International Conference on Theoretical Physics 2010

(Dubna-Nano 2010)

Journal of Physics: Conference Series Volume 248

Dubna, Russia
5-10 July 2010

Editors:

Vladimir Osipov
Valentin Nesterenko
Yury Shukrinov

ISBN: 978-1-61782-077-9

International Conference on Theoretical Physics 2010

(Dubna-Nano 2010)

Journal of Physics: Conference Series Volume 248

Dubna, Russia
5-10 July 2010

ISBN: 978-1-61782-077-9

IOP Institute of Physics

Printed from e-media with permission by:

Curran Associates, Inc.
57 Morehouse Lane
Red Hook, NY 12571

Some format issues inherent in the e-media version may also appear in this print version.

Printed by Curran Associates, Inc. (2010)

For permission requests, please contact the Institute of Physics
at the address below.

Institute of Physics
Dirac House, Temple Back
Bristol BS1 6BE UK

Phone: 44 1 17 929 7481
Fax: 44 1 17 920 0979

techtracking@iop.org

Additional copies of this publication are available from:

Curran Associates, Inc.
57 Morehouse Lane
Red Hook, NY 12571 USA
Phone: 845-758-0400
Fax: 845-758-2634
Email: curran@proceedings.com
Web: www.proceedings.com

TABLE OF CONTENTS

CARBON NANOSYSTEMS

QUANTUM DOTS

SPECTROSCOPY AND DYNAMICS OF ATOMIC CLUSTERS

QUANTUM TRANSPORT

JOSEPHSON JUNCTIONS

MODELLING

APPLICATIONS AND PERSPECTIVES

PREFACE

The International Conference on Theoretical Physics 'Dubna-Nano2010' was held on 5–10 July 2010, at the Bogoliubov Laboratory of Theoretical Physics, Joint Institute for Nuclear Research, Dubna, Moscow region, Russia. The previous conference of this series was at Dubna in 2008.

The conference provided the opportunity for the presentation and discussion of theoretical and experimental advances in the rapidly growing area of nanophysics, with the accent on its theoretical aspects. The multidisciplinary character of the conference allowed an effective exchange of ideas between different areas of nanophysics. The following topics were covered: carbon nanosystems (graphene, nanotubes, fullerenes), quantum dots, quantum transport, spectroscopy and dynamics of atomic clusters, Josephson junctions, modelling, applications and perspectives.

Approximately 120 scientists from 26 countries participated in the conference. The program included 63 oral talks and 70 posters. The 62 contributions are included in these proceedings.

We would like to express our gratitude to all participants for their presentations and discussions, which made the conference indeed successful. We are deeply indebted to the members of the International Advisory Committee (Professors T Ando, J Fabian, F Guinea, P Hawrylak, K Kadowaki, T Koyama, Yu I Latushev, Yu E Lozovik, M Machida, B K Nikolic, N F Pedersen, P-G Reinhard, J M Rost, A Ya Vul') and the Local Organizing Committee for their fruitful work.

The financial support of BLTP JINR, Russian Foundation for Basic Research, Heisenberg–Landau Program and Bogoliubov–Infeld Program was of a great importance.

Additional information about 'Dubna-Nano2010' is available at the homepage http://theor.jinr.ru/~nano10.

Vladimir Osipov, Valentin Nesterenko and Yury Shukrinov
Editors

PEER REVIEW STATEMENT

All papers published in this volume of *Journal of Physics: Conference Series* have been peer reviewed through processes administered by the proceedings Editors. Reviews were conducted by expert referees to the professional and scientific standards expected of a proceedings journal published by IOP Publishing.

Aharonov-Bohm effect on columnar defects in thin graphite and graphene

Yu.I. Latyshev[1], A.P. Orlov[1], E.G. Shustin[2], N.V. Isaev[2], W. Escoffier[3], P. Monceau[4], C.J. van der Beek[5], M. Konczykowski[5], I. Monnet[6]

[1]Kotelnikov Institute of Radio-Engineering and Electronics RAS, Mokhovaya 11-7, 125009 Moscow, Russia

[2]Kotelnikov Institute of Radio-Engineering and Electronics RAS, Fryazino 141190, Moscow region, Russia

[3]Laboratoire National des Champs Magnétiques Intenses (LNCMI-T), 143 Av. de Rangueil, 31400, Toulouse, France

[4]Institut Néel, CNRS, 25 Rue des Martyrs, 38042 Grenoble, France

[5]Laboratoire des Solides Irradiés, Ecole Polytechnique, Palaiseau, 91128 France

[6]Centre de Recherche sur les Ions, les Materiaux et la Photonique (CIMAP, GANIL), Caen, 14050 France

E-mail: yurilatyshev@yahoo.com

Abstract. We studied Aharonov-Bohm effect on thin graphite single crystals containing columnar defects (nanoholes) produced by irradiation with heavy ions. The thickness of irradiated single crystals has been varied from 50 nm down to less then 1 nm by thinning the thicker samples with beam-plasma etching. For irradiated samples we found the existence of oscillating, field-periodic contribution of magnetoresistance with a periodicity corresponding to the flux quantum *hc/e* per area of nanohole independently of sample thickness. The results point to the significant contribution of Dirac fermions to the oscillating part of magnetoresistance and also to the existence of the edge states located near the nanohole perimeter.

1. Introduction

Shortly after a discovery of Dirac fermions in graphene [1, 2], they were also found in thin single crystals of graphite by far infrared (FIR) magneto-transmission measurements [3], ARPES [4], STM [5] and interlayer tunneling [6] techniques. The specific features of Dirac fermions: (1) conic spectrum $E(k) = \pm v_F\hbar|k|$, where v_F is the Fermi energy, and (2) specific Landau quantization with the energy of Landau levels E_n proportional to $(n\ H)^{1/2}$, have been used for their identification in graphite. The presence of Dirac fermions in graphite was attributed either to the linear dispersion at the H-point of the graphite Brillouin zone [3] or to the existence of graphene layer at the surface of graphite [7, 8].

Another interesting property of Dirac fermions is the absence of backscattering [9, 10] and the related absence of the weak localization effects. That feature was confirmed on graphene mesoscopic ring samples, where oscillations of magnetoresistance with a period, corresponding to flux quantum $\Phi_0 = hc/e$ per area of the hole have been observed, while a contribution of *hc*/2*e* oscillations, due to weak localization, was strongly suppressed [11].

In a present paper we study A-B effect on columnar defects produced by irradiation of thin single crystals of graphite and graphene with heavy ions. Columnar defects serve as nanoholes (antidots) for studies of magnetic flux quantization. We found the same periodicity of oscillating part of magnetoresistance of irradiated sample *hc/e*, as for graphene ring indicating contribution of Dirac fermions [11]. We show that this periodicity does not depend on graphite thickness with its variation from about 50 graphene layers down to bi-layer graphene. That points to the surface contribution of Dirac fermions in graphite.

2. Experimental

We used in experiments high quality single crystals of natural graphite. The in-plane conductivity was 30–50 μΩ cm and conductivity anisotropy along and across the layers was $4\ 10^3$ at room temperature. At liquid helium temperatures it reached $3\ 10^4$. The high quality of the crystals was also confirmed by studies the FIR magneto-transmission [13] and Shubnikov–de Haas oscillations of magnetoresistance [14].

Thin graphite single crystals with thickness of 30-100 nm and with lateral sizes of several hundred microns have been detached out of fixed thick graphite single crystal using adhesive tape. The glue layer of the tape has been solved in acetone leaving thin crystal to float freely in a solvent and then it has been picked up onto the substrate.

At the second stage thin graphite crystal has been thinned by soft plasma etching [15]. We used original beam-plasma reactor [16] to provide low energy ions. Electron beam excites Ar-plasma in a reactor and induces a flow of low energetic monochromatic ions in a direction perpendicular to the beam direction. This flow has been used for etching. The energy of ions was 60 - 80 eV providing soft defectless etching with a rate of about 10 nm per hour. A thickness of the thinned crystal has been controlled by measuring of its resistance *in situ*. The etching was terminated when the sheet resistance of the sample achieved few kΩ corresponding to 1-2 layer graphene.

Thin graphite/graphene crystals have then been masked and patterned in oxygen discharge to get Hall bar geometry. The width of the patterned part of the crystals was 10-20 μm and the length between potential probes was about 100 μm (Fig.1). Low ohmic electric contacts have been made by pressing of In-stubs into the crystal, as shown in Fig.1.

Fig. 1. Microphotograph of the patterned graphite single crystal of 30 nm thickness on sapphire substrate with indium contact pads.

We measured 8 samples of graphite single crystals irradiated with Xe-ions with energy of 90 MeV along the *c*-axis with doses corresponding to 5 10^8 - 10^{10} def/cm^2. Two of them have been preliminary thinned by soft plasma etching. We analyzed the size of columnar defects using scanning AFM, as a size of hillocks of the amorphous substance squeezed out of them [17] (Fig. 2). This effect is known and is attributed to the relaxation of force stresses accompanying the formation of columnar defects. The hillock density corresponds to the radiation dose. The columnar defect diameter D over the hillock flat part is estimated as $D \leq 24$ nm at its height of about 6 nm [12]. Magnetoresistance of graphite crystals has been measured in resistive magnet (up to 32 T) in Grenoble National High Magnetic Field Lab. (LNCMI-G) or in pulsed fields (up to 55 T) in LNCMI-T, Toulouse, using standard techniques. The lowest achievable temperature was 1.2 K.

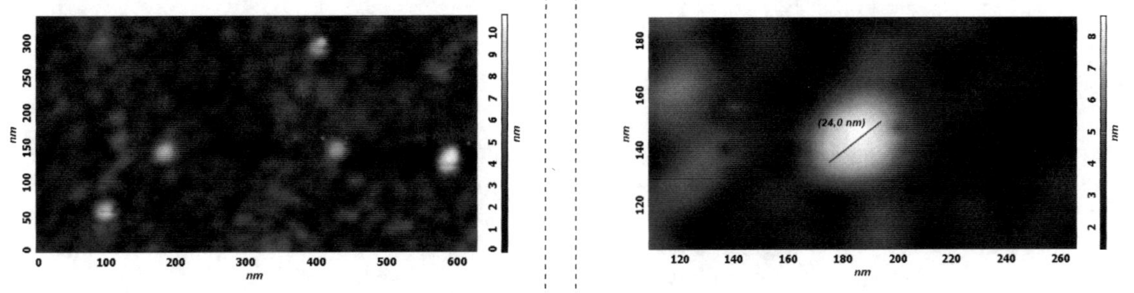

Fig. 2. AFM images of (a) a group of columnar defects and (b) a single columnar defect in sample # 3 irradiated by Xe-ions with energy of 90 MeV with a dose corresponding to 10^9 def/cm^2. The columnar defects are AFM recognized as hillocks of an amorphous substance squeezed out of them (see the main text).

3. Experimental results

Fig.3 (left panel) shows the full scale dependence of resistance of low dose irradiated sample #1 on magnetic field *H*//*c* at several specified temperatures. In the region below 8 T, the Shubnikov–de Haas oscillations are clearly seen (right panel of Fig. 3). Their period is inversely proportional to the field and equals to 0.22 T^{-1} , in agreement with previous data [14, 18]. At higher fields new, field periodic oscillations appear. Three oscillations with the minima at 15, 22.5, and 30 T are pronounced in the region of high magnetic fields exceeding 10 T. The period of these oscillations corresponds to 7.5 T and a much weaker minimum at 7.5 T is also seen. The $R(H)$ curves almost coincided in the field of increasing and decreasing regimes as well as in reversing the field direction, $H \rightarrow -H$. Specific features of these oscillations are their big amplitude (up to 15%) and their existence up to quite high temperatures. Fig. 2a shows the $R(H)$ curves at 1.2, 4.2, 10, 20, and 32 K. It is seen that the oscillation amplitude decreases with increasing temperature, however, they are still observable up to temperatures of 30 K.

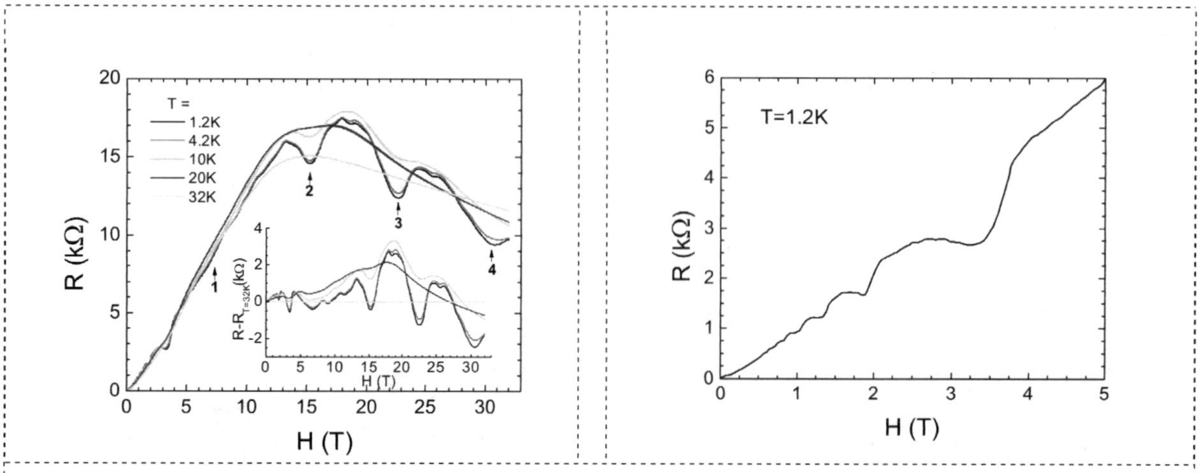

Fig. 3. The in-plane resistance of the irradiated graphite single crystal (sample # 3) versus magnetic field $H \parallel c$ at 1.2, 4.2, 10, 20, and 32 K. The sample is 50-nm thick, the columnar defect density is $c = 10^9$ defects/cm^2. The arrows mark the group of minima $R(H_n)$, where $H_n = nH_0$ (n is an integer, $H_0 = 7.5$ T). The right panel shows the enlarged fragment of the $R(H)$ curve up to 5 T at $T = 1.2$ K, which illustrates the Shubnikov–de Haas magnetoresistance oscillations.

Similar large scale, field-periodic oscillations were observed in the samples at the radiation doses of $3\ 10^9$ and 10^{10} defects/cm^2, but with lower amplitudes. The reference measurements of the unexposed samples demonstrate the absence of such oscillations. On average, the $R(H)$ curves of the unexposed samples and low dose exposed samples are similar in their shape to those of bulk samples of high quality graphite single crystals [19]. They have a wide maximum near 20 T, then slowly droop to a minimum at 28–30 T. Near the maximum, the normalized magnetoresistance $R(H_{max})/R(0)$ reaches $5\ 10^3$. At a defect density of 10^9 defects/cm^2, it is almost invariable; however, with further increase of defect density c the magnetoresistance at $H = 15$–20 T decreases (by an order of magnitude at $c = 10^{10}$ defects/cm^2), and the $R(H)$ curves become monotonic with saturation at $H > 20$ T.

To extend the range of effect observation we used pulsed magnetic field up to 55T. Fig. 4 shows R(H) dependence and its oscillating part of our thinnest sample #5 obtained by soft plasma etching. This sample was characterized by local Raman spectroscopy as a bi-layer graphene [15]. Fig. 3 clearly shows 6 oscillations of magnetoresistance above 10 T observed on this thinnest sample. The period of oscillations, 7.6 ± 0.3 T, is very close to what was observed in sample #3 of 50 nm thick under the same irradiation conditions.

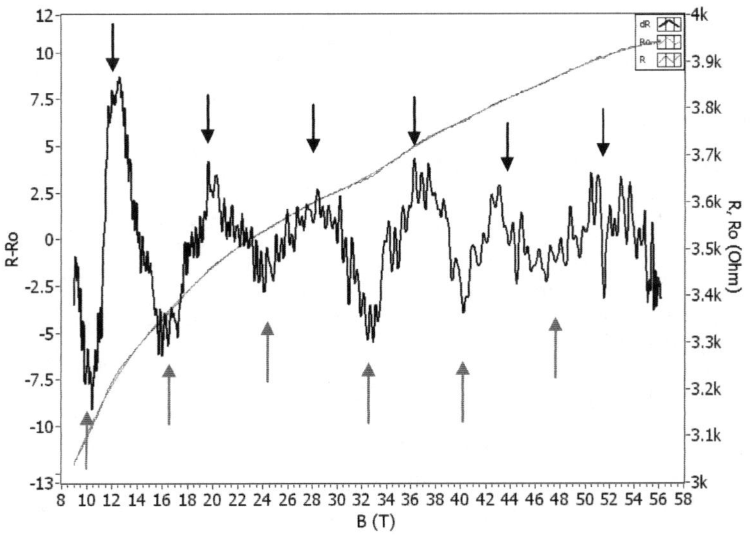

Fig.4. Magnetoresistance and its oscillating part of bi-layer graphene sample #5 with columnar defects at T=1.5 K. Black and red arrows indicate correspondingly maximums and minimums of the oscillating part of magnetoresistance. B//c.

4. Data analysis and discussion.

We begin analysis of the experimental data with an estimation of the magnetic flux through the area of columnar defect corresponding to the period of magnetoresistance oscillations. The field period for one oscillation, ΔH, obtained from experiment, is $\Delta H = 7.5$ -7.6 T. Under the assumption that the paths close to the columnar defect perimeter make the main contribution, the corresponding magnetic flux is $\Delta\Phi = \Delta H\pi D2/4 = 3.5 \pm 0.8$ 10–8 mT cm^2. Despite a large uncertainty in the columnar defect size, this value is much closer to hc/e = 4.14 10–8 mT cm^2 than to hc/2e = 2.07 10–8 mT cm^2. We should also take into account that the effective diameter of the orbit of carriers around a columnar defect is larger than its geometric diameter. This shifts the estimate of ΔΦ closer to hc/e. Thus we can conclude that the period of magnetoresistance oscillations corresponds to the flux quantum per area of columnar defect. This result agrees well with the experiments on graphene rings [7] and implies that for both objects, thin graphite crystals and bi-layer graphene with columnar defect, we observed Aharonov-Bohm interference of Dirac fermions.

The fact that effect does not depend on sample thickness indicates that effect is very likely related with the surface layers of graphite similarly as it was recently observed for topological insulators [20]. That conclusion is supported by recent STM [7] and FIR magneto-transmission [8] measurements which showed that the surface layer of graphite is often represents as a graphene layer lying on graphite substrate.

There is another interesting point following from our experiment. Normally, Aharonov-Bohm effect is observed in ring geometry samples. This geometry lets fix the quantized orbits between inner and outer boundary of the ring. In our experiment A-B type effect was observed on non-ring geometry and as shown is determined by the orbits located close to the hole. We consider that the restricted orbits near the hole can exist due to the edge states located near graphene boundary. Otherwise, the averaging contribution of all possible orbits around hole would smear out interference effect. Existence of the edge states in graphene has been recently predicted theoretically [21]. Our observations of Aharonov-Bohm effect on nanoholes (antidots) strongly support the existence of the edge states in graphene.

Acknowledgments
This work was supported by the Russian Foundation for Basic Research (project # 08-02-01093a); by Agence Nationale pour la Recherche (grant # ANR_07_BLAN_0136); by the European Associated Laboratory (LEA) between the Institute of Radio-Egeneering and Electronics, Russian Academy of Sciences, and the Institut Néel, CNRS; and by the Russian Academy of Sciences (programs "Strongly Correlated Electrons in Solids and Structures" and "Physics of New Materials and Structures"). The authors are also acknowledged the support of the European Commission from the 7th framework programme "Transnational Access", contract N° 228043 – EuroMagNET II – Integrated Activities.

References

[1] Novoselov K S et al. 2005 *Nature* **438** 197
[2] Zhang Y B et al. 2005 *Nature* **438** 201
[3] Orlita M *et al.* 2008 *Phys. Rev. Lett.* **100** 136403
[4] Zhou S Y et al. 2007 *Nature Physics* **2** 595
[5] Li G and Andrei Eva Y 2007 *Nature Phys* **3** 623
[6] Latyshev Yu I et al. 2008 *J. of Physics: Conf. Series* **129,** 012032
[7] Li G et al. 2009 *Phys. Rev. Lett.* **102** 176804
[8] Nuegebauer P et. al. 2009 *Phys. Rev. Lett.* **103** 136403
[9] Katsnelson M I, K S Novoselov, and A K Geim 2006 *Nature Phys.* **2**, 620
[10] Lozovik Yu E, S P Merkulov, and A A Sokolik 2008 *Usp. Fiz. Nauk* **178**, 757 [2008 *Phys. Usp.* **51**, 727]
[11] Russo S et al. 2008 *Phys. Rev. B* **77** 085413
[12] Latyshev Yu I et al. 2009 *JETP Lett.* **90**, 480
[13] Orlita M et al. 2008 *cond-mat.*: 0805.0553v1
[14] Scneider J. V. et al. 2009 *Phys. Rev. Lett.* **102** 166403
[15] Latyshev Yu I, E G Shustin et al., *Proceedings of the* 2nd *Moscow Intern. Forum on Nanotechnologies, Moscow*, 6-8 Oct., 2009, p. 465-467
[16] Shustin E G, N V Isaev, M P Temiryazeva, Yu V Fedorov 2009 *J.Vacuum* **83**, 1350
[17] Skuratov V A et al. 2003 *Nucl. Instrum. Methods Phys. Res.* **B 203** 136
[18] Lukyanchuk I A and Y Kopelevich 2006 *Phys. Rev. Lett.* **97** 256801
[19] Ochimizu H et al. 1992 *Phys. Rev. B* **46** 1986
[20] Ihn T 2010 *Nature Materials* **9** 187
[21] Volkov V A, I V Zagorodnev 2009 *Low Temperature Physics* **35** 2, Volkov V A and I V Zagorodnev 2009 *Journal of Physics: Conference Series* **193** 012113

Journal of Physics: Conference Series **248** (2010) 012002 doi:10.1088/1742-6596/248/1/012002

Superconducting proximity effect in graphene nanostructures

M. Hayashi[1,4], **H. Yoshioka**[2] **and A. Kanda**[3,4]

[1]Faculty of Education and Human Studies, Akita University, Akita 010-8502, Japan
[2]Department of Physics, Nara Women's University, Nara 630-8506, Japan
[3]Institute of Physics, University of Tsukuba, Tsukuba 305-8571, Japan
[4]JST-CREST, 4-1-8 Honcho, Kawaguchi, Saitama 332-0012, Japan

E-mail: m-hayashi@ed.akita-u.ac.jp

Abstract. The superconducting proximity effect is studied theoretically for graphene(G)-superconductor(S) hybrid structures. Especially S-G-S junction is considered and the critical current is argued focusing on its temperature and junction-length dependences. The free energy of the total system is calculated by use of the tunneling approximation and the critical current is estimated from it. It is reported that the critical current can oscillate as a function of temperature and junction-length in bi-layer junctions. Some detailed comparison with ordinary S-normal metal-S junction is also given and both monolayer and bilayer junctions show some characteristic deviations.

1. Introduction

Graphene has been attracting much interest because of their unique electronic properties as well as potential application[1, 2, 3]. From the very beginning of the research several intriguing phenomena originating from peculiar band structure of monolayer and bilayer graphenes have been reported [4, 5]. Their superconducting properties are also studied theoretically and experimentally. It has been shown that, in spite of the vanishing density of states at the Fermi energy, a finite supercurrent can actually flow through graphene[6, 7, 8]. One of the present authors (AK) has fabricated microstructures consist of ultrathin graphite films and investigated superconducting proximity effect[9, 10, 11] and it has been found that the critical supercurrent depends largely on the gate voltage.

In this paper we calculate the critical current of superconductor(S)-graphene(G)-superconductor(S) Josephson junction using tunneling approximation and path integral formalism [12]. A similar calculation for monolayer graphene has been done by several authors [13, 14, 15]. In this paper we especially pay attention to the difference between monolayer and bilayer junctions focusing on the behaviors at non-zero temperatures. We argue the proximity length characteristic to Josephson junctions and examine the superconducting state within graphene.

The tunneling between superconductors and graphene can be treated in several different methods. In most of the preceding studies, the superconducting lead attached to graphene is treated by introducing non-vanishing superconducting gap to the graphene in the contact area [7, 16, 17]. In this paper we adopt the tunneling approximation. In this method, we treat perturbatively the tunneling matrix elements between leads and graphene. Usually the junctions look like (a), however we consider the point-contact-like junction as depicted in Fig. 1 (b) for simplicity. We note that, even in junction (a), the critical current is mostly determined by the shortest part of the junction, and therefore the behavior of (a) and (b) may not change significantly. This point will be discussed elsewhere in greater detail.

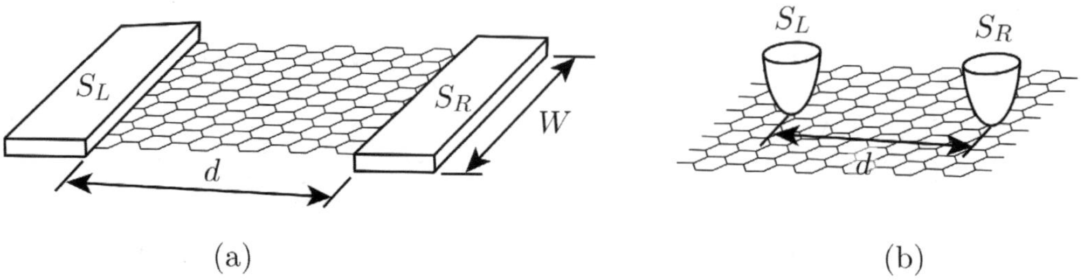

Figure 1. (a) A conventional S-G-S junction. S_L and S_R indicate superconducting leads, whose separation and width are d and W, respectively. (b) Point-contact type S-G-S junction. Here we have shown only one layer of the graphene for simplicity.

2. Model

2.1. Hamiltonian and Band Structure of Graphene

We start from the following tight-binding Hamiltonian which describes the electrons in left and right superconducting leads and monolayer (or bilayer) graphene (for simplicity we hereafter indicate them by L, R and G, respectively):

$$H_L = \sum_{\vec{k}\sigma} \left(\xi_{\vec{k}} c^{L\dagger}_{\vec{k}\sigma} c^{L}_{\vec{k}\sigma} + \Delta^L c^{L\dagger}_{\vec{k}\uparrow} c^{L\dagger}_{-\vec{k}\downarrow} + \Delta^{L*} c^{L}_{-\vec{k}\downarrow} c^{L}_{\vec{k}\uparrow} \right), \tag{1}$$

$$H_R = H_L \ (L \to R), \tag{2}$$

$$H_G = -t \sum_{\langle j,k\rangle\sigma} \left(a^\dagger_{j\sigma} b_{k\sigma} + h.c. \right)$$

$$- t \sum_{\langle j,k\rangle\sigma} \left(\bar{a}^\dagger_{j\sigma} \bar{b}_{k\sigma} + h.c. \right) - t_\perp {\sum_{\langle j,k\rangle\sigma}}' \left(a^\dagger_{j\sigma} \bar{b}_{k\sigma} + h.c. \right) \tag{3}$$

$$\simeq \sum_{\vec{k}\sigma} E_{\vec{k}} \left(\alpha^\dagger_{\vec{k}\sigma} \alpha_{\vec{k}\sigma} - \beta^\dagger_{\vec{k}\sigma} \beta_{\vec{k}\sigma} \right). \tag{4}$$

Here $c^{L(R)}_{\vec{k}\sigma}$ is the annihilation operator of electron with wave number $\vec{k}$ and spin σ, and $\Delta^{L(R)}$ the superconducting gap. The gap is assumed to be spatially constant in both leads. $\xi_{\vec{k}} = \hbar^2 |\vec{k}|^2/(2m)$ with m being electron mass in the leads. We also assume that $\xi_{\vec{k}}$ is common in both leads. In Eq. (3) and (4), we have given the graphene part; $a_{j\sigma}$ and $b_{j\sigma}$ are the annihilation operator of an electron located at A and B sublattice of the graphene, respectively, and, t is the hopping matrix element. The second line of Eq. (3) is for bilayer graphene. The annihilation operators in the second layer (we also call it lower layer) is given by $\bar{a}_{j\sigma}$ and $\bar{b}_{j\sigma}$, and $t_\perp$ is the inter-layer hopping matrix element. Since A-B stacking is stable for bilayer system we have introduced only A-B hopping. The summation $\sum'$ is taken over interlayer pairs. Here, we limit the electron hopping in the graphene to the nearest neighbors, which makes the electron dispersion symmetric about the zero energy. The quasiparticle annihilation operators with positive ($E_{\vec{k}}$) and negative ($-E_{\vec{k}}$) energy, obtained after diagonalization, are denoted by $\alpha_{\vec{k}\sigma}$ and $\beta_{\vec{k}\sigma}$, respectively. In case of monolayer, the dispersion is given by

$$E_{\vec{k}} = |\gamma_{\vec{k}}| \equiv |te^{i\vec{k}\cdot\vec{\delta}}(1 + e^{i\vec{k}\cdot\vec{a_1}} + e^{i\vec{k}\cdot\vec{a_2}})|, \tag{5}$$

where $\vec{\delta}$, $\vec{a}_1$ and $\vec{a}_2$ are shown in Fig. 2 (a). Near the K point, $|\gamma_{\vec{k}}|$ is approximated by $\hbar v_F|\vec{k} - \vec{K}|$ with $\hbar v_F = 3ta/2$ and the same holds for K'. ($\vec{K}$ shows the position of K-point.) These structures are usually described as Dirac cones. In case of bilayer, Dirac cones of the two monolayers split into high and low energy bands, and the energy of the latter is given by $E_{\vec{k}} = |\gamma_{\vec{k}}|^2/t_\perp$, which has a parabolic form. The higher energy band is usually about 0.2 eV above, and is neglected hereafter.

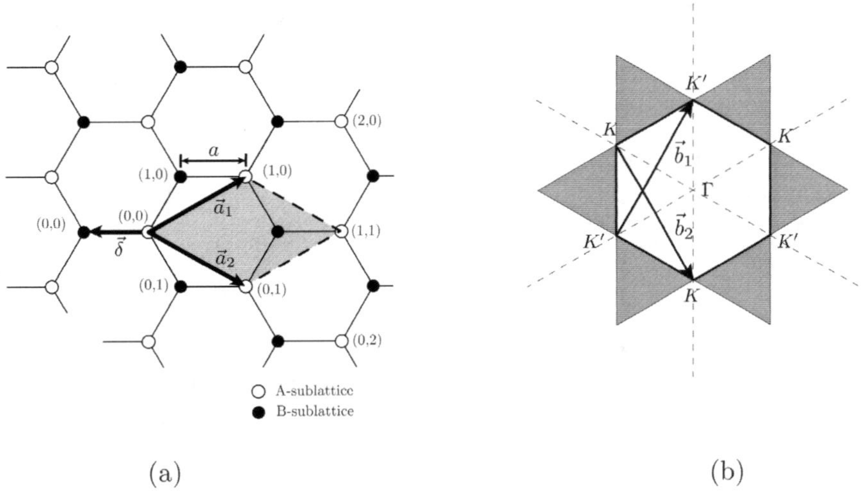

Figure 2. (a) Honeycomb lattice of graphene and its unit cell (shaded region). Two lattice vectors ($\vec{a}_1$ and $\vec{a}_2$) are indicated. Open and filled circles, respectively, show the A and B sublattice of graphene. (b) Reciprocal lattice of graphite sheet. The first (hexagon in the center) and the second Brillouin zone (shaded regions) are shown with two reciprocal lattice vectors ($\vec{b}_1$ and $\vec{b}_2$). K and K' are the points where two Fermi points exists.

Let us consider the tunneling of electrons between graphene and left leads. We assume the following tunneling Hamiltonian,

$$H_T^{L(A)} = -t'\sum_{j\in L}(c_{j\sigma}^{L\dagger}a_{j\sigma} + h.c.), \quad H_T^{L(B)} = -t'\sum_{j\in L}(c_{j\sigma}^{L\dagger}b_{j\sigma} + h.c.) \tag{6}$$

where $H_T^{L(A)}$ and $H_T^{L(B)}$ describe the tunneling of electron from left superconductor to A and B site of the graphene, respectively, with t' being the tunneling matrix element. For right leads, we similarly define $H_T^{R(A)}$ and $H_T^{R(B)}$. In case of bilayer system, tunneling occurs only to the upper layer, denoted by $a_{j\sigma}$ and $b_{j\sigma}$. In Eq.(6), $j \in L$ means that the j-th A (or B) site is in the left junction area. In this paper, the junction area is finally reduced to a lattice point.

2.2. Calculation of Free Energy

We assume $\Delta^R = \Delta_0 e^{i\theta^R}$ and $\Delta^L = \Delta_0 e^{i\theta^L}$, where Δ_0 and $\theta^{R(L)}$ are constants, and the free energy $F(\theta)$, where $\theta = \theta^R - \theta^L$, can be calculated by imaginary-time path integral formalism in the following way,

$$F(\theta) = -\beta^{-1}\ln Z(\beta),\quad Z(\beta) = \int \mathcal{D}[\Psi_{\vec{k}}]\, e^{-S/\hbar}, \tag{7}$$

$$S = \int_0^{\beta\hbar} d\tau \Big[\sum_{\vec{k}\sigma} \Psi^\dagger_{\vec{k}} \cdot \hbar\frac{\partial}{\partial\tau}\Psi_{\vec{k}} + H - \mu_L N_L - \mu_G N_G - \mu_R N_R\Big], \tag{8}$$

where $\Psi_{\vec{k}} = {}^t(c^R_{\vec{k}\uparrow}, c^{R*}_{-\vec{k}\downarrow}, \alpha_{\vec{k}\uparrow}, \alpha^*_{-\vec{k}\downarrow}, \beta_{\vec{k}\uparrow}, \beta^*_{-\vec{k}\downarrow}, c^L_{\vec{k}\uparrow}, c^{L*}_{-\vec{k}\downarrow})$ are Grassmann variables, and $H = H_L + H_R + H_G + H_T$. $\beta = 1/(k_B T)$ and, T and k_B are temperature and Boltzmann constant, respectively. μ_a and N_a $(a = L, R, G)$ are the chemical potential and number operator in L, R and G region, respectively. We set $\mu_L = \mu_R \equiv \mu$ in this paper. μ_G is kept in the following calculation, although finally we discuss only $\mu_G = 0$ case. The action in the Fourier space is given by

$$S = \hbar \sum_{\vec{k},\vec{k}',\omega_n} \Psi^\dagger_{\vec{k},\omega_n} \cdot \Pi(\vec{k},\vec{k}',\omega_n)\cdot \Psi_{\vec{k}',\omega_n} \tag{9}$$

$$\Pi = \begin{pmatrix} -\{G^R\}^{-1} & \Gamma^{R\alpha} & \Gamma^{R\beta} & 0 \\ \Gamma^{R\alpha\dagger} & -\{G^\alpha\}^{-1} & 0 & \Gamma^{L\alpha\dagger} \\ \Gamma^{R\beta\dagger} & 0 & -\{G^\beta\}^{-1} & \Gamma^{L\beta\dagger} \\ 0 & \Gamma^{L\alpha} & \Gamma^{L\beta} & -\{G^L\}^{-1} \end{pmatrix} \tag{10}$$

$$G^{R(L)} = \delta_{\vec{k},\vec{k}'}\begin{pmatrix} g^{R(L)}_{\vec{k},\omega_n} & f^{R(L)}_{\vec{k},\omega_n} \\ f^{R(L)*}_{\vec{k},\omega_n} & -g^{R(L)*}_{\vec{k},\omega_n}\end{pmatrix},\quad G^{\alpha(\beta)} = \delta_{\vec{k},\vec{k}'}\begin{pmatrix} g^{\alpha(\beta)}_{\vec{k},\omega_n} & 0 \\ 0 & -g^{\alpha(\beta)*}_{\vec{k},\omega_n}\end{pmatrix} \tag{11}$$

where the arguments ($\vec{k}$, $\vec{k}'$ and ω_n) are suppressed accordingly. The Green's functions are given as

$$g^{R(L)}_{\vec{k},\omega_n} = -\frac{i\omega_n + (\xi_{\vec{k}} - \mu)/\hbar}{\omega_n^2 + \{(\xi_{\vec{k}}-\mu)^2 + \Delta_0^2\}/\hbar^2},\quad f^{R(L)}_{\vec{k},\omega_n} = -\frac{\Delta^{R(L)}}{\omega_n^2 + \{(\xi_{\vec{k}}-\mu)^2 + \Delta_0^2\}/\hbar^2} \tag{12}$$

$$g^{\alpha(\beta)}_{\vec{k},\omega_n} = \frac{1}{i\omega_n - (\pm E_{\vec{k}} - \mu_G)/\hbar} \tag{13}$$

where $\omega_n = (2n+1)/(\beta\hbar)$ is the Matsubara frequency. In Eq. (13), $\pm$ sign should be taken $+$ for α and $-$ for β. The tunneling terms are defined as

$$\Gamma^{L\alpha} = \Gamma^L_A + \Gamma^L_B,\quad \Gamma^{L\beta} = -\Gamma^L_A + \Gamma^L_B \tag{14}$$

$$\Gamma^L_{A\vec{k}\vec{k}'} = -\frac{t'}{\sqrt{2}\hbar}\frac{1}{\sqrt{n_G n_L}} v^L_A(\vec{k}-\vec{k}')e^{-i\theta_{\vec{k}'}}\sigma_z,\quad \Gamma^L_{B\vec{k}\vec{k}'} = -\frac{t'}{\sqrt{2}\hbar}\frac{1}{\sqrt{n_G n_L}} v^L_B(\vec{k}-\vec{k}')\sigma_z, \tag{15}$$

$$v^L_{A(B)}(\vec{k}) \equiv \sum_{j\in L} e^{-i\vec{k}\cdot\vec{r}^{A(B)}_j} \tag{16}$$

where σ_z is the Pauli matrix, dagger (†) means $\Gamma^{L\dagger}_{A\vec{k}\vec{k}'} = \{\Gamma^L_{A\vec{k}'\vec{k}}\}^*$, and $\vec{r}^{A(B)}_j$ denotes the A (B) site of the j-th unit cell. The phase factor $e^{i\theta_{\vec{k}}}$ denotes $\gamma_{\vec{k}}/|\gamma_{\vec{k}}|$. The right-lead terms, $\Gamma^{R\alpha}$ etc., can be given in the same manner. n_a $(a = L, R, G)$ are the number of lattice sites in L, R and G region, respectively.

Here we divide Π into two parts as $\Pi = -\tilde{G}^{-1} + \tilde{\Gamma}$, where $\tilde{\Gamma} \propto t'$, and then perform the perturbative expansion with respect to t'. The lowest order terms that contribute to the

Journal of Physics: Conference Series **248** (2010) 012002 doi:10.1088/1742-6596/248/1/012002

Josephson coupling energy F_J of two superconductors are the fourth order terms. These terms are further classified by the tunneling sites: 1) the tunneling to left and right superconductors occurs within the same sublattice (A or B), which we denote by F_J^{AA}, 2) the tunneling to left and right superconductors occur at the different sublattice, which we denote by F_J^{AB}. Then we obtain

$$\beta F_J^{AA} = -2 \sum_{\vec{k},\vec{k}',\omega_n} \mathrm{Re}\{\chi^L_{A\vec{k}\vec{k}'} \chi^{R*}_{A\vec{k}\vec{k}'}\} g^{\alpha\beta}_{\vec{k}\omega_n} g^{\alpha\beta}_{\vec{k}'\omega_n}, \quad \beta F_J^{AB} = -2 \sum_{\vec{k},\vec{k}',\omega_n} \mathrm{Re}\{\chi^L_{A\vec{k}\vec{k}'} \chi^{R*}_{B\vec{k}\vec{k}'}\} \bar{g}^{\alpha\beta}_{\vec{k}\omega_n} \bar{g}^{\alpha\beta}_{\vec{k}'\omega_n} \tag{17}$$

where $g^{\alpha\beta} = g^\alpha + g^\beta$ and $\bar{g}^{\alpha\beta} = g^\alpha - g^\beta$. Here $\chi^{R(L)}_{\vec{k},\vec{k}'}$ can be further calculated as,

$$\chi^{L(R)}_{A\vec{k},\vec{k}'} = \left(\frac{t'}{\sqrt{2}\hbar}\right)^2 \frac{1}{n_G n_{L(R)}} \sum_{\vec{l}} v_A^{L(R)}(\vec{k}-\vec{l}) v_A^{L(R)}(\vec{l}-\vec{k}') e^{i(\theta_{\vec{k}}-\theta_{\vec{k}'})} f^{L(R)}_{\vec{l}}, \tag{18}$$

$$\chi^{L(R)}_{B\vec{k},\vec{k}'} = \left(\frac{t'}{\sqrt{2}\hbar}\right)^2 \frac{1}{n_G n_{L(R)}} \sum_{\vec{l}} v_B^{L(R)}(\vec{k}-\vec{l}) v_B^{L(R)}(\vec{l}-\vec{k}') f^{L(R)}_{\vec{l}}. \tag{19}$$

After some calculation, we obtain F_J^{AA} as

$$\beta F_J^{AA} = -2 \left(\frac{t'}{\sqrt{2}\hbar}\right)^4 \left(\frac{s_1 m k_F^s}{2\pi}\right)^2 \sum_{\omega_n} \sum_{\vec{r}_j \in R, \vec{r}_{j'} \in L} \frac{\Delta_0^2 \cos(\theta^L - \theta^R)}{(\hbar\omega_n)^2 + \Delta_0^2} |g^{\alpha\beta}(\vec{r}_j^A - \vec{r}_{j'}^A, \omega_n)|^2, \tag{20}$$

where s_1 and k_F^s are the unit cell area and the Fermi wave number of the lead superconductors. We have introduced the real space representation of $g^{\alpha\beta}_{\vec{k},\omega_n}$. The A-B term F_J^{AB} is obtained by changing $g^{\alpha\beta}(\vec{r}_j^A - \vec{r}_{j'}^A, \omega_n)$ to $\bar{g}^{\alpha\beta}(\vec{r}_j^A - \vec{r}_{j'}^B, \omega_n)$. The supercurrent flowing through the junction can be calculated from the Josephson energy F_J by

$$I(\theta) = \frac{2e}{\hbar} \frac{dF_J}{d\theta}. \tag{21}$$

Usually contact between superconductor and graphene is macroscopic, including many lattice points. Therefore F_J is given by the average of F_J^{AA} and F_J^{AB}. However it is also possible to consider point contacts, such as STM using superconducting tips, and we also discuss later the behavior of F_J^{AA} and F_J^{AB} separately in this paper.

Bearing in mind that there are two Fermi points in graphene dispersion, the real-space Green's function can be calculated as follows,

$$\begin{aligned} g^{\alpha(\beta)}(\vec{r}, \omega_n) &= s_0 \int_{\mathrm{BZ}} \frac{d^2\vec{k}}{(2\pi)^2} g^{\alpha(\beta)}_{\vec{k},\omega_n} e^{i\vec{k}\cdot\vec{r}} = s_0 \int_{\mathrm{BZ}} \frac{d^2\vec{l}}{(2\pi)^2} \left\{ g^{\alpha(\beta)}_{\vec{K}+\vec{l},\omega_n} e^{i\vec{K}\cdot\vec{r}} + g^{\alpha(\beta)}_{\vec{K}'+\vec{l},\omega_n} e^{i\vec{K}'\cdot\vec{r}} \right\} e^{i\vec{l}\cdot\vec{r}} \\ &= s_0 \left(e^{i\vec{K}\cdot\vec{r}} + e^{i\vec{K}'\cdot\vec{r}}\right) \int_{\mathcal{R}^2} \frac{d^2\vec{l}}{(2\pi)^2} \frac{1}{i\omega_n - (\pm E_{\vec{l}} - \mu_G)/\hbar} e^{i\vec{l}\cdot\vec{r}}, \end{aligned} \tag{22}$$

where $s_0 = 3\sqrt{3}a^2/2$ is the unit cell area of the graphene with a being the smallest in-plane carbon atom spacing, and BZ denotes the first Brillouin zone. In the last line, the integral domain is expanded to infinity for simplicity.

3. Critical Current of Superconductor-Graphene-Superconductor junction

3.1. Monolayer case

In monolayer case, we obtain

$$\left|g^{\alpha\beta}(\vec{r}^{A}_{jj'},\omega_n)\right|^2 = s_0^2\Omega_{\vec{r}^{A}_{jj'}}\frac{|\omega_n - i\mu_G/\hbar|^2}{(\pi v_F^2)^2}\left|K_0\left(|\vec{r}^{A}_{jj'}|\frac{\omega_n - i\mu_G/\hbar}{v_F}\mathrm{sign}(\omega_n)\right)\right|^2 \tag{23}$$

$$\Omega_{\vec{r}^{A}_{jj'}} = \left|e^{i\vec{K}\cdot\vec{r}_{jj'}} + e^{i\vec{K}'\cdot\vec{r}_{jj'}}\right|^2 = 2\left\{1 + \cos(\vec{K} - \vec{K}')\cdot\vec{r}^{A}_{jj'}\right\} \tag{24}$$

where $\vec{r}^{A}_{jj'} = \vec{r}^{A}_{j} - \vec{r}^{A}_{j'}$. Here we used the formula

$$\int_0^\infty \frac{tJ_0(ta)}{t^2 + z^2}dt = K_0(za), \quad (a > 0, \mathrm{Re}(z) > 0), \tag{25}$$

where $J_n(z)$ and $K_n(z)$ are the n-th order Bessel function and modified Bessel function of the second kind, respectively. $\Omega_{\vec{r}^{A}_{jj'}}$ gives a atomic scale modulation of the Josephson energy as a function of the tunnel site, whose average over lattice sites is 2. By substituting this to Eq. (20), we obtain F_J^{AA}. In a similar way, F_J^{AB} can be obtained in a slightly different functional form. However numerical difference between these two is almost negligible and we discuss only F_J^{AA} in this section.

In Fig. 3, we have plotted the junction length d and temperature dependence of the critical current in monolayer system. We set $\mu_G = 0$ in the following plots. Fig. 3 (a) is the logarithmic plot of I_c as a function of d. It can be seen that I_c decay exponentially as a function of d, when $d > l_0$, where $l_0 = (\hbar v_F)/\Delta_0(0)$, and $\Delta_0(0)$ is the gap at $T = 0$. (Note that l_0 is not exactly the coherence length of superconducting leads, since v_F here is that of graphene.) Therefore we can define the proximity length $\xi_g(T)$ by $I_c \sim \exp(-d/\xi_g(T))$. Fig. 3 (b) shows the logarithmic plot of I_c vs. $\sqrt{T}$, T and T^2. Remarkably, the curve is most linear for T^2-plot, which means that $\xi_g(T)$ behaves like $\propto 1/T^2$. This is somewhat unusual, since in the S-conventional metal-S junction, we usually obtain $\xi_g(T) \propto 1/T$ for clean-metal case, and $\xi_g(T) \propto 1/\sqrt{T}$ for dirty-metal case. A similar behavior is experimentally observed for junction using gated thin graphite films [18], although its relation to present result is not conclusive at this moment.

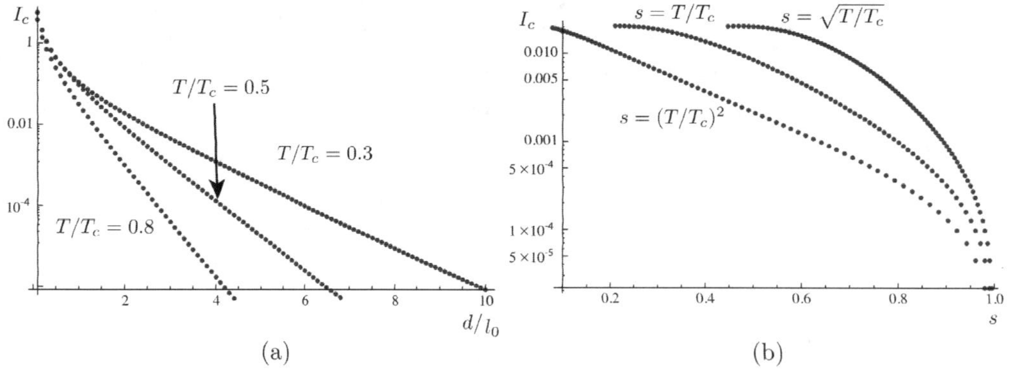

Figure 3. (a) Logarithmic plot of the critical current in monolayer system as a function of the junction length d. (b) Three different plots of the temperature dependence of I_c where $d = 2l_0$. The parameter s denotes $\sqrt{T/T_c}$, T/T_c and $(T/T_c)^2$ for three curves.

3.2. Bilayer case

In bilayer case, we obtain

$$\left|g^{\alpha\beta}(\vec{r}^{A}_{jj'},\omega_n)\right|^2 = s_0^2\Omega_{\vec{r}^{A}_{jj'}}\left(\frac{m_b}{\pi\hbar}\right)^2 4\,\mathrm{Im}\left[K_0\left(|\vec{r}^{A}_{jj'}|\frac{\sqrt{2m_b(i\hbar\omega_n+\mu_G)}}{\hbar}\right)\right]^2 \tag{26}$$

$$\left|\bar{g}^{\alpha\beta}(\vec{r}^{B}_{jj'},\omega_n)\right|^2 = s_0^2\overline{\Omega}_{\vec{r}^{B}_{jj'}}\left(\frac{m_b}{\hbar\pi}\right)^2 4\,\mathrm{Re}\left[K_1\left(|\vec{r}^{B}_{jj'}|\frac{\sqrt{2m_b(i\hbar\omega_n+\mu_G)}}{\hbar}\right)\right]^2 \tag{27}$$

$$\overline{\Omega}_{\vec{r}^{B}_{jj'}} = \left|e^{i\vec{K}\cdot\vec{r}^{B}_{jj'}-i\theta_j} - e^{i\vec{K}'\cdot\vec{r}^{B}_{jj'}+i\theta_j}\right|^2 \tag{28}$$

where $m_b = 2v_F^2/t_\perp$, $\vec{r}^{A}_{jj'} = \vec{r}^{A}_{j} - \vec{r}^{A}_{j'}$, $\vec{r}^{B}_{jj'} = \vec{r}^{A}_{j} - \vec{r}^{B}_{j'}$, and θ_j is the angle between $\vec{r_j}^A$ and the x-axis. (The sign of the square root in the arguments of K_0 and K_1 should be determined so that the real part is positive.) By use of Eq. (20), we obtain the Josephson energy of bilayer system. In Fig. 4 (a), we have plotted the contribution of F_J^{AA} and F_J^{AB} to I_c separately. One can see, in contrast to monolayer case, F_J^{AA} and F_J^{AB} contributions show oscillation as a function of d. However, as seen from Fig. 4 (b), this oscillation is washed out when we average the two contributions. The d dependence of I_c is then simple exponential decay as in the case of monolayers. In Fig. 4 (c), the temperature dependence of I_c is examined. In this case, $\xi_g(T) \propto 1/\sqrt{T}$ is a good approximation, which probably originates from the square root in the argument of K_0 and K_1.

4. Discussion

In this paper we have studied the critical current of S-G-S josephson junction. It was found that the behavior of critical current as a function of junction length d and temperature is largely affected by the band structure of graphene.

In monolayer case, we found an anomalous temperature dependence of the proximity length $\xi_g(T) \propto 1/T^2$. The origin of this behavior is not clear at this stage, however contributions from ω_n-sum is important, since taking account of only the most dominant terms gives ordinary $\xi_g(T) \propto 1/T$ behavior as one can see from Eq. (24). Therefore $\xi_g(T) \propto 1/T^2$ comes from the mixed contributions from many modes and is probably difficult to be reduced to a simple picture.

In bilayer case, we found a peculiar oscillation of the critical current as a function of junction length. (The oscillation can also be seen in temperature domain.) The oscillation disappears if we average over contributions of tunneling at A- and B-sublattice of graphene. From Eq. (28) the origin of oscillation can be attributed to the interference between electrons in upper and lower band at the Dirac points. However the curvature of the band is also important, since we do not see any corresponding phenomena in monolayer case. If we understand detailed origin of this oscillation, it may shed a new light on the subtle quantum interference effect in bilayer graphene, which however is a future problem.

Acknowledgments

This work was supported by Grants-in-Aid for Scientific Research (KAKENHI) 22540329. Some parts of the calculations were carried out on computers in YITP, Kyoto University.

References

[1] Novoselov K S, Geim A K, Morozov S V, Jiang D, Zhang Y, Dubonos S V, Grigorieva I V and Firsov A A 2004 Science **306** 666

[2] Novoselov K S, Geim A K, Morozov S V, Jiang D, Katsnelson M I, Dubonos S V, Grigorieva I V and Firsov A A 2005 Nature **438** 197

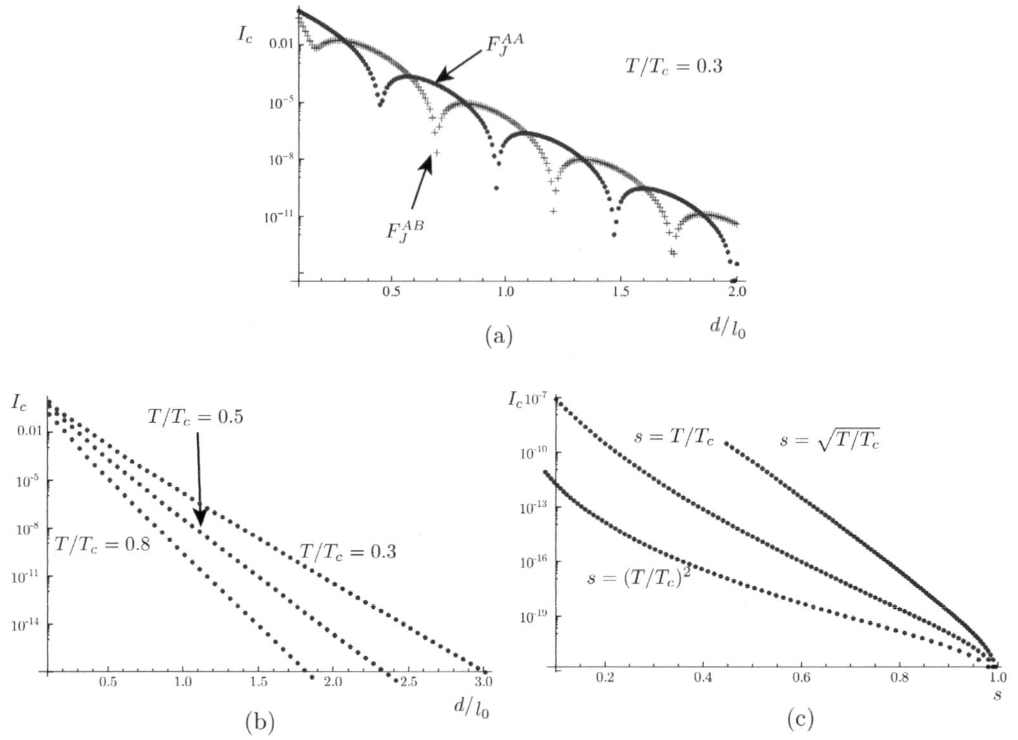

Figure 4. (a) Logarithmic plot of the contributions of F_J^{AA} and F_J^{AB} to I_c as a function of the junction length d. ($T = 0.3T_c$) (b) Three plots with different temperatures of the d-dependence of I_c, where both contribution from F_J^{AA} and F_J^{AB} are included. (c) Three different plots of the temperature dependence of I_c. The parameter s denotes $\sqrt{T/T_c}$, T/T_c and $(T/T_c)^2$ for three curves. ($d = 2l_0$)

[3] Zhang Y, Tan Y T, Stormer H L and Kim P 2005 Nature **438** 201
[4] Slonczewski J C and Weiss P R 1958 Phys. Rev. **109** 272
[5] Koshino M and Ando T 2006 Phys. Rev. B **73** 245403
[6] Beenakker C W J 2006 Phys. Rev. Lett. **97** 067007
[7] Titov M and Beenakker C W J 2006 Phys. Rev. B**74** 041401
[8] Heersche H B, Herrero P J, Oostinga J B, Vandersypen L M K and Morpurgo A F 2007 Nature **446** 56
[9] Moriki T, Kanda A, Sato T, Miyazaki H, Odaka S, Ootsuka Y, Aoyagi Y and Tsukagoshi K 2007 Physica E**40** 241
[10] Sato T, Moriki T, Tanaka S, Kanda A, Miyazaki H, Odaka S, Ootsuka Y and Aoyagi Y 2008 Physica E **40** 1495
[11] Sato T, Kanda A, Moriki T, Goto H, Tanaka S, Ootsuka Y, Miyazaki H, Odaka S, Tsukagoshi K and Aoyagi Y 2008 Physica C **486** 797
[12] Mori M, Hikino S, Takahashi S and Maekawa S 2007 J. Phys. Soc. Jpn., **76** 054705
[13] González J and Perfetto E 2007 Phys. Rev. B**76** 155404
[14] González J and Perfetto E 2008 J. Phys.: Condens. Matter **20** 145218
[15] Hayashi M, Yoshioka H and Kanda A 2008 J. of Phys.: Conference Series **109** 012014
[16] Burset P, Levy Yeyati A and Martín-Rodero A 2008 Phys. Rev. B**77** 205425
[17] Black-Schaffer A M and Doniach S 2009 Phys. Rev. B**79** 064502
[18] Kanda A, in preparation.

Journal of Physics: Conference Series **248** (2010) 012003 doi:10.1088/1742-6596/248/1/012003

Electromagnetic response of the composites containing chemically modified carbon nanotubes

A M Nemilentsau[1], M V Shuba[1], P N D'yachkov[2], G Ya Slepyan[1], P P Kuzhir[1] and S A Maksimenko[1]

[1] Institute for Nuclear Problems, Belarus State University, Bobruiskaya 11, 220030 Minsk, Belarus

[2] Kurnakov Institute of General and Inorganic Chemistry, Russian Academy of Sciences, Leninskii Prospekt 31, 119991 Moscow, Russia

E-mail: andrei.nemilentsau@gmail.com

Abstract. We demonstrate theoretically that the chemical modification of the bundles and composites containing mixture of semiconducting and metallic single-wall carbon nanotubes (SWNTs) leads to the substantial enhancement of the characteristics of their electromagnetic response in the terahertz (THz) frequency range. Boron and nitrogen doping is used to illustrate the effect. In particular, frequencies of antenna resonances in the bundles containing doped SWNTs are shifted to the blue in comparison to the ones in the bundles containing the same number of pristine SWNTs. Moreover, doping increases the resonant values of the bundles polarizability. Enhancement of the conductivity of the composite containing doped SWNTs is also demonstrated. The origin of the behavior is the metallization of the chemically modified semiconducting SWNTs in the bundles and composites due to the injection of the additional charge carriers.

1. Introduction

It is known that chemical modification of carbon nanotubes can lead to substantial improvement of their electric transport characteristics due to the enrichment by charge carriers [1]. Particularly, appreciable decrease of the dc resistivity has been reported [2] for the boron-doped multi-walled nanotubes. Moreover, increase of the dc electrical conductivity and the metallic to semiconductor nanotubes ratio has been experimentally demonstrated in case of the nitrogen substitutional doping of the composites containing mixture of metallic and semiconductor single-wall carbon nanotubes (SWNTs) [3]. Note that the metallization of semiconducting SWNTs due to nitrogen doping could serve as a tool for percolation threshold reducing in SWNT-based composites; normally this threshold is attained at the pristine SWNT weight concentration of about 1 % [4].

By analogy with the quasi-static regime one can expect that the high-frequency electromagnetic response of carbon nanotubes can also be controlled by the chemical modification. The enhancement of the terahertz electromagnetic response of SWNT bundles and composites is of particular interest due to the rapid progress in the field of the teraherz imaging and sensing [5]. The SWNT-based terahertz detectors have already been demonstrated experimentally [6, 7]. Metallic SWNT antennas [8, 9, 10, 11] can be utilized for the terazertz radiation manipulation as they are predicted to possess antenna resonances in the frequency rage from 1 to 30 THz [9]. In contrast, semiconducting SWNTs weakly interact with electromagnetic radiation in this frequency range and therefore are not suitable for terahertz applications. Unfortunately, fabrication of uniform composites containing only metallic SWNTs is still a challenging task [12] and typical SWNT bundles and composites consist of mixtures of metallic and semiconducting SWNTs.

In this work we demonstrate theoretically that SWNTs doping by either boron (B) or nitrogen (N) atoms improve electromagnetic response of bundles and composites comprising mixture of metallic and semiconducting SWNTs. Substitution of carbon atoms in the SWNT crystalline lattice by the third and fifth valence group atoms injects additional charge carriers to the SWNT. Weakly affecting the metallic SWNT conductivity this leads to the semiconductor SWNTs metalization. In its turn, this leads to the enhancement of the antenna efficiency of the bundles and conductivity of the composites comprising chemically modified SWNTs in comparison with the pristine SWNT based bundles and composites.

The paper is organized as follows. The band structures of both pristine and chemically modified SWNTs are presented in section 2. In section 3 we present results of calculations of electromagnetic response of chemically modified SWNT based bundles and composites. Brief conclusions are presented in section 4.

2. Band structure

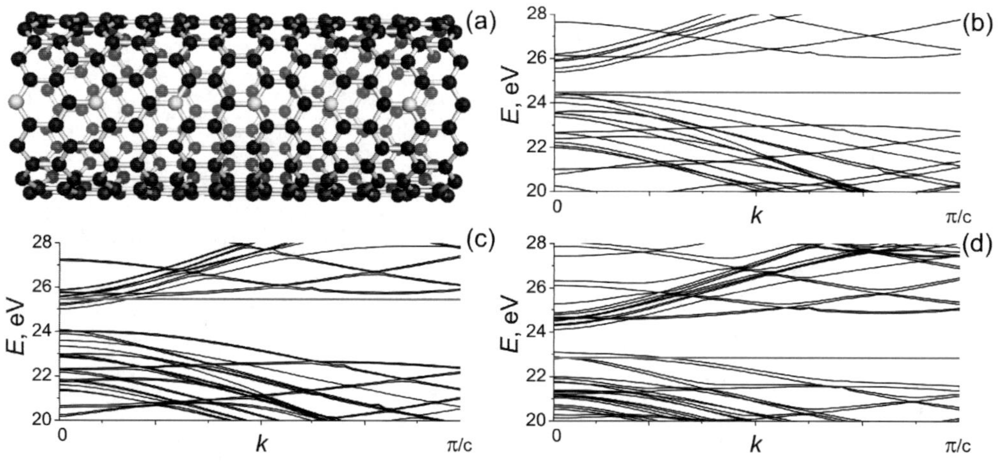

Figure 1. (a) Fragment of the chemically modified (13,0) SWNT containing six unit cells. Carbon atoms are designated by blue color. A single carbon atom in each unit cell of a SWNT is replaced by the dopant atom (yellow color). (b)-(d) Band structure of (13,0) SWNT in the vicinity of the Fermi level (red line): (b) pristine, (c) N doped, (d) B doped. $c = 3b$ is the SWNT translational vector length, where $b = 1.42$ Å is the interatomic distance in the SWNT.

In order to demonstrate the SWNT doping influence on the SWNT electromagnetic response we restrict our consideration to the case when only a single carbon atom in each unit cell of a SWNT is replaced by the dopant atom (Fig. 1(a)). Band structures of both pristine and chemically modified SWNTs are calculated using the muffin-tin potential approximation [13] and the linear augmented cylindrical waves (LACW) method [14, 15]. The LACW method is the well known in the solid state physics augmented plane-wave method [13] transferred to the systems possessing cylindrical symmetry. The SWNT Bloch functions are the linear superpositions of the augmented cylindrical waves that obtained by matching the solutions of the Schrödinger equation within each muffin-tin sphere with cylindrical-wave solutions in the interstitial region [14]. The Bloch functions expansion coefficients and electron dispersion law are defined by the variational method. We construct augmented cylindrical waves assuming that the electron motion is restricted to the cylindrical layer by two infinitely high cylindrical potential barriers of radii $R_l - 1.38$ Å and $R_l + 1.27$ Å, where R_l is the radius of the SWNT of type l. The LACW method has been already successfully implemented to calculate band structure of the pristine SWNTs [15]. Results of calculations of band structures of both pristine and chemically modified (13,0) SWNTs are presented on Fig. 1. As one can see, carbon atoms replacement by N (B) atoms leads to the Fermi energy increasing (decreasing) due to the additional electron (holes) incorporation. As the Fermi level crosses either the conductivity (1c) or the valence band (1d) one can make a conclusion that chemical modification by N and B atoms leads to the metalization of semiconducting SWNT.

3. Electromagnetic response of chemically modified SWNTs

To study electromagnetic wave scattering by SWNT-based bundles and composites we use the integral-equation technique presented earlier for single [9] SWNTs and for bundles [10] of SWNTs. To apply integral equations formalism [9], the linear axial surface conductivity of a SWNT should be calculated first. In the terahertz frequency range the dominant contribution to the SWNT conductivity is provided by the electron intraband motion [16]. Thus, the conductivity of pristine semiconducting SWNTs is negligibly small in comparison with the conductivity of metallic ones and is assumed to be zero in further calculations. To evaluate the axial conductivity of metallic SWNTs and chemically modified semiconducting SWNTs (revealing metallic behavior accordingly to Fig. 1), we use the Drude model of the conductivity[16]. The electron relaxation time for both pristine and doped SWNTs is assumed to be the same since the electron mean-free path corresponding to the scattering on the optical phonons is much less than the mean-free path corresponding to the scattering on impurities [17]. This assumption is in a good agreement with the experimental results [18] on the electromagnetic waves absorption by SWNT-based composites in the THz range: The THz absorption peak width is only slightly affected by the SWNT doping. Further we assume the relaxation time to be equal to 3.3×10^{-14} s [19].

As an example, let us consider the electromagnetic response of an approximately cylindrical bundle with the diameter 3.8 nm, containing three (13,0), two (12,0) and two (11,0) SWNTs (further we refer to it as M-bundle). The polarizability spectra of both pristine and N-doped M-bundles of two different lengths L are depicted in Figs. 2a,b. The electromagnetic response of B-doped SWNTs is similar to that of N-doped SWNTs and is not discussed further. As one can see, polarizability spectra of both pristine and doped SWNT bundles contain resonance peaks, the resonance frequency of the doped SWNT bundles being blue-shifted in comparison with the pristine ones. These resonances are the antenna resonances of surface waves in the SWNT bundle. The angular resonance frequency ω is defined by the condition [9]

$$L\omega = \pi c(2s-1)\mathrm{Re}\beta, \tag{1}$$

where s is an integer and c is the speed of light in the vacuum. In single SWNT the slow-wave coefficient Reβ can be as small as [16] Re$\beta \approx 0.01$. However, in the bundle the slow-wave coefficient is increased due to the electromagnetic interaction of the metallic SWNTs [10] resulting in the resonance frequency blue shift in comparison to the single SWNT resonance frequency. In its turn, the doping-caused metallization of semiconducting SWNTs in the M-bundle increases the number of metallic SWNTs in the bundle in comparison to the pristine SWNT bundle and thus leads to the additional resonance frequency blue-shift.

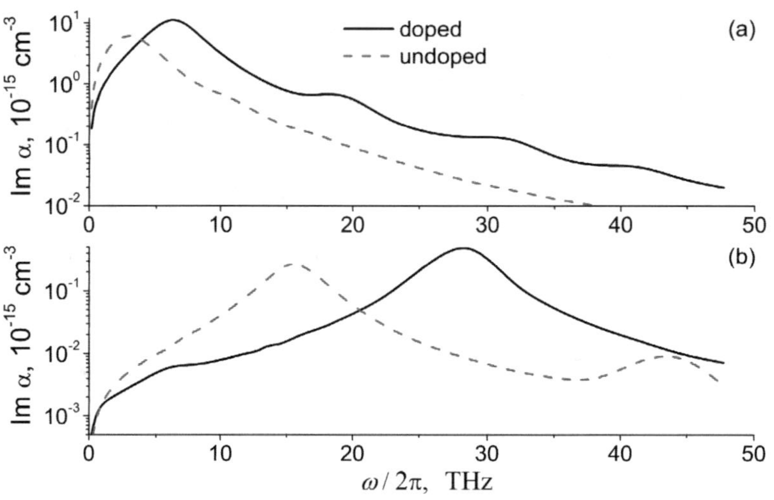

Figure 2. Frequency dependence of the imaginary part of polarizability of SWNT bundle comprising three (13,0), two (12,0) and two (11,0) SWNTs before (solid line) and after (dashed line) the SWNTs chemical modification by the nitrogen atoms (Fig. 1). SWNTs length in the bundle is equal to (a) $L = 1000$ nm and (b) $L = 200$ nm. Doped SWNTs contain approximately 2 percent of nitrogen impurities.

Let us now consider a dilute composite material comprising different types of randomly dispersed and oriented SWNTs in the matrix with dielectric permeability close to 1. The distribution function $N_\ell(L)$ describes the number of SWNTs of type ℓ and length L per unit volume. All orientations are equiprobable, and f denotes the volume fraction occupied by the SWNTs (conceived as cylinders of volume $\pi R_\ell^2 L$). To calculate the effective conductivity $\sigma_{eff}(\omega)$ of the composite material in case when $L \ll \lambda$, where $\lambda = 2\pi/k$ is the free-space wavelength, we use the simple Waterman-Truell formula [20]

$$\sigma_{eff}(\omega) \cong \frac{\omega}{4i\pi}\left[1 + \frac{4\pi f}{3}\sum_{\ell}\int_0^\infty \alpha_\ell(\omega, L)N_\ell(L)dL\right], \qquad (2)$$

where $\alpha_\ell(\omega, L)$ is the polarizability of an isolated SWNT calculated using the integral-equation technique [9].

We calculate the conductivity of the composite containing SWNTs of length $L = 1000$ nm of three different chiralities (with chiral vectors (11,0), (12,0), (13,0)). For simplicity we assume that the number of SWNTs of different types in the composite is the same. Terahertz peak in the conductivity spectrum of undoped SWNT composite depicted on Fig. 3(a) is due to the antenna resonance in the metallic (12,0) SWNTs. The nitrogen doping leads to the substantial enhancement of the conductivity of semiconducting (11,0) and (13,0) SWNTs while the conductivity of the (12,0) metallic SWNT changes only slightly. As a result, the antenna resonances emerge also in semiconducting SWNTs and the effective composite conductivity is increased over the whole low-frequency range, as shown in Fig. 3(a). It should be noted, that doping of the composite containing isolated SWNTs leads to the increase of the resonant peak amplitude and width while the resonance frequency remains practically unchanged. Similar behavior of the terahertz conductivity peak for SWNT-based composite doped by the tetrafluorotetracyanoquinodimethane is observed experimentally in Ref. [18].

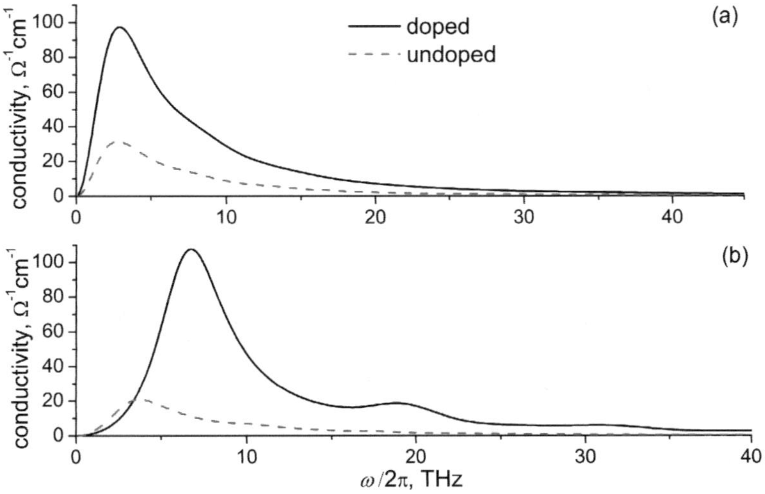

Figure 3. Frequency dependence of conductivity $\mathrm{Re}(\sigma_{eff})$ of the SWNT-based composite, containing (a) (11,0), (12,0), (13,0) SWNTs and (b) M-bundles (Fig. 2) before (solid line) and after (dashed line) doping. $L = 1000$ nm. Doped SWNTs contain approximately 2 percent of nitrogen impurities. $f = 0.005$.

Conductivity frequency dependence of the composite containing identical M-bundles (Fig. 3) is presented on Fig. 3(b). In this case doping not only increases the resonant peak amplitude but also shifts to the blue the low-frequency resonance in correspondence with the blue-shift of the resonance frequency of the single bundle polarizability (Fig. 2(a)).

4. Conclusions

Concluding, we demonstrate theoretically that the chemical modification of the bundles and composites containing mixture of the semiconducting and metallic SWNTs is an efficient way of the control of their electromagnetic response in the terahertz frequency range. In particular, in CNT bundles we predict the doping-caused increase of the antenna resonance frequency and amplitude. The enhancement of the conductivity of the doped SWNT composites resulting in the improvement of the absorbance and reflectance of the SWNT-based samples has also been demonstrated. Extrapolating these results to the microwave range, we can propose doping as a tool for the enhancement of the electromagnetic shielding of composite materials with embedded SWNTs and nanotube bundles.

Acknowledgments

This research was partially supported by the IB BMBF (Germany) under project BLR 08/001, EU FP7 CACOMEL project FP7-247007, ISTC project B-1708, and NATO Collaborative Linkage Grant (Reference No. CBP. EAP.CLG983910). The work of A. M. N. was supported by the BRFFR (Belarus) young scientists grant F09M-071. The work of P. N. D. was supported by the RBRF (Russia) grant 08-03-00262.

References

[1] Lee R S, Kim H J, Fisher J E, Lefebvre J, Radosavljevic M, Hone J and Johnson A T 2000 *Phys. Rev.* B **61** 4526
[2] Ishii S, Watanabe T, Ueda S, Tsuda S, Yamaguchi T and Takano Y 2008 *Appl. Phys. Lett.* **92** 202116

[3] Villalpando-Paez F *et al* 2006 *Chem. Phys. Lett.* **424** 345
[4] Liu L, Matitsine S, Gan Y B, Chen L F, Kong L B and Rozanov K N 2007 *Appl. Phys. Lett.* **101** 094106
[5] Chan W L, Deibel J and Mittleman D M 2007 *Rep. Prog. Phys.* **70** 1325
[6] Fu K, Zannoni R, Chan C, Adams S H, Nicholson J, Polizzi E and Yngvesson K S 2008 *Appl. Phys. Lett.* **92** 033105
[7] Kawano Y, Uchida T and Ishibashi K 2009 *Appl. Phys. Lett.* **95** 083123
[8] Hanson G W 2005 *IEEE Trans. Antennas Propag.* **53** 3426
[9] Slepyan G Ya, Shuba M V, Maksimenko S A and Lakhtakia A 2006 *Phys. Rev.* B **73** 195416
[10] Shuba M V, Maksimenko S A and Lakhtakia A 2007 *Phys. Rev.* B **76** 155407
[11] Rutherglen C and Burke P 2009 *Small* **5** 884
[12] Hersam M C 2008 *Nat. Nanotech.* **3** 387
[13] Ashcroft N W and Mermin N D 1976 *Solid State Physics* (New York: Holt-Saunders)
[14] D'yachkov P N 2003 *Encyclopedia of Nanoscience and Nanotechnology* vol 1, ed H.S. Halwa (New York: Am. Sci. Publ.) p 191
[15] D'yachkov P N and Hermann H 2004 *J. Appl. Phys.* **95** 399
[16] Slepyan G Ya, Maksimenko S A, Lakhtakia A, Yevtushenko O and Gusakov A V 1999 *Phys. Rev.* B **60** 17136
[17] Yao Z, Kane C L and Dekker C 2000 *Phys. Rev. Lett.* **84** 2941
[18] Akima N *et al* 2006 *Adv. Mater.* **18** 1166
[19] Slepyan G Ya, Shuba M V, Maksimenko S A, Thomsen C and Lakhtakia A 2010 *Phys. Rev.* B **81** 205423
[20] Waterman P C and Truell R 1961 *J. Math. Phys.* **2** 512

Journal of Physics: Conference Series **248** (2010) 012004 doi:10.1088/1742-6596/248/1/012004

ELASTIC PROPERTIES OF GRAPHENE-GRAPHANE NANORIBBONS

O E Glukhova[1], **I N Saliy**[1], **R Y Zhnichkov**[2], **I A Khvatov**[2], **A S Kolesnikova**[1] **and M M Slepchenkov**[1]

[1]Department of Physics, Saratov State University, 410012, Saratov, Russia
[2]Povolzhsky Regional Center of New Information Technologies, Network and Telecommunication Systems, Saratov State University, 410012, Saratov, Russia

E-mail: glukhovaoe@info.sgu.ru

Abstract: The results of theoretical investigation of the atomic structure, deformations, and elastic properties of graphene-graphane nanoribbons (GGN) are represented here. To study the properties of GGN we applied the empirical method based on the bond-order potential developed by Brenner and the tight-binding method. We calculated the Young's pseudo-modulus of GGN and the strain energy of GGN subject to axial tension and compression. The curve of the strain energy collapse occurs at the axial compression of 0.03-0.04. Plane atomic network subject to axial compression becomes wave-like. This is a so-called phase transition. Elasticity of armchair-graphene nanoribbons is greater than elasticity of armchair nanotubes and graphane nanoribbons with the same width and length.

Introduction

Discovery of new materials leads obligatory to discovery of their new applications and properties.

Synthesis of graphene-graphane nanoribbons (GGN) has stimulated very broad and intense research both of theoretical investigation of GGN and their applications. This is due to their unique properties. Graphane is the simplest organic compound made only of carbon and hydrogen atoms. All the known hydrocarbons, until now, are molecules that consist of a carbon backbone with hydrogen atoms attached. Graphane is a semiconductor and it provides a playground for fundamental science and technological applications. The first-principles total energy calculations show that this compound has a very favorable formation energy, quite comparable with other hydrocarbons.

As it is known, carbon nanotubes demonstrate excellent mechanical properties. So it is possible that the other carbon nanostructures such as GGN may also be of high elasticity.

In the present work two methods were used to investigate the elastic properties: the empirical method based on the bond-order potential developed by Brenner and the tight-binding method. To describe the interactions between bonded and non-bonded atoms a multiscale topological network around each atom was introduced.

1. Empirical study

1.1. Computational method

To study the elastic properties and deformations of GGN we applied the empirical method based on the bond-order potential developed by Brenner [1]. The entire system energy is described by the sum of the binding energy E_b, the torsional energy E_{tors} and the van der Waals energy E_{vdW}:

$$E_{tot} = E_b + E_{tors} + E_{vdW} . \tag{1}$$

To describe the interaction between the atom and its environment we introduce three different regions in the topological network (see figure 1).

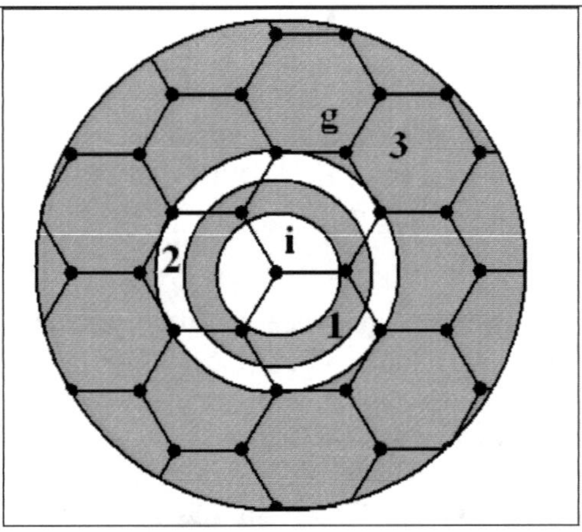

Figure 1. The multiscale topological network around each an atom

As shown in figure 1, there are three regions around atom i: the first (near), the second (intermediate) and the third (far). Atoms from the near region are covalently bonded with atom i, atoms from the other regions are non-bonded with atom i. The far region has no borders.

Each pair of covalently bonded atoms interacts via potential-energy:

$$E_b = \frac{1}{2}\sum_{i=1}^{Nat}\left(\sum_{j(\neq i)}\left(V_R(r_{ij}) - B_{ij}V_A(r_{ij})\right)\right) \tag{2}$$

This is binding energy. Here V_R is the repulsive pair term, V_A is the attractive pair term, r_{ij} is the distance between the atom with number i and atom j from the near region. The function B_{ij} is a many-body term. This term was introduced to describe the specificity of σ–π interaction. So, the value of the binding energy depends on the position and chemical identity of atoms.

The original bond-order potential by Brenner lacked any torsional interactions around a single bond. To describe the non-bonded interaction between the atoms we introduce the torsional interaction energy to the entire system energy. The torsional potential is given by

$$E^{tors} = \frac{1}{2}\sum_{i=1}^{Nat}\left(\sum_{j\neq i}\left(\sum_{k\neq i,j}\left(\sum_{l\neq i,j,k} V_{tors}(\omega_{ijkl})\right)\right)\right) \tag{3}$$

The torsional potential $V_{tors}(\omega_{ijkl})$ is given as a function of dihedral angle ω. The torsion angle ω_{ijkl} is defined in the usual way as the angle between the plane defined by the vectors $\boldsymbol{r}_{ik}$ and $\boldsymbol{r}_{ij}$ and that of defined by $\boldsymbol{r}_{ij}$ and $\boldsymbol{r}_{jl}$. Here atoms j and k are given from the intermediate region (second) and atom l is given from the third (far) region.

The van der Waals energy E_{vdW} defines the interaction between non-bonded atoms:

$$E_{vdW} = \frac{1}{2}\sum_{i=1}^{Nat}\left(\sum_{j(\neq i)} V_{vdW}(r_{ij})\right) \tag{4}$$

As known, it is advisable to incorporate the van der Waals potential only if the short-distance potential becomes zero. This is to prevent an artificial reaction barrier from forming due to the steep repulsive wall of the van der Waals potential. To achieve a seamless coupling between a short-distance potential and a long-distance potential we apply our multiscale definition of the topological network around the atom.

The van der Waals interaction energy may be described by the Lennard-Jones, Morse, Buckingham potentials and so on. We use the Morse potential that is given by

$$V_{Morse}(r_{ij}) = D_e\left(\left(1-\exp\left(-\beta(r_{ij}-r_e)\right)\right)^2-1\right)+E_r\cdot\exp\left(-\beta' r_{ij}\right), \tag{5}$$

where D_e is the average bond energy, E_r is the repulsion nucleus energy, β, β' [2]. The interatomic distance in graphite calculated by the empirical method is equal to 0.1421 nm.

Using the empirical method we studied some GGN with length-to-width ratio $\frac{L}{D}$ from 2 to 3.22. Nanostructures with ratio $\frac{L}{D}$ less than 3 are classified as nanoparticles, nanostructures with ratio greater than 3 are classified as nanoribbons. The study of deformations and elastic properties of nanoparticles and nanoribbons was implemented with the help of the following algorithm:

1) Optimization of atomic structure by entire system energy minimization (see equation (1)) on atomic coordinates (the atomic structure obtained from the previous optimization);
2) Tension or compression of the atomic network of a nanoribbon and reoptimization of atomic structure with fixed atoms on the nanoribbon ends;
3) Calculation of the Young's pseudo-modulus for the elastic tension of a nanoribbon by 1% according to the formula:

$$Y_p = \frac{F}{D}\frac{L}{\Delta L}, \tag{6}$$

where a deformation force is given by $F = \frac{2\Delta E}{\Delta L}$. Here ΔE is the strain energy, namely, the total energy at a given axial strain minus the total.

1.2. Results of empirical study

We calculated the Young's pseudo-moduli of the armchair- and zigzag-nanoribbons with different length and width. The dependences of the Young's pseudo-modulus are shown in figure 2. We can see that the pseudo-modulus increases with the length and decreases with the width. When the length-to-width ratio $\frac{L}{D}$ becomes 3 and greater than 3 the value of the Young's pseudo-modulus stops to increase.

We also studied the axial tension and compression of nanoribbons. At first the length is increased or decreased by 1-10 %, then the entire energy is minimized but without affecting the atoms at the both ends. Strain energy of nanoribbons subject to axial tension is presented in figure 3.

The reported results show that the stress-strain curve may be fitted by a parabola. So, at the respective strains from 0 to 0.12 we have the region of elastic deformation. We studied nanoribbons subject to axial compression. The strain energy collapse occurs at the respective strain 0.04.

We compared the elastic properties of the carbon nanotube and the nanoribbon of the same width. We calculated the Young's pseudo-modulus of the nanotube (5,5) with the perimeter about 2.1 nm and one of the armchair-nanoribbon with the width 2.3 nm. The Young's pseudo-modulus of the armchair-nanoribbon is greater than the modulus of the nanotube by 27 %.

The dependences of the strain energy and the Young's pseudo-modulus for graphane-nanostructures are given in figure 4.

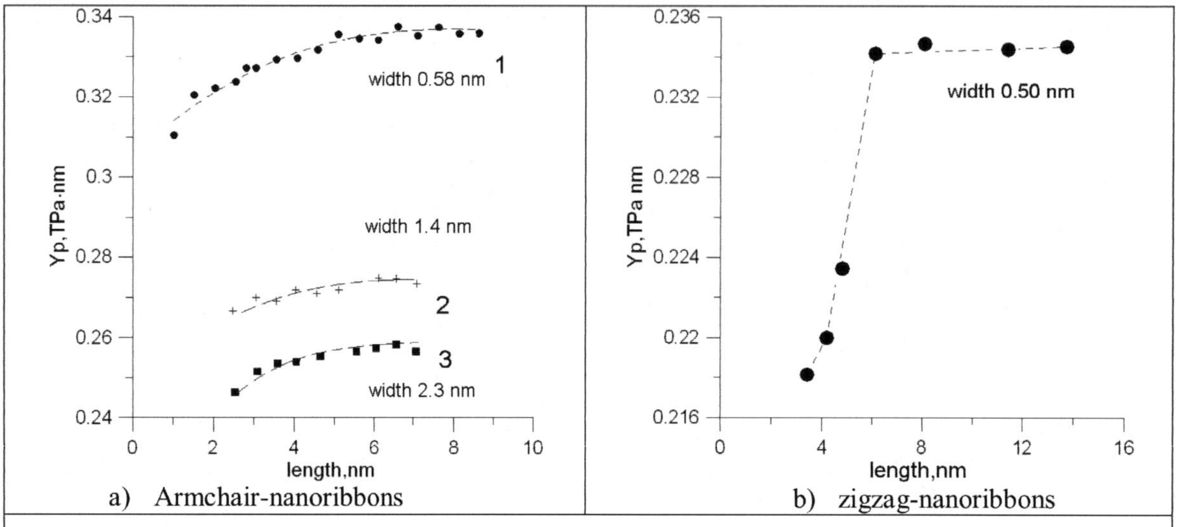

Figure 2. The Young's pseudo-moduli of nanoribbons

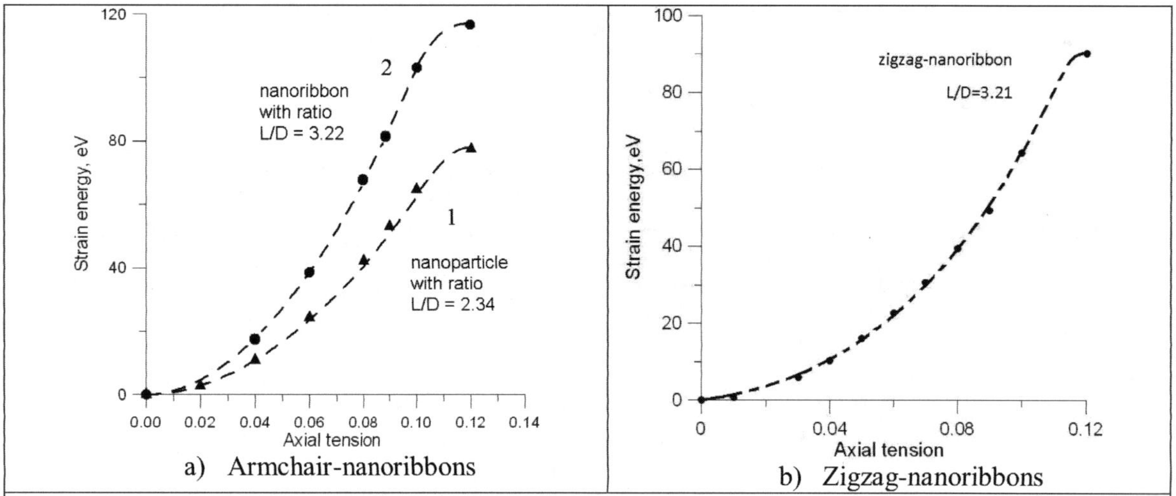

Figure 3. Strain energy of nanoribbons subject to axial tension

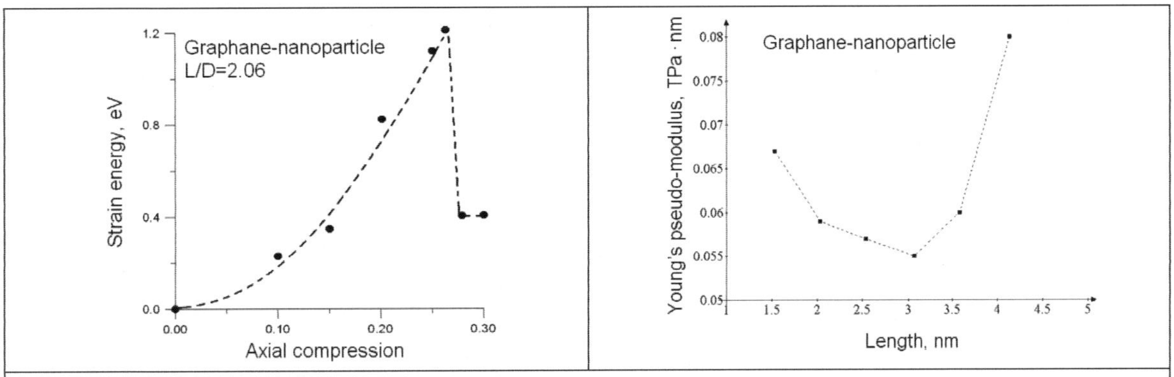

Figure 4. Parameters of elasticity of graphane-nanostructures

2. Quantum-mechanical study

2.1. Computational scheme

Let us consider the unusual process of the axial compression of the nanoribbon. In this work TB was applied to study this process. To study stability of carbon nanoclusters the TB method was earlier presented in [3] and it was successfully implemented. The energy of a system of ion cores and valence electrons is written as

$$E_{tot} = E_{bond} + E_{rep} \tag{7}$$

Here the term E_{bond} is the bond structure energy that is calculated as the sum of energies of the single-particle occupied states. Those single-particle energies become known by solving the following equation

$$\mathbf{H}\,|\psi_n\rangle = \varepsilon_n\,|\psi_n\rangle, \tag{8}$$

where $\mathbf{H}$ is the one-electron Hamiltonian, ε_n is the energy of the nth single-particle state. The wave functions $|\psi_n\rangle$ can be approximated by linear combination

$$|\psi_n\rangle = \sum_{l\alpha} C^n_{l\alpha}\,|\varphi_{l\alpha}\rangle, \tag{9}$$

where $\{\varphi_{n\alpha}\}$ is an orthogonal basis set, l is the quantum number index and α labels the ions. The matrix elements in equation (8) are calculated after fitting a suitable database obtained from the experiments and were presented in [3]. For carbon compounds the matrix elements are calculated by

$$V_{\alpha\beta}(r) = V^0_{\alpha\beta}\left(\frac{p_3}{r}\right)^{p_1} \exp\left\{p_1\left[-\left(\frac{r}{p_2}\right)^{p_4} + \left(\frac{p_3}{p_2}\right)^{p_4}\right]\right\}, \tag{10}$$

where r is the distance between atoms.

Term E_{rep} in equation (9) is phenomenon energy that is a repulsive potential. It can be expressed as a sum of two-body potentials as

$$E_{rep} = \sum_{\alpha,\beta\rangle\alpha} V_{rep}\left(r_{\alpha\beta}\right), \tag{11}$$

where V_{rep} is a pair potential between atoms at α and β. This two-body potential describes an interaction between bonded and non-bonded atoms and is presented in [3]

$$V_{rep}(r) = p_5\left(\frac{p_3}{r}\right)^{p_6} \exp\left\{p_6\left[-\left(\frac{r}{p_2}\right)^{p_4} + \left(\frac{p_3}{p_2}\right)^{p_4}\right]\right\}, \tag{12}$$

where i and j are orbital moments of wave function, γ presents the bond type (σ or π). The values of the parameters $V^0_{\alpha\beta}$, the atomic terms and p_n for carbon compounds are given in table 1 [3].

Optimization of atomic structure is implemented by entire system energy minimization on atomic coordinates (see equation (9)). The study of the compression process was implemented with the algorithm presented earlier. The parameters were fitted from the experimental data for fullerenes and carbon nanotubes. Transferability to other carbon compounds was tested by comparison with *ab initio* calculations and experiments.

Our transferable tight-binding potential can reproduce changes correctly in the electronic configuration as a function of the local bonding geometry around each carbon atom.

Table 1. Values of the parameters

ε_s, эВ	ε_p, эВ	$V^0_{ss\sigma}$, эВ	$V^0_{sp\sigma}$, эВ	$V^0_{pp\sigma}$, эВ	$V^0_{pp\pi}$, эВ
-10,932	-5,991	-4,344	3,969	5,457	-1,938
p_1	p_2, Å	p_3, Å	p_4	p_5, эВ	p_6
2,796	2,32	1,54	22	10,92	4,455

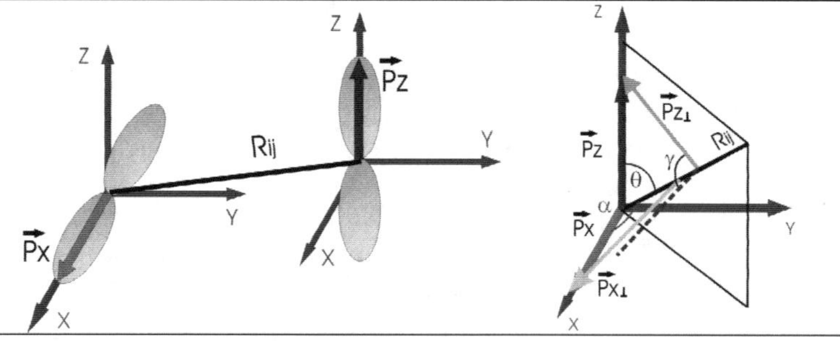

Figure 5. Schematic representation of the interaction of P_z- and P_x- orbital.

All *S*- and *P*-orbitals are given in the real Cartesian coordinates system. To reproduce changes in the electronic configuration of the local bonding geometry around each atom correctly we defined *P*-orbital as an axial vector (see figure 5). Each axial vector makes the angle with direction R_{ij} (α, β, θ) and it may be written as the geometrical sum of the two vectors:

$$\vec{P}_x = \vec{P}_{xD} + \vec{P}_{x\perp},\ \vec{P}_y = \vec{P}_{yD} + \vec{P}_{y\perp},\ \vec{P}_z = \vec{P}_{zD} + \vec{P}_{z\perp} \tag{13}$$

Here $\vec{P}_{xD}$, $\vec{P}_{yD}$, $\vec{P}_{zD}$ are projections to an interatomic direction, $\vec{P}_{x\perp}$, etc are projections to an orthogonal direction.

So, to describe the interaction between P_z and P_x (see figure 5) we must write:

$$\vec{P}_x \cdot \vec{P}_z = \underbrace{\vec{P}_{xD} \cdot \vec{P}_{zD}}_{\sigma\text{ - bonding}} + \underbrace{\vec{P}_{x\perp} \cdot \vec{P}_{z\perp}}_{\pi\text{ - bonding}} \tag{14}$$

The angle between projections $\vec{P}_{xD}$ and $\vec{P}_{zD}$ is equal to zero, but the other angle between projections to an orthogonal direction is not zero and it is equal to γ (see figure 5). As a result of some mathematical transformations we can write the expressions for $\cos\gamma$ and the energy of the interaction between P_z and P_x in the following way:

$$\cos\gamma = -\frac{\cos\alpha \cdot \cos\theta}{\sin\alpha \cdot \sin\theta},$$

$$V_{PxPz}(r_{ij}) = \cos\alpha \cdot \cos\theta \left(V^{\sigma}_{PxPz}(r_{ij}) - V^{\pi}_{PxPz}(r_{ij})\right) \tag{15}$$

As well known, the expression for the energy of the interaction between S and P-orbitals can be defined simply:

$$V_{SPz}(r_{ij}) = V^{\sigma}_{SPz}(r_{ij}) \cdot \cos\theta$$

The presented scheme provides the consideration and calculation of the rehybridization between σ- and π-orbitals to reproduce the electronic configuration and the local bonding geometry around each atom. In figure 6 we can see that the atom in sp^2 hybridization becomes one in $sp^{2+\Delta}$ hybridization because of a curvature of the topological network. The degree of rehybridization is defined by the pyramidalization angle. This angle is calculated by the following formula:

$$\theta_p = \theta_{\sigma\pi} - \frac{\pi}{2} \tag{16}$$

Our own program was used to research the nanoribbons with the help of the tight-binding method. It provides the calculation of the total energy of nanostructures, which consist of 50-5000 atoms. We adapted our TB method to be able to run the algorithm on a parallel computing machine (a computer cluster).

So, the presented transferable tight-binding method and the described scheme to reproduce the electronic configuration and the local bonding geometry around each atom are well suited for computer simulations. We tested our scheme by comparison with the experiments for fullerene and some carbon nanotubes.

2.2. Results of quantum-mechanical study

To simulate axial compression the atoms on the ends were fixed on the plates (see figure 7). The plates were moved towards each other to decrease the length by some percent.

The plane atomic network subject to axial compression becomes wave-like. It is a so–called phase transition. The calculation of density of states demonstrates the absence of changes in the electronic structure. However, the topology has nonzero pyramidalization angles. The amplitude of a wave and its period are not constant and they change along the length (see figure 8a). The strain energy collapse occurs at the strain of axial compression 0.03-0.04.

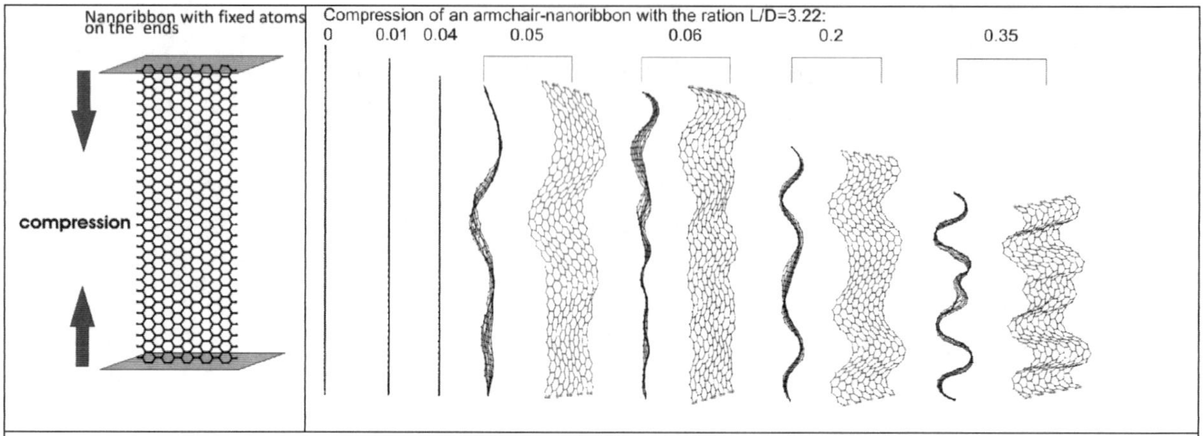

Figure 7. Simulation of axial compression

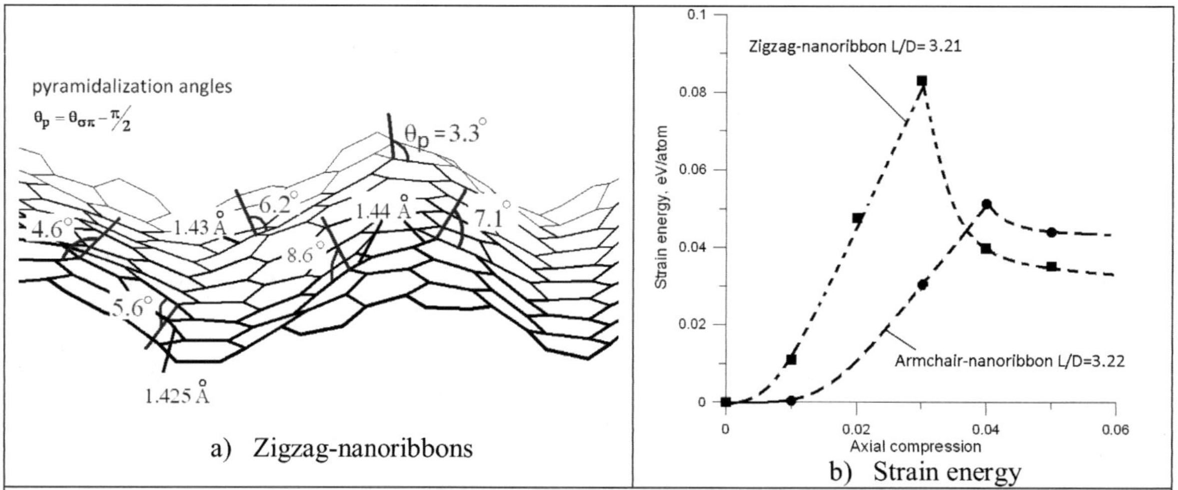

Figure 8. The results of tight-binding study of nanoribbons subject to axial compression

3. Conclusions

- Elasticity of armchair-graphene nanoribbons is greater than elasticity of armchair nanotubes and graphane nanoribbons with the same width and length.
- The curve of the strain energy collapse occurs at the axial compression of 0.03-0.04. Plane atomic network subject to axial compression becomes wave-like. This is a so-called phase transition.
- The curvature of the atomic network because of its compression is suggested to increase the reactivity of the nanoribbon.

References

[1] Brenner D W 1990 Empirical potential for hydrocarbons for use in similating the chemical vapor deposition of diamond film Phys.Rev. B. 42 15 9458-71

[2] Wang Y, Tomanek D and Bertsch G F 1991 Stiffness of a solid composed of C_{60} clusters *Phys.Rev.* B. **44** 12 6562-65

[3] Glukhova O E and Zhbanov A I 2003 Equilibrium state of C_{60}, C_{70} and C_{72} nanoclusters and local defects of the molecular skeleton *Solid State Phys.* **45** 189-96

Anomalous thermal conductivity in multiwalled carbon nanotubes with impurities and short-range order

V E Egorushkin, N V Melnikova, A N Ponomarev, A A Reshetnyak[1]

Institute of Strength Physics and Material Science, Siberian Branch of the Russian Academy of Sciences, 2/4, prospect Academicheskii, Tomsk, 634021, Russia

[1]E-mail: reshet@ispms.tsc.ru

Abstract. Low-temperature thermal conductivity and thermopower of multiwalled carbon nanotubes considered within a bundle of nanotubes are calculated taking into account multiple scattering of electrons on the atoms of impurities (like single carbon atoms) and short-range order regions arising due to a some distribution of the impurities. The calculations are realized on a base of the temperature diagrammatic Feynman techniques and the results of our research are in a good quantitative and qualitative agreement with the corresponding experimental data for multiwalled carbon nanotubes with diameter less or equal 10 nm at T<50K.

1. Introduction

Carbon nanotubes (CNTs) have unusual electron transport properties in a low temperature range ($T<50K$), which are similar to ones for amorphous metals and alloys with short-range order regions [1 -3]. So, the low-temperature resistivity $\rho(T)$ of CNT decreases, but does not increase, when T rises [4]. As the consequence, the low-temperature conductivity $\sigma(T)$ increases with grows of temperature approximately as $T^{1/2}$ [5]. In its turn, the thermal conductivity $k(T)$ rises sharply in this temperature regime up to its maximum value and then decreases (see Fig. 1 [6]). At last, the thermopower $S(T)$, being by more sensitive to a nature of electron transport, to be considered as a function of temperature is linear, positive, and well described by the Mott formula at $T<50K$. Near $50K$ the curve $S(T)$ has a break or maximum (see Fig. 2 [6-8]) following by a change in the slope with respect to the T axis. At the same time, both the value of thermopower and that of $S(T)$ slope, i.e. $\partial S(T)/\partial T$, have extra-ordinarily large magnitudes. Furthermore, in single-walled CNTs (SWNTs) exposed to air or oxygen, the positive values of thermopower becomes by negative ones for the SWNTs stripped of oxygen [8].

Conventionally, it is assumed that at $T>50K$ the temperature behavior of the thermopower is determined by both the phononic subsystem and by the influence of defects on the symmetric electron-hole structure of CNTs. The latter influences on values of $S(T)$, as well as of thermal conductivity $\rho(T)$, at T<50K unfortunately have been insufficiently understood at present [9,10]. As it was above-mentioned the amorphous metals and alloys with short-range order regions and CNTs have the same discriminating peculiarity to be concerned in the fact that the structure of all these systems includes the domains with short-range order. In particular, the defects of surface in CNTs are stipulated not only by possible declination from an ideal hexagonal structure, but also by the catalyst impurities, retained atoms, attached radicals (OH, CO and etc.), which change the configurations of valence and conduction bands.

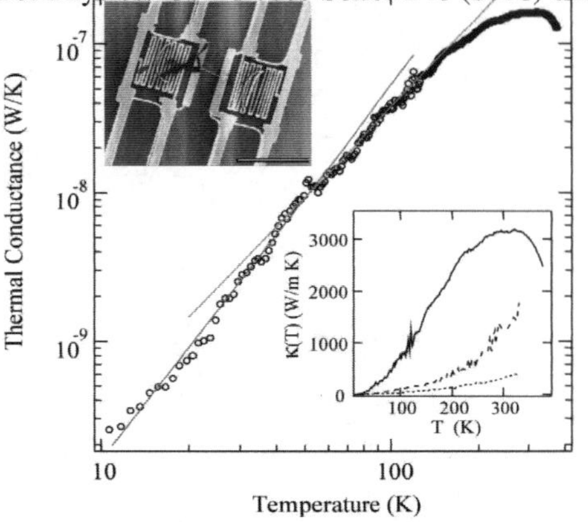

Figure 1. The thermal conductance of an individual MWNT of a diameter 14 nm see, Ref. [6]. Lower inset: Solid line represents $\chi(T)$ of an individual MWNT (d=14 nm). Broken and dotted lines represent small (d=80 nm) and large bundles (d=200 nm) of MWNTs, respectively. Upper inset: SEM image of the suspended islands with the individual MWNT. The scale bar represents 10 μm.

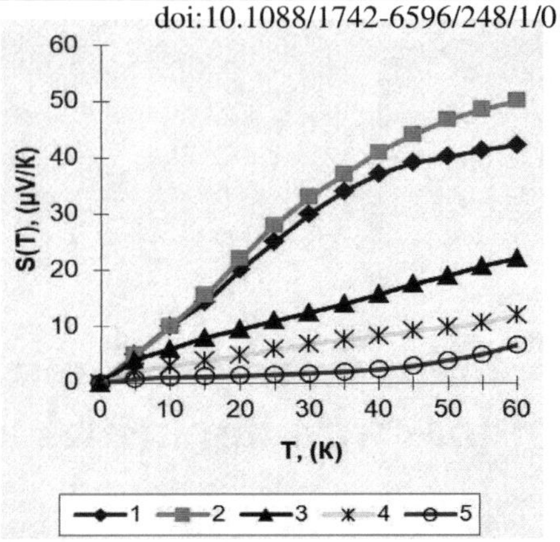

Figure 2. Temperature dependence of thermoelectric power of SCNT bundles to a diameter of 5•10^{-6} m and a length of 10^{-3} m (curves 1,2 from Ref. [7]), bundles of CNTs with a diameter of 10^{-8} m and a length of 10^{-3} m (curve 3) , MCNTs with a diameter of 10^{-8} m and a length of 2.5•10^{-6} m (Curve 4 from Ref.[7]) and bundles of CNT diameter of 10^{-8} m and a length of 10^{-3} m (curve 5 in Ref.[6]).

Thus, the CNTs (not annealed to a purification) may be considered as "dirty" metallic systems where electrons are directly scattered by the structure defects: impurities and new chemical bonds forming the short-range order regions [11]. Within the such consideration we shall calculate relaxation time of electrons multiply elastically scattered on the structural inhomogeneities and then apply this result to obtain the corresponding contribution to electron thermal conductivity following to Wiedemann-Franz law. We shall also list the result of our calculation of the contributions to the low-temperature thermopower [11] where the one-electronic contribution is calculated by means of multiple elastic electronic scattering on the defects of the hexagonal CNT structure.

2. Electron relaxation time

Let us consider a solid (bundle of CNTs) with stochastically distributed defects such as catalyst impurities, retained atoms and so on (unconditionally of their origin). Then, we introduce microconcentrations (see, Ref.[12] for details) – the occupation numbers $c_s(\vec{R}_i)$ of the nodes denoted by the coordinates $\vec{R}_i$ =(x_i, y_i, z_i), $i=1,\ldots,N$ by doped atoms of the type $s=1,2,\ldots$. In absence of long-range order $c_s(\vec{R}_i)=c_s+\delta c_s(\vec{R}_i)$. Here c_s is the macroconcentration of type s atoms in the bundle of CNTs and $\delta c_s(\vec{R}_i)$ is the fluctuation of concentration. The average expectation-value of microconcentration with respect to configurations of the atoms is equal to, $\langle c_s(\vec{R}_i)\rangle = c_s$, whereas the average expectation-values for the fluctuation vanishes $\langle \delta c_s(\vec{R}_i)\rangle$=0, and second order moment of $c_s(\vec{R}_i)$: $\langle \delta c_s(\vec{R}_i)\delta c_s(\vec{R}_j)\rangle$, $i,j=1,\ldots,N$, will determine either new chemical bonds or new short-range order [12] in a system with defects. Its Fourier image $\langle \delta c_s(R_i)\delta c_s(R_j)\rangle \sim \langle |c_k|^2\rangle$ determines the structure of short-range order region, so that, (for simplicity we restrict ourselves by the case of binary system, when $s=1$, $c_s = c$) $\langle |c_k|^2\rangle = N^{-1}c(1-c)\sum_{i=1}^{N}\alpha_i \cos \vec{k}\vec{R}_i$, where α_i are the short-range order

Journal of Physics: Conference Series **248** (2010) 012005 doi:10.1088/1742-6596/248/1/012005

parameters and $\alpha_0 = 1$. The quantity $\left\langle |c_k|^2 \right\rangle_0 = N^{-1}c(1-c)$ determines an "ideal solid solution" for all $\alpha_{i\neq 0} = 0$ or presence of impurities under consideration at $c \le 0,1\,,1$.

Then, we introduce a stochastic field of "new" atoms of unique type, for $s=1$

$$V(\vec{R}) = \sum\nolimits_i U(\vec{R} - \vec{R}_i), \tag{1}$$

where potential field $U(\vec{R} - \vec{R}_i)$ (till not completely determined) corresponds to one atom with $\vec{R}_i$. We will describe the electron interaction with new field $V(\vec{R})$ by the following Hamiltonian

$$H_{int} = \sum\nolimits_i c(\vec{R}_i) \int \psi^+(\vec{R}) U(\vec{R} - \vec{R}_i) \psi(R) d\vec{R}, \tag{2}$$

where ψ^+, ψ are the field operators of electrons.

For simplicity, we suppose that $U(\vec{R} - \vec{R}_i) = U_0 \delta(\vec{R} - \vec{R}_i)$ because of the radius of action of the potential U is enough small in comparison with the distance between atoms (sparse system). In the case under consideration, instead of the site potential we may pass to site t-matrices representing the amplitude of multiple scattering on the site. So U_0 may be considered as an effective potential t_0. Furthermore, we assume U_0 to be small what allows us to use only the first orders of perturbation theory series for amplitude of scattering with respect to U_0. For one-particle Green function (GF) G the series looks as follows

$$G = G_0 + G_0 V G_0 + G_0 V G_0 V G_0 + \ldots, \tag{3}$$

or, equivalently

$$\begin{aligned} G(\vec{r},\vec{r}') &= G_0(\vec{r},\vec{r}') + \sum_i c(\vec{R}_i) \int G_0(\vec{r},\vec{R}) U(\vec{R}-\vec{R}_i) G_0(\vec{R},\vec{r}') d\vec{R} + \sum_{i,j} c(\vec{R}_i) c(\vec{R}_j) \\ &\times \int\int G_0(\vec{r},\vec{R}) U(\vec{R}-\vec{R}_i) G_0(\vec{R},\vec{R}') U(\vec{R}'-\vec{R}_j) G_0(\vec{R}',\vec{r}') d\vec{R} d\vec{R}' + \ldots = G_0 + \sum_i G_i^{(1)} + \sum_{i,j} G_{i,j}^{(2)} + \ldots . \end{aligned} \tag{4}$$

As to the rest terms of perturbation series one should be noted that after dividing in the each perturbation theory order one-node ($i = j = k\ldots$) and two-node ($i \neq j \neq k\ldots$) terms, which may be taken into account by including of t – matrix instead of U, i.e. $U(\vec{R} - \vec{R}') = t_0 \delta(\vec{R} - \vec{R}')$. The rest more complicated scattering events should be neglected because the corresponding terms are proportional to c^3 or $< \delta c \delta c >^2$ being much less than c^2 or $< \delta c \delta c >$.

Now, we consider separately the first-order with respect to U_0 terms in the Eq.(4)

$$G_i^{(1)} = c(\vec{R}_i) \int G_0(\vec{r},\vec{R}) U(\vec{R} - \vec{R}_i) G_0(\vec{R},\vec{r}') d\vec{R}. \tag{5}$$

In momentum representation, we have (with omitting the vector sign "$\rightarrow$" over $\vec{p}, \vec{R}, \vec{k}$)

$$G(r,R) = \int G(p) e^{ip(r-R)} dp, \qquad U(R-R') = \int U(k) e^{ik(R-R')} dk. \tag{6}$$

Substituting expressions (6) into Eq. (5) we get the expression for propagator of one-particle GF

$$G_i^{(1)}(p,p') = c(R_i) e^{i(p-p')R_i} G_0(p) U(p-p') G_0(p'). \tag{7}$$

Finally, averaging $G_i^{(1)}$ with respect to configurations R and summarizing over i, we get

$$< G^{(1)}(p,p') >= c_S U_0 \delta(p-p') \left| G_0(p) \right|^2. \tag{8}$$

Note, that in the Eq.(8) $U = U(0) = U_0$ without using the approximation $U(x) = U_0 \delta(x)$. In the latter case, $U(p-p') = U_0$. Expression (8) coincides for small c with the analogous quantity $< G^{(1)} >$ for the case of "dirty" metals [12].

Let us turn to the next term in the Eq.(4)

$G^{(2)}(r,r') =$ 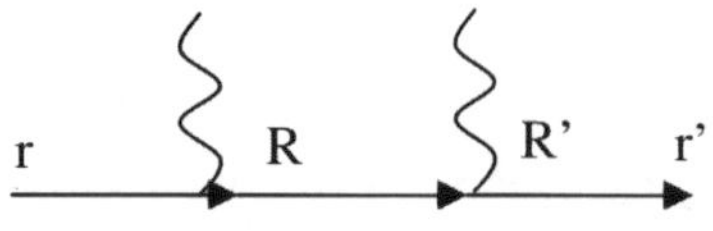

or, equivalently, in analytical form,

$$G^{(2)}_{i,j} = \iint G_0(r-R)c(R_i)U(R-R_i)G_0(R-R')c(R_j)U(R'-R_j)G_0(R'-r')dRdR'.$$

In p – representation after the change $p'' \leftrightarrow p'$ we have

$$\begin{aligned} G^{(2)}_{i,j}(p,p') &= c(R_i)c(R_j)G_0(p)G_0(p')\int U(p-p_1)G_0(p_1)U(p_1-p')e^{-i(p-p_1)R_i}e^{i(p'-p_1)R_j}dp_1 \\ &= G_0(p)G_0(p')\int U(p-p_1)G_0(p_1)U(p_1-p')c(R_i)e^{-i(p-p_1)R_i}c(R_j)e^{i(p'-p_1)R_j}dp_1 \end{aligned} \qquad (9)$$

Eq. (9) describes "double" scattering by one site and "single" scattering by two sites (the quotes assume the possibility of including t –matrix that means the repeated multiple scattering by one site and by two sites, correspondently). It is useful estimate those terms in (9):

1) for $i = j \Rightarrow c^2(R_i) = c(R_i)$ and therefore

$$\begin{aligned} G^{(2)}_{i,i}(p,p') &= G_0(p)G_0(p')\int U(p-p_1)G_0(p_1)c(R_i)e^{-i(p-p')R_i}U(p_1-p')dp_1 \\ &= c(R_i)e^{-i(p-p')R_i}G_0(p)G_0(p')\int U(p_1-p')G_0(p_1)U(p-p_1)dp_1 \end{aligned}. \qquad (10)$$

After averaging of the propagator of GF $G^{(2)}_{i,i}(p,p')$ in (10) with respect to configurations of impurities and then summation over index i [$\langle c(R_i)\rangle = c$; $\sum_i e^{-i(p-p')R_i} = \delta(p-p')$] we have

$$< G^{(2)}(p,p') >_{ii} = cU_0^2\delta(p-p')G_0^2(p)\int G_0(p_1)dp_1 . \qquad (11)$$

2) for $i \neq j$

$$G^{(2)}_{i\neq j} = G_0(p)G_0(p')\int U(p-p_1)G_0(p_1)U(p_1-p')c(R_i)e^{-i(p-p_1)R_i}c(R_j)e^{-i(p_1-p')R_j}dp_1.$$

Averaging and summation over i,j lead to the representation

$$\begin{aligned} <G^{(2)}(p,p')>_{i\neq j} &= G_0(p)G_0(p')\int U(p-p_1)G_0(p_1)U(p_1-p')\Big\{\sum\nolimits_{i\neq j} e^{-i(p-p_1)R_i}e^{-i(p_1-p')R_j}\Big\{\langle c(R_i)c(R_j)\rangle \\ &+c^2\Big\}\Big\}dp_1 = U_0^2G_0(p)\Big\{c^2G_0^2(p)\delta(p-p')+G_0(p')\int\langle c(p-p_1)c(p_1-p')\rangle G_0(p_1)dp_1\Big\} \end{aligned} .(12)$$

One should be noted that the first term in the Eq.(12), $U_0^2c^2G_0^3(p)\delta(p-p')$, coincides with the corresponding term in the GF for the "dirty" metal [13].

In its turn, from the analysis of the second term,

$$U_0^2G_0(p)G_0(p')\int < c(p-p_1)\, c(p_1-p') > G_0(p_1)dp_1, \qquad (13)$$

under the suggestion that the electron scattering on the defects in disordered system to be elastic it follows $p = p'$ and the Eq. (13) takes the form

$$U_0^2G_0^2(p)\int < |c(p-p_1)|^2 > G_0(p_1)dp_1. \qquad (14)$$

Collecting the resulting terms (8), (11), (12) together with use of the Eqs. (13), (14) we get

$$\begin{aligned} \langle G(p,p')\rangle &= G_0 + \langle G^{(1)}\rangle + \langle G^{(2)}\rangle_{ii} + \langle G^{(2)}\rangle_{i\neq j} = G_0 + cU_0\delta(p-p')|G_0(p)|^2 \\ &+ U_0^2G_0^2(p)\delta(p-p')\Big[c\int G_0(p_1)dp_1 + c^2G_0(p) + \int\langle|c(p-p_1)|^2\rangle G_0(p_1)dp_1.\Big] \end{aligned}, \qquad (15)$$

The representation (15) permits one to find self-energy part Σ, electron relaxation time $\tau^{-1} = -2\mathrm{Im}\Sigma$ and properly the GF averaged over disorder. To this end, we start from the Dyson equation

$$G(p) = G_0(p) + G_0(p)\Sigma G(p), \qquad (16)$$

or, in the first order with respect to Σ,

$$G(p) = G_0(p) + G_0^2(p)\Sigma\,, \tag{17}$$

From the Eqs. (15), (17) we may to derive the expression for Σ as follows,

$$\Sigma = cU_0 + c^2U_0^2G_0(p) + cU_0^2\int G_0(p_1)dp_1 + U_0^2\int\left\langle\left|c(p-p_1)\right|^2\right\rangle G_0(p_1)dp_1. \tag{18}$$

The first two summands do not give contribution to the imaginary part of Σ, $\operatorname{Im}\Sigma$. However, they determine the contribution to a change of chemical potential (i.e. of the Fermi level), $\mu^* = \mu - cU_0$, and contributions of the following higher order infinitesimal.

Therefore, to calculate the electron relaxation time τ we have to consider two last terms in the Eq. (18). As the result for $\operatorname{Im}\Sigma = \operatorname{Im}\Sigma_1^{(2)} + \operatorname{Im}\Sigma_2^{(2)}$, we have, firstly,

$$\Sigma_1^{(2)} = cU_0^2\int G_0(p_1)dp_1\,.$$

Pass to integrating with respect to scalar ξ, $\xi = p^2/2m - \mu$, and omitting the factor $(2\pi)^{-3}$ we get

$$\int G_0(p_1)d\vec{p}_1 = \frac{\nu_0}{4\pi}\int\limits_{|n|=1} d\vec{n}\int d\xi G_0(\xi) = \nu_0\int\frac{d\xi}{\varepsilon - \xi + i0 sign\varepsilon} = -i\pi\nu_0 sign\varepsilon\,,$$

with angle vector $\vec{n}$. Thus, we obtain for $\Sigma_1^{(2)}$ the expression,

$$\Sigma_1^{(2)} = -icU_0^2\pi\nu_0 sign\,\varepsilon = -\frac{i}{2\tau_{imp}}sign\,\varepsilon, \tag{19}$$

where ν_0 is the density of electron states on the Fermi level, and

$$\left(\tau_{imp}\right)^{-1} = 2\pi cU_0^2\nu_0\,. \tag{20}$$

The last expression coincides with the corresponding expression for electron relaxation time calculated within the theory of "dirty" metals with impurities [13, 14].

Secondly, for the quantity $\Sigma_2^{(2)}$ from (18), $\Sigma_2^{(2)} = U_0^2\int\left\langle\left|c(p-p_1)\right|^2\right\rangle G_0(p_1)dp_1$. and validity of the formula $\left\langle\left|c(k)\right|^2\right\rangle = \frac{c(1-c)}{N}\sum\limits_{i=0}^{N}\alpha_i\cos\vec{k}R_i$ (see formula before Eq.(1)) [12] we derive,

$$\Sigma_2^{(2)} = U_0^2\frac{c(1-c)}{N}\sum_{i=0}^{N}\alpha_i\int_0^\infty\frac{\cos(\vec{p}-\vec{p}_1)R_i d\vec{p}_1}{\varepsilon - \xi(p_1) + i0\,sign\,\varepsilon}\,.$$

The standard Euler formula for $\cos y$, $y = (\vec{p}-\vec{p}_1)\vec{R}$, $\cos y = \frac{1}{2}\left(e^{iy_i} + e^{-iy_i}\right)$ leads to the representation

$$\int_0^\infty\frac{e^{\pm i\vec{p}_1\vec{R}_i}R_i d\vec{p}_1}{\varepsilon - \xi(p_1) + i0 sign\varepsilon} = \frac{4\pi}{R}\int_0^\infty\frac{\sin(p_1R_i)p_1dp_1}{\varepsilon - \xi(p_1) + i0 sign\varepsilon}\,. \tag{21}$$

Therefore, $\Sigma_2^{(2)}$ has the form

$$\Sigma_2^{(2)} = -\frac{4\pi^2U_0^2mc(1-c)}{N}\sum_{i=0}^{N}\alpha_i\frac{\cos(\vec{p}R_i)}{R_i}e^{ixR_i sign\varepsilon}, \tag{22}$$

where we have introduced the notation for variable x, $x = \sqrt{2m(\varepsilon + \varepsilon_F + i0 sign\varepsilon)}$

Substituting expressions (19) and (22) into Eq. (18) we obtain the self-energy part of GF,

$$\Sigma(p,\varepsilon) = -\frac{i\sin\varepsilon}{2\tau_{np}} - \frac{4\pi^2U_0^2mc(1-c)}{N}\sum_{i=0}^{N}\alpha_i\frac{\cos(\vec{p}R_i)}{R_i}e^{ixR_i sign\varepsilon}\,. \tag{23}$$

It is easily to extract the real and imaginary parts of $\Sigma(p,\varepsilon)$ where the former term will correspond to the renormalization of the chemical potential $\mu \to \mu^* = \mu - \Sigma_{real}$ and, the latter one for small momentum of electrons: $|p| \to 0$ takes the form

$$\operatorname{Im}\Sigma(p,\varepsilon)_{|p|\to 0} = \operatorname{Im}\Sigma(\varepsilon) = -i\left(\frac{1}{2\tau_{np}} + \frac{4\pi^2 U_0^2 mc(1-c)}{N}\sum_{i=0}^{N}\alpha_i \frac{\sin xR_i}{R_i}\right) sign\varepsilon .$$

Therefore, for small electron energy ε we have:

$$\operatorname{Im}\Sigma(\varepsilon) \approx -\tfrac{i}{2}(s\ ign\ \varepsilon)\ \tau^{-1}, \tag{24}$$

where,

$$\frac{1}{2\tau} = \frac{1}{2\tau_{imp}} + 4\pi^2 U_0^2 mc(1-c)\frac{x}{N}\sum_{i=1}^{N}\alpha_i . \tag{25}$$

The parameter of short-range order α_i in (25) is negative in the first coordinate sphere ($\alpha_1 < 0$) and positive in the second one ($\alpha_2 > 0$). It is important that the coordinate sphere parameters α_i may change a sign in dependence of atoms which generate the pairs. The first term in (25), τ_{imp}^{-1} represents the inverse relaxation time of electrons scattered by impurities. The second one corresponds to the inverse relaxation time of electron scattered by structural inhomogeneities of short-range order type.

At last, the one-electron GF averaged with respect to configurations in "dirty" metallic CNT has the form (see, for details Refs.[12-14]:

$$\langle G(\varepsilon, p)\rangle = \left(\varepsilon - \xi(p) + \tfrac{i}{2}\tau^{-1}(p) sign\ \varepsilon\right)^{-1}, \tag{26}$$

where $\xi(p) = p^2/2m - \mu^*$, $\vec{p}$ and m – momentum and mass of electron, μ^* is the renormalized electron chemical potential. We shall use expressions (25) and (26) to calculate the contribution of multiple elastic electronic scattering by the short-range order regions.

3. Thermal conductivity

As far as we consider model CNTs representing at low-temperature region the metallic systems with elastic electron scattering we may use the Wiedemann - Franz law to determine a thermal conductivity

$$\chi(T) = \tfrac{1}{3}\left(e^{-1}\pi k_B\right)^2 T\sigma(T) . \tag{27}$$

where k_B, e, n are respectively the Boltzman constant, charge of the electron and concentration of electrons with $\sigma(T) = m^{-1}e^2\tau n$ being by Drude conductivity. Because of the electron relaxation time is given by the Eq. (25), we have for the first coordination sphere, after the following substitution $\varepsilon \to \varepsilon_l = \pi T(2l+1)$ (for vanishing integer l)

$$\sigma(T) = \frac{ne^2}{4\pi^2 cU_0^2 m\nu_0}\frac{1}{1-BT^{1/2}}, \tag{28}$$

where $B = (\nu_0 N)^{-1} 2\sqrt{2\pi}(1-c)m^{3/2}$. Finally, we obtain

$$\chi(T) = \frac{k_B{}^2 n}{12cU_0^2 m\nu_0}\frac{T}{1-BT^{1/2}}, \tag{29}$$

It is useful to estimate the above contribution to thermal conductivity of multi-walled CNTs. Doing so, we have for the concentration of electrons in CNT, n=10^{28}-10^{29}m^{-3}; e=1,6·10^{-19}C; m= 9,1·10^{-31} kg; ν_0=1eV^{-1}=0,625·10^{19} J^{-1}; c=0,1; U_0^2=(0,05eV)^2=5·10^{-4}·(1,6·10^{-19})^2 J^2. Hence, we get

$\left(4\pi^2 cU_0^2 mv_0\right)^{-1} ne^2\hbar = 7,1\cdot10^6 S\cdot m^{-1}$ and $B=\left(Np_0\right)^{-1}4\pi^2\sqrt{2\pi}(1-c)m^{3/2}k_B^{1/2}=1,4K^{-1/2}$ (with restored dependence of Plank constant $\hbar$ and momentum on the Fermi level p_0). Therefore, the estimation holds

$$\left(12cU_0^2 mv_0\right)^{-1} k_B{}^2 n\hbar \approx 4,6\ W\cdot K^{-2}\cdot m^{-1}, \tag{30}$$

that permits one to derive for T~50K the value for thermal conductivity, $\chi(T)\propto -26W\cdot K^{-1}\cdot m^{-1}$. The last result is to be in a good qualitative and quantitative agreement with the experimental data [6] for multi-walled CNTs with impurities with diameter less or equal 10 nm and we present on Figure 3 the low-temperature dependence of the calculated thermal conductivity.

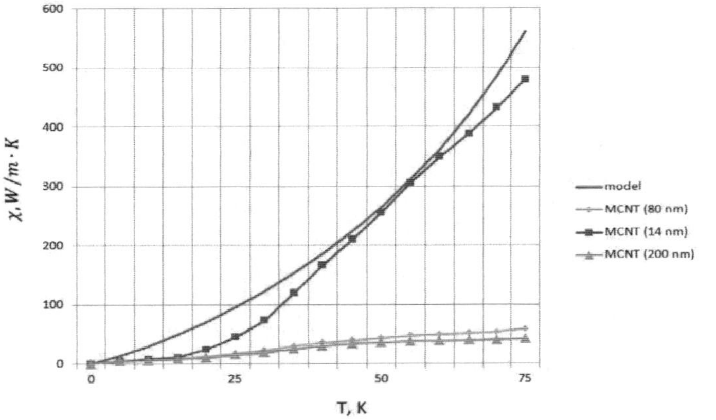

Figure 3. Thermal conductivity of a model and single MCNT with diameter 200 нм (curve 1), 80 нм (curve 2) and 14 нм (curve 3). (Obtained from the experimental data in Ref.[6]). Theoretical temperature dependence of the input of the multiple elastic scattering of the electrons on the short-range ordered regions in $\chi(T)$ (curve 4 as red type).

4. Thermopower

In our recent paper [11] we have proposed a theory of low-temperature thermopower in CNTs. On a base of the Mott formula

$$S(T)=\tfrac{1}{3}|e|^{-1}k_B^2\pi^2T\left(\tau\partial\tau^{-1}/\partial\varepsilon\right)_{\varepsilon=\varepsilon_F}, \tag{31}$$

we have calculated the corresponding one-particle contribution to thermopower in the form,

$$S(T)=-\left(18|e|N\right)^{-1}k_B^2\pi m(1-c)\sum_i\alpha_iR_i^2T,\quad R_0=2\pi\left(k_0\right)^{-1}, \tag{32}$$

where R_i, α_i are the radius of coordinate sphere and the parameters of short-range order, which includes the coordinate numbers [14] and k_0 is the wave-vector on Fermi level.

From (33) we would like to make the following conclusions:

- At low T the magnitude and curve slope of $S(T)$ significantly depend on the size of inhomogeneity regions and do not determined by the parameters of electronic structure.
- The change of thermopower sign for MCNTs in the oxygenated and deoxygenated states related to the geometry of a new structure of MCNTs, so that for $\alpha_i>0$ we have $S(T)<0$, and vice versa ([10]).
- The value of thermopower correlates to the size of samples of CNTs (their diameter through R_i).
- With grows of the impurity concentration c the magnitude $|S(T)|$ decreases in accordance with [10].
- The break or maximum of $S(T)$ around 50K may be due to forming of a stable region of structural inhomogeneities in CNTs (the so-called "mesoscopic first order phase transition" [6]), which depends on the change of number of the parameters α_i in eq. (32) part of which can be positive or negative in the neighborhood of a critical point T_0, hence changing the form of $S(T)$.

The estimation of one-electron contribution to thermopower described by the Eq. (32) permits to derive that $S(T)\propto -2\cdot 10^{10}N^{-1}(1-c)\sum \alpha_i R_i^2 T$. For $N^{-1}(1-c)\sum_i \alpha_i R_i^2 \propto -0.5\cdot 10^{-16}$ m^2, $\alpha_i R_i^2<0$, $\alpha_i \propto 10^{-2}$, $R_i \propto 10^{-7}$ m, $S(T)\propto T$ mkV/K. Therefore, for $T=10K$ we have $S(T)$~10 mkV/K. This result is in a good agreement with the experimental data [6, 7] (see, for details Ref. [11]).

5. Conclusion

We have suggested the new model of multi-walled CNT with impurities within bundle of CNTs by means of introducing special distribution of the atoms of impurities and short-range order regions arising due to those impurities and defects of hexagonal lattice structure. The results of our first-principle research of the contributions of multiple elastic electron scattering on the impurities and short-range order regions in the framework of above model of CNT and temperature Green functions approach to thermal conductivity and thermopower are presented by the Eqs.(29),(32) and shown to be in a good quantitative and qualitative agreement with the corresponding experimental data for T<50K for multiwalled carbon nanotubes with diameter less or equal 10 nm. Hence, our model may describe unusual low-temperature behaviour of electron transport properties of metallized carbon CNTs.

In order to construct the complete theory of electron transport in CNTs at low temperatures we have to take into account inelastic electron-electron interaction (that is especially important for the electron resistance) and its interference with multiple elastic electron scattering on the defects. From the other hand, the enlargement of the results of temperature dependence of thermal conductivity and thermopower into region of T>50K require to turn on the phononic subsystem of CNT into consideration. The resolution of the above two points are within scopes of our research.

Acknowledgement

The work is supported in part by the RFBR (Project № 09-02-992). The authors thanks the organizers of the DubnaNANO2010 Conference for support and hospitality.

References

[1] Mizutani U 1993 *Phys. Stat. Sol. (b)* **176** pp. 9-30
[2] Gyunterodt G I and Bek G (eds.) 1986 *Metallic Glasses, Vol. 2: Atomic Structure and Dynamics, Electronic Structure, and Magnetic Properties* (Mir, Moscow)
[3] Zallen R 1998 *The physics of amorphous solids* (Wiley Classic Library Edition)
[4] Jang J W, Lee D K, Lee C E, Lee T J, Lee C J, et al. 2002 *Solid State Comm.* **124** p.147
[5] Postma H W Ch et al 2000 *Phys. Rev.* B **62** R10653 - R10656
[6] Kim P, Shi, McEuen P L 2001 *Phys. Rev. Lett.* **87** No 21 215502
[7] Kong, W.J. Lu, L. H W Zhu, H.W. Wei, B.Q. and Wu, D.H. 2005 *J. Phys. Cond. Matt.*, **17**, 1923-1928
[8] Bradley K, Jhi S-H, Collins Ph G, Hone J, et al. 2000 *Phys. Rev. Lett.* **85** No 20 4361-4364
[9] Grigorian L, Sumanasekera G U, et al. 1999 *Phys. Rev.* B **60** No 16 R11309-R11312
[10] Avouris Ph, Martel R, et al. 2000 *Electrical properties of carbon nanotubes: spectroscopy, localization and electrical breakdown. Science and Application of Nanoubes* ed. by Tomanek and Enbody (Kluwer Academic/Plenum Publishers, New York) pp 223–237
[11] Egorushkin V E, Melnikova N V, Ponomarev A N 2009 *Russ. Phys. J.* **52** № 3 252–264
[12] Levitov L S and Shitov A V 2003 *Green's Functions. Problems and Solutions*, 2nd ed. (Fizmatlit, Moscow), in Russian
[13] Abrikosov A A, Gorkov L P, and Dzyaloshinski I E 1963 *Methods of Quantum Field Theory in Statistical Physics* (Dover, New York)
[14] Iveronova V I and Katsnelson A A 1977 *Short-range Order in Solid Solutions* (Nauka, Moscow)

Model of fluid flow in nanotube: classical and quantum features

I Yu Popov[1], **S.A.Chivilikhin**[1], **V V Gusarov**[2]

[1] St.-Petersburg State University of Information Technologies, Mechanics and Optics, Kronverkskiy, 49, St.-Petersburg, 197101, Russia

[2] St.-Petersburg State Technological Institute (Technical University), Moskovskiy, 26, St.-Petersburg, 190000, Russia

popov@mail.ifmo.ru

Abstract. Different applications of nanostructures are related with fluid flows through these systems. In its turn, experiments show that flows in nanostructures have many specific features, which can't be explained in classical terms only. Particularly, flow through nanotube is extremely fast in comparison with the analogous classical one. We suggest a model based on the possibility of existence of molecular clusters (Frenkel crystallites) in the fluid. There are some experimental evidences of such phenomenon. Under this assumption one needs to take into account quantum effects. Particularly, the boundary condition, which plays the crucial role for the flow in nanostructures, takes the form of sliding condition instead of the adhesion condition for the classical flow. The parameters of the boundary conditions are determined by solving of quantum scattering problem for the particle of the fluid by the wall potential. Main features of the flow are described in the framework of the model. For very narrow nanotubes another phenomena have an influence on the flow- possibility of existence of solitons in nanotube walls. These soliton solutions are similar to Davidov's solitons in molecular chains. This model of flow is also described.

1. Introduction

Last years, fluid and gas flows in nano-sized domains are intensively studied [1]. At present, there are no general equations of nano-hydrodynamics. Usually, the molecular dynamics is used for computations [2]. As for analytical approaches, the simplest one involves introducing the slip condition at the boundary [3]. There is also a hybrid method, incorporating the continuous approach and the molecular dynamics (which is used to analyze the fluid structure and determine the rheological law) [4]. In [5], fluctuations are taken into account when deriving the hydrodynamic equations. In [6] viscous-elastic behavior of water in nano-sized gaps was experimentally studied. Great increase of effective viscosity of water in a channel of nano-width is observed. For nano-channel flow the molecular structure of the medium plays a crucial role. It is similar to Brownian motion [7] and body motion through rarefied gas [8]. More precisely, flows in nano-channels are influenced by local heterogeneity of molecular structure of the liquid if its size is compared with the channel width. A hypothesis about the existence of locally-ordered structures in liquid is in [9]. During several decades scientists discuss whether there are local domains in liquid (crystallites) in which the molecular structure is similar to that for crystal [10]. Investigations of fluid flows in nano-sized domains show that it is strongly influenced by local ordering of nano-sized scale. Experiments ([6], [11]) show that the effective viscosity of water in nano-channel with hydrophilic walls is essentially greater than the corresponding macroscopic value. Calculations in the framework of molecular dynamics [11] show that there are ordered structures (like periodic) of sizes less than nanometer. Computing experiments in [12] lead to appearance of ice-like state in nanotubes of small diameters under normal conditions. Experimental and theoretical investigations of water state in carbon nanotube [13 – 14] show that there is ice-like envelope with liquid water inside in the nanotube. Increasing of effective fluid viscosity via channel diameter was marked in [15] for channels of few micrometers diameters. Influence of walls isn't evident. In some experiments there are increasing of effective viscosity, but in others the opposite effect takes place. Particularly, the experiments in [16] show that the fluid flow speed through carbon nanotube (of few nanometers diameter) is essentially greater than one calculated in the framework of classical fluid theory. Possible model of such phenomenon – superfluidity in nano-channel – is suggested in [17]. As it was mentioned earlier, the effective viscosity can be both extremely small in some cases and extremely great in other situations (in comparison with its classical value). Such non-trivial correlation between the nanotube diameter and the viscosity, possibly, is related with local ordering in the liquid. The first variant of such approach is in [18], where the model of nanotube flow close to crystallite model is suggested. In the present paper the relation of the effective viscosity of a fluid in nanotube with the dynamics of locally ordered nano-sized clusters is described. The number of such clusters increases if the temperature tends to the crystallization point. The sizes of such clusters are close to the diameters of hydro-silicate nanotubes. For this reason, it is of importance to study the existence of crystal clusters of this kind in a liquid inside nanotubes and the influence of clusters on the mass transport through the tube. Computations based on molecular dynamics (with the Lennard-Jones potential) show that solid nano-sized clusters can exist both in the nanotube and outside it. The clusters (crystallites) can have a size of order 1 nm, which is close to the internal diameter of silicate nanotube. Note that the computed size of crystallite depends on parameters of the interaction potential, i.e., on the chemical composition of the liquid. Moreover, in unbounded domains, the clusters are also nano-sized. In the case of narrow nanotube with elastic walls (e.g., carbon nanotube) there are elastic waves in walls [19]. It has great influence on the flow. We suggest a solvable model of such flow. Particularly, it is shown that wall soliton induces a flow.

2. Flow of liquid with crystallites

We suggest models taking into account the existence of crystallites in the nanotube. The first model takes into account energy transformations. Namely, we determine the power of pressure drop and compare it with the energy loss due to viscosity and melting.

We use quantum mechanical treatment to explain the type of the boundary condition on the inner surface of nanotube

$$v_x\big|_s - v_s = L_s \left.\frac{\partial v_x}{\partial n}\right|_s ,$$

where v_s and L_s are the characteristic velocity and length (so-called sliding velocity and sliding length).

The character of the flow in the nanotube depends on the correlation between the equilibrium size of the crystallite and the nanotube diameter. We try to find the effective viscosity μ_{eff} of the fluid in the nanotube which is defined by the following way. Consider the correlation between our nanotube (possibly, containing crystallites) and classical tube (with the Poiseille flow) of the same size and with the same pressure drop and flow rate. The viscosity of the Poiseille flow having the same parameters is called the effective viscosity of the flow in the nanotube. Note that there is some space between the crystallite and the nanotube wall. It is occupied by so-called non-autonomous phase (of width h)

having the properties of a fluid with another viscosity μ_0 [20]. Size correlations leads to a few particular cases. Introduce some notations. Let D be the nanotube diameter, L be its length, H_e be the equilibrium size of the crystallite and h_e be the equilibrium width of the non-autonomous phase.

The character of the flow depends on the existence of the crystallite inside the tube. For small nanotube diameter ($D < 2h_e$) there is no crystallite. If $2h_e \le D < H_e + 2h_e$ then there exists a crystallite inside the nanotube but its diameter is less than the equilibrium one. If $H_e + 2h_e \le D < 2H_e + 3h_e$ then there are not greater than one crystallite in a cross-section of the nanotube. Correspondingly, one obtains three different expressions for the effective viscosity:

$$\mu_{ef} = \begin{cases} \dfrac{qD^2}{32\nu L} + \mu_0 \left[\dfrac{0.5D}{0.5D + L_s} \left(1 - \dfrac{v_s}{v}\right) \right]^2 & , D < 2h_e \\ \dfrac{qD^2}{32\nu L}\left[1 - \left(\dfrac{D-h_e}{H_e}\right)^2\right] + \mu_0 \dfrac{h_e(D-h_e)}{(h_e + L_s)^2}\left(1 - \dfrac{v_s}{v}\right)^2 & , 2h_e \le D < H_e + 2h_e \\ \mu_0 \dfrac{h(D-h)}{(h+L_s)^2}\left(1 - \dfrac{v_s}{v}\right)^2 & , H_e + 2h_e \le D < 2H_e + 3h_e \end{cases}$$

Here

$$h = \begin{cases} h_e, & H_e + 2h_e < D < \alpha H_e + 2h_e, \\ 0.5(D - \alpha H_e), & \alpha H_e + 2h_e \le D < 2H_e + 3h_e. \end{cases}$$

We assume that for $H_e + 2h_e < D < \alpha H_e + 2h_e$ the width of the non-autonomous phase preserves (h_e) and the cross-section of the crystallite increases from H_e to αH_e (α is some dimensionless parameter). For larger values of D ($\alpha H_e + 2h_e \le D < 2H + 3h_e$), the width of the crystallite preserves ($H = \alpha H_e$) and the width of the non-autonomous phase h increases: $h = 0.5(D - \alpha H_e)$.

The character of obtained viscosity via diameter dependence is in good correlation with the experimental data [15], [16].

3. Sliding boundary condition and surface waves

Boundary conditions play crucial role in nanotube flow. The character of this condition is determined by the quantum interaction between liquid and walls. Consider the corresponding quantum problem. Let one have a nanotube having a periodic set of atoms (centers) at the wall (a cylinder of radius R with OZ axis. These centers are at N vertices of polygons. Polygons are parallel, and the distance between neighbor polygons is h. To describe the system we use zero range potential method [24]. Using of δ-interaction allowing one to simplify considerably the procedure of spectral analysis [25, 26]. Correct mathematical description of point-like interaction is given in the framework of the theory of self-adjoint extensions of symmetric operators. Formally, the Hamiltonian H of the system is a perturbation of free Hamiltonian (i.e. the Laplace operator $-\Delta$ by periodic system of zero range potentials which are in the vertices of the above mentioned polygons (nodes):

$$H = -\Delta + \sum_{n \in \mathbb{Z}_N} \sum_{s \in \mathbb{Z}} \alpha \delta(r - r_{ns}),$$

where α is the intensity of the perturbation, r_{ns} is the radius-vector of a node having the following cylindrical coordinates: $r_{ns} = (R\cos(\theta n), R\sin(\theta n), sh)$, $s \in \mathbb{Z}$, $n \in \mathbb{Z}_N$, $\theta = 2\pi / N$.

The construction of our model is as follows. First, we restrict the Laplace operator onto the set of functions vanishing at the nodes. The restricted operator is symmetric and non-self-adjoint. It has self-adjoint extensions which gives us the model operator [25]. To choose the particular extension it is necessary to satisfy the condition at the nodes. The continuous spectrum of the model operator contains values corresponding to modes of waveguide type concentrated near the wall of the nanotube. The corresponding wave function has the following form:

$$\psi(r)=\sum_{n,s} c_{ns}G(r_{ns},r;E),$$

where the coefficients c_{ns}, $c_{ns}\in\mathbb{C}$, are determined by the "boundary" conditions at the nodes, E is the energy of the mode, $G(r_{ns},r;E)$ is the Green function of the Laplace operator:

$$G(r,r';E)=G(r-r';E)=\frac{\exp(-\sqrt{-E}\,|r-r'|)}{4\pi|r-r'|}.$$

Wave function has the following asymptotics in a neighborhood of each point r_{ns}:

$$\psi(r)=\frac{a_{-1}}{|r-r_{ns}|}+a_0+o(|r-r_{ns}|),\ r\to r_{ns}.$$

The "boundary" condition at the node r_{ns} gives us the following correlation: $a_0=\alpha a_{-1}$ $\Im\alpha=0$. This condition means that for each centre r_{ns}:

$$\lim_{|r-r_{ns}|\to 0}\frac{1}{|r-r_{ns}|\psi}\frac{\partial}{\partial|r-r_{ns}|}(|r-r_{ns}|\psi)=\alpha.$$

Using well-known asymptotic expansion for $G(r_{ns},r;E)$ in a neighborhood of each point r_{ns}, one obtains the following system:

$$(-\alpha-\sqrt{-E})c_{ns}+4\pi\sum_{n',s'(n',s')\neq(n,s)} c_{n's'}G(r_{ns}-r_{n's'};E)=0,\ \ s=0,\pm1,\ldots,\ n\in\mathbb{Z}_N.$$

Then, the dispersion equation takes the form

$$\alpha=-\sqrt{-E}-\widehat{G}(j,q;E),$$

$$\widehat{G}(j,q;E)=\sum_{n,s:(n,s)\neq(0,0)}\frac{\exp(-iqs-i\theta jn)\exp(-\sqrt{-E}\sqrt{2R^2(1-\cos(\theta n))+h^2s^2})}{\sqrt{2R^2(1-\cos(\theta n))+h^2s^2}}$$

Now, it is simple to describe the spectrum. Particularly, there are bands corresponding to solutions having the type of surface waves concentrated near the nanotube wall and moving along it. To explain fast flow in nanochannels, which is observed in experiments, some authors (see, e.g., [27]) formally replace the no-slip condition by the sliding one. The parameter, speed of sliding, is chosen empirically. We have shown above that there exist waveguide modes concentrated near the channel wall and have found its parameters. The existence of these surface waves leads to the sliding condition. One can determine the sliding speed by the following way. Consider the scattering of a particle by the wall potential to determine the momentum transmitted to the wall. We use a way analogous to that in [37].

Let H_0 be the unperturbed Hamiltonian, H_1 be the perturbation (wall potential V), ψ_a be the solution of the scattering problem by the wall potential V corresponding to the particle with fixed momentum p_a. The momentum operator $\hat{p}$ commutes with H_0, Φ_a is the eigenfunction of $\hat{p}$ ($\hat{p}\Phi_a=p_a\Phi_a$). Solution ψ_a corresponds to Φ_a, i.e. satisfies the equation:

$$\psi_a = \Phi_a + (E - H_0 + i0)^{-1} H_1 \psi_a \quad (E - H_0)\psi_a = H_1 \psi_a .,$$

One can determine the mean value of the operator $(i\hbar)^{-1}[\hat{p}, H] = (i\hbar)^{-1}[\hat{p}, H_1]$:

$$(\psi_a, (i\hbar)^{-1}[\hat{p}, H_1]\psi_a) = 2\hbar^{-1} p_a \operatorname{Im}(\Phi_a, H_1\psi_a) + 2\pi\hbar^{-1} \sum_b p_b \left|(\Phi_b, H_1\psi_a)\right|^2 \delta(E_a - E_b). \qquad (1)$$

If one takes the identical operator instead of $\hat{p}$, then (1) transforms to the form

$$0 = 2\hbar^{-1} \operatorname{Im}(\Phi_a, H_1\psi_a) + 2\pi\hbar^{-1} \sum_b \left|(\Phi_b, H_1\psi_a)\right|^2 \delta(E_a - E_b). \qquad (2)$$

One gets the expression for $\operatorname{Im}(\Phi_a, H_1\psi_a)$ from (2) and inserts it into (1). Then,

$$(\psi_a, (i\hbar)^{-1}[\hat{p}, H_1]\psi_a) = 2\pi\hbar^{-1} \sum_b (p_b - p_a) \left|(\Phi_b, H_1\psi_a)\right|^2 \delta(E_a - E_b). \qquad (3)$$

Note that the right hand side of (3) is the mean momentum transmitted during the scattering of ψ_a. Summation over all states ψ_a gives us the full mean transmitted momentum. We are interested in the longitudinal transmitted momentum δp: $\delta p = (\psi_a, (i\hbar)^{-1}[\widehat{p_\tau}, H_1]\psi_a)$. Using (3), one obtains (in our case $H_1 = V$)

$$\delta p = \gamma (p_s - p_a), \quad \gamma = 2\pi\hbar^{-1} \sum_b \left|(\Phi_b, H_1\psi_a)\right|^2 \delta(E_a - E_b),,$$

$$p_s = \frac{\sum_b p_b \left|(\Phi_b, H_1\psi_a)\right|^2 \delta(E_a - E_b)}{\sum_b \left|(\Phi_b, H_1\psi_a)\right|^2 \delta(E_a - E_b)}$$

Using the value of transmitted momentum one can obtain the boundary condition which takes the form

$$\frac{\partial v}{\partial n} = \frac{1}{L_s}(v - v_s), \quad v = \frac{p}{m}, \quad v_s = \frac{p_s}{m}, \quad L_s = \frac{a}{\tau\gamma},$$

where τ is the mean time of molecule transition between two equilibrium states ("Frenkel transition") [28], a is the distance between these equilibrium positions, m is the mass of liquid particle.

4. Wall soliton and flow in nanostructure

For the case of narrow channel another effect can play an important role. It should be stressed that boundary condition, wall structure and profile play crucial role. Namely, it is necessary to take into account vibration and waves in molecular chains forming the nanotube wall. There are experimental evidences of such wall vibration [19]. The most interesting is solitary wave. There are different models describing such solutions (see, e.g., [29]). We shall consider so-called "Davidov soliton", stable solution of this type in long molecular chain. The analogous waves are also in nanotubes [30].

Consider the soliton influence on the flow. The soliton moves in the nanotube wall and looks like a moving local extension (or constriction) of the tube. Let the area of the cross-section of the tube be $S = S(x - Vt)$, where x is the longitudinal coordinate, V is the soliton velocity. Using the continuity equation

$$\frac{\partial S}{\partial t} + \frac{\partial (Su)}{\partial x} = 0,$$

where u is velocity of the flow ($u = V + (v - V)S_0 / S$, v and S_0 are the flow velocity and the area of the cross-section outside the domain occupied by the soliton), one can obtain the following rough estimation

$$v = \overline{v} + 2\left(1 - \frac{<S>}{S_0}\right)\frac{V - \overline{v}}{1 + v_s/\overline{v}},$$

where $<S>$ is the average area of the nanotube cross-section, $\overline{v}$ is the flow in the same nanotube without soliton. To obtain better result it is necessary to describe the flow in more details (see below). The description of the wall soliton is quantum mechanical problem. Nanotube wall can be treated as a system of molecular chains (or molecular net) [29-31]. Let us consider an exciton of effective mass m interacting with displacement $u(x,t)$ of molecules having masses M from the equilibrium positions at points $x = na$. The corresponding Hamiltonian is [31]

$$H = \frac{1}{a}\int \left(\frac{\hbar^2}{2m}|\psi_x|^2 + \frac{1}{2}Mu_t^2 + MV_0^2 U(\rho) - \vartheta\rho|\psi|^2\right)dx,$$

where ρ is the relative decreasing of the equilibrium distance a between neighbor molecules, $\rho = -u_x$, V_0 is the longitudinal sound speed in the linear approximation of the chain, ϑ is the energy which characterizes the interaction of the exciton and the displacement, $U(\rho)$ is the dimensionless potential of intermolecular interaction having minimum at , $\rho = 0$, ψ - normalized exciton wave function. The Hamiltonian leads to the following system of equations:

$$i\hbar\psi_t + \frac{\hbar^2}{2m}\psi_{xx} + \vartheta\rho\psi = 0, \quad u_{tt} - V_0^2 U_{\rho\rho} u_{xx} = \frac{\vartheta}{M}(|\psi|^2)_x.$$

The system has a stable solution having the form of solitary wave of the profile Φ depending on U. Particularly, for cubic nonlinearity, $U(\rho) = \frac{1}{2}\rho^2 + \frac{\gamma}{3}\rho^3$:

$$\Phi(\xi) = \frac{1}{2}\left(\frac{3\sigma}{4}\right)^{1/3}\left(\frac{\tau}{\gamma}\right)^{1/3} sch^2\left(\frac{1}{2}\left(\frac{\tau\sigma^2}{6\gamma}\right)^{1/3}\xi\right).$$

The speed V of this soliton is as follows:

$$V = V_0 \frac{1 + \frac{\tau M}{\gamma m}\left(1 - \frac{3}{5}(\gamma\sigma)^{1/3}(\tau/6)^{2/3}\right)}{1 + \frac{\tau M}{\gamma m}}, \quad \sigma = \frac{2\vartheta m a^2}{\hbar^2}, \quad \tau = \frac{\vartheta}{MV_0^2}$$

Geometrically, this solution seems as a local extension (or constriction) of the tube, which moves with velocity V. This motion leads to the motion of liquid in the neighborhood of this extension (restriction). The estimation of the Reynolds number shows that the Stokes approximation is appropriate. To describe the flow we use a model in which the local perturbation of the boundary is replaced by the point-like one (stokeslet approximation) [32, 33]. The mathematical background of the model is the theory of self-adjoint extensions of symmetric operators [34]. Consider two-dimensional straight channel (strip). The result for cylindrical channel is absolutely analogous. The Stokes flow for this case is described by the stream function ψ satisfying the biharmonic equation (for axisymmetric flow in a cylindrical domain the Laplace operator is replaced by the Stokes operator). As for the boundary conditions, it is more convenient to study moving boundary and fixed singularity, i.e. we assume that the normal derivative of the stream function isn't zero (it is equal to the wall velocity).

Let us describe briefly the model of point-like perturbation for the Stokes flow. The starting point is the operator Δ_0^2 which is the closure of the restriction of Δ^2 onto the set of smooth functions vanishing in a neighborhood of zero. The domain of the operator is

$$D(\Delta_0^2)=\{u,\ u\in L_2(\Omega),\Delta^2 u\in L_2(\Omega),u(0)=u'_{x_i}(0)=u''_{x_ix_j}(0)=0,i,j=1,2\}.$$

A model operator Δ_e^2 is obtained as a self-adjoint extension of symmetric operator Δ_0^2. Due to the correlation $\Delta_0^2\subset\Delta_e^2\subset\Delta_0^{2*}$ one can search the extension as a restriction of the adjoint operator. The domain of the operator Δ_0^{2*} consists from the elements of the following form:

$$u(x)=\sum_{i,j=1}^{2}c_{ij}^u g_{x_ix_j}(x)+\sum_{i=1}^{2}c_i^u g_{x_i}(x)+c_0 g(x)+\xi(x)(a_0+\sum_{i=1}^{2}a_i^u x_i+\sum_{i,j=1}^{2}a_{ij}^u\beta_{ij}x_ix_j)+u_0(x).$$

Here $u_0\in D(\Delta_0)$, $\beta_{ij}=1, i\neq j$, $\beta_{ii}=2^{-1}$, $\xi(x)-$ smooth cutting function: $\xi(x)=1,|x|\le 1$, $\xi(x)=0,|x|\ge 2$. To construct the model operator one should establish some correlation between these coefficients in the asymptotic expansion, for example, $U_0=AU_1$, $A=A^*$, $U_1=(c_0^u,c_1^u,c_2^u,c_{11}^u,c_{12}^u,c_{22}^u)$, $U_0=(a_0^u,a_1^u,a_2^u,a_{11}^u,a_{12}^u,a_{22}^u)$. As for detailed description of all classes of extensions, see [35]. The model allows one to obtain the streamfunction (and, consequently, the picture of the flow) in an explicit form. Calculations show that there is an eddy which follows the soliton, i.e. the soliton produces a flow in nanotube (see Fig. 1)

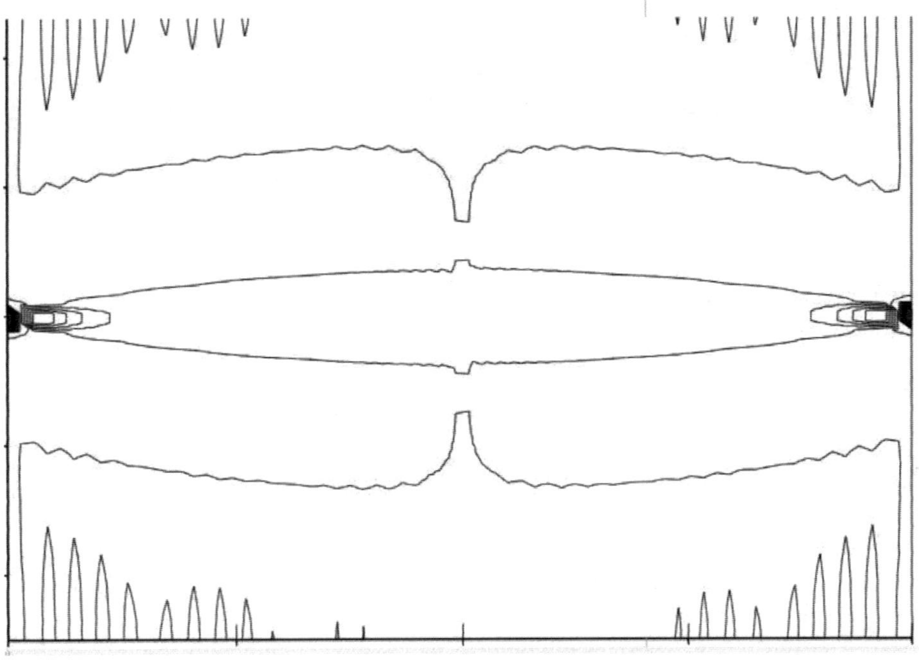

Fig. 1.

The work was supported by the Program "Development of the Scientific Potential of High School of Russia" (project 2.1.1/4215) and the Program "Scientific Staff of Innovative Russia 2009-2013" (project NK-526P_24).

References

[1] Li D 2008 *Encyclopedia of Microfluids and Nanofluids* (NY: Springer)
[2] Kadau K, Germann T C, Hadjiconstantinou N G, Lomdahl P S and Dimonte G 2004 *Proc. Nat.*

Acad .Sci .USA **101** 5851
[3] Joly L, Ybert C and Bocquet L 2006 *Phys. Rev. Lett.* **96** 046101
[4] Nie X B, Chen S Y and Robbins M O 2004 *J. Fluid Mech.* **500** 55
[5] Kang W and Landman U 2007 *Phys. Rev. Lett.* **98** 064504
[6] Antognozzi M Humphris A D L and Miles M J 2001 *Appl. Phys. Lett.* **78** 300
[7] Einstein A and Smolukhovski 1936 *Brownian Motion* (Leningrad: ONTI)
[8] Kogan M N 1967 *Dynamics of Rare Gas (Kinetic Theory)* (Moscow: Nauka)
[9] Frenkel Ya I 1941 *Uspekhi Fiz Nauk* **25** 1
[10] Zatsepina G N 1974 *Structure and Properties of Water* (Moscow: MSU)
[11] Li T D, Gao J, Szoszkeiwicz A Landman U and Riedo E 2007 *Phys. Rev.* **B75** 115415
[12] Mashl R J, Joseph S, Aluru N R and Jacobsson E 2003 *Nano Lett.* **3** 589
[13] Kolesnikova A I, Loong Ch-K de Souza N R, Burnhamb C J and Moravskyc A P 2006 *Physica B* **385-386** 272
[14] Kolesnikova A I, Zanotti J-M, Loong Ch-K and Thiyagarajan P 2002 *Phys. Rev. Lett.* **93** 035503
[15] Deryagin B V, Ovcharenko F D and Churaev V N 1989 *Water in Disperse Systems* (Moscow: Chemistry)
[16] Holt J K, Park H G, Wang Y, Stadermann M, Artyukhin A B, Grigoropoulos C P and Noy A 2006 *Science* **312** 1034
[17] Maslov V P 2007 *Theor. Math. Phys.* **153** 1677
[18] Hanasaki I and Nakatani A 2006 *Model. Simul. Mater. Sci. Eng.* **14** 9
[19] Insepov Z, Wolf D and Hassanein A 2006 *Nano Lett.* **6** 1893
[20] Gusarov V V and Popov I Yu 1996 *Nuovo Cim.* **18D** 799
[21] Thomas J A and McGaughey A J H 2008 *Nano Lett.* **8** 2788
[22] Liu Y and Wang Q 2005 *Phys. Rev.* **B72** 085420
[23] Joseph S and Aluru N R 2008 *Nano Lett.* **8** 452
[24] Demkov Yu N and Ostrovskii V N 1975 *Zero-range Potential Method in Atomic Physics* (Leningrad: LSU)
[25] Albeverio S, Gesztesy F, Hoegh-Krohn R and Holden H 1989 *Solvable Models in Quantum Mechanics* (Berlin: Springer)
[26] Pavlov B S and Shushkov A A 1989 *Math. Sb.* **137** 147
[27] Chivilikhin S A, Gusarov V V, Popov I Yu and Svitenkov A I 2008 *Russian J. Math. Phys.* **15** 410
[28] Frenkel Ya I 1975 *Kinetic Theory of Liquid* (Moscow: Nauka)
[29] Khusnutdinova K R, Samsonov A M and Zakharov A S 2009 *Theor. Math. Phys.* **159** 475
[30] Brizhik L, Eremko A, Piette B and Zakrzewski W 2010 *Int. J. Quant. Chem.* **110** 10
[31] Davidov A S 1985 *Solitons in Molecular Systems* (Dordrecht: Reidel)
[32] Popov I Yu 1993 *Phys. Scripta* **47** 682
[33] Gugel Yu V, Popov I Yu and Popova S L 1996 *Fluid Dyn. Res.* **18** 199
[34] Pavlov B S 1987 *Uspekhi Mat. Nauk* **42** 99
[35] Popov I Yu 1996 *Rev. Mat. Univ. Compl. Madrid* **9** 235
[36] Gerjuoy E 1965 *J. Math. Phys.* **1** 993

International Conference on Theoretical Physics Dubna-Nano 2010 IOP Publishing
Journal of Physics: Conference Series **248** (2010) 012007 doi:10.1088/1742-6596/248/1/012007

Sub-Poissonian phonon distributions in suspended carbon nanotubes

G Piovano[1], F Cavaliere[1] and M Sassetti[1]

[1] Dipartimento di Fisica and CNR-SPIN, Università di Genova, Via Dodecaneso 33, 16146 Genova

E-mail: giulia.piovano@gmail.com

Abstract. We study a quantum dot embedded into a vibrating suspended carbon nanotube. The out-of-equilibrium dynamics of the vibrational degrees of freedom is analyzed introducing the phonon Fano factor $\mathcal{F}$ to characterize different phonon distributions. *Sub-Poissonian* distributions with $\mathcal{F} < 1$ are found in selected transport regimes and are explained in terms of a *selective* population of phonon states due to a peculiar transport mechanism.

1. Introduction

Recent improvements in building nanoelectromechanical systems (NEMS) open a new scenario for studying the interaction between mechanical and electronic degrees of freedom at the nanoscale [1]. The electron-phonon coupling is responsible for such peculiar transport features as the Franck-Condon blockade [2]. A notable class of NEMS is represented by freely vibrating carbon nanotubes (CNTs), seamless cylinders of graphene [3] suspended between two metallic contacts at different voltages. Capacitive coupling to a gate electrode provides the possibility to modulate the effective charge in the system. The interest on such systems has increased considerably in the last years [4, 5, 6]. Signatures due to different vibrational modes, from the stretching ones [4, 5] to the radial breathing mode [6] have been detected in transport experiments and analyzed theoretically [7, 8, 9]. The properties of the out-of-equilibrium phonon populations of these systems have attracted however less attention [10]. In this paper we will theoretically study the vibron dynamics, which can exhibit a non-classical *sub-Poissonian* population of phonons, with fluctuations smaller than the average phonon number [11, 12].

2. Model

The CNT, of length $\mathcal{L}$, is suspended between two metallic gates, see Fig. 1. Here, it is assumed that the CNT vibrates according to its fundamental stretching mode. Within the CNT, a quantum dot of length L is formed due to structural defects at $x_1 = 0$ and $x_2 = L$, which act as tunnel barriers, while the CNT portions at $x < 0$ ($x > L$) act as source (drain) CNT contacts. A symmetric bias V, applied across the system forces electrons to flow from the source to the drain lead via the dot. Additionally, the dot is capacitively coupled to an external gate, at a voltage V_g, which tunes its excess charge.

Journal of Physics: Conference Series **248** (2010) 012007 doi:10.1088/1742-6596/248/1/012007

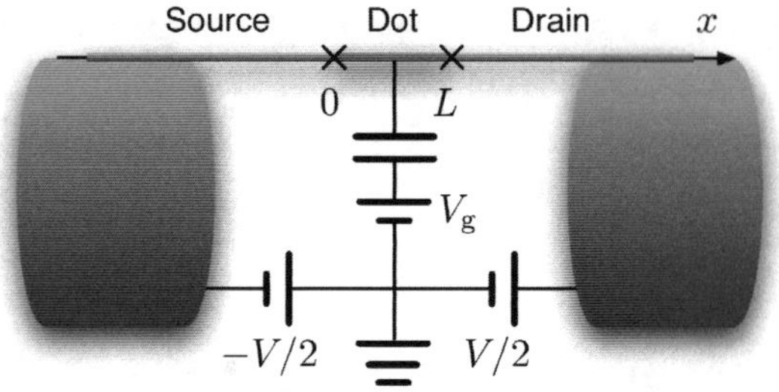

Figure 1. Model setup: a suspended CNT (blue) connected to two metallic gates. Along the CNT two tunneling barriers create a quantum dot (red) between $x_1 = 0$ and $x_2 = L$, capacitively coupled to a gate electrode.

The Hamiltonian of the CNT dot reads [13]

$$H_{\rm d} = \frac{E_{\rm c}}{2}(N - N_{\rm g})^2 + \sum_{\eta,\sigma,j} \varepsilon_{\eta,\sigma,j} n_{\eta,\sigma,j}\,, \tag{1}$$

where $E_{\rm c}$ is the charging energy, $n_{\eta,\sigma,j} = d^\dagger_{\eta,\sigma,j} d_{\eta,\sigma,j}$ is the occupation number of the j-th orbital electron state with spin $\sigma = \pm 1$ in the branch $\eta = 1, 2$. The spectrum is $\varepsilon_{\eta,\sigma,j} = j\varepsilon_0 + (\eta - 1)\Delta\varepsilon$, with ε_0 the electronic level spacing and $\Delta\varepsilon$ an energy mismatch between the two branches. Additionally, $N = \sum_{\eta,\sigma,j} n_{\eta,\sigma,j}$ is the dot total charge while $N_{\rm g}$ is the charge induced by the gate voltage $V_{\rm g}$. The lowest quantized stretching mode (vibron) is modeled by (here and in the following, $\hbar = 1$) $H_{\rm v} = \Omega b^\dagger b$, where Ω is the vibron frequency whose occupation is increased (decreased) by $b^\dagger$ (b). In this work we consider a dot *smaller* than the vibron ($L < \mathcal{L}$). In such a situation, the Anderson-Holstein model provides an appropriate description of the coupling between electrons and the vibron. Deviations from this model, occurring for $L \gtrsim \mathcal{L}$, have been recently considered [9]. The vibron couples with the total dot charge via the Hamiltonian

$$H_{\rm d-v} = \Omega\lambda(b^\dagger + b)(N - N_{\rm g})\,, \tag{2}$$

with λ a coupling parameter. The dot is also connected with the external CNT leads via

$$H_{\rm t} = \sum_\alpha t_\alpha \sum_{\eta,\sigma,k,j} c^\dagger_{\alpha,\eta,\sigma,k} d_{\eta,\sigma,j} + {\rm H.c} \tag{3}$$

with t_α the tunneling amplitude on barrier $\alpha = {\rm S, D}$ and $c^\dagger_{\alpha,\eta,\sigma,k}$ the creation operator for an electron with spin σ and longitudinal momentum k on the branch η of the lead α. The latter are treated as a noninteracting electron gas.

In order to diagonalize the linear coupling between the dot and the vibron we perform the Lang-Firsov canonical transformation [14] with generator $\mathcal{U} = e^{\lambda(b^\dagger - b)(N - N_{\rm g})}$. The transformed total Hamiltonian $\tilde{H} = \mathcal{U}(H_{\rm d} + H_{\rm v} + H_{\rm d-v})\mathcal{U}^\dagger$ and tunneling Hamiltonian $\tilde{H}_{\rm t} = \mathcal{U} H_{\rm t} \mathcal{U}^\dagger$ are expressed in terms of the original operators as

$$\tilde{H} = \frac{E_{\rm c}}{2}(N - N_{\rm g})^2 + \sum_{\eta,\sigma,j} \varepsilon_{\eta,\sigma,j} n_{\eta,\sigma,j} + \Omega b^\dagger b \tag{4}$$

$$\tilde{H}_{\rm t} = \sum_{\alpha=S,D} t_\alpha e^{\lambda(b - b^\dagger)} \sum_{\eta,\sigma,k,j} c^\dagger_{\alpha,\eta,\sigma,k} d_{\eta,\sigma,j} + {\rm h.c.}\,. \tag{5}$$

In the following, we are interested in the low-energy dynamics of the vibrational degrees of freedom. Since $\varepsilon_0 \gg \Omega$, as pointed out in recent experiments [5, 9], we can neglect the plasmonic excitations of the quantum dot. In this case, the dot states in branch η with spin σ are compactly filled with $N_{\eta,\sigma}$ electrons and the $\tilde{H}$ can be labeled as $|\{N_{\eta,\sigma}\}, l\rangle$ with l the occupation number of the vibron. Treating $\tilde{H}_{\rm t}$ to lowest order in t_α and tracing out the leads degrees of freedom, one obtains a master equation for the occupation probability of the dot-vibron states $P(\{N_{\eta,\sigma}\}, l)$

$$\dot{P}(\{N_{\eta,\sigma}\}, l) = \sum_{\alpha,\eta,\sigma} \Gamma^{(\alpha)}_{N'_{\eta,\sigma};l' \to N_{\eta,\sigma};l} P(\{N'_{\eta,\sigma}\}, l') - \sum_{\alpha,\eta,\sigma} \Gamma^{(\alpha)}_{N_{\eta,\sigma};l \to N'_{\eta,\sigma};l'} P(\{N_{\eta,\sigma}\}, l) \quad (6)$$

where

$$\Gamma^{(\alpha)}_{N_{\eta,\sigma};l \to N'_{\eta,\sigma};l'} = \Gamma_0 \frac{|t_\alpha|^2}{|t_{\rm S}|^2} X_{ll'} \{ f_\alpha(\Delta E)\delta_{N',N+1} + [1 - f_\alpha(\Delta E)]\delta_{N',N-1} \} \quad (7)$$

with $\Gamma_0 = 2\pi\nu|t_{\rm S}|^2$, $N = \sum_{\eta,\sigma} N_{\eta,\sigma}$ ($N' = \sum_{\eta,\sigma} N'_{\eta,\sigma}$), ν the density of states of the leads (assumed constant and equal on both contacts), ΔE the energy difference between the final and the initial states and $f_{\rm S/D}(E) = f(E \pm eV/2)$. Here, $f(E)$ is the Fermi function at temperature T and e is the electron charge. In Eq. 7, we have introduced the Franck-Condon factors [2] $X_{ll'} = |\langle l| \exp\left[\lambda(b - b^\dagger)\right] |l'\rangle|^2 = e^{-\lambda^2} \lambda^{2|l'-l|} (l_<!/l_>!) L_{l_<}^{|l'-l|}(\lambda^2)$, with $l_< = \min\{l, l'\}$, $l_> = \max\{l, l'\}$ and $L_a^b(x)$ the generalized Laguerre polynomials. It is also useful to introduce the asymmetry between the barriers with the source and the drain contacts $A = |t_{\rm D}|^2/|t_{\rm S}|^2$. In the following, we restrict the numerical solution of Eq. (6) in the steady state near the resonance between the two charge states of the dot $N = 4\mathcal{N}$, with $N_{\eta,\sigma} = \mathcal{N}\ \forall\ \eta, \sigma$, and $N + 1$ with one extra electron in the lowest unoccupied charge states, which occurs at $N_{\rm g} = 1/2 + 4\mathcal{N} + \mathcal{N}\varepsilon_0/E_{\rm c}$.

3. Results

In order to characterize the vibrational dynamics of the system, we introduce the vibron Fano factor $\mathcal{F} = {\rm var}(l)/\langle l\rangle$ [15], with $\langle l\rangle$ the average vibron occupation number and ${\rm var}(l) = \langle l^2\rangle - \langle l\rangle^2$ its variance[1]. In analogy to the photon case, we can distinguish between sub-, super- and Poissonian distributions according to $\mathcal{F} < 1$, $\mathcal{F} > 1$ and $\mathcal{F} = 1$. For conventional boson distributions such as e.g. photons in a classical light field, a super-Poissonian Fano factor is typically expected [15].

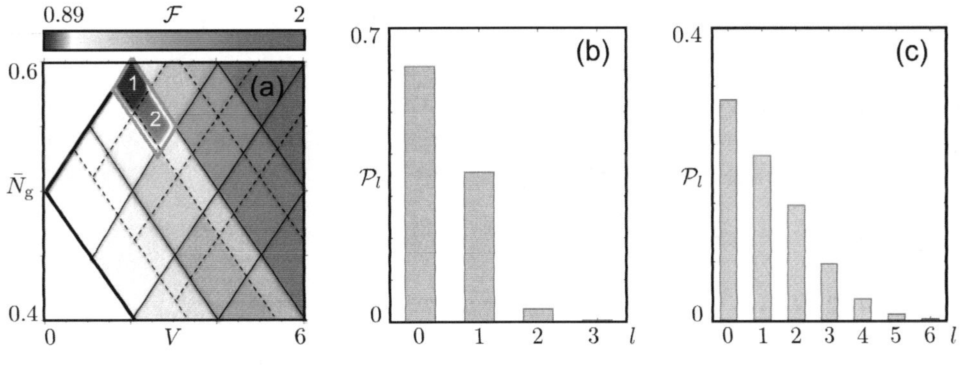

Figure 2. (a) Density plot of $\mathcal{F}$ as a function of V (units $\Omega/|e|$) and $\bar{N}_{\rm g} = N_{\rm g} - 4\mathcal{N} - \mathcal{N}\varepsilon_0/E_{\rm c}$. White: $\mathcal{F} = 1$. (b,c) Phonon populations in different points of the $(V, \bar{N}_{\rm g})$ plane: (b) $|e|V = 2\ \Omega$, $\bar{N}_{\rm g} = 0.54$ (zone 1, with $\mathcal{F} = 0.89$); (c) $|e|V = 5.7\ \Omega$, $\bar{N}_{\rm g} = 0.5$, with $\mathcal{F} = 1.7$. In all plots, $k_{\rm B}T = 0.02\ \Omega$, $\lambda^2 = 0.6$, $A = 1/20$, $E_{\rm c} = 20\ \Omega$, $\Delta\epsilon = 0.6\Omega$.

Figure 2a shows a density plot of $\mathcal{F}$ in the plane $(V, \bar{N}_{\rm g})$, where $\bar{N}_{\rm g} = N_{\rm g} - 4\mathcal{N} - \mathcal{N}\varepsilon_0/E_{\rm c}$

[1] Both quantities contain contributions due to the dot charge state through the electron-vibron coupling [11].

obtained solving numerically the master equation (6) in the stationary regime. The solid thick lines denote transitions between states with N and $N+1$ charges and the vibron in its ground state. These lines separate regions at low bias where transport is interdicted (Coulomb blockade) from areas where it is allowed. Thin solid (dashed) lines, with energy spacing Ω, mark the onset of vibrational excitations triggered by tunneling events on dot states with $\eta = 1$ ($\eta = 2$).
For the non-equilibrated dynamics of the vibron in our system, a super-Poissonian Fano factor is typically obtained. However, in selected parameter regimes, $\mathcal{F} < 1$ can be achieved: this occurs in the regions of the $(V, \bar{N}_{\rm g})$ plane denoted as "1" and "2" contoured in green in Fig. 2a. This phenomenon is linked with a peculiar *selective* population of the vibron states [11], in which $\mathcal{P}_0, \mathcal{P}_1 \gg \mathcal{P}_{l\geq 2}$, with $\mathcal{P}_l = \sum_{N_{\eta,\sigma}} P(\{N_{\eta,\sigma}\}, l)$. This population is achieved, in selected regimes of bias and gate voltages, if $\lambda \lesssim 1$ and if $A \neq 1$ [11]. Figure 2b shows $\mathcal{P}_l$ calculated in the center of zone 1: here $\mathcal{P}_0 \approx 0.6$ and $\mathcal{P}_1 \approx 0.35$, with $\mathcal{P}_{l\geq 2} \ll 0.1$.
Let us analyze in more details the transport mechanisms leading to this kind of population. In zone 1, tunnel in events occur with $\Delta l = l_{\rm f} - l_{\rm i} \leq 1$ (with $l_{\rm f/i}$ the vibron number of the final/initial state), while tunnel out transitions imply $\Delta l \leq 0$. Due to the above constraints, no direct population of the vibron states with $l_{\rm f} \geq 2$ from the vibrational ground state $l_{\rm i} = 0$ is allowed. Indeed the population of a state with given $l_{\rm f} \geq 2$ implies at least $l-1$ tunnel-out events starting from $l_{\rm i} = 0$. As a result, the population of vibron states with $l \geq 2$ scales as $\mathcal{P}_l \propto A^{l-1}$. For $\lambda \lesssim 1$ and $A < 1$, this leads to a "filtering" of the probability distribution with $\mathcal{P}_0 \lesssim \mathcal{P}_1 \ll \mathcal{P}_{l\geq 2}$ and an ensuing $\mathcal{F} < 1$ [11]. In zone 2, tunnel out events from the state with $\eta = 2$ may occur with $\Delta l \leq 1$: this results in a stronger de-population of the $l = 1$ state and a weaker sub-Poissonian Fano factor.
For higher bias voltage, super-Poissonian distributions are generally achieved. Due to the large number of transport channels supporting vibron excitations, one finds *broad* phonon distributions with $\mathcal{F} > 1$. An example is shown in Fig. 2c.

4. Conclusions

In this work we have studied a quantum dot embedded into a freely-vibrating suspended CNT. We have analyzed the out-of-equilibrium occupation probability of vibron states, which in selected transport regimes exhibits a *sub-Poissonian* phonon distribution. We have explained this non-classical behavior as due to a *selective* population of vibron states.

Acknowledgments

Financial support by the EU-FP7 via ITN-2008-234970 NANOCTM and by INFM-CNR via "Progetto giovani" are gratefully acknowledged.

5. References

[1] Blencowe M P 2005 *Contemporary Physics* **46** 249
[2] Koch J, von Oppen F, and Andreev A V 2006 *Phys. Rev.* B **74** 205438
[3] Saito R and Dresselhaus M S 1998 *Physical Properties of Carbon Nanotubes* (Imperial College Press)
[4] Sapmaz S, Jarrillo-Herrero P, Blanter Y M, and Dekker C 2006 *Phys. Rev. Lett.* **96** 026801
[5] Leturcq R et al. 2009 *Nature Phys.* **5** 327
[6] Leroy B J, Lemay S G, Kong J, and Dekker C 2004 *Nature* **432** 371
[7] Zazunov A, Feinberg D, and Martin T 2006 Phys. Rev. B **73** 115405
[8] Haupt F, Cavaliere F, Fazio E, and Sassetti M 2006 *Phys. Rev. B* **74**
[9] Cavaliere F, Mariani E, Leturcq R, Stampfer C, and Sassetti M 2010 *Phys. Rev.* B **81** 201303(R)
[10] Mitra A, Aleiner I, and Millis A J, Phys. Rev. B **69** 245302
[11] Merlo M, Haupt F, Cavaliere F, and Sassetti M 2008 *New J. Phys.* **10** 023008
[12] Cavaliere F, Piovano G, Paladino E, and Sassetti M 2008 *New J. Phys.* **10** 115004
[13] Oreg Y, Byczuk K, and Halperin B I 2000 *Phys. Rev. Lett.* **85** 365
[14] Mahan G D 2000 *Many-Particle System* (New York: Kluwer Academic/Plenum Publisher)
[15] Mandel L and Wolf E 1995 *Optical Coherence and Quantum Optics* (New York: Cambridge University Press)

International Conference on Theoretical Physics Dubna-Nano 2010 IOP Publishing
Journal of Physics: Conference Series **248** (2010) 012008 doi:10.1088/1742-6596/248/1/012008

The electronic spectra of double-wall zig-zag carbon nanotube affected by the magnetic field.

M. Pudlak[1], **R. Pincak**[1,2]

[1] Institute of Experimental Physics, Slovak Academy of Sciences, Watsonova 47,043 53 Kosice, Slovak Republic

[2] Joint Institute for Nuclear Research, BLTP, 141980 Dubna, Moscow region, Russia

E-mail: pudlak@saske.sk, pincak@saske.sk

Abstract. We study the electronic spectra of (9,0)-(18,0) double walled carbon nanotubes influenced by the external magnetic field. We choose the orientation of the magnetic field parallel to the axis of nanotube. We found the strong change of the electronic spectra due to external magnetic field. That means gap between valence and conductive bands in DWN will be changed.

1. Introduction

A single-wall carbon nanotube can be described as a graphene sheet rolled into a cylindrical shape so that the structure is one-dimensional with axial symmetry and in general exhibiting a spiral conformation called chirality. Of special interest is the prediction that the calculated electronic structure of a carbon nanotube can be either metallic or semiconducting, depending on its diameter and chirality. The energy gap for a semiconductor nanotube, which is inversely proportional to its diameters, can be directly observed by scanning tunneling microscopy measurements. We now consider the electronic structure of carbon nanotubes in a uniform external magnetic field. There are two high symmetry cases for the direction of the magnetic field: one with the magnetic field parallel to the nanotube axis ($B \parallel z$) and the other with the magnetic field perpendicular to the nanotube axis, ($B \perp z$). Hereafter the nanotube axis is taken along the z-axis. In this paper we consider the case of $B \parallel z$. Especially we are interested in the zigzag $(9,0) - (18,0)$ double-wall nanotubes(DWNs) affected by the magnetic field. The synthesis of DWNs has been reported recently [1, 2]. Their electronic structure was investigated by the local density approximation [3, 4] and the tight-binding model [5, 6, 7, 8, 9]. When the magnetic field is parallel to the nanotube axis, electrons moving within the nanotube surface will feel a force perpendicular to the surface. As far as we consider only the transfer integral between two atoms within the nanotube surface, the electronic structure would appear to be unaffected by the magnetic field. This, however, is not correct. The wavefunction will change its phase factor and thus its momentum, k, will shift depending on the magnetic flux penetrating the cross section of the carbon nanotube. This phenomenon is generally known as the Aharonov-Bohm effect, discussed often in the case of cylindrical geometry. Since the carbon nanotube can be a metal or a semiconductor, depending on whether there is an allowed wavevector k in the circumferential direction that has the value of the K. point in the two-dimensional Brillouin zone, this Aharonov-Bohm effect will modify the energy gap of a carbon nanotube as a function of magnetic field.

2. $(9,0)-(18,0)$ zigzag tubules

The π electronic structures are calculated from the tight-binding Hamiltonian

$$H=\sum_i \epsilon|\varphi_i^{out}\rangle\langle\varphi_i^{out}|+\sum_{i,j}\gamma_{ij}\left(|\varphi_i^{in}\rangle\langle\varphi_j^{in}|+h.c\right)+\sum_i\widetilde{\epsilon}|\varphi_i^{in}\rangle\langle\varphi_i^{in}|+\sum_{i,j}\widetilde{\gamma}_{ij}\left(|\varphi_i^{in}\rangle\langle\varphi_j^{in}|+h.c\right)$$
$$+\sum_{l,n}W_{ln}\left(|\varphi_l^{in}\rangle\langle\varphi_n^{out}|+h.c\right), \qquad (1)$$

ϵ and $\widetilde{\epsilon}$ are Fermi energies of the outer and inner nanotubes; $|\varphi_i^{out}\rangle$, $|\varphi_i^{in}\rangle$ are π orbitals on site i at the outer and inner tubes; γ_{ij}, $\widetilde{\gamma}_{ij}$ are the intratube hopping integrals; W_{ij} are the intertube hoping integrals which depends on the distance d_{ij} and angle θ_{ij} between the π_i and π_j orbitals.

$$W_{ij}=\frac{\gamma_0}{8}\cos(\theta_{ij})e^{(\xi-d_{ij})/\delta}, \qquad (2)$$

where θ_{ij} is an angle between the ith atom of the inner shell and the jth atom of the outer shell, d_{ij} is the interatom distance and ξ is a intertube distance. The characteristic length $\delta=0.45\mathring{A}$. We assume the symmetric geometry of zig-zag DWNT [7]. It was considered that hopping between shells takes place only between atoms which occupy position directly each above other. It means we take into account only the interactions

$$W_{ii}=\frac{\gamma_0}{8}. \qquad (3)$$

We look for solution in the form

$$\Psi=\sum_{i=1}^{12}c_i\psi_i. \qquad (4)$$

We have secular equations

$$\sum_{j=1}^{12}\langle\psi_i|H|\psi_j\rangle c_j=\tilde{E}c_i, \qquad (5)$$

where

$$\langle\psi_i|H|\psi_j\rangle=\delta_{ij}E_i, \qquad (6)$$

for $i,j=1,...8$ and $i,j=9,...12$, and the interaction between shells is described by the terms $\langle\psi_i|H|\psi_j\rangle$ for $i=1,...8$; $j=9,...12$, see [10]. The vector potential A for B|| in this coordinate system is given by

$$A=\left(\frac{\Phi}{L},0\right), \qquad (7)$$

where Φ is the magnetic flux penetrating the cross section of a carbon nanotube, and L is the diameter of tube. Thus the shift k is

$$k_x=k_x+\frac{\Phi}{L\Phi_0}, k_x=k_x+\frac{\Phi}{2L\Phi_0}, \qquad (8)$$

for inner and outer tube, respectively. To construct the Hamiltonian, we use only the valence and conductive states of individual nanotubes in the absence of intertube interaction. The electronic structures can be calculated from the Hamiltonian

$$H=\begin{pmatrix}\Delta+b2 & 0 & H_{3,11} & H_{3,12}\\ 0 & \Delta-b2 & H_{4,11} & H_{4,12}\\ H_{11,3} & H_{11,4} & -\Delta+b6 & 0\\ H_{12,3} & H_{12,4} & 0 & -\Delta-b6\end{pmatrix} \qquad (9)$$

where

$$b2 = \gamma_0(1 - 2\xi \cos\frac{\sqrt{3}ka}{2} + \xi^2)^{\frac{1}{2}}, \quad (10)$$

$$b6 = \gamma_0(1 - 2\tilde{\xi} \cos\frac{\sqrt{3}ka}{2} + \tilde{\xi}^2)^{\frac{1}{2}}, \quad (11)$$

$$H_{3,11} = H_{4,12} = \frac{1}{4\sqrt{2}}\frac{\gamma_0}{8}\left(1 + e^{-i(\varphi_6-\varphi_2)}\right), \quad (12)$$

$$H_{4,11} = H_{3,12} = \frac{1}{4\sqrt{2}}\frac{\gamma_0}{8}\left(1 - e^{-i(\varphi_6-\varphi_2)}\right), \quad (13)$$

$$e^{i(\varphi_6-\varphi_2)} = \gamma_0^2 \frac{1 - \xi e^{i\frac{\sqrt{3}ka}{2}} - \tilde{\xi} e^{-i\frac{\sqrt{3}ka}{2}} + \xi\tilde{\xi}}{b2\ b6}. \quad (14)$$

$2\Delta(\approx\ 0.21eV)$ is a Fermi energy difference of the outer and inner nanotubes, $\Psi(k) = (d_3, d_4, d_{11}, d_{12})$ and d_i is an amplitude to find electron in state ψ_i. The wave functions ψ_3, ψ_4 are conductance and valence states of outer nanotube and ψ_{11}, ψ_{12} are conductance and valence states of inner nanotube in the absence of the intertube interaction [10]. The parameter $\gamma_0(\simeq 3eV)$ is the hoping integral in the graphene.

The parameters ξ, $\tilde{\xi}$ can be expressed in the form

$$\tilde{\xi} = 2\tilde{\beta}\cos(\frac{m\pi}{9} + \frac{\Phi}{9\Phi_0}), \quad (15)$$

and

$$\xi = 2\beta\cos(\frac{m\pi}{9} + \frac{\Phi}{18\Phi_0}), \quad (16)$$

where the parameters $\tilde{\beta}$ and β are the same as in [10, 12]

The eigenvalues of Hamiltonian (1) for some values of $\sqrt{3}ka/2$ near the point $k = 0$ are depicted on Fig.1. and was computed in [12]. The Fig.2. shows broadening of the gap between valence and conductive bends due to the influence of the magnetic field parallel to the z axes.

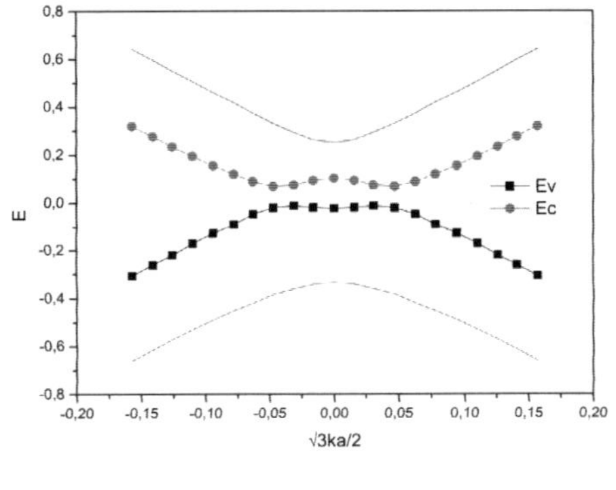

Figure 1. Spectra of zigzag DWN where E_c and E_v are conductive and valence band.

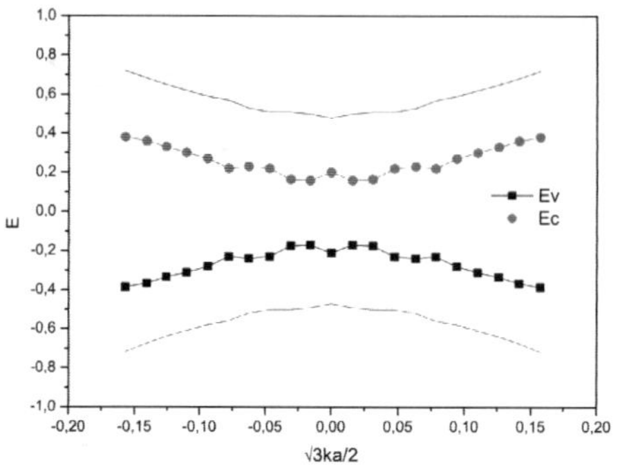

Figure 2. Spectra of zigzag DWN affected by the magnetic field with value $\Phi/\Phi_0 = 0.5$.

3. Conclusion

As was described in [10, 12] the Fermi level of the outer shell is about 0.21 eV higher than the Fermi level of the inner shell in the case of zig-zag SWNs. In the case of zig-zag DWNs the curvature do not shift the k_F of the individual nanotubes. The result is that these DWNs are semiconductors. Generally we can say that the conductivity depends on the relative position of the Fermi points k_F of individual nanotubes. If there is no shift the DWN is a semiconductor. If there is a shift in the dependence on Fermi levels and the energy gaps of individual nanotubes the DWN can be semimetal or semiconductor. Since the presence of an energy gap in a carbon nanotube is determined by whether or not the one-dimensional energy bands cross or do not cross at the K or K' points at the corners of the 2D Brillouin zone, a semiconducting carbon nanotube can become metallic in a parallel magnetic field at certain values of the phase shift, and conversely a metallic nanotube can become semiconducting in a parallel magnetic field, the energy gap thus oscillates. The significance in a carbon nanotube is that the semiconducting or metallic nature of the nanotube can be altered only by applying a magnetic field parallel to the nanotube axis. This is because the distinction between a semiconducting and a metallic carbon nanotube arises from a quantum effect in which discrete wave numbers in the circumferential direction distinguish between metallic and semiconducting properties. It is noted that the one-dimensional energy dispersion relations for carbon nanotubes at the top of the valence band and the bottom of the conduction band follow a linear t relation only when the carbon nanotube is metallic. When the carbon nanotube has a semiconducting energy gap, the energy dispersion is quadratic at the top of the valence band and at the bottom of the conduction band. Thus the effective mass of an electron contributing to the transport properties of a carbon nanotube is a function of magnetic field.

ACKNOWLEDGEMENTS — The work was supported by the Slovak Academy of Sciences in the framework of CEX NANOFLUID, and by the Science and Technology Assistance Agency under Contract No. APVV 0509-07 and by VEGA Grant No. 2/0069/10.

References

[1] T.Sugai,H.Yoshida,T.Shimada,T.Okazaki,H.Shinohara 2003 Nano Lett. **3** 769
[2] Z.Zhou *et al.* 2003 Carbon **41** 337
[3] Y.Kwon,D.Tomanek 1998 Phys.Rev. B **58** R16001
[4] S.Okada,A.Oshiyama 2003 Phys.Rev. Lett. **91** 216801
[5] Y.H.Ho,C.P.Chang,F.L.Shyu,R.B.Chen,S.C.Chen,M.F.Lin 2004 Carbon **42** 3159
[6] R.Saito,G.Dresselhaus,M.S.Dresselhaus 1993 J.Appl.Phys. **73** 494

[7] Y.H.Ho,G.W.Ho,S.J.Wu,M.F.Lin 2006 J.Vac.Sci.Technol. B **24** 1098
[8] Ph.Lambin, V.Meunier,A.Rubio 2000 Phys.Rev.B **62** 5129
[9] J.-C.Charlier, X.Blase,S.Roche 2007 Review of Modern Physics **79** 677
[10] M. Pudlak, R. Pincak 2009 Eur.Phys.J.B **67** 565
[11] T.Frankel, *The Geometry of Physics*(Cambridge University Press,Cambridge, 1999)
[12] M. Pudlak, R. Pincak 2008 Journal of Physics, Conf. Ser. **129** 012011
[13] C.H. Lee *et al.* 2008 J.Phys.:Condens.Matter **20** 075213
[14] A.G.Souza Filho *et al.* 2007 NANO LETTERS **7** 2383
[15] B.Shan,K.Cho 2006 Phys.Rev.B **73** 081401(R)
[16] P.N.Dyachkov, D.V.Makaev 2006 Phys.Rev. B **74** 155 442

Journal of Physics: Conference Series **248** (2010) 012009 doi:10.1088/1742-6596/248/1/012009

Conductance through the disclination dipole defect in metallic carbon nanotubes

D.V. Kolesnikov[1] **and V.A. Osipov**[2]

[1,2]BLTP JINR, Dubna, Moscow region, Russia

E-mail: [1]kolesnik@theor.jinr.ru, [2]osipov@theor.jinr.ru

Abstract. The electronic transport properties of a metallic carbon nanotube with the five-seven disclination pair characterized by a lattice distortion vector are investigated. The influence of the disclination dipole includes induced curvature and mixing of two sublattices. Both these factors are taken into account via a self-consistent perturbation approach. The conductance and the Fano factor are calculated within the transfer-matrix technique.

PACS: 73.63.Fg, 72.80.Rj, 72.10.Fk

1. Introduction

Transport properties of variously shaped carbon nanostructures and graphene are of great practical and theoretical interest. In particular, the conductivity of carbon nanotubes is currently the subject of wide investigations. It is known that disclinaton topological defects (fivefolds) can convert planar graphene surface into the conical one with a marked modification of the electronic states [1]. In the nanotubes, however, the presence of an isolated pentagonal ring is not allowed and the simplest topological defect is the 5-7 disclination dipole (DD). The simple model for the junction between two metallic tubes was investigated in [2].

In this paper, we study the electronic transport properties of a metallic nanotube containing closely spaced 5-7 disclination dipole (i.e., the junction composed of two semi-infinite *metallic* nanotubes).

2. Geometry

A metallic or semiconducting character of the tube is governed by the translational vector $\vec{T_0}(n,m)$, where (n,m) are the numbers of steps along two unit cell vectors in the honeycomb lattice. The boundary conditions for the tube with a translational vector $\vec{T}$ transforms to the angular vector field a_φ which dependis on chirality as $(n+m)\ mod\ 3$. This field, along with the angular momentum, subsequently generates the mass-type term in the effective Dirac equation (see [3] for detail).

As for the fivefold-sevenfold pair in the nanotube, it appears to be the source of translational-type holonomy [4], therefore it should produce additional gauge vortex field. However, in this paper we restrict our consideration to a special case when the condition $(n+m)\ mod\ 3 = 0$ is fulfilled at *both* sides. Additionally, at low energies below the first gap value ($|E| << hV_F/R_0$, where V_F is the Fermi velocity and R is the tube radius) one can take into account for the

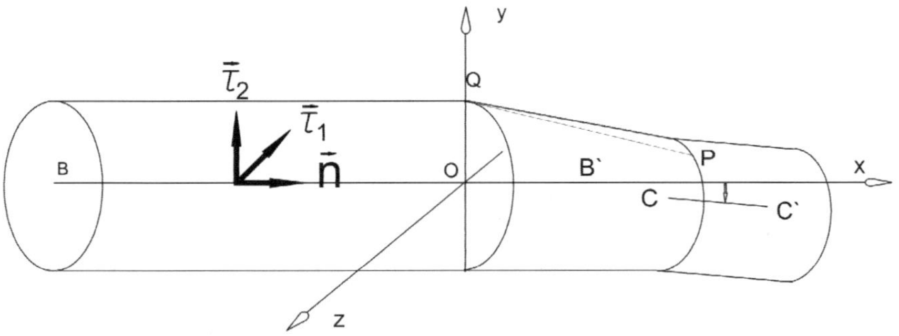

Figure 1. A schematic picture of the nanotube surface. The fivefold is located at the point Q, and the sevenfold - at the point P. The thick tube axis BB' coincides with the x-axis. The thin tube axis CC' is shifted due to the disclination dipole QP.

estimation of first-order perturbation only the main conducting channel corresponding to the lowest angular momentum.

As is known, the dislocation-type defect touches both the chirality and the radius of the carbon nanotube. In the general case, the tube with DD can be characterized by the distortion vector $\vec{b}$: for the translational vector $\vec{T}$ one can find, that it has the value $\vec{T_0}$, $T_0 = 2\pi R_0$ on the one side and $\vec{T_0} + \vec{b}$ on the other. Trying to describe a shape of the nanotube, it is useful to perform a development of the junction region onto a 2D plane.

Let us associate the development with the xy-plane and choose the axis along the tube on the left side as the x-axis. In this case, one has $\vec{T_0}(0, T_0, 0)$ and $\vec{T}(\xi) = \vec{T}(T_x, T_y, 0)$. It is convenient to introduce the new frame with the vector $\vec{n} = (\vec{T} \times \vec{e}_z)/T$ ($\vec{e}_i$ $(i = x, y, z)$ are the orthogonal basis vectors) which follows the tube axis (see Fig. 1).

The orthogonal to $\vec{n}$ vectors are $\vec{\tau}_1 = -\vec{e}_z$ and $\vec{\tau}_2 = \vec{\tau}_1 \times \vec{n} = \vec{T}/T$. To take into account the disclination dipole, which is situated at the top of the cylinder, one needs to describe the skewed conical surface between Q and P points. Actually, the DD has two effects: it changes the radius and shifts the surface of the tube in the direction normal to the tube axis. Since the radius $R(\xi) = T(\xi)/(2\pi)$ changes its value on ΔR, the ring will be shifted in the direction of $\vec{\tau}_2$ by ΔR, so that we must add the shift $\vec{\tau}_2 T/(2\pi)$.

Finally, the surface is parametrized as following:

$$\vec{r} = \vec{n}\xi + \vec{\tau}_2 \frac{T}{2\pi} + \vec{\tau}_1 \frac{T}{2\pi} \cos\varphi + \vec{\tau}_2 \frac{T}{2\pi} \sin\varphi, \tag{1}$$

where $-\infty < \xi < \infty$ and $0 \leq \varphi < 2\pi$ are the normal and the transversal coordinates, respectively.

The form of the tube with 5-7 rings is supposed to be

$$\vec{r}(\xi\frac{T_y}{T} + \frac{T_x}{2\pi}(1+\sin\varphi), -\xi\frac{T_x}{T} + \frac{T_y}{2\pi}(1+\sin\varphi), -\frac{T}{2\pi}\cos\varphi) \tag{2}$$

where we assume

$$\vec{T}(\xi) = \vec{T_0} + \frac{\vec{b}}{2}(\tanh\frac{\xi}{\xi_0} + 1) \tag{3}$$

to be the tube translation vector, depending on the coordinate ξ, the distortion vector $\vec{b}$ and an effective parameter ξ_0, which corresponds to the half-width of the surface region curved by the DD.

3. Conductance

From (2, 3) one can find metrical tensor, metrics connection coefficients $\Gamma^{\kappa}_{\mu\lambda}$, tetradic coefficients $e^{i}_{\ \mu}$ which are determined as $g_{\mu\nu} = e^{i}_{\ \mu} e^{k}_{\ \nu}\delta_{ik}$, and the spin connection coefficients which are determined as $(\omega_\mu)^{ab} = e^{a}_{\ \nu} D_\mu e^{b\nu}$, where $D_\mu := \partial_\mu + \Gamma_\mu$. From now on we set $E_F = 0$, $V_F = \hbar = 1$. The effective Dirac equation on the curved surface is defined as

$$-i\sigma_i e_i^{\ \mu}(\nabla_\mu - iA_\mu)\psi = E\psi, \tag{4}$$

with σ_i being the Pauli matrices ($i = 1, 2$), ψ is a 2-component spinor wavefunction, and A_μ ($\mu = \xi, \varphi$) being the DD field. The derivative includes a spin connection term Ω_μ and it is written as $\nabla_\mu = \partial_\mu + \Omega_\mu$, $\Omega_\mu = \frac{1}{8}\omega_\mu^{\ ab}[\sigma_a, \sigma_b]$. In the linear in $\vec{b}$ approximation and for the momentum equal to zero, the unperturbed Dirac operator has the form $-i\sigma_1\partial_\xi$, and the perturbation has the form

$$\hat{\mathcal{V}} = i\sigma_1 \frac{(\vec{b}\times\vec{T_0})_z}{8\pi^2 R_0 \xi_0 \cosh^2(\xi/\xi_0)}\partial_\xi - \sigma_1 < A_\xi >_\varphi - \sigma_1 \frac{\vec{b}\cdot\vec{T_0}}{16\pi^2 R_0^2 \xi_0 \cosh^2(\xi/\xi_0)} - \frac{\sigma_2}{R_0} < A_\varphi >_\varphi . \tag{5}$$

Here $<>_\varphi$ denotes the averaging over φ and $(\vec{b}\times\vec{T_0})_z = b_x T_{0y} - T_{0x} b_y$. Now the perturbation (5) is localized near $\xi = 0$ with a half-width ξ_0. In order to remove the differentiation in (5) one can multiply the operator $\hat{\mathcal{D}}_0 + \hat{\mathcal{V}}$ by $\left(1 + \frac{(\vec{b}\times\vec{T_0})_z}{8\pi^2 R_0 \xi_0 \cosh^2(\xi/\xi_0)}\right)$. In this case, the additional potential $IE(\vec{b}\times\vec{T_0})_z/(8\pi^2 R_0\xi_0\cosh^2(\xi/\xi_0))$ appears, where I is a unit matrix.

The field $\vec{A}$ describes the DD-induced valley mixing. It may be written as $A_j = \mu\epsilon_{jk}\partial_k G$ where $j, k = \xi, \varphi$, μ is an effective "charge" of each disclination, G is a dipole potential, which satisfies the equation

$$\nabla^2 G(\xi,\varphi) = \frac{2\pi}{R}[d_\xi \delta'(\xi)\delta(\varphi) + d_\varphi \delta(\xi)\delta'(\varphi)/R_0], \tag{6}$$

$\nabla^2 = \partial_\xi^2 + \partial_\varphi^2/R_0^2$, ϵ_{jk} is a unit antisymmetric tensor, $\delta(x)$ and $\delta'(x)$ are the Dirac delta function and it's derivative, and $\vec{d}$ is a vector connecting five- and sevenfold (it has the same order as $\vec{b}$ [5]). One can find G with the Fourier expansion $G = \sum_n e^{in\varphi}G_n(\xi)$. For the zero momentum $n = 0$, the field $\vec{A}$ has the form

$$A_\xi = 0,\ A_\varphi = \frac{4\pi^2 \mu b_\xi}{R_0}\delta(\xi), \tag{7}$$

where $\delta(\xi)$ is a Dirac delta function. Finally, the perturbation operator looks as following:

$$\hat{\mathcal{V}} = -I\frac{(\vec{b}\times\vec{T_0})_z E}{8\pi^2 R_0\xi_0\cosh^2(\xi/\xi_0)} - \sigma_1\frac{\vec{b}\cdot\vec{T_0}}{16\pi^2 R_0^2\xi_0\cosh^2(\xi/\xi_0)} - \sigma_2\frac{4\pi^2\mu d_\xi}{R_0^2}\delta(\xi). \tag{8}$$

Notice that the last term in (8) corresponds to the ξ-dependent mass term $\sigma_2 m(\xi)$ in the Dirac equation. Let us calculate the transfer matrix and the conductance according to the method described in [6] and [7]. To transform the initial Dirac equation to the diagonal form, the unitary rotation is performed: $\psi \to \mathcal{L}\psi$, $\hat{\mathcal{V}} \to \mathcal{L}\hat{\mathcal{V}}\mathcal{L}^\dagger$, $\sigma_i \to \mathcal{L}\sigma_i\mathcal{L}^\dagger$, $\mathcal{L} = (\sigma_1 + \sigma_3)/\sqrt{2}$. For zero angular momentum, the equation for the transfer matrix is

$$\mathcal{T}(\frac{L}{2}, -\frac{L}{2}) = \mathcal{T}^{(0)}(\frac{L}{2}, -\frac{L}{2}) - i\int_{-\frac{L}{2}}^{\frac{L}{2}} dx \mathcal{T}^{(0)}(\frac{L}{2}, x)\sigma_3\hat{\mathcal{V}}(x)\mathcal{T}(x, -\frac{L}{2}), \tag{9}$$

where $\mathcal{T}^{(0)}(x_2, x_1) = \exp(i\sigma_3 E(x_2 - x_1))$ and L is the length of the tube.

Taking into account $\mu = \tau_2/4$,one can find the transfer matrix in the form

$$\mathcal{T} = \exp(i\sigma_3 EL) + i[\frac{\vec{b}\cdot\vec{T}_0}{8\pi^2 R_0^2} + \sigma_3 \frac{(\vec{b}\times\vec{T}_0)_z E}{4\pi^2 R_0}]\exp(i\sigma_3 EL) + \mu d_\xi \sigma_1 \tau_2 \frac{4\pi^2}{R_0^2}. \tag{10}$$

The scattering matrix $(t^\dagger)^{-1}$ is defined by the upper-left element of the transfer matrix, so that the last term in (10) is neglected. The conductance is defined as $G = (4e^2/h)Tr[t^\dagger t]$, where the trace is taken over the channel space. It takes the form

$$G = \frac{4e^2}{h}[1 - \left(\frac{\vec{b}\cdot\vec{T}_0}{2T_0^2} + \frac{(\vec{b}\times\vec{T}_0)_z \epsilon}{T_0^2}\right)^2], \tag{11}$$

where $\epsilon = ER_0$ is a dimensionless energy (in the absolute units of hV_F/R_0). Correspondingly, the Fano factor $F = 1 - Tr[(t^\dagger t)^2]/Tr[t^\dagger t]$ reads

$$F = \left(\frac{\vec{b}\cdot\vec{T}_0}{2T_0^2} + \frac{(\vec{b}\times\vec{T}_0)_z \epsilon}{T_0^2}\right)^2. \tag{12}$$

4. Conclusion

In this paper, we have investigated the electronic transport properties of metallic nanotubes with the disclination dipole defect. The case of disclination dipoles preserving the chirality (metallic character of the tube) was considered. This assumption simplifies the problem under consideration and allows us to construct a self-consistent expansion scheme for the effective Dirac equation. The perturbation includes both the curvature term, which appear as the Gaussian-like smooth localized potential, and the DD-induced term. The standard transfer-matrix approach is used to calculate the conductance and the Fano factor of the structure. As the marked effect of the curvature, the energy-dependent term in both the conductance and the Fano factor appears, depending on the angle between vectors $\vec{b}$ and $\vec{T}_0$. For the angle less than $\pi/2$ the conductance was found to decrease with energy, and vice versa.

This work has been supported by the Russian Foundation for Basic Research under grant No. 08-02-01027.

[1] Lammert P and Crespi V. 2004 *Phys. Rev. B* **69** 035406
[2] Tamura R and Tsukada M. 2000 *Phys. Rev. B* **61** 8548
[3] Kane C L and Mele E J. 1997 *Phys. Rev. Lett* **78** 1932
[4] Vozmediano M A H, Katznelson M I, and Guinea F. 2010 *Preprint* arXiv:1003.5179
[5] For the planar case, $|\vec{d}| = |\vec{b}|$; see Yazyev O V and Louie S G. 2010 *Preprint* arXiv:1004.203
[6] Titov M. 2007 *Europhys. Lett* **79** 17004
[7] Schuessler A et al. 2009 *Phys. Rev. B* **79** 075405

Journal of Physics: Conference Series **248** (2010) 012010 doi:10.1088/1742-6596/248/1/012010

The effect of the orientation of the Stone-Wales defects on the bands structure of carbon nanotubes

P Partovi-Azar[1] **and A Namiranian**[1,2]

[1] Computational Physical Science Laboratory, Department of Nano-Science, Institute for research in fundamental sciences (IPM), Tehran, Iran.

[2] Department of Physics, Iran University of Science and Technology, Narmak, 16345, Tehran, Iran.

E-mail: p.partovi@nano.ipm.ac.ir, afshinn@iust.ac.ir

Abstract. We have studied the effect of orientation of Stone-Wales (SW) defects formed on the armchair single-wall carbon nanotubes (SWCNTs) on their metal-semiconductor behaviour by means of the density functional formalism. It has been shown that two third of possible orientations of SW defects, at high concentrations, open a $\approx$ 40 meV gap at Fermi level while in the other one third orientations tube remains metallic.

A single-wall carbon nanotube (SWCNT), a graphene sheet warped around a specific axis, is supposed to be the building block for future nano-electronic technologies [1, 2]. The surprising feature of a perfect SWCNT, as a cluster of carbon atoms, is the dependence of its electrical conductivity on its geometrical structure alone: almost one third of these structures depending on their warping (chiral) vector are conductors, while the others are semiconductors [3]. However, presence of any kind of imperfection, may change the conductivity of a SWCNT drastically [4]. Now it has been a decade since the first proposition that addition of impurity atoms, like Nitrogen or Boron, on metallic SWCNTs, may reduce their conductance up to 50 percents [5]. An important class of imperfections on SWCNT, are the structural disorders. A common structural disorder, so-called Stone-Wales (SW) defect [6], appears when a carbon-carbon bond rotates 90 degrees, and consequently four hexagons around that reconstruct to form two pentagon-heptagon pairs. Usually a SW defect is formed on nanotubes under high tensile strain [7]. The existence of a SW defect alters many mechanical and electrical properties of CNTs [8, 9, 10, 11].

Here we have focused on possible metal-simiconductor transition in armchair nanotubes, which are believed to be metallic, in the presence of a high concentration of the SW defects. Depending on the mentioned rotating carbon bond, two types of SW defects are recognized on an armchair SWCNT.

Each carbon atom on the surface of a SWCNT has three bonds with three neighbors. In armchair tubes, when a SW defect occurs, rotating bond forms a 60 degree angle or it lies parallel to the tube axis. As such, there are two distinct orientation for a SW defect on a SWCNT (figure 1). Hereafter we call these two types 60 degree and parallel types respectively.

We have used a density functional formalism based on Troullier-Martins [12] norm-conserving pseudo-potentials using pseudo-atomic orbitals as implemented in SIESTA code [13]. Real space mesh grid is chosen in such a way that there is a mesh point approximately at each 0.15 Å on

Journal of Physics: Conference Series **248** (2010) 012010 doi:10.1088/1742-6596/248/1/012010

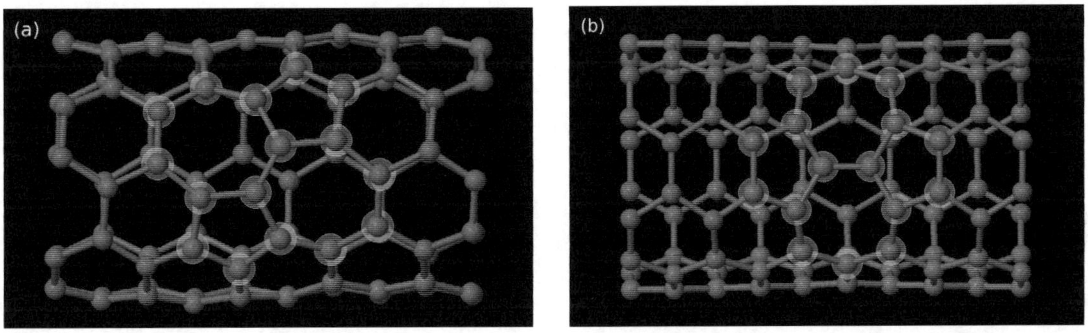

Figure 1. Relaxed positions of carbon atoms in a SW defected (5, 5) tube. Rotated bond in SW defect form a 60 angle with tube axis in (a) and its parallel to tube axis in (b). Samples contain 100 carbon atoms

three lattice vectors. We have used the generalized gradient approximation for the exchange-correlation functional and the structure is relaxed until we reach a maximum force per atom less than 0.04 eV/Å. Histograms of the tube radius and the C-C bond-length in a (5, 5) SWCNT are shown in figure 2 (a) and (b) for two types of SW defects, respectively. The range and the standard deviation, σ, in both (a) and (b) for the case of parallel rotating bond in the SW defect is broader. Moreover, radius and bond-length tend to higher values in comparison with the assumed value in a clean sample.

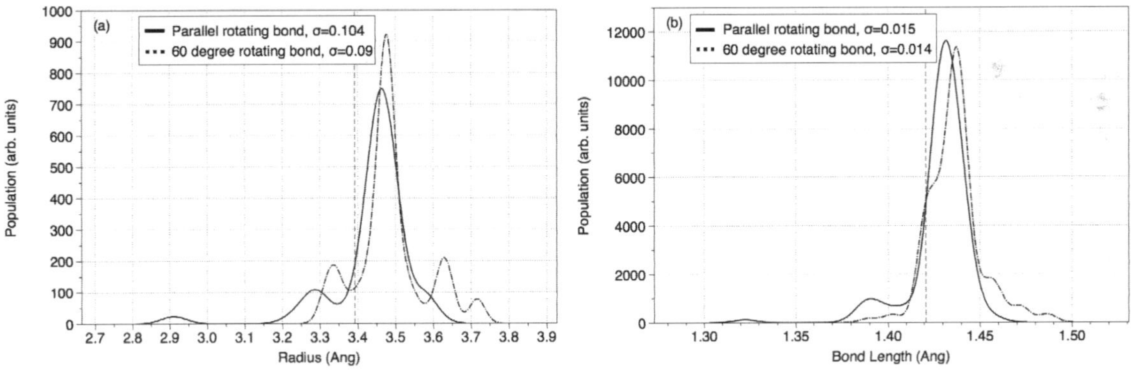

Figure 2. Histogram of (a) radius and (b) bond-length in a relaxed sample of a (5, 5) SWCNT with two types of SW defects (solid and dashed lines). Vertical dashed lines in (a) and (b) are the tube diameter and C-C bond-length in a clean sample, respectively.

Figure 3 shows the band structure of the above sample with two type of the SW defects. The results do not show a gap opening due to the SW defect in (a), while there is an $\approx$ 40 meV gap in (b) – predicting a semiconducting behaviour for the SWCNT. The K point in the figure is the point at which the two bands are closest to one another. Figure 3 shows a strong dependence of the electronic structure on the orientation of the SW defect at high concentrations. Figure 4 shows the total density of states (DOS) for two samples studied in figures 1 to 3 near the Fermi level.

In figure 4 (a) there is no gap at the Fermi level, while in (b), which corresponds to the 60 degree SW defect, there is a gap opening of about 40 meV.

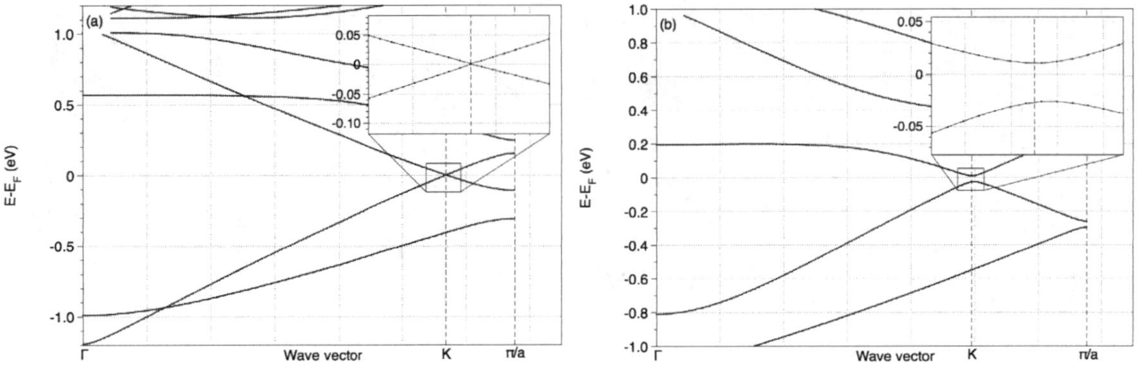

Figure 3. Band structure of a (5, 5) SWCNT with parallel (a) and 60 degree SW defect (b). While the SWCNT remains metallic in (a), there is an ≈ 40 meV gap in energy at "K" point in (b).

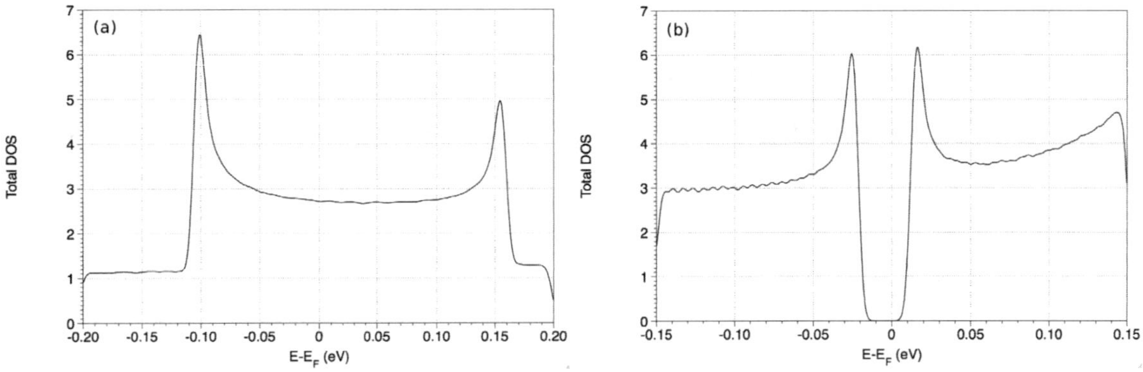

Figure 4. Band structure of a (5, 5) SWCNT with parallel (a) and 60 degree SW defect (b). While the SWCNT remains metallic in (a), there is an ≈ 40 meV gap in energy at "K" point in (b).

In summary, the effect of orientation of Stone-Wales defect on the electronic structure of the armchair single-wall carbon nanotubes is studied at high concentrations using a density functional formalism. Results show a strong dependence of the electronic structure on the orientation of the Stone-Wales defect. In two third of the orientations of the defect, the armchair tube, which in its clean form is believed to be a metal, undergoes a metal-semiconductor transition as a gap in energy opens near the Fermi level.

References

[1] Javey A, Guo J, Wang Q, Lundstrom M and Dai H 2003 *Nature (London)* **424** 654
[2] Anantram M P and Leonard F 2006 *Rep. Prog. Phys.***69** 507
[3] Hamada N, Sawada S I and Oshiyama A 1992 *Phys. Rev. Lett.* **68** 1579
[4] Charlier J Ch, Blase X and Roche S 2007 *Rev. Mod. Phys.* **79** 677
[5] Choi H J, Ihm J, Louie S G and Cohen M L 2000 *Phys. Rev. Lett.* **84** 2917
[6] Stone A J and Wales D J 1986 *Chem. Phys. Lett.* **128** 501
[7] Terrones H and Terrones M 1997 *J. Phys. Chem. Solids* **58** 1789
[8] Choi H J and Ihm J 2000 *Phys. Rev. Lett.* **84** 2917
[9] Andriotis A N, Menon M and Srivastava D 2002 *J. Chem. Phys.* **117** 2836
[10] He Y, Zhang C, Cao C and Cheng H 2007 *Phys. Rev. B.* **75** 235429

[11] Tekleab D, Carroll D L, Samsonidze G and Yakobson B I 2001 *Phys. Rev. B.* **64** 035419
[12] Troullier N and Martins J L 1991 *Phys. Rev. B* **43** 1993-2006
[13] Soler J, Artacho E, Gale J D, García A, Junquera J, Ordejón P and Sánchez-Portal D 2002 *J. Phys.: Condens. Matter* **14** 2745

Numerical simulation of Platonic hydrocarbons and fullerenes

K P Katin, D A Lobanov[1] and M M Maslov

National Research Nuclear University "MEPhI", Kashirskoe Sh. 31, Moscow 115409, Russia

E-mail: LobanovDen@yandex.ru

Abstract. Thermal stability of small cage clusters known as Platonic hydrocarbons (tetrahedrane, cubane) was studied over a wide temperature range using tight-binding molecular dynamics simulation. Activation energies and frequency factors in the Arrhenius equation were obtained for these clusters. *Ab initio* calculations using HF, and B3LYP methods with 6-31G* basis set were performed on C_{20}, C_{36} and C_{60} fullerenes to compute the lowest energy barriers preventing their decomposition. Possible decomposition products were also analyzed in detail.

1. Introduction

Since the discovery of C_{60} fullerene in 1985 [1] cage-like carbon clusters have been widely studied both theoretically and experimentally. The fact that C_{60} is the smallest fullerene, that obeys the isolated pentagon rule [2], probably explains why it can be synthesized in large amounts [3]. The smallest possible C_{20} fullerene (see figure 1(a)) was synthesized and experimentally identified in 2000 [4], carbon atoms in its shell are arranged in the network of pentagons only. Fullerene C_{36} (see figure 1(b)) attracts a special fundamental interest [5]. It occupies an intermediate position between C_{20} and C_{60} fullerenes and possesses uncommon physicochemical properties. Great attention is focused on the study of the macroscopic crystalline forms based on C_{36} fullerenes similar to C_{60} fullerites [5, 6]. Both C_{20} and C_{36} fullerenes have adjacent pentagons, which makes them more reactive than C_{60}. This fact enables them to form a covalently bound solids with fullerenes playing the role of elemental building blocks [6, 7].

Platonic hydrocarbons are named after Platonic solids (five regular polyhedrons). Tetrahedrane C_4H_4 (see figure 1(c)) is the smallest and most strained hydrocarbon because of its 60-degree bonding angle of carbon atoms. Due to molecule's high strain it was not synthesized yet, but there exist some of its compounds where hydrogen atoms are substituted by organic groups [8, 9]. Cubane C_8H_8 (see figure 1(d)) [10] is the most rigorously studied cluster among the Platonic hydrocarbons. Since the carbon bond angle of 90° is closer to the equilibrium value for the tetrahedral bond angle of 109.5°, this molecule is less strained than tetrahedrane. After Eaton and Cole synthesized it for the first time in 1964, cubane is widely studied as a promising high-energy fuel and a starting material for novel chemical compounds [11, 12].

Since many practical applications of these molecules depend on their stability, it is important to investigate under what conditions they transit to different structures. Generally when exposed

[1] Corresponding author

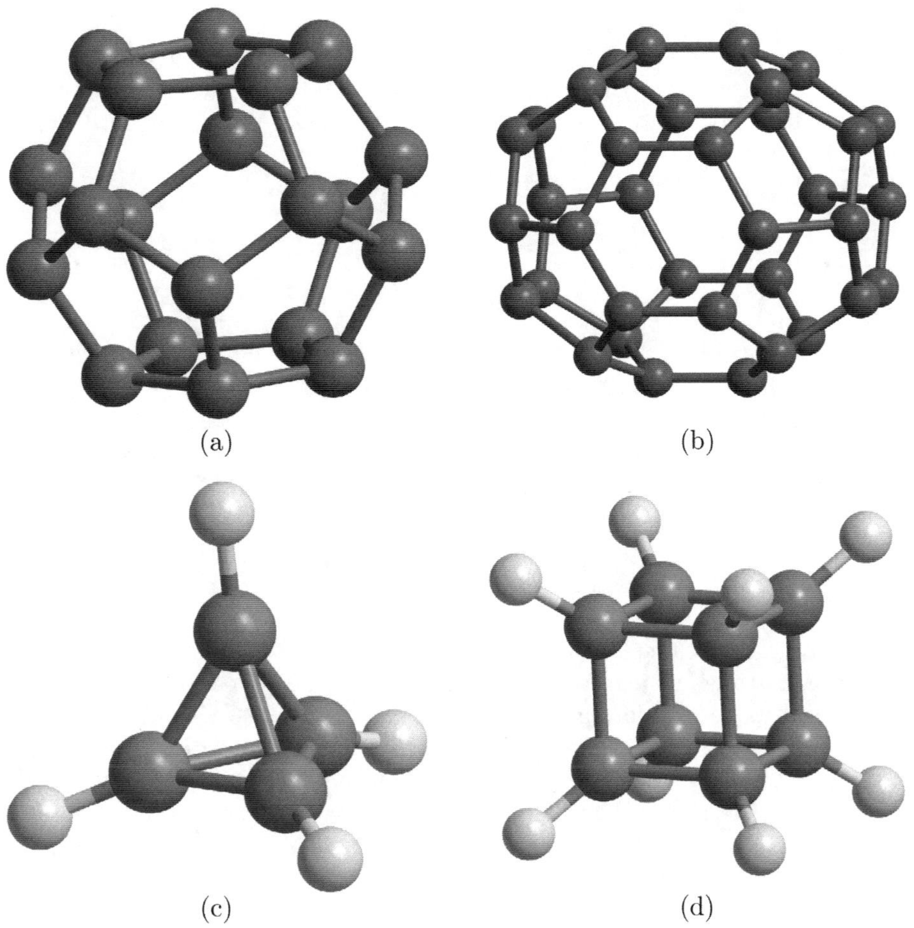

Figure 1. Fullerenes and Platonic hydrocarbons studied in this work: (a) fullerene C_{20}; (b) fullerene C_{36}; (c) tetrahedrane C_4H_4; (d) cubane C_8H_8. Dark grey spheres are carbon atoms, light grey spheres are hydrogen atoms.

to high temperature molecules begin to evolve by repeatedly crossing energy barriers from one local minimun to another, with rate τ^{-1} described by the Arrhenius equation:

$$\tau^{-1}(T) = A\exp(-E_{\mathrm{a}}/k_{\mathrm{B}}T), \tag{1}$$

where k_{B} is the Boltzmann constant. Two quantities determining the rate are activation energy E_{a} and frequency factor A. In this paper we report calculations on stability of Platonic hydrocarbons C_4H_4 and C_8H_8 and fullerenes C_{20}, C_{36} and C_{60}. We found the lowest energy transition states of fullerenes binding the initial cluster with the next isomer and simulated thermal dissociation of Platonic hydrocarbons over a wide temperature range, allowing us to find frequency factors and activation energies.

2. Computational details

Calculations on C_4H_4 and C_8H_8 were performed with the nonorthogonal tight-binding potential (NTB) for hydrocarbons [13], for fullerenes NTB and the orthogonal tight-binding potential (OTB) [14] were used. A molecular dynamics (MD) simulation was performed on hydrocarbons.

During simulation full energy of the system was held constant, and temperature was defined as a mean kinetic energy of the atoms. Tetrahedrane was simulated at about 20 different temperatures in the range of 500-800 K. In computations on cubane the system was sampled at about 50 different temperatures evenly distributed across the range of 1050-2000 K. In this study we used Newton method to optimize geometries of the first transition states. Fullerenes were also treated using the HF and DFT methods with 6-31G* basis set. At DFT level B3LYP functional was used.

3. Results

3.1. Thermal stability of the tetrahedrane and cubane molecules

At the beginning we optimized initial molecular geometries. Bond lengths in the tetrahedrane are $l_{\mathrm{C-C}} = 1.525$ Å and $l_{\mathrm{C-H}} = 1.068$ Å and in cubane are $l_{\mathrm{C-C}} = 1.570$ Å and $l_{\mathrm{C-H}} = 1.082$ Å. The values for cubane are close to the ones obtained experimentally in [15] ($l_{\mathrm{C-C}} = 1.571$ Å and $l_{\mathrm{C-H}} = 1.097$ Å). Since tetrahedrane was not synthesized yet, our resuls can only be compared to the results of previous computations. The lengths computed in [16] at B3LYP/6-31G* level are $l_{\mathrm{C-C}} = 1.480$ Å and $l_{\mathrm{C-H}} = 1.073$ Å. To access relative stabilities binding energies E_{b} were calculated:

$$E_{\mathrm{b}} = \frac{n \cdot E(\mathrm{C}) + m \cdot E(\mathrm{H}) - E(\mathrm{C_n H_m})}{n+m}, \tag{2}$$

where $E(\mathrm{C_n H_m})$ is potential energy of the molecule, $E(\mathrm{C})$ and $E(\mathrm{H})$ are energies of isolated carbon and hydrogen atoms, respectively. Binding energy of tetrahedrane is 3.90 eV/atom and for cubane E_{b} equals to 4.42 eV/atom (experimental value is 4.47 eV/atom [15]).

We performed an MD simulation to identify the sequence of transitions during thermal decomposition of Platonic hydrocarbons. The tetrahedrane decomposition starts from breaking of two opposite carbon bonds and transition to cyclobutadiene ($E_{\mathrm{b}} = 4.29$ eV/atom). Next follows a fast (<1 ps) "splitting" of cyclobutadiene into two acetylene molecules ($E_{\mathrm{b}} = 4.54$ eV/atom). This increase in E_{b} shows that stored potential energy of 5.12 eV/molecule is released in the process. The cubane decomposition starts from the transition to *syn*-tricyclooctadiene ($E_{\mathrm{b}} = 4.47$ eV/atom), which is followed by a fast (0.1-1 ps) transition to cyclooctatetraene ($E_{\mathrm{b}} = 4.73$ eV/atom) or to bicyclooctatriene ($E_{\mathrm{b}} = 4.65$ eV/atom). Finally, bicyclooctatriene transforms into cyclooctatetraene or dissociates to benzene and acetylene ($E_{\mathrm{b}} = 4.75$ eV/atom). The values of mechanical energy released in transition to cyclooctatetraene (4.96 eV/molecule) and in dissociation to benzene and acetylene (5.28 eV/molecule) are very close to that in tetrahedrane. The decomposition in both cases results in lowering the strain by increasing the carbon bond angle.

Direct calculation of the tetrahedrane and cubane lifetimes τ at different temperatures allowed us to find the values of the activation energies E_{a} and the frequency factors A by fitting the data to the Arrhenius equation (1). For the tetrahedrane these values are $E_{\mathrm{a}} = 0.7 \pm 0.1$ eV and $A = 10^{14.52\pm0.59}$ s^{-1}, and for the cubane they are $E_{\mathrm{a}} = 1.9 \pm 0.1$ eV and $A = 10^{16.03\pm0.36}$ s^{-1}. The values of activation energies and frequency factors can be used to estimate tetrahedrane and cubane lifetimes in the range of high and low temperatures, where experimental data are unavailable.

3.2. The values of energy barriers preventing the dissociation of fullerenes

We addressed the problem of fullerenes' stability by optimizing geometries and calculating the heights U of the lowest energy transition states bridging ideal fullerene with next energy minimum using both *ab initio* approaches and tight-binding methods [13, 14]. The detailed results for energy barriers are shown in table 1. Symbols F_i^N and S_{i-j}^N denote the first local minimum and corresponding saddle point, respectively, indices i and j denote the isomer ("0" is the ideal or "undistorted" cluster and "1" is the next minimum) and N is the number of atoms

Table 1. Potential energy (eV) of the metastable and saddle structures of fullerenes.

	Method			
	B3LYP/6-31G*	HF/6-31G*	OTB	NTB
F_1^{20}	5.86	4.00	3.83	3.02
S_{0-1}^{20}	5.98	4.40	5.00	3.29
F_1^{36}	-0.15	-0.83	-0.33	-0.14
S_{0-1}^{36}	6.30	5.73	5.74	4.52
F_1^{60}	1.31	2.01	1.42	1.33
S_{0-1}^{60}	7.37	8.58	6.48	5.23

in the molecule. The energies in the table are given relative to the energies of ideal fullerenes i.e. F_0^{20}, F_0^{36} and F_0^{60}.

4. Concluding remarks

Thermal stability of the tetrahedrane and cubane molecules was analyzed using tigh-binding and *ab initio* approaches. Temperature dependencies of their lifetimes were obtained and main products of dissociation were identified. During this process strain is released by increasing bond angles. The problem of fullerenes' stability was addressed by calculating heights of the lowest energy barriers preventing the decomposition of fullerenes. The obtained results are essential in practical applications of Platonic hydrocarbons, fullerenes and their derivatives.

Acknowledgments

The work was supported financially by Federal Target Program "Research and Scientific-Pedagogical Cadres of Innovative Russia" for 2009-2013 years (Appl. Number NK-102P; Grant P416 and Appl. Number NK-494P; Grant P286). The authors also wish to thank L A Openov and A I Podlivaev for valuable discussion of the results during the course of this work.

References

[1] Kroto H W, Heath J R, O'Brien S C, Curl R F and Smalley R E 1985 *Nature* **318** 162–3
[2] Kroto H W 1987 *Nature* **329** 529–31
[3] Krätschmer W, Lamb L D, Fostiropoulos K and Huffman D R 1990 *Nature* **347** 354–8
[4] Prinzbach H, Weiler A, Landenberger P, Wahl F, Wörth J, Scott L T, Gelmont M, Olevano D and Issendorff B 2000 *Nature* **407** 60–3
[5] Piskoti C, Yarger J and Zettl A 1998 *Nature* **393** 771–3
[6] Fowler P W, Mitchell D and Zerbetto F 1999 *J. Am. Chem. Soc* **121** 3218–9
[7] Davydov I V, Podlivaev A I and Openov L A 2009 *Russian Physics Journal* **52** 1199–204
[8] Maier G, Pfriem S, Schäfer U and Matusch R 1978 *Angew. Chem. Int. Ed. Engl.* **17** 520–1
[9] Maier G and Born D 1989 *Angew. Chem. Int. Ed. Engl.* **28** 1050–2
[10] Eaton P E and Cole T W 1964 *J. Am. Chem. Soc.* **86** 962–4
[11] Eaton P E 1992 *Angew. Chem. Int. Ed. Engl.* **31** 1421–36
[12] Li Z and Anderson S L 2003 *J. Phys. Chem.* A **107** 1162–74
[13] Maslov M M, Podlivaev A I and Openov L A 2009 *Phys. Lett.* A **373** 1653–7
[14] Xu C H, Wang C Z, Chan C T and Ho K M 1992 *J. Phys.: Condens. Matter* **4** 6047
[15] Hedberg L, Hedberg K, Eaton P E and Nodari N end Robiette A G 1991 *J. Am. Chem. Soc.* **113** 1514–7
[16] Ren X Y, Jiang C Y, Wang J and Liu Z Y 2008 *J. Mol. Graphics Modell.* **27** 558 – 62

THE STUDY OF FORMATION CONDITIONS OF CARBON FILMS WITH DIFFERENT STRUCTURE DIMENSION

G.G. Kosakovskii[1], **Z.Ya. Kosakovskaya**[2] ,
A.M. Smolovich[2], **M.M. Simunin**[1], **and V.K. Nevolin**[1]
[1]Moscow Institute of Electronic Technology, 124498, Moscow, Zelenograd, Russia
[2]Institute Radioenginering and Electronics of RAS, 125009, Moscow, Mochovaya str. 11/7, Russia

E-mail: german_kos@mail.ru

Abstract: The influence of thermal and geometrical properties of a substrate on the structure of carbon films synthesized by the high temperature methods was studied theoretically and experimentally. It–is shown that the carbon films deposited on various substrates had the identical morphology of structure at equality of the temperature conductivity-thickness ratio in condition of identical the other technological parameters of growth films.

It is known that the physical and geometrical properties of substrates strongly influence on the growth mechanism of the films. For example, in identical technological conditions the nanotube film and the diamond-like film were grown on a surface of the quartz plates with various thicknesses. There is no criterion for the optimum substrate choice, which would take into account its physical and geometrical parameters.

In our work the influence of the physical and geometrical parameters of the substrates on the structure of carbon films, which were synthesized in a single high temperature technological process, is studied. Thus the temperature and the density of carbon vapor flow to substrates and also surface temperature of substrate were equal for all substrates on the nucleation stage of films.

One of the basic factors determining the nucleation mechanism of films is the efficiency of energy equipartition between the substrate's surface and carbon particles of vapor flow named accommodation coefficient $\alpha = (T_v - T_r)/(T_v - T_s)$, where T_v is a temperature of carbon vapor flow; T_r is temperature of adatoms; T_s is surface temperature of substrate. As the carbon films on various substrates are formed at single process, T_s has dominant influence on the growth mechanism. It is known that the significant parameters for forming a carbon structure such as the coefficients of adsorption and desorption of carbon particles, their diffusions on the substrate's surface exponentially depends on the surface temperature of the substrate.

In our experiments the flat substrates from Si, KBr, NaCl, LiF, polycor and cover glass with thickness from 0.01 mm to 5 mm were used. The carbon films with thickness 120 nm on various substrates were grown at a single technology process by the thermic evaporation of graphite in vacuum in the pulse-periodic regime (time length of impulse – 2 s, frequency – 0.5 Hz and film growth rate v = 5-6 nm/sec). The films structure was researched by the SEM, STM, and AFM methods.

The calculation of change of the surface temperature was proceeded for our flat substrates. The heat conduction equation (1) was solved for the one-dimensional case (flat substrate)

$$\frac{\partial T_s}{\partial \tau} = a_{ï} \frac{\partial^2 T_s}{\partial x^2} \qquad (1)$$

where $\grave{a}_{\ddot{i}}$ is the temperature conductivity; $\frac{\partial T_s}{\partial \tau}$ and $\frac{\partial T}{\partial x}$ are change of the surface temperature during time and along thickness of substrate, respectively. The model of thermal stresses allowing to take a surface temperature at any moment of time [1] was used for solution of equation (1). At boundary conditions $\lambda \frac{\partial T}{\partial x}|_{x=0} = -q$ and $\frac{\partial T}{\partial x}|_{x=d} = 0$ the solution of equation (1) according to [2] is

$$T_{\ddot{i}} = T_0 + \frac{2qa_{\ddot{i}}}{\lambda_{\ddot{i}}}\sqrt{\frac{\tau}{\pi}}\left[1 + 2\sum_{n=1}^{\infty}\left\{\exp(-n^2\alpha^2) - \sqrt{\pi}n\alpha\left[1 - \Phi(n\alpha)\right]\right\}\right] \tag{2}$$

According to [2] at the nucleation stage $\alpha = \frac{d}{\grave{a}_{\ddot{i}}\sqrt{\tau}} > 1{,}2$ the solution of equation is

$$\grave{O}_s = \grave{O}_0 + \left(\frac{2qa_{\ddot{i}}}{\lambda d}\right)\sqrt{\frac{t}{\pi}} \quad . \tag{3}$$

Using an equation (3) the surface overheat curves (Ts-T0) were built by step-by-step method for the first impulse at conditions q = 4 kW-and t = 2s (Figure 1). From figure1 we can see, that all substrates may be divided on three groups having the near value of the substrates overheating. In each group the flat substrates were made from the various materials with different thicknesses but had the approximate equal value of $\kappa = \frac{d}{\grave{a}_{\ddot{i}}}$, where d – thickness of substrate, $\boldsymbol{a_n}$ – coefficient of its temperature conductivity. Thus the different substrates had the identical temperature conditions for the nucleation of carbon films.

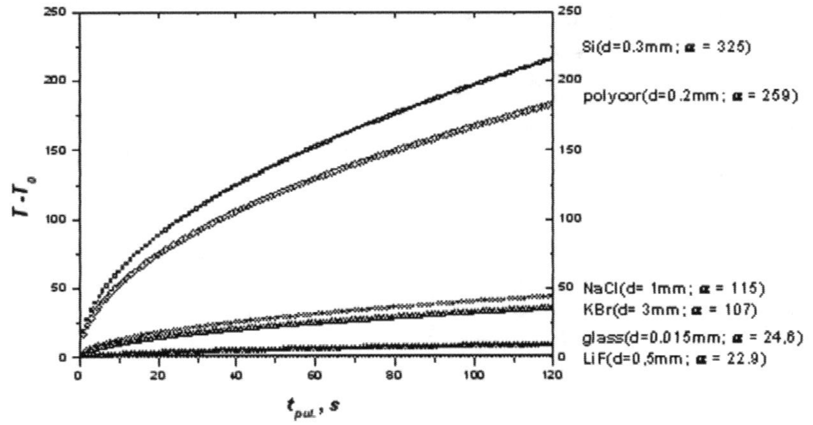

Figure 1. Overheat of substrate surface at the nucleation stage.

On figure 2 the images of four different structures of four carbon films from the various substrate groups (see figure 1), which were synthesized in the one technological process are shown. Probably, the structures of films on figures 2a and 2b may be classified as 3D and 2D allotropic carbon forms, which had nucleation on the substrate surface. The carbon films on figures 2c and 2d

Journal of Physics: Conference Series **248** (2010) 012012 doi:10.1088/1742-6596/248/1/012012

may be classified as 1D carbon nanostructure, which had nucleation in volume of carbon vapor flow near the substrate surface.

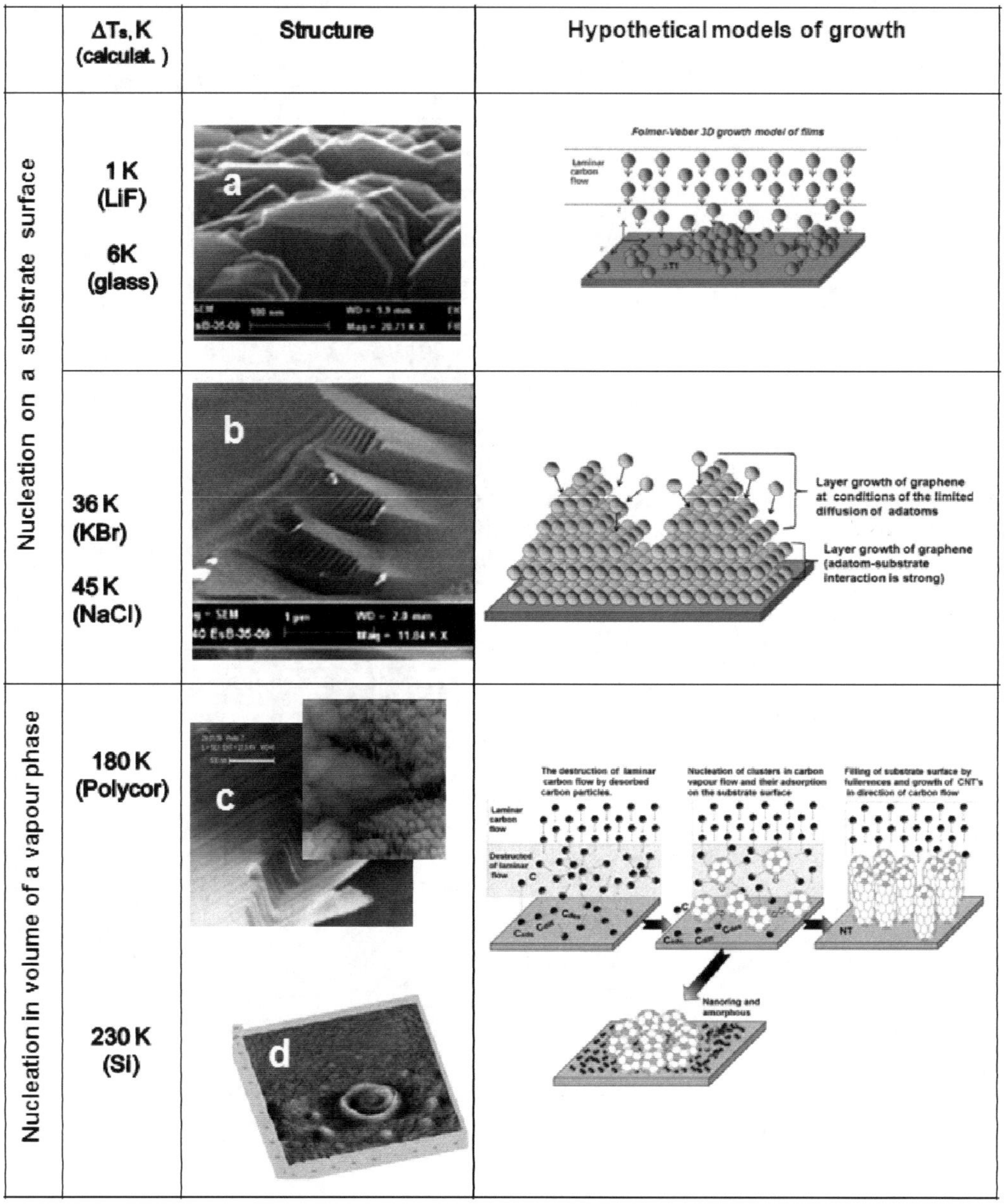

Figure 2. Structure and hypothetical models of films growth on various substrates formed in single technology process:
a) SEM image of carbon films on glass. d=0.015mm
b) SEM image of carbon films on NaCl. d=1mm
c) SEM and STM images of carbon films on polyco.r d=0.2mm
d) AFM image of carbon films on Si. d=0.3mm

The first group of substrates (LiF and cover glass; k~23) had a minor overheat. The image of structure of these films is shown on figure 2a. The films have a polycrystalline structure with size of crystallites near 120 nm. Crystallites have crystallographic faceting with angle, representative for diamond. Initiation of the such germs completely agrees with the Folmer-Veber scheme of the films growth [3], which expects three-dimensional growth (figure 2a). This is possible if the binding energy between carbon adatoms of the vapour flow near substrate surface and a carbon nucleus is more than theirs bonds with the substrate surface.

The second group of substrates (KBr and NaCl; k~110) had overheat ($T_s - T_0$) near 45K. The film on NaCl (Figure 2b) has wavy structure, generated by the parallel planes of, hypothetically, graphite. This structure can grow by the Stranskogo-Krastanova mechanism. The third group of substrates (polycor and Si; k ~ 250 and k~325 respectively) had overheat (Ts – T0) near 180K for polycor and 250 for Si. The images of the carbon films structure on polycor and Si are shown on fig. 2c and 2d. The films have the various structure (aligned 1D carbon structures, nanorings, and amorphous). The carbon film on the polycor has oriented columnar structure with diameter 2-4 nm (figure 2c). The growth of graphite fibrous is impossible at such high temperatures of the substrates surface because it is necessary bordering according to diffusion of the graphite`s adatoms. We suggest the following mechanism of growth of 1D carbon structure. The coefficient of adatoms` desorption is high because the substrate`s surface has high temperatures. The part of carbon adatoms were evaporated diffusely from substrate to a hot laminar flow. The part of cold carbon adatoms were evaporated from substrate to a hot laminar flow diffusely. There are the conditions for the clusters (fullerenes) forming in the vapor carbon flow near the substrate surface. The cluster's mass is very big. So, the clusters have not possibility to evaporate from the substrate. The fullerenes like the adatoms diffuse on a substrate surface and form nucleuses. Growth of the nanotubes in the carbon flow direction begins after the complete filling of surface by the fullerenes. The formation of nanorings (figure 2d) may be caused by the high desorption coefficient of carbon adatoms and the high diffusion coefficient of fullerenes on the substrates surface.

The authors thank A. Koz'min for help in study of the films structure by the SEM and AFM methods. In part this work was supported by grant RFBR 08-02-01093-a

References:

1. Gofman R.U. Physic of thin films. – *M., Nauka,* 1968.- 225c
2. Kozlov D.V. Analytical methods of decision of equalization of heat conductivity. Rudiments of hydrophysics. Learning aids.- *M. MSUN*, 2004. – 246 pp.
3. L.I.Maissel and R.Grang. Handbook of thin film technology. - *Mc Graw Hill Hook Company*, 1970. -634c

Nanocones rolling in hydro-thermal medium and flows in conical domains

I.V.Blinova, K.N.Kyzyurova, I.Yu.Popov

St.-Petersburg State University of Information Technologies, Mechanics and Optics, Kronverkskiy, 49, St.-Petersburg, 197101, Russia

popov@mail.ifmo.ru

Abstract. In many applications of tubular nanostructures it is necessary to describe fluid flows through these systems, which have many specific features. Moreover, these flows play an important role in the processes of nanotubes formation. Our report is devoted to the description of these processes in conical nanostructures. Hydro-thermal method of nanotube formation is very effective. But it is often leads to formation of nanostructures of different morphology, particularly, nanocones. Process of nanocone formation due to rolling from plane structure in hydro-thermal medium is considered. The system comes to equilibrium if there is lattices correspondence between neighbor layers. Due to this reason the conical angle depends on the lattice parameters. This dependence is analyzed. Movement of the conical surface during the formation process leads to the appearance of flow in conical domain which has an influence on the formation process.

1. Introduction

Recently, the increasing interest was attracted to the problem of synthesis, studying of structural features, behaviour and properties of substance in nano-sized state. It is related, first of all, with unusual physical and chemical, optical, mechanical and other properties of such systems (see, e.g., [1]), and also with great prospects of practical use of nanocrystal constructional. The unusual behaviour of such objects in that case when at least one of characteristic sizes of particles does not exceed 10 nanometers is most expressed in nanotubes, owing to minimally admissible radius, capillary forces reach extremely possible value that opens unlimited possibilities for intercalation and synthesis of new composite nanomaterials and nanodevices with operated properties. It is worth noting, that so far there are no appropriate mathematical models of processes of nanotube formation. It limits the possibilities of forecasting of synthesis dynamics and its results.

The problem of nanotube formation is closely related with the problem of flows in nano-sized structures. Simple estimations [2] show that the Reynolds number for such flow is very small. Consequently, it is possible to use the Stokes approximation. Low Reynolds number flows in conical domains are intensively investigated [3], [4] starting from the well-known Moffatt work [5] It is possible to use two models for our structure: the domain between two cones with common vertex and the domain between two parallel cones. The first model allows one to obtain the solution in an analytical form (in this case one can make separation of the variables). For the second model one obtains the result by approximate calculations only.

2. Nanotube formation

The process of nanotube formation consists of several stages. The initial state is sandwich structure of stressed double plane layers. Final result is the ensemble of nanotubes [7] in hydro-thermal medium (Fig. 1).

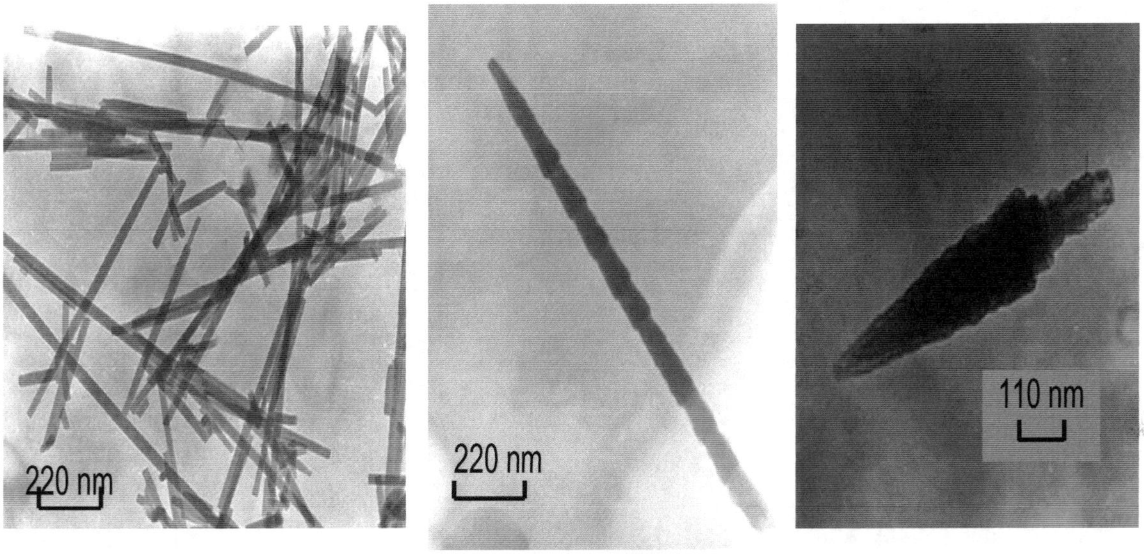

Fig. 1. Non-organic nanotubes $Mg(Fe,Co,Ni)_3Si_2O_5(OH)_4$

The first stage is a separation of one double layer from the sample. The second stage is rolling of this layer and formation of tubular (cylindrical or conical) structure [2]. And the final stage is evolution of the ensemble of nanotubes in the hydro-thermal medium [6]. In the present paper we deal with the second stage. The case of cylinder was considered in [2]. To describe the dynamics of nanocone formation we use the model of stressed elastic layer. Taking into account the elastic forces, adhesion and liquid viscous friction, one obtains the corresponding dynamical equation. As a result, the average angular velocity of nanocone rolling is given by the following formula

$$\omega = \frac{E\delta^4}{144\mu \sin^2\gamma}\left[\frac{2\cot\gamma(r_2 - r_1) - \cot^2\gamma \ln\left(\frac{r_2}{r_1}\right)}{(r_2 + r_1)(r_2^2 + r_1^2)}\right],$$

Here E is the Young's modulus for nanocone, δ is the width of the double layer, γ is the nanocone angle, r_1 and r_2 are radii of two cone boundaries, μ is the viscosity.

As for the length of the nanocone and the cone angle it depends not only on the parameters of the initial nanoplate, but also on the lattice constant for the layer. There is lattice correspondence of neighbor layers in equilibrium state. Absence of such correspondence leads to transformation of the structure into one closer to equilibrium. The majority of nanotubes in the set of experimentally obtained nanorolls have parameters close to the equilibrium one. Correlation between the lattice constants and the interlayer distance predetermines the angle and length of the nanocone. For given nanotube parameters it is obtained by simple calculations in MathCad. It is in correlation with the experiment results: it is shown that these parameters aren't arbitrary. There are preferable angles and lengths (one can see it on Fig.1).

3. Flow in nanocone

It is necessary to describe the flow in more details to determine the parameters of the rolling process. We consider axisymmetric Stokes flow in the space between two cones with common vertex and semivertex angles ϑ_1, ϑ_2, $\vartheta_1 < \vartheta_2$. A Stokes streamfunction $\psi(r,\vartheta)$ exists and is related to the velocity components (u,v) along the direction of the unit vector (e_r, e_ϑ) according to

$$u = \frac{1}{r^2 \sin\vartheta}\frac{\partial\psi}{\partial\vartheta}, \quad v = -\frac{1}{r\sin\vartheta}\frac{\partial\psi}{\partial r}$$

The fluid motion is governed by the fourth-order equation

$$\left(\frac{\partial^2}{\partial r^2} + \frac{1-\xi^2}{r^2}\frac{\partial^2}{\partial\xi^2}\right)^2 \psi = 0,$$

where $\xi = \cos\vartheta$. Let us introduce new function by

$$\psi(r,\vartheta) = \Re\left(r^{\mu+3/2}(1-\xi^2)f_\mu(\xi)\right).$$

where $\Re(\mu) > 1/2$. Then f_μ is easily determined by successively solving the equations

$$\frac{d}{d\xi^2}((1-\xi^2)g_\mu) + (\mu-1/2)(\mu-3/2)g_\mu = 0,$$

$$\frac{d}{d\xi^2}((1-\xi^2)f_\mu) + (\mu-1/2)(\mu-3/2)f_\mu = g_\mu,$$

To obtain the general solution, for $\mu \neq 3/2$,

$$f_\mu = AP'_{\mu+1/2}(\xi) + BP'_{\mu-3/2}(\xi) + CQ'_{\mu+1/2}(\xi) + DQ'_{\mu-3/2}(\xi),$$

where a prime denotes differentiation of a Legendre function. Impermeable and no-slip boundary conditions apply at each solid surface bounding the fluid. Hence, one obtains

$$f_\mu(\xi_1) = f_\mu(\xi_2) = 0, \quad f'_\mu(\xi_1) = f'_\mu(\xi_2) = 0,$$

where $\xi_1 = \cos\vartheta_1$, $\xi_2 = \cos\vartheta_2$. Correspondingly, μ is determined by the following equation

$$\begin{vmatrix} P'_{\mu+1/2}(\xi_1) & P'_{\mu-3/2}(\xi_1) & Q'_{\mu+1/2}(\xi_1) & Q'_{\mu-3/2}(\xi_1) \\ P'_{\mu+1/2}(\xi_2) & P'_{\mu-3/2}(\xi_2) & Q'_{\mu+1/2}(\xi_2) & Q'_{\mu-3/2}(\xi_2) \\ P''_{\mu+1/2}(\xi_1) & P''_{\mu-3/2}(\xi_1) & Q''_{\mu+1/2}(\xi_1) & Q''_{\mu-3/2}(\xi_1) \\ P''_{\mu+1/2}(\xi_2) & P''_{\mu-3/2}(\xi_2) & Q''_{\mu+1/2}(\xi_2) & Q''_{\mu-3/2}(\xi_2) \end{vmatrix} = 0.$$

For each mode $\mu = \mu_n$ $(n = 1,2,3,...)$, the radial and polar velocities are

$$u(r,\xi) = \Re\left(-r^{\mu-1/2}\left((1-\xi^2)f_\mu\right)'\right),$$

$$v(r,\xi) = (1-\xi^2)^{1/2}\Re\left(-(\mu+3/2)r^{\mu-1/2}f_\mu\right),$$

4. Discussion

The stream function for the flow caused by a source, for example, posed at the cone vertex is obtained as a series in the obtained eigenfunctions. For the case of point source one gets the solution in an explicit form. It is possible also to consider line circle source at the cone boundary (it gives the axisymmetric flow too and reduces to point source in our two-dimensional problem). Rigorous theory of point-like sources for the Stokes flow [8] is based on the theory of self-adjoint extensions of symmetric operators (see, e.g., [9]). It is shown that the there are eddies in the flow. It has an influence on the nanotube formation process. In the case of moving boundary (rolling nanocone) one should replace two homogeneous boundary conditions by the corresponding inhomogeneous ones. If we assume that the nanotube wall is slightly penetrable, then it is necessary to replace the last pair of the boundary conditions by inhomogeneous ones.

The work was supported by the Program "Development of the Scientific Potential of High School of Russia" (project 2.1.1/4215) and the Program "Scientific Staff of Innovative Russia 2009-2013" (project NK-526P_24).

References

[1] Gleiter H 2000 *Acta Mater.* **48** (1) 1
[2] Chivilikhin S A, Popov I Yu and Gusarov V V 2007 *Doklady Physics* **52** (1) 60
[3] Malhotra C P, Weidman P D and Davis A M J 2005 *J.Fluid Mech.* **522** 117
[4] Hall O, Gilbert O D and Hills C P 2009 *Fluid Dyn. Res.* **41** 011402
[5] Moffatt H K 1964 *J.Fluid Mech.* **18** 1
[6] Chivilikhin S A, Popov I Yu, Svitenkov A I, Chivilikhin D S and Gusarov V V 2009 *Doklady Physics* **54** (11) 491
[7] Sharikov F Yu, Korytkova E N and Gusarov V V 2007 *Phys. Chem. Glass* **33** (3) 710
[8] Pozrikidis C 1992 *Boundary Integral and Singularity Methods for Linearized Viscous Flow* (Cambridge: Cambridge Univ. Press)
[9] Gugel Yu V, Popov I Yu and Popova S L 1996 *Fluid Dyn. Res.* **18** 199

Journal of Physics: Conference Series **248** (2010) 012014 doi:10.1088/1742-6596/248/1/012014

Spin transport through a triangular graphene flake

R Farghadan[1], A Saffarzadeh[2, 3], E Saievar Iranizad[1],

[1] Department of Physics, Tarbiat Modares University, Tehran, Iran

[2] Department of Physics, Payame Noor University, Tehran, Iran

[3] Computational Physical Sciences Laboratory, Department of Nano-Science, Institute for Research in Fundamental Sciences (IPM), P.O. Box 19395-5531, Tehran, Iran

E-mail: Farghadan@modares.ac.ir

Abstract. We theoretically investigate the coherent spin polarized transport through a graphene flake by focusing on the magnetic effect of the zigzag edges. The graphene flake with triangular shape is sandwiched between two nonmagnetic one dimensional electrodes. The semi-infinite electrodes cannot effectively alter the ferromagnetic spin configuration at the zigzag edges. The electronic structure and magnetic properties of this junction are simulated self consistently by use of the single-band tight-binding approximation and the mean-field Hubbard model. Our numerical results show that the spin transport strongly depends on the magnetic configuration of the graphene flake and hence, a fully spin-polarized conductance without external field can be produced. In addition, the current-voltage characteristic shows a high splitting between spin-up and spin-down currents due to the effect of edge-induced magnetism on the electronic transport through the graphene flake. The results may be useful for carbon-based spintronic applications.

1. Introduction

The control and manipulation of spin degree in carbon-based systems have attracted much attention duo to experimental possibility to fabrication and the special charge and spin transport properties [1]. In such structures the spin-polarized currents conserved due to the low spin orbit interaction and weak hyperfine coupling [2]. Various experimental and theoretical works are applied to investigate the feasibility of generation of carbon-based spin-dependent device. [3, 4]

The honeycomb lattice of graphene is a bipartite lattice which is formed by two triangular sublattices A and B [5, 6]. For a bipartite lattice and the Hubbard repulsive parameter the imbalance between the total number of atoms in the two sublattices can produce a total spin $2S= N_A-N_B$ [6]. Also, the zigzag edge of graphene flake at the ground state has an antiferromagnetic spin alignment that is mostly localized at the zigzag edge and has its largest value of the magnetic moment at the edge.

In this paper, we theoretically investigate the spin-polarized transport through a triangular graphene flake sandwiched between two one-dimensional nonmagnetic electrodes which paying special attention to the magnetic effects of atoms in the scattering region. To reduce the effects of electronic structure of the electrodes on the scattering region, we consider the case that only one carbon of the graphene flake is connected to each electrode, as shown in Figure 1(a). The graphene flake which consists of 78 carbon atoms has ferromagnetic spin configuration due to its special design at the zigzag edges. In this case the three sides of triangular channel belong to the same sublattice and hence, same magnetic moments localized at the edges. The total spin value at the scattering region, according to the Lieb's theorem, reaches S=3 and we found that the semi-infinite electrodes cannot effectively alter this spin configuration.

2. Formalism and Model

Here the Hamiltonian of the left and right one-dimensional electrodes is described within the single-band tight-binding approximation and takes the form

$$H_\alpha = \sum_{i_\alpha} \varepsilon_0 \, d^\dagger_{i_\alpha} d_{i_\alpha} - t \sum_{ij} d^\dagger_{i_\alpha} d_{i_\alpha} \tag{1}$$

where $d^\dagger_{i_\alpha}$ (d_{i_α}) creates and (destroys) an electron at site i in electrode $\alpha\,(=R,L)$, t is the hopping integral between nearest-neighbor atoms and ε_0 is the on site energy of the electrodes. The electron conduction is mainly determined by triangular junction. In order to obtain the magnetic moment of each atom in the scattering region, we use the Hubbard model within the unrestricted Hartree-Fock approximation and tight binding model. Therefore, the central part of the device is described using the mean-field Hubbard model for the π orbitals with nearest neighbor hopping t and the on-site Coulomb interaction U. Therefore, the channel Hamiltonian can be written as [7,8]

$$H_C = -t \sum_{ij,\sigma} d^\dagger_{i\sigma} d_{j\sigma} + U \sum_i \{\langle n_{i\downarrow}\rangle n_{i\uparrow} + \langle n_{i\uparrow}\rangle n_{i\downarrow} - \langle n_{i\downarrow}\rangle\langle n_{i\uparrow}\rangle\} \tag{2}$$

where $d^\dagger_{i\sigma}$ $(d_{i\sigma})$ creates (annihilates) an electron with spin σ at site i, $n_{i\sigma} = d^\dagger_{i\sigma} d_{i\sigma}$ is the number operator and $\langle n_{i\sigma}\rangle$ denotes the spin-dependent expectation value for the number operator at site i. We solve self consistently the mean field Hamiltonian by an iteration method. Therefore the Green's function and the spin density on each atom in the channel should be iteratively calculated until a convergence of the charge density is reached. The Green's function of the graphene flake (channel) is written as [9]

$$G_C(E,V) \equiv [EI - H_C - \Sigma_L(E,V) - \Sigma_R(E,V)] \tag{3}$$

here $\Sigma_\alpha(E) = -t\, \exp(ik_\alpha a)$ is the self-energy which contains the influence of the electronic structure of the semi-infinite electrodes on the channel.

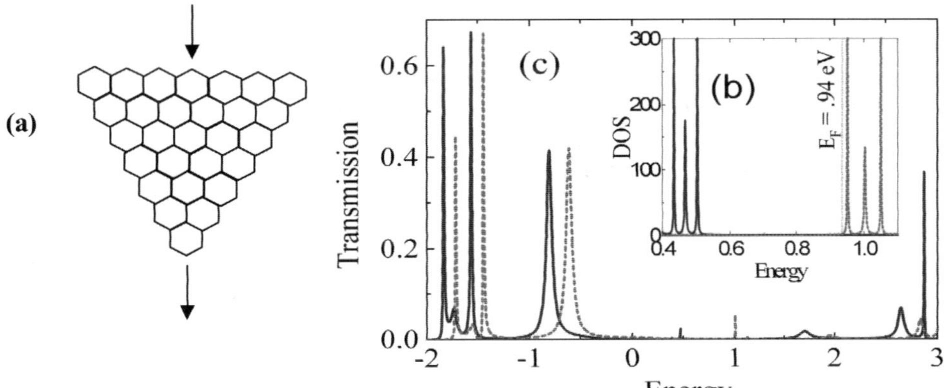

Figure 1. (a) The triangular graphene flake with two nonmagnetic one dimensional electrodes (narrow arrows). (b) DOS for the channel in contact with the electrodes. (c) Transmission coefficients as a function of energy at zero gate voltage. The solid (dashed) line is for majority (minority) spin electron.

We assume that the hopping parameter and lattice constant of the electrodes are same as the graphene flake.

Using the equation (3) for the non-equilibrium Green's function (NEGF), the spin-dependent density of states and the expectation value for the number operator of electron with spin σ at site i in the central region, are given by

$$\rho_{i\sigma}(E) = -\frac{1}{\pi}\mathrm{Im}(< i\sigma \,|\, G_C(E) \,|\, i\sigma >), \qquad (4)$$
$$\langle n_{i\sigma} \rangle = \int_{-\infty}^{E_F} \rho_{i\sigma}(E)\, dE$$

The magnetic moment on each carbon atoms can be written as $m_i = (\langle n_{i\uparrow}\rangle - \langle n_{i\downarrow}\rangle)\,\mu_B$. The transmission probabilities of spin-up and spin-down channels are independent and the electronic transport can be decoupled into two spin currents: one for spin-up and the other for spin-down. Therefore, the spin-dependent currents for a constant bias voltage are calculated by the Landauer-Buttiker formalism [9]:

$$I_\sigma(V_a) = \frac{e}{\hbar}\int_{-\infty}^{+\infty} T_\sigma(\varepsilon, V_a)[f(\varepsilon-\mu_L) - f(\varepsilon-\mu_R)]d\varepsilon \qquad (5)$$

where $f(\varepsilon)$ is the Fermi-Dirac distribution function, $\mu_{L,R} = E_F \pm eV_a/2$ are the chemical potentials of the electrodes and $T_\sigma(\varepsilon, V_a) = Tr[\hat{\Gamma}_L \hat{G} \hat{\Gamma}_R \hat{G}^\dagger]$ is the spin- energy- and voltage-dependent transmission function. Using Σ_α, the coupling matrices Γ_α can be expressed as $\Gamma_\alpha = -2\mathrm{Im}(\Sigma_\alpha)$

3. Result and discussion

We study the results of electronic density of states (DOS), the transmission coefficients, and the current-voltage characteristics. The DOS for the triangular graphene flake has been shown in Figure 1(b). In the figure, the highest occupied molecular orbital (HOMO) and the lowest unoccupied molecular orbital (LUMO) are clear. The complete separation between the majority and the minority electrons in most of the energy ranges explains the localized magnetization at the graphene flake. The interaction between the molecular levels of graphene flak and the surface states of semi infinite electrodes has a week broadening effect on the electronic structure of nanodisk. Therefore, the DOS show sharp peaks as shown in Figure 1(b). In the presence of the electrodes, the difference between two spin-polarized densities of states at each atomic site doesn't change notably except at the contact sites and the maximum value of the magnetic moment reaches $0.24\mu_B$. The Fermi energy shifts from zero to $E_F = 0.94\,eV$ due to the Hubbard repulsive parameter $(U = 2.82\ eV)$.

Transport through molecules is different with respect to the bulk materials, i.e., standard metals or semiconductors, due to the collapse of Fermi energy into a single energy level HOMO [10]. The transport properties are mainly determined by HOMO and LUMO levels. In the case of small bias voltage, the probability of transmission for spin-down electrons is more than the spin-up ones. Figure 1(c) shows the transmission coefficients as a function of energy for spin-up and spin-down electrons. We see that the transmission coefficient at zero gate voltage has a number of sharp peaks and spin-down peak appears near the Fermi energy. It is clear that, in most of the energy ranges, the values of transmission are completely separated. This is due to the spin-dependent scattering at the Fermi energy.

We have studied the degree of spin polarization (P) for electrons traversing the channel which can be defined as $P(\varepsilon) = [T_\uparrow(\varepsilon) - T_\downarrow(\varepsilon)]/[T_\uparrow(\varepsilon) + T_\downarrow(\varepsilon)]$. From the transmission spectrum it is clear that, we can obtain a spin-polarization as high as 100%. This spin polarization of the electron conduction can play an important role on the spin-dependent transport through the graphene flake. For this reason, we have plotted in Figure 2(a) the spin currents in terms of the applied voltage. From the I-V characteristic, we see the step like behavior for two spin subbands. The minority peak at the Fermi energy is responsible for spin down electron conduction at low bias. Also, the applied voltage can change the difference between both the spin-up and spin-down currents significantly.

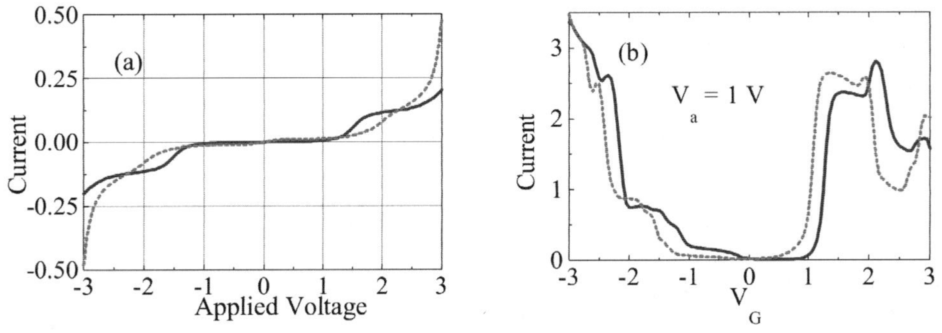

Figure 2. Current-voltage characteristics: (a) current as a function of applied voltage at V_G=0.0 V and (b) at V_a=1 V. The solid (dashed) line is for majority (minority) spin electron.

The influence of gate voltage shifts the molecular levels relative to the Fermi energy of electrodes, and hence the transmission coefficient may significantly vary [11]. Since the currents depend on the molecular levels laying between the left and right chemical potential, by altering the gate potential the molecular levels shift inside the energy window and the intensity and polarization of current for two spin currents change. To show the effect of gate voltage, we plot the current verses the gate voltage in Figure 2(b). In some ranges of gate voltage there isn't any peak of DOS and the current verses the gate voltage vanishes.

4. Conclusion

In this work, the coherent spin-polarized transport through triangular junction was investigated based on the non-equilibrium Green's function technique, and the mean-field Hubbard model. We have shown that the graphene flake whit zigzag edge is strongly spin selective and perfect spin polarization can be produced without applying magnetic field or magnetic materials.

The difference between the transmission coefficients of two spin subbands in most of energy ranges is very large and there is a great spin splitting between two transmission coefficients even in the range of energy which is far from the Fermi energy. Consequently, high spin-polarized currents can be produced. In addition, one can manipulate the degree of spin polarization by applying a gate voltage and a bias potential on the channel and electrodes, respectively. These results may be useful for spin injection into the semiconducting spintronic devices.

References

[1] Guo J and Ouyang Y 2009 *Applied Physics Letter* **94** 24104
[2] Zutic I, Fabian J and Sarma S D 2004 *Rev. Mod. Phys.* **76** 323
[3] Bery L and Fertig H A 2007 *Phys. Rev. B* **76** 205435
[4] Wimmer M, Adagideli I, Berber S, Toma´nek D and Richter K 2008 *Phys. Rev. Lett.* **100** 177207
[5] Rossier J F and Palacios J J 2007 *Phys. Rev. Lett.* **99** 177204
[6] Lieb E. H 1989 *Phys. Rev. Lett.* **62** 1201
[7] Fujita M, Wakabayashi K and Nakada K 1996 *J. Physical Society of Japon* **65** 1920
[8] Sancho M P L, Juan F D and Vozmediano M A H 2009 *Phys. Rev. B* **79** 075413
[9] Datta S 2006 *Quantum Transport atom to transistor.* (New York:Cambridge University press)
[10] Sanvito S and Rocha A R 2006 *J. Comput. Theor. Nanosci.* **3** 624
[11] Saffarzadeh A 2008 *J. Appl. Phys.* **103** 083705

Journal of Physics: Conference Series **248** (2010) 012015 doi:10.1088/1742-6596/248/1/012015

Rabi oscillations and saturable absorption effect in single-wall carbon nanotubes

A M Nemilentsau, G Ya Slepyan, and S A Maksimenko

Institute for Nuclear Problems, Belarus State University, Bobruiskaya 11, 220030 Minsk, Belarus

E-mail: andrei.nemilentsau@gmail.com

Abstract. We have studied theoretically a single-wall carbon nanotube (SWNT) interaction with the short high-intensive electromagnetic pulse whose carrier frequency is in the resonance with the frequency of the electron interband transitions in the SWNT. For this purpose kinetic equations describing evolution of the density matrix of the electron sub-system have been solved numerically and spectra of the surface current density induced in the SWNT by the external electromagnetic field have been calculated. We have demonstrated that Rabi oscillations of the population of the SWNT electronic bands occur for the pulse intensity exceeding 10^{10} W/cm^2.

1. Introduction

Carbon nanotubes are known to possess strongly nonlinear optical response [1]. In particular, it has been demonstrated that single-wall carbon nanotubes (SWNTs) have high values of the third-order nonlinear optical susceptibility [2, 3, 4], may be used for effective generation of the high-order harmonics [5, 6] and as saturable absorbers for passive mode-locking in lasers [7, 8]. Among the different nonlinear processes Rabi oscillations of the inversion that can occur in two-level [9] or two-bands [10] systems illuminated by a high-intensive resonant electromagnetic field are of a great interest. We expect that due to the SWNTs strong optical nonlinearity Rabi oscillations will also occur in SWNTs illuminated by the external electromagnetic field.

2. System under the consideration

Let us consider single-wall carbon nanotube (SWNT) illuminated by an electromagnetic pulse (see Fig. 1) of the form:

$$\mathbf{E}(t) = \mathbf{e}_z \begin{cases} E_0 \sin(\omega_0 t), & 0 \le t \le \sigma_0 \\ 0, & t < 0,\, t > \sigma_0 \end{cases}, \tag{1}$$

where E_0 is the pulse amplitude, σ_0 is the pulse duration, $\mathbf{e}_z$ is the unit vector along the z-axis. The direction of pulse propagation $\mathbf{k}$ is normal to the SWNT axis and the vector of electric field strength $\mathbf{E}(t)$ is polarized along the SWNT axis (see Fig. 1a). The pulse carrier frequency ω_0 is in the resonance with the frequency of electron interband transitions $\omega_{cv}(p,s) = (\mathcal{E}_c(p,s) - \mathcal{E}_v(p,s))/\hbar$ (see Fig. 1b) for the electrons with quasi-momentum $p = 0$ in the sub-band $s = 9$, where

$$\mathcal{E}_{c,v}(p,s) = \pm\gamma_0 \sqrt{1 + 4\cos\left(\frac{3pb}{2\hbar}\right)\cos\left(\frac{\pi s}{m}\right) + 4\cos^2\left(\frac{\pi s}{m}\right)} \tag{2}$$

is the electrons dispersion law, $\gamma_0 = 2.7$ eV, p is the electron quasi-momentum, indices c, v stand for the conduction and valence bands, respectively. In $(m, 0)$ SWNT both conduction and valence bans are split on $2m$ sub-bands [11], designated by index $s = 1, .., 2m$. First Brillouin zone in the SWNT is defined by the condition $-\pi\hbar/3b < p \leq \pi\hbar/3b$, $b = 0.142$ nm.

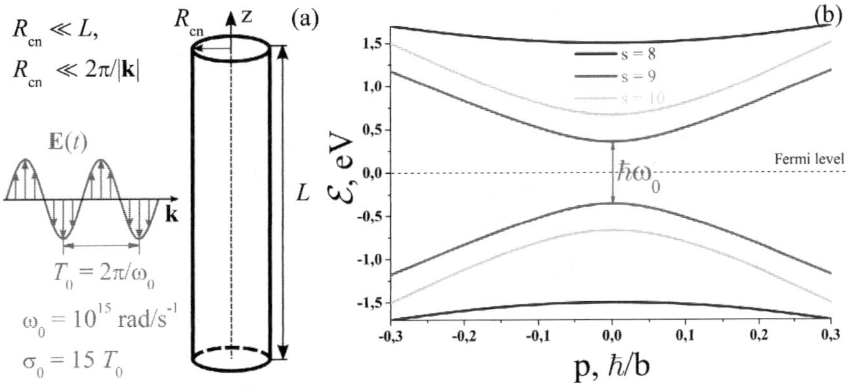

Figure 1. (*a*) Schematic representation of (14,0) SWNT (black cylinder) illuminated by the electromagnetic pulse $\mathbf{E}(t)$ of the form defined by Eq. (1). (*b*) Band structure of (14,0) SWNT in the vicinity of the Fermi level.

3. Evolution of the dynamical inversion

To study the SWNT response on the applied electromagnetic field we solve numerically kinetic equations for the density matrix $\rho_{\alpha\beta}(t, p, s)$ of π-electrons in SWNT[3, Eq. (1)]

$$\frac{\partial \rho_{\rm inv}(t,p,s)}{\partial t} + eE(t)\frac{\partial \rho_{\rm inv}(t,p,s)}{\partial p} = -\frac{\rho_{\rm inv}(t,p,s) - \rho^{\rm eq}(p,s)}{T_1} + \frac{2ie}{\hbar}E(t)R_{cv}(p,s)(\rho_{vc}(t,p,s) - \rho_{cv}(t,p,s)), \quad (3)$$

$$\frac{\partial \rho_{cv}(t,p,s)}{\partial t} + eE(t)\frac{\partial \rho_{cv}(t,p,s)}{\partial p} = -\left(\frac{1}{T_2} + i\omega_{cv}(p,s)\right)\rho_{cv}(p,s) - \frac{ie}{\hbar}E(t)R_{cv}(p,s)\rho_{\rm inv}(t,p,s), \quad (4)$$

where $T_{1,2} = 40$ fs is the electron relaxation time, $R_{cv}(p, s)$ is the normalized matrix elements of the dipole momentum operator, $\rho_{\rm inv}(t, p, s) = \rho_{cc}(t, p, s) - \rho_{vv}(t, p, s)$ is the dynamical inversion.

Results of calculations are presented in Fig. 2 for (14,0) SWNT illuminated by the electromagnetic field defined by Eq. (1). We restrict our consideration to sub-band $s = 9$ and only small part $(-0.3 \leq p \leq 0.3)$ of first Brillouin zone of the SWNT corresponding to the electron interband transitions frequencies $\omega_{cv}(p, s)$ that are close to the carrier frequency ω_0 of the pulse. As one can see, for low intensities of the driving field (Fig. 2a,b) the dynamical inversion increases while the pulse is switched on and decreases exponentially after the pulse passage due to the relaxation effects. The slight inversion oscillations with the frequency equals to the carrier frequency of the pulse is due to the intraband electrons motion. With the pulse intensity increase (Fig. 2c,d) we observe that high frequency oscillations with the period T_0 are superimposed by the low frequency oscillations with the oscillations period T_R. These are so-called Rabi oscillations of the inversion that are due to the electron transitions between the valence and conduction bands.

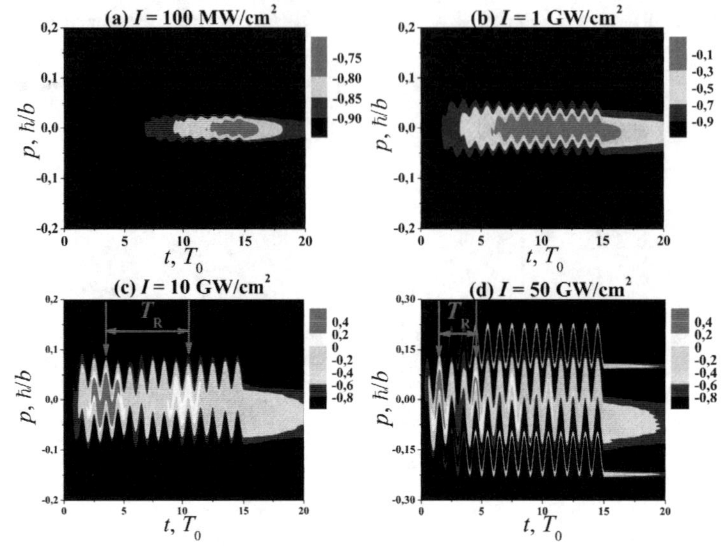

Figure 2. Evolution of the dynamical inversion $\rho_{\rm inv}(t,p,s=9)$ in (14,0) SWNT illuminated by the electromagnetic field defined by Eq. (1) (see also Fig. 1a) for different field intensities I. T_R is the period of Rabi oscillations.

4. Surface current density

We also calculated the surface current density induced in the SWNT by the external electromagnetic field defined by Eq. (1)

$$j(t)=\frac{e}{2\pi^2\hbar R_{cn}}\sum_{s=1}^{2m}\int\left(\frac{\partial\mathcal{E}_c(p,s)}{\partial p}\rho_{\rm inv}(t,p,s)+i\omega_{cv}(p,s)R_{cv}(p,s)(\rho_{vc}(t,p,s)-\rho_{cv}(t,p,s)\right)dp, \quad (5)$$

where integration is performed over the first Brillouin zone of the SWNT. The spectra of the induced current density $j(t)$ are presented on Fig. 3. As one can see, three different regimes of the SWNT interaction with the external electromagnetic field can be indicated. In the linear regime (Fig. 3a) the value of the induced current is approximately proportional to the electric field strength, i.e. $|j(\omega_0)|\sim E_0$. For the driving field intensities exceeding 10^9 W/cm^2 the interaction between the SWNT and external electromagnetic field occurs in the saturation regime (Fig. 3b). In this regime the value of the induced electric $|j(\omega_0)|$ varies very slow with the electric field intensity increase, i.e. the SWNT-electromagnetic field interaction is saturated. Further increase of the intensity leads to the appearance of the additional spectral lines in the induced current density spectra with the frequencies $\omega_0\pm\Omega_R$. The origin of these lines are the Rabi oscillations of the dynamical inversion presented on Fig. 2c,d and Ω_R is the Rabi frequency. As one can see from Fig. 3c the Rabi frequency increases with the electromagnetic pulse intensity increase. In more details the intensity dependence of the spectral component of the induced current density $|j(\omega_0)|$ is presented on Fig. 3d.

5. Conclusions

Concluding, we have studied the single-wall carbon nanotube (SWNT) interaction with the short high-intensive electromagnetic pulse whose carrier frequency is in the resonance with the frequency of the electron interband transitions in the SWNT. To describe the dynamics of the electron sub-system in the SWNT we use kinetic equations for the electron density matrix. These

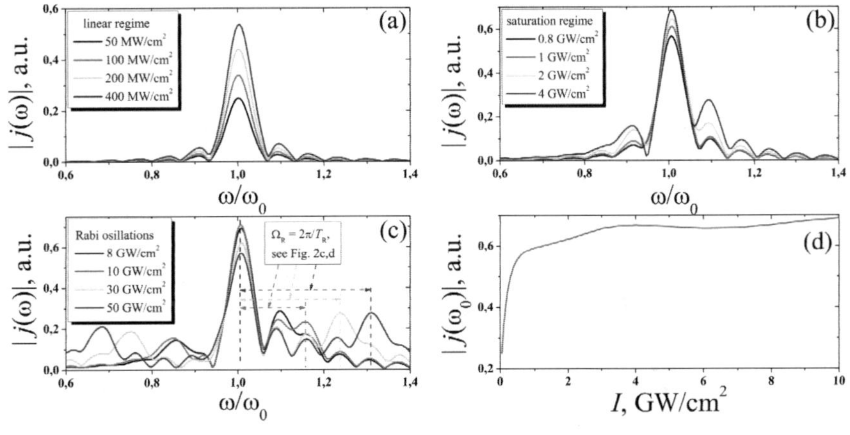

Figure 3. (a)-(c) Spectra $|j(\omega)|$ of the surface current density $j(t)$ induced in (14,0) SWNT by electromagnetic field defined by Eq. (1); Ω_R is the Rabi frequency. (d) Intensity dependence of the spectral component $|j(\omega_0)|$ of the induced current.

equations have been solved numerically and the density of the surface current induced on the SWNT surface by the external electromagnetic field has been calculated. We have demonstrated that for the pulse intensities of order 10^9 W/cm^2 SWNTs interaction with pulse occur in the saturation regime. However, for the pulse intensities higher that 10^{10} W/cm^2 Rabi oscillations of the inversion have been predicted. The triplet Mollow in the spectra of the induced current density has been demonstrated.

Acknowledgments

This research was partially supported by the Belarus Republican Foundation for Fundamental Research (BRFFR) under projects F10R-002, F10CO-020 and young scientists grant F09M-071, and EU FP7 under projects FP7-230778 TERACAN, FP7-247007 CACOMEL and FP7-266529 BY-NanoERA.

References

[1] Wang J, Chen Y and and Blau W J 2009 *J. Mat. Chem.* **19** 7425
[2] Margulis Vl A and Sizikova T A 1998 *Physica* B **245** 173
[3] Nemilentsau A M, Slepyan G Ya, Khrutchinskii A A and Maksimenko S A 2006 *Carbon* **44** 2246
[4] Zarifi A, Fisker C and Pedersen T G 2007 *Phys. Rev.* B **76** 045403
[5] Slepyan G Ya, Maksimenko S A, Kalosha V P, Herrmann J, Campbell E E B, and Hertel I V 1999 *Phys. Rev.* A **60** R777
[6] Slepyan G Ya, Khrutchinskii A A, Nemilentsau A M, Maksimenko S A and Herrmann J 2004 *Int. J. Nanosci.* **3** 343
[7] Rozhin A G, Sakakibara Y, Kataura H, Matsuzaki S, Ishida K, Achiba Y and Tokumoto M 2005 *Chem. Phys. Lett.* **405** 288
[8] Solodyankin M A, Obraztsova E D, Lobach A S, Chernov A I, Tausenev A V, Konov V I, Dianov E M 2008 *Opt. Lett.* **33** 1336
[9] Tritschler T, Mücke O D and Wegener M 2003 *Phys. Rev.* B **68** 033404
[10] Golde D, Meier T and Koch S W 2008 *Phys. Rev.* B **77** 075330
[11] Reich S, Thomsen C and Maultzsch J 2004 *Carbon Nanotubes: Basic Concepts and Physical Properties* (Weinheim: Wiley-VCH)

Journal of Physics: Conference Series **248** (2010) 012016 doi:10.1088/1742-6596/248/1/012016

Linear conductance oscillations in quantum wires and stripes with Rashba interaction

M. Magdalena Gelabert[1] **and Llorenç Serra**[1,2]

[1] Departament de Física, Universitat de les Illes Balears, E-07122 Palma de Mallorca, Spain
[2] Institut de Física Interdisciplinària i Sistemes Complexos IFISC (CSIC-UIB), Universitat de les Illes Balears, E-07122 Palma de Mallorca, Spain

E-mail: magdalena.gelabert@uib.es, llorens.serra@uib.es

Abstract. An analysis of the linear conductance of 2D quantum wires and stripes with Rashba spin-orbit interaction and attached to spin polarized leads is presented. Differences and similarities between both systems are highlighted. We discuss the dependence of the conductance with both energy and Rashba intensity.

1. Introduction

Almost twenty years after the theoretical proposal of a spin transistor by Datta and Das [1] the underlying physical principle of conductance oscillation induced by electrical tuning of the spin-orbit coupling has been proved in experiment [2]. This spin-orbit coupling is due to the asymmetry of the vertical (z) confinement felt by a 2D electron gas lying in the xy plane. It is known as Rashba interaction [3] and its tunability with electric gates has been demonstrated in quantum wells [4] and quantum wires [5]. Although the subject of conductance oscillations induced by Rashba coupling has been a central topic in spintronics for some time (see Ref. [6] for a review) the experiment by Koo *et al* spurred on a renewed interest [7, 8, 9, 10].

The physics of quantum wires with spin-orbit coupling differs depending on whether the coupling is uniform or localized in space. With an extended and uniform coupling the system is characterized by a modified subband structure, as compared to the case without Rashba coupling. A characteristic of this modified subband structure is the existence of anticrossing points in the presence of in-plane magnetic fields [11, 12]. An interesting consequence of the anticrossings is the existence of anomalous conductance plateaus due to the subband maxima and minima. Indeed, whenever the Fermi energy crosses a local extremum the number of propagating modes increases or decreases by one unit, with a corresponding increase or decrease in conductance. Quite remarkably, these anomalous conductance variations have been recently measured in a GaAs quantum wire with transport carried by holes [13].

A localized Rashba interaction, restricted to a small region of a quantum wire, is predicted to generate peculiar conductance dips due to the existence of quasibound states. This physical behavior was dubbed the Fano-Rashba effect in Ref. [14] due to the similarity with the well-known Fano resonances in electron scattering by atoms [15]. Since in real devices the spin-orbit channel is usually attached to ferromagnets an inhomogenous Rashba coupling is a more realistic description than an extended one for small-sized devices.

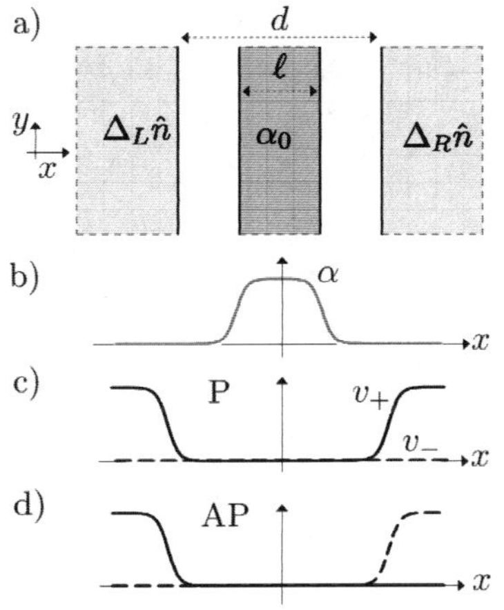

Figure 1. (Color online) Sketch of the physical system (a) and of the spatial variation of Rashba intensity $\alpha(x)$ (b) and of the spin-dependent potentials $v_\pm(x)$ (c,d).

In this work we discuss differences and similarities in the conductance oscillations of quantum wires and stripes with localized Rashba interaction, two distinct cases that we treated separately in Refs. [8, 9]. Other related works dealing with spin-orbit wires and stripes are Refs. [16, 17, 18, 19, 20, 21, 22, 23]. While a quantum wire is characterized by a transverse confinement, which we assume to be parabolic, a stripe has a vanishing confinement in the transverse direction. Transport in a quantum wire is carried by a discrete set of transverse modes that are coupled by the Rashba interaction. On the contrary, a stripe has a continuum of transverse momenta q, uncoupled with each other since the Rashba interaction preserves this quantum number. We shall discuss the physical implications of these two behaviors on the linear conductance and the polarization of the transmitted current.

2. Model

Our model considers a 2D electron gas in the xy plane with transport occurring along x and with the possibility of a transverse confining potential $v_t(y)$. We call *stripe* the system described by a vanishing $v_t(y)$, while $v_t(y) = m\omega_0^2 y^2/2$ corresponds to a parabolic *wire* characterized by a confinement energy $\hbar\omega_0$. The reader is addressed to Refs. [8, 9] for full details of the model while here it is just sketched for the sake of completeness. The system Hamiltonian reads

$$\mathcal{H} = -\frac{\hbar^2}{2m_0}\left(\frac{d^2}{dx^2} + \frac{d^2}{dy^2}\right) + v_t(y) + \Delta(x)\,\hat{n}\cdot\vec{\sigma} + |\Delta(x)| + \mathcal{H}_R, \tag{1}$$

where $\mathcal{H}_R$ is the Rashba Hamiltonian,

$$\mathcal{H}_R = \frac{1}{\hbar}\left(\alpha(x)p_y\sigma_x - \frac{1}{2}\{\alpha(x), p_x\}\sigma_y\right)\,. \tag{2}$$

Polarized leads in the direction of $\hat{n}$ are described by means of a Zeeman field $\Delta(x)$ that couples to the spin vector $\vec{\sigma}$. The functions determining the Hamiltonian are $\Delta(x)$ and $\alpha(x)$. These quantities take a constant value in the three parts of our system: left contact (L), central region and right contact (R), and they vary smoothly, described by a Fermi-type function, at the interfaces. We denote by m_0 the conduction-band effective mass of the semiconductor and by α_0 the Rashba intensity of the central region. The Zeeman field in contact c, where $c = L, R$, is denoted by Δ_c, respectively. The case of parallel polarized contacts (P) corresponds to $\Delta_L = \Delta_R \equiv \Delta_0$, while the case of antiparallel polarizations (AP) corresponds to $\Delta_L = -\Delta_R \equiv \Delta_0$, where the total Zeeman splitting is $2\Delta_0$. For simplicity, Δ_0 is assumed equal in both contacts. We use the notation $\hat{n}$P and $\hat{n}$AP to indicate parallel and antiparallel configurations along a certain direction $\hat{n}$. Figure 1 shows a sketch of the model with the variation of the Rashba intensity $\alpha(x)$. It also shows the potentials v_s, for $s = \pm$ spins, defined as

$$v_s(x) = s\Delta(x) + |\Delta(x)| \, , \tag{3}$$

where $|\Delta(x)|$ is a scalar gate potential aligning the band bottom of the different regions. Notice that in the P configuration the $s = -$ spin sees no potential at all while $s = +$ is confined by a potential well of width d. On the contrary, in the AP configuration both spins feel a potential step, but in opposite contacts. As we will discuss below, these differences in potential landscape for $+$ and $-$ spins greatly influence the transport properties of the stripe with polarized contacts.

3. Linear conductance and polarization

We calculate the linear conductance G from left to right contacts when an infinitesimal potential bias is applied along x. In the Landauer formalism of scattering G is given by the quantum transmission T at the Fermi energy [24]. In practice, however, we have to differentiate the quantum wire and the stripe systems due to the discrete and continuous character of the transverse modes, respectively. In terms of the conductance quantum $G_0 = e^2/h$, Landauer formula for the quantum wire reads

$$\text{(wire)} \quad G = G_0 \sum_{ns,n's'} T_{n's',ns} \, , \tag{4}$$

where ns and $n's'$ are labeling transverse mode and spin in L and R contacts, respectively. On the other hand, for the stripe the corresponding formula for the conductance per unit of transverse length L_y reads

$$\text{(stripe)} \quad \frac{G}{L_y} = \frac{G_0}{4\pi} \sum_{ss'} \left\{ \kappa_{Ls} \int_{-\pi/2}^{\pi/2} d\theta \, |\cos\theta| \, T_{s's}(\kappa_{Ls},\theta) + \kappa_{Rs} \int_{\pi/2}^{3\pi/2} d\theta \, |\cos\theta| \, T'_{s's}(\kappa_{Rs},\theta) \right\} \, , \tag{5}$$

where κ_{cs} is the Fermi wavevector in contact $c = L, R$ and spin s, while θ gives the polar angle of the vectorial momentum for the incident electron. We shall also calculate the relative polarization p of the transmitted current, defined as $p = G_p/G$, where the *polarized conductance* G_p takes into account the spin of the current in the R terminal. Its expression for the quantum wire and the stripe are straightforward extensions of Eqs. (4) and (5).

4. Results

This section presents our results for the linear conductance of wires and stripes, Eqs. (4) and (5), using the method of Refs. [8, 9]. We shall discuss the dependence with energy and Rashba intensity for various orientation of the Zeeman fields in the contact regions.

4.1. Energy dependence

Figures 2 and 3 display a comparison of $G(E)$ for a stripe (S) and a wire (W). Black symbols correspond to the results without Rashba coupling while gray (red in color) symbols are for an intensity of $\alpha_0 = 0.3E_U L_U$, where E_U and L_U denote energy and length unit, respectively. As in Refs. [8] and [9] we choose a units system in which $\hbar = m_0 = 1$. Additionally, for the stripe we choose the energy unit $E_U = \Delta_0$ while for the wire we take $E_U = \hbar\omega_0$. In terms of these units it is $L_U = \sqrt{\hbar^2/m_0 E_U}$.

In the quantum stripe (left panels in Figs. 2 and 3) the energy $2\Delta_0$ signals the threshold from only one propagating spin ($E < 2\Delta_0$) to both spins propagating in the contacts ($E > 2\Delta_0$).

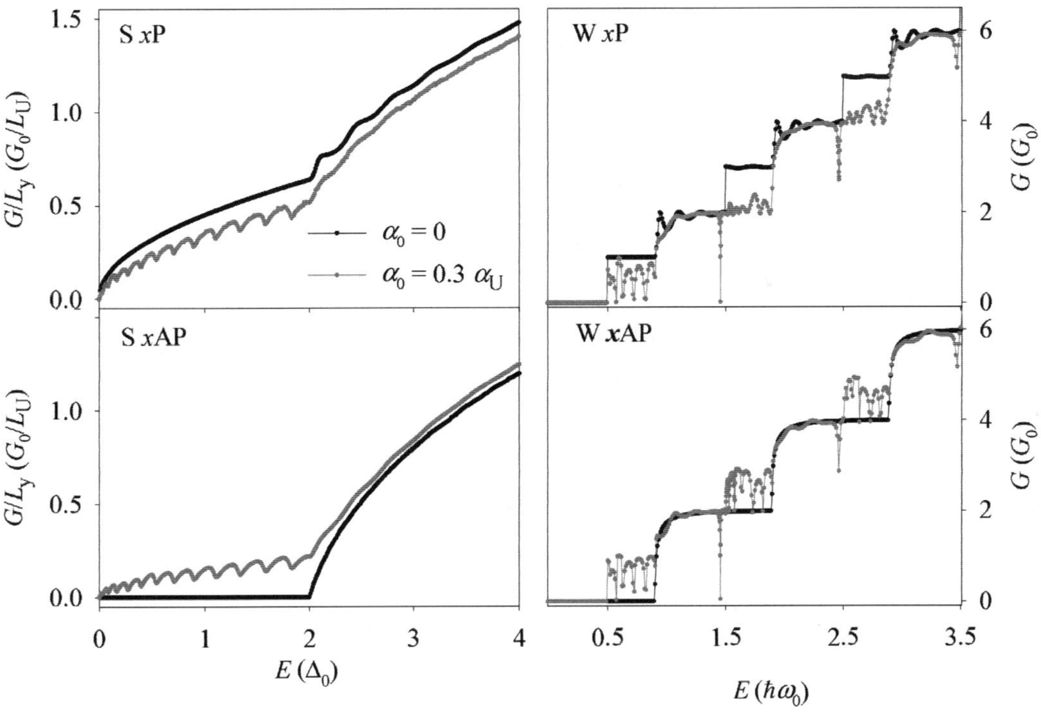

Figure 2. (Color online) Conductance of the stripe (S, left) and wire (W, right) systems for polarization of the contacts along x. Upper and lower rows correspond to parallel (P) and antiparallel (AP) orientations, respectively. We take $\ell = 8L_U$ (length of the Rashba region), and $\Delta_0 = 0.2\hbar\omega_0$ in the wire case. Black symbols are for vanishing Rashba coupling while gray (red) ones correspond to $\alpha_0 = 0.3\alpha_U$. The unit of Rashba coupling is given by $\alpha_U = \sqrt{\hbar^2\Delta_0/m_0}$ and $\sqrt{\hbar^3\omega_0/m_0}$ for the quantum stripe and wire, respectively.

This transition reflects in an abrupt enhancement of conductance when the energy exceeds the threshold. In addition, there are conductance oscillations of two types: for x-oriented contacts the system displays Fano oscillations below the threshold $2\Delta_0$ due to the Rashba-induced coupling between the propagating spin and the quasibound states of the opposite evanescent spin. Above threshold there are Ramsauer oscillations due to the potential steps in the polarized leads. Notice, however, that in AP configurations the Ramsauer oscillations are greatly reduced and, also, that in y-polarization the Fano oscillations are absent. Quite remarkably, while the conductance of P configurations is enhanced by the Rashba coupling with respect to the spin-orbit-free case, i.e., red-gray symbols are higher than black ones, this situation is reversed in AP configurations. This behavior can be interpreted as a destruction of the spin-valve behavior due to the mixing induced by the Rashba coupling.

Let us now focus on the wire conductances shown in the right panels of Figs. 2 to 3. In the absence of spin-orbit coupling the conductance is characterized by a staircase appearance; each step corresponding to the activation of additional transverse modes. In P configurations steps for up and down spins are shifted by an energy $2\Delta_0$ and, therefore, the conductance jumps by G_0 from one step to the next. In the AP case the corresponding increments are doubled, $2G_0$. With the addition of the Rashba coupling (red-gray symbols) we see that in general the

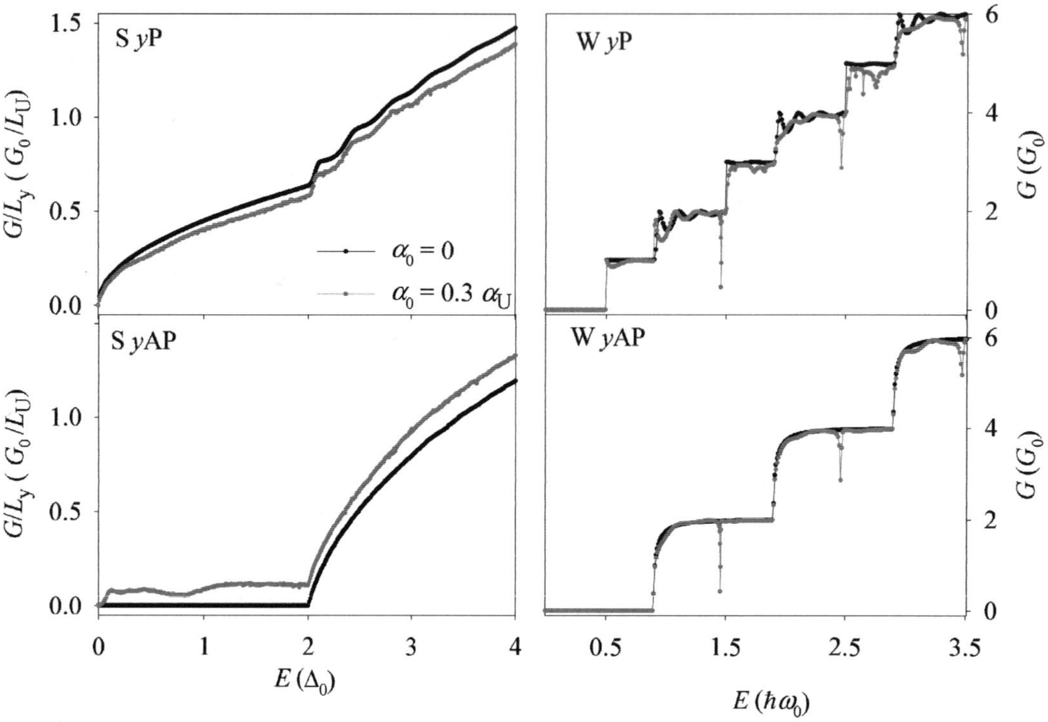

Figure 3. (Color online) Same as Fig. 2 for y configuration of the contacts.

conductance displays more oscillation. As a reminiscence of the perfectly clean wires we refer to the energy interval $[n - 1/2, n + 1/2]\hbar\omega_0$ as the *n-th conductance plateau.* Comparing with the stripe case (left panels) we notice that the wire conductance for each conductance plateau resembles the result of the stripe. Looking, for instance, the result in Fig. 2 for the second plateau $[1.5, 2.5]\hbar\omega_0$ we see at the beginning Fano oscillations followed by a sudden increase in conductance for $E \approx 1.9\hbar\omega_0$, quite similar to the stripe behavior. Interestingly, a qualitative difference with the stripe can be seen at the end of the plateau as a pronounced conductance dip. This is again a Fano resonance but is qualitatively different from those seen at the beginning of the plateau. It originates in a quasibound state shifted by a negative energy from the next plateau by the Rashba coupling. Notice that this is different with respect to the Fano resonances at the beginning of the plateau which stem from quasibound states induced by the polarized contacts. These wire conductance structures are repeated almost regularly in each plateau.

The results for other configurations of the polarized contacts show analogous similarities between wire and stripe conductances for each conductance plateau. Notice, in particular that in y orientation the Fano oscillations at the beginning of the plateaus are absent while those at the end are still present. It is also worth mentioning that the yAP configuration displays the less-structured conductance of all, the Rashba interaction only inducing dips at energies $E \approx (n + 1/2)\hbar\omega_0$ in this case.

Figure 4 displays the energy dependence of the polarization p for selected cases to illustrate the polarization mechanism of the localized Rashba coupling. The stripe is characterized by its full polarization $|p| = 1$ for energies below the Zeeman gap $2\Delta_0$. When exceeding this threshold, the polarization decreases in absolute value and tends to zero for high enough energies. Ramsauer oscillations can be clearly seen for energies slightly above threshold. Comparing black and gray

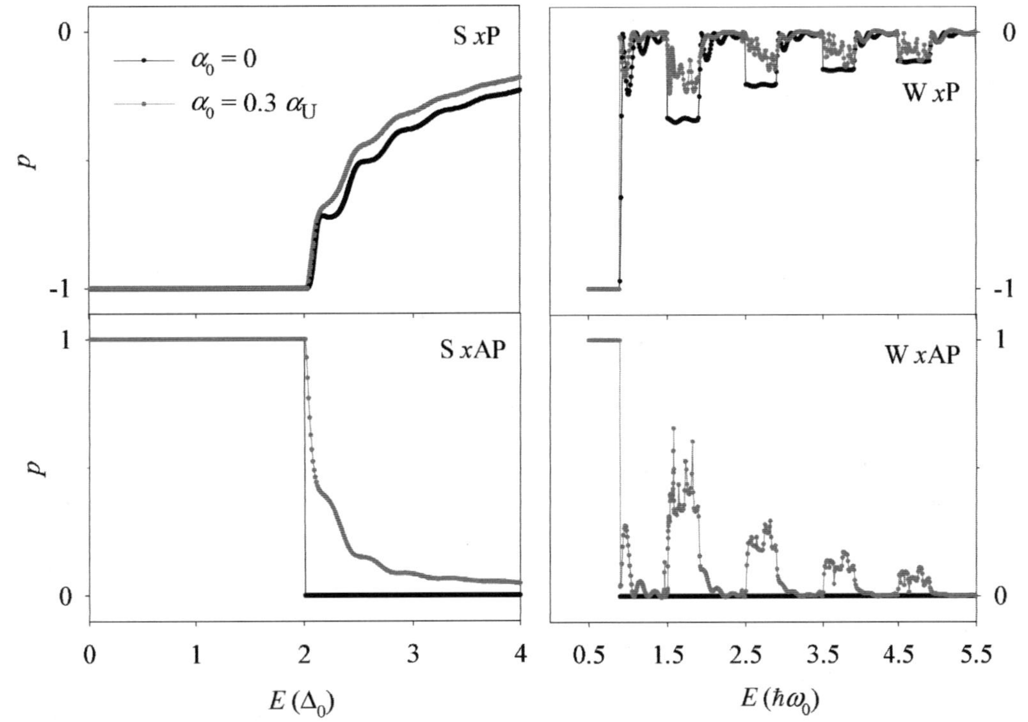

Figure 4. (Color online) Polarization of the transmitted current, defined as $p = G_p/G$, where $G_p = \sum_{s'} s' G_{s's}$ is the polarized conductance. As in Fig. 2, black and gray (red) symbols correspond to the absence and presence of Rashba interaction, respectively. The configuration of the contacts for each panel is also labeled as in Fig. 2.

(red) symbols, we notice that the main effect of the Rashba interaction (red-gray symbols) in the stripe geometry is to smooth the transition from full to vanishing polarization in AP configuration.

Focusing now on the wire polarization of the transmitted current, displayed in the right panels of Fig. 4, we notice that each conductance-plateau-region, $[n-1/2, n+1/2]\hbar\omega_0$, is again a qualitative repetition of the stripe behavior. The P configuration is characterized by a transition from high absolute polarization at the beginning of the plateau to low polarization towards the plateau end. There is an overall tendency to decrease the polarization when the energy increases and conspicuous oscillations are superimposed on the general trend. In agreement with our preceding analysis of the conductance, we can associate the oscillations in the high-$|p|$ part of each plateau with Fano resonances due to quasibound states and those on the vanishing tail with Ramsauer potential oscillations. The Fano resonances are quite narrow and, although not seen in the discrete set of plotted symbols of Fig. 4, polarization can be actually reversed with respect to the main trend when the energy is very close to some resonances. An approximate expression for the polarization, not taking into account conductance oscillations, is obtained from the number of up (n_+) and down (n_-) propagating channels as $|p| \approx (n_+ - n_-)/(n_+ + n_-)$. This simplified formula is in qualitative agreement with the behavior shown in the right panels of Fig. 4 when counting the number of propagating modes for a given energy.

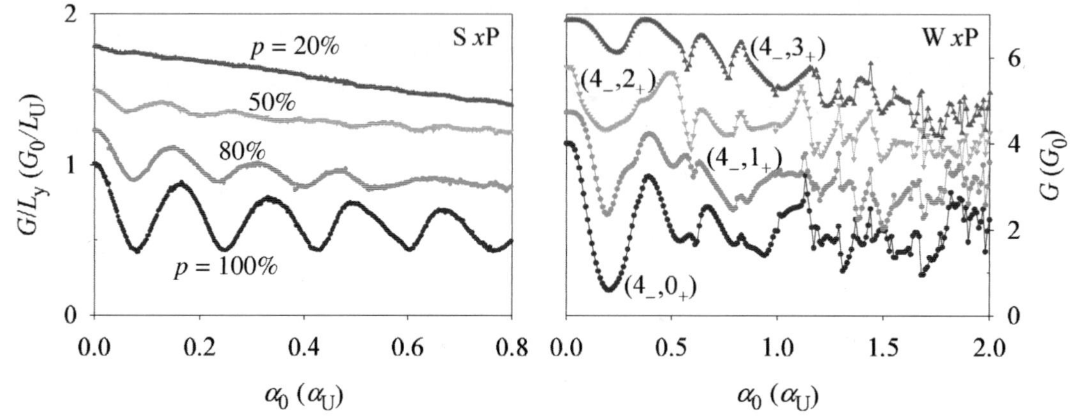

Figure 5. (Color online) Conductance as a function of α_0 in stripe (left) and wire (right) geometries. The polarization of the contacts is indicated for the stripe, while for the wire (n_-, n_+) gives the number of propagating down and up spin modes.

4.2. Dependence on Rashba coupling intensity

The Datta-Das spin transistor relies on the oscillatory character of the conductance as a function of Rashba intensity. The present subsection analyzes this dependence of the conductance, at a fixed energy, in quantum wires and stripes. In particular, we explore the robustness of the oscillations when $E > 2\Delta_0$ and both spins can propagate in the contacts, i.e., the regime of partially polarized contacts. Figure 5 displays the conductance in both systems. In the stripe the incident-current polarization can be varied in a continuous way simply by increasing the energy above the Zeeman barriers of the contacts. For the wire, however, we always have an integer number of spin-down and spin-up propagating modes (n_-, n_+) and, as mentioned above, approximate polarizations $|p| \approx (n_+ - n_-)/(n_+ + n_-)$. Of course, partial reflections and transmissions can lead to deviations from this simplified expression.

For the stripe, shown in the left panel of Fig. 5, at full polarization ($p = -1$) the conductance displays damped oscillations with α_0. Remarkably, when the polarization of the leads is reduced the oscillating behavior is greatly quenched and it is fully washed out for polarizations below 20%. In the wire geometry (right panel) the behavior is less regular. At full polarization there is a clear initial oscillation, but as α_0 increases the conductance exhibits an irregular or *disordered* behavior. When the polarization is reduced the initial oscillation is heavily distorted, but is more robust than in the stripe case. At large α_0's the region of irregular conductance is not qualitatively modified when the polarization is reduced.

5. Conclusions

In this work we have compared the linear conductance of a parabolic wire and a stripe with a localized Rashba interaction and polarized contacts. The spin splitting of the polarized contacts is modeled by means of Zeeman fields. For energies below the Zeeman energy gap the spin-selective barriers of the contacts induce the formation of quasibound states. These quasibound states couple with propagating states via the Rashba coupling and manifest as Fano resonances of the conductance for both stripe and wire systems. In the case of the wire these Fano resonances appear at the beginning of each conductance plateau. At the end of each plateau we also find a second class of Fano resonances, conductance dips, due to quasibound states which are not induced by the polarized leads, but they originate in the Rashba interaction alone. For energies above the Zeeman gap the conductance shows Ramsauer oscillations due to the underlying potential. We have also discussed the variations for x and y orientation of the polarized contacts in both parallel and antiparallel configurations.

As a function of α_0 the stripe conductance shows a damped oscillating behavior which is not robust for partial polarizations; i.e., it becomes monotonous for polarizations below $\approx 20\%$. The wire conductance is qualitatively similar but with a more robust initial oscillation. At high values of α_0 the wire conductance is characterized by an irregular behavior that we attribute to the existence of many resonances.

Acknowledgments

We acknowledge R. López and D. Sánchez for valuable discussions. This work was supported by the Grant No. FIS2008-00781.

References

[1] Datta S and Das B 1990 *Appl. Phys. Lett.* **56** 665
[2] Koo H C, Kwon J H, Eom J, Chang J, Han S H and Johnson M 2009 *Science* **325** 1515
[3] Rashba E I 1960 *Fiz. Tverd. Tela* (Leningrad) **2**, 1224 [1960 *Sov. Phys. Solid State* **2**, 1109]
[4] Nitta J, Akazaki T, Takayanagi H and Enoki T 1997 *Phys. Rev. Lett.* **78**, 1335
[5] Engels G, Lange J, Schäpers Th and Lüth H 1997 *Phys. Rev. B* **55**, R1958
[6] Zutic I, Fabian J and Das Sarma S 2004 *Rev. Mod. Phys.* **76** 323
[7] Agnihotri P and Bandyopadhyay S 2010 *Physica E* **42**, 1736
[8] Gelabert M M, Serra Ll, Sánchez D and López R 2010 *Phys. Rev. B* **81**, 165317
[9] Gelabert M M and Serra Ll 2010 *ArXiv*:1005.2480 (unpublished)
[10] Zainuddin A N M, Hong S, Siddiqui L and Datta S 2010 *Arxiv*:1001.1523 (unpublished)
[11] Nesteroff J A, Pershin Yu V and Privman V 2004 *Phys. Rev. B* **69**, 121306(R)
[12] Serra Ll, Sánchez D and López R 2005 *Phys. Rev. B* **72**, 235309
[13] Quay C H L, Hughes T L, Sulpizio J A, Pfeiffer L N, Baldwin K W, West K W, Goldhaber-Gordon D and de Picciotto R 2010 *Nature Phys.* **6**, 336
[14] Sánchez D and Serra Ll 2006 *Phys. Rev. B* **74**, 153313
[15] Fano U 1961 *Phys. Rev.* **124**, 1866
[16] Matsuyama T, Hu C M, Grundler D, Meier G and Merkt U 2002 *Phys. Rev. B* **65**, 155322
[17] Pala M G, Governale M, König J and Zülicke U 2004 *Europhys. Lett.* **65**, 850
[18] Shelykh I A and Galkin N G 2004 *Phys. Rev. B* **70**, 205328
[19] Nikolić B and Souma S 2005 *Phys. Rev. B* **71**, 195328
[20] Zhang L, Brusheim P and Xu H Q 2005 *Phys. Rev. B* **72**, 045347
[21] Jeong J S and Lee H W 2006 *Phys. Rev. B* **74**, 195311
[22] Perroni C A, Bercioux D, Marigliano Ramaglia V and Cataudella V 2007 *J. Phys.: Condens. Matter* **19**, 186227
[23] Wan J, Cahay M and Bandyopadhyay S 2007 *J. App. Phys* **102**, 034301 [erratum 2007 *J. App. Phys* **102**, 099902]
[24] Datta S 2002 *Electronic transport in mesoscopic systems* (Cambridge UP)

International Conference on Theoretical Physics Dubna-Nano 2010 IOP Publishing
Journal of Physics: Conference Series **248** (2010) 012017 doi:10.1088/1742-6596/248/1/012017

Interaction effects in quantum dots in a vertical magnetic field

A Puente[1], M Pons[1] and R G Nazmitdinov [1,2]

[1] Departament de Física, Universitat de les Illes Balears, 07122 Palma de Mallorca, Spain
[2] BLTP, Joint Institute for Nuclear Research, 141980 Dubna, Russia

E-mail: rashid@theor.jinr.ru

Abstract. The Hartree-Fock approximation has been used to compute the properties of parabolic two-dimensional quantum dots under a perpendicular magnetic field. The addition energies for vertical quantum dots up to $N = 20$ electrons is analysed at different strengths of the effective electron-electron interaction. A remarkable agreement between experimental data and results for the addition energy for $N = 2, 3, 4, 5$ electrons exhibits an important role of the screening of the effective interaction for interpretation of the ground state transitions in the magnetic field. We suggest a recipe to determine the orbital momenta of the QD ground states. We show that quantum shell structure of quantum dots survives only at small magnetic fields.

1. Introduction

Quantum dots (QDs) attract a considerable experimental and theoretical attention during last decade (cf [1, 2, 3, 4]). In particular, this interest is due to the fact that QDs may provide a natural realization of quantum bit. On the other hand, QDs can be considered as a tiny laboratory allowing direct investigation of fundamental properties of charge and spin correlations at the atomic scale.

Mesoscopic systems such as QDs exhibit many common properties specific to a finite Fermi system. A prominent feature is a regular single–particle motion which can manifest itself in pronounced shell effects. In recent experiments [5, 6] it was found that the energy needed to place an extra electron (addition energy) into a vertical QD at zero magnetic field has characteristic maxima which correspond to the sequence of magic numbers of a two-dimensional harmonic oscillator. If a QD is placed under a perpendicular magnetic field, one observes the orbital momentum and spin oscillations of the ground state of a QD with alteration of the field strength [1]. At certain field range the oscillations disappear and it is believed that the electrons form a finite-size analog of infinite integer quantum Hall liquid (with filling one) [2]. In QDs this fully polarized state is called the maximum density droplet (MDD). Further, for a low enough electron density, Wigner [7] predicted that electrons should localize, creating an ordered spatial structure that breaks the complete translational symmetry of the homogeneous electron gas. According to a general opinion, the Wigner crystallization in QDs, whose localized states are referred to as Wigner molecules, should occur at significantly larger densities than in bulk. This is based on the argument that in QDs potential-energy contributions can easily exceed the kinetic terms and, therefore, electronic motion can be effectively quenched by manipulating the external confinement and/or an applied magnetic field.

The existence of different phases in QDs has become a topical subject in mesoscopic physics [2, 3]. Such studies shed light on various aspects of strongly correlated *finite* systems, which are different from bulk and can be controlled experimentally. It gives a hope to understand the transition from finite-size effects to a conventional condensed matter picture of a many-body quantum system, as a function of electron-electron interaction and external fields. In fact, the high controllability of QDs creates a remarkable opportunity to study in detail novel strongly correlated phenomena in various experimental setups.

In theoretical studies of QDs the underlying lattice of the semiconductor material is taken into account by using an effective mass for the conduction electrons, and a static dielectric constant, reducing the Coulomb repulsion. For vertical QDs, the resulting confining potential is, to a good approximation, harmonic [1, 5, 6]. Various approaches, based on such approximations, including ab initio calculations such as diffusion or path integral Monte Carlo methods, spin-density functional methods have been applied with some degree of success to analyze the ground state energies of N-electron QD [2]. In many cases, a convenient starting point to treat finite systems is a mean field description like the Hartree-Fock (HF) approach. Self-consistency between the mean field and the single-particle orbitals and total energy minimization are the basic conditions at this level. However, the total energy minimization procedure being of nonlinear nature is still a real challenge for many-body quantum theory.

Recently we developed an efficient, numerically stable HF procedure. In the present communication we compare our results with the state of art calculations [8] for few-electron vertical QDs. The main focus of our report is, however, the analysis of quantum size effects (like shell structure) and transition to the classical limit of localized electronic states in vertical QDs, with the increase of the electron-electron interaction and/or the magnetic field.

2. The Model

We consider a system of N interacting electrons whose motion is restricted to the xy plane by a circular parabolic potential. The system is also subject to an external magnetic field applied in the vertical direction (z). The full Hamiltonian thus reads

$$H = \sum_{i=1}^{N} \left[\frac{1}{2m} \left(\mathbf{p} - \frac{e}{c}\mathbf{A} \right)^2 + \frac{1}{2} m\omega_0^2 r^2 \right]_i + \sum_{i>j=1}^{N} \frac{e^2}{\kappa\, r_{ij}} + g^* \mu_B B S_z \ . \tag{1}$$

Here, m, $\kappa (= \epsilon_r)$ and g^* are the electron's effective mass, the dielectric constant and the effective gyromagnetic factor, respectively, and we have used planar polar coordinates ($r^2 = x^2 + y^2$). The two contributions within the square brackets are, respectively, the generalized kinetic energy in terms of the vector potential $\mathbf{A}$, and the external confinement. Within the so-called symmetric gauge one has $\mathbf{A}(x, y) = B/2(-y, x)$. The next contribution is the Coulomb interaction and, finally, the last term is the Zeeman energy involving the total spin operator S_z and the universal Bohr's magneton $\mu_B = |e|\hbar/2m_e c$.

In the chosen gauge the magnetic field contributions can be recast into the form of an effective parabolic confinement of frequency $\Omega = \sqrt{\omega_0^2 + \omega_c^2/4}$, where $\omega_c = |e|B/mc$ is the cyclotron frequency, and an addition orbital-momentum-dependent term $(\hbar\omega_c/2)\ell_z$ (cf [4]). By introducing the length $\ell_\Omega \equiv \sqrt{\hbar/(m\Omega)}$ and the energy $E_0 = \hbar\Omega$ units we can rewrite Hamiltonian (1) in terms of only two adimensional parameters: $R_{mp} = e^2/(\kappa\ell_\Omega)/\hbar\Omega$ and $W_{mp} = \omega_c/\Omega$, – the magneto-parabolic units (mpu) [9]. In the absence of a magnetic field R_{mp} coincides with the so-called Wigner parameter R_W (cf [3]). Note also that W_{mp} has a maximal value $W_{mp} = 2$ that corresponds to a zero confinement $\omega_0 = 0$ or an infinite magnetic field. Within the mpu system the Hamiltonian reads

$$\mathcal{H} = \frac{H}{\hbar\Omega} = \sum_{i=1}^{N} \left[-\frac{1}{2}\nabla^2 + \frac{1}{2}r^2 + \frac{W_{mp}}{2}\ell_z \right]_i + R_{mp} \sum_{i>j=1}^{N} \frac{1}{r_{ij}} + \frac{g^* m^*}{2} W_{mp} S_z \,, \tag{2}$$

where $m^* = m/m_e$ is the adimensional ratio of effective to bare electron mass, and $r \to r/\ell_\Omega$. The passage from the mpu system, with a given R_{mp} and W_{mp}, to physical units requires the knowledge of the effective mass m and the dielectric constant κ. More specifically, with fixed m and κ one can invert R_{mp} and W_{mp} for the effective confinement Ω and cyclotron ω_c frequencies or, equivalently, for the external confinement ω_0 and the magnetic field B. In terms of these results the physical values of the energy ($\hbar\Omega$) and length (ℓ_Ω) units are readily found (see Fig.1 in [9]). The R_{mp} parameter is a measure of the relative importance of electron-electron interaction to confinement potential strength. It is used usually as a measure to define the different phases in QDs and, in particular, a formation of Wigner molecules [3, 9]. Note that we use the effective electron-electron interaction, where the parameter R_{mp} is associated with the effective charge (a screening parameter). Such a form can be justified due to, for example, the QD thickness [10].

Introducing the basis of oscillator and spin states that diagonalizes the first term in Eq. (2), $\{\,|a\eta\rangle;\ a = 1, \ldots \mathcal{N};\ \eta = \uparrow, \downarrow\,\}$, where a labels the orbital part and η the spin, an arbitrary single-particle orbital $|i\rangle$ is expanded as $|i\rangle = \sum_{a\eta} B^{(i)}_{a\eta}\,|a\eta\rangle$. The oscillator (Fock-Darwin) states $|a\rangle$ are characterized by radial (n_a) and orbital momentum (m_a) quantum numbers (cf [4]). In our calculations ground states are eigenstates of the S_z and of the $\hat{\ell}_z$ operators. In the chosen basis, the HF equations are written as a system of nonlinear eigenvalue equations for the matrix of B coefficients.

3. Shell structure and classical limit

The most simple theoretical approach to analyse the experimental data from transport studies of correlated QD states is to calculate the addition energy

$$E_{add}(N) = E_{N+1} - 2E_N + E_{N-1}, \tag{3}$$

where E_N is the energy of the an N-electron dot state [1]. The energy $E_{add}(N)$ is proportional to the difference between the gate voltages for two successive current peaks in transport experiments.

Recently Maksym *et al* [8] developed a model of a vertical QD and reproduced quite well experimental addition energies for $N = 2, 3, 4$. Although the three-dimensional nature of QDs is accounted for, the real calculations are based on the two-dimensional parabolic potential. The authors constructed the effective two-dimensional electron-electron interaction (the screened Coulomb interaction) which does not depend on the magnetic field. They diagonalized the model Hamiltonian in the many-electron basis consisted of Slater determinants. To compare with the results [8], we have used the following parameters for InGaAs QDs: $\hbar\omega_0 = 4.5\,\text{meV}$, $m^* = 0.0653$, dielectric constant $\epsilon_r = 12.7$. These values provide for the strength of the bare Coulomb interaction the following estimation: $R_{mp}(\omega_c = 0) \equiv R_W \approx 1.56$. Since we use the effective two-dimensional interaction, the effective value $R_W = 0.78$ was taken to reproduce the available experimental data. Our calculations demonstrate a good agreement with the experimental data as well as with the theoretical results obtained in [8].

For N=2, we found that the singlet $(L_z, S_z) = (0, 0)$ state is replaced by fully polarized triplet $(-1, 1)$ state (MDD phase) at $B_{HF} = 3.8$ T (L_z and S_z are the quantum numbers of the z-component of the total orbital momentum and of the total spin, respectively). The experimental transition takes place at $B_{exp} = 4.2$ T, while the theoretical calculations of Maksym *et al* [8] give $B_{theor} = 3.8$ T.

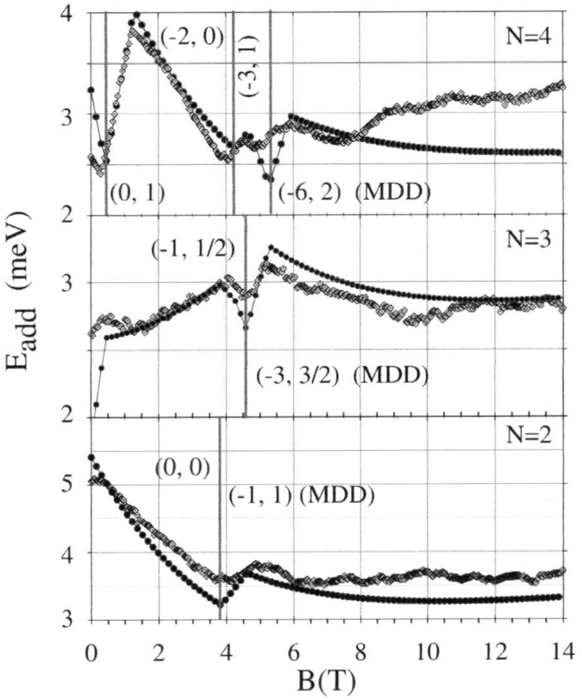

Figure 1. Experimental and theoretical addition energies $E_{add}(N)$ for $N = 2, 3, 4$ as a function of the magnetic field B. A symbol $\diamondsuit$: experimental data [8]; a symbol $\bullet$: Hartree-Fock results. Ground states are labeled by (L_z, S_z) (see text). The vertical line indicates the transition between different ground states.

For $N = 3$ our calculations predict the state $(-1, 1/2)$ to be the lowest state until $B_{HF} = 4.72$ T. Two electrons occupy the lowest available states with $m = 0$ but with different $S_z = \pm 1/2$ quantum numbers. The next available state is characterized by $m = -1$, and the largest possible value for this configuration has $S_z = 1/2$. At $B_{HF} = 4.72$ T we obtain the onset of the MDD phase $(-3, 3/2)$.

At small magnetic field $B_{HF} < 0.46$ T, for $N = 4$ we obtain the ground state $(0, 1)$. A partially polarized state is favourable due to the HF exchange term which supports the configuration with two aligned spins. At $B_{HF} = 0.46$ T ($B_{exp} = 0.33$ T, $B_{theor} = 0.3$ T) the magnetic field overcomes the exchange contribution and we obtain $(-2, 0)$ state. This state persists till $B_{HF} < 4.42$ T. At $B_{HF} = 4.42$ T ($B_{exp} = 4.06$ T, $B_{theor} = 4.0$ T) the magnetic filed polarizes the system partially, and, with the aid of the exchange term, the HF calculations produce the ground state $(-3, 1)$. At $B_{HF} \geq 5.33$ T we obtain the onset of the MDD phase $(-6, 2)$. At $B_{HF} = 14.31$ T the MDD state is replaced by the ground state $(-10, 2)$.

We reproduce a double peak structure for N=3 and N=4 around 4-6 T. The spike observed for $N = 3$ at $B_{HF} = 3.8$ T is due to the transition to the MDD phase in two-electron system (see also Eq.(3)). The following dip is caused by the transition $(-1, 1/2)$ to $(-3, 3/2)$ in $N = 3$, while the transition from $(-2, 0)$ to $(-3, 1)$ in $N = 4$ changes only the slope of the former transition. The next spike observed for $N = 3$ at $B_{HF} = 5.33$ T is due to the transition $(-3, 1) \rightarrow (-6, 2)$ in $N = 4$. The large peak at $B_{HF} = 1.3$ T for $N = 4$ is due to the ground state transition $(-1, 1/2) \rightarrow (-4, 1/2)$ in $N = 5$ (not shown). The next transition in $N = 5$: $(-4, 1/2) \rightarrow (-6, 3/2)$ at $B_{HF} = 4.80$ T,– has a little effect. However, the transition to the MDD phase $(-6, 3/2) \rightarrow (-10, 5/2)$ in $N = 5$ is responsible for the small spike in the addition energy of $N = 4$ at $B_{HF} = 5.87$ T.

Although the agreement is good in general till $B_{exp} \leq 6$ T, there are some discrepancies at high magnetic field for $N = 4$ case (see Fig.1). Similar problem appears in more sophisticated calculations of Maksym *et al* [8]. The analysis of finite-thickness effect in two-electron QDs

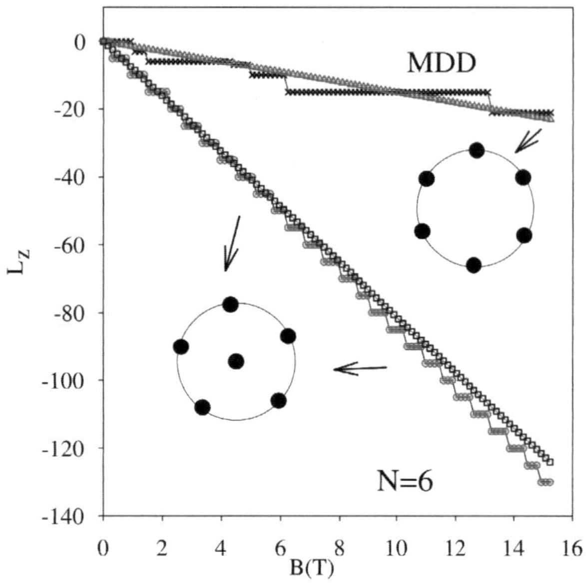

Figure 2. The ground state orbital momentum for six electrons as function of the magnetic field. The symbol: × (black) corresponds $R_W = 0.8$; ○ (green) corresponds to $R_W = 10$. In both cases the MDD phase has $L_z = 15$ (the longest plato for $R_W = 0.8$). For $R_W = 0.8(10)$, starting from the MDD phase, the ground state orbital momentum follows the sequence of magic numbers $L_z = 15, 21...$ ($L_z = 15, 20, 25, ...$). The symbol △ (□) corresponds to the values of L_z from Eq.(6) for $R_W = 0.8(10)$. The equidistant points on the circles are associated with the equilibrium configurations at a given R_W. The equilibrium classical configurations is $N = 5 + 1$ ($R_W = 10$).

reveals a magnetic-field dependence of the screening of the effective interaction [10]. The stronger is the magnetic field the lesser is expected the screening effect in QDs. Indeed, we reproduce the decrease and the increase of the addition energies for N=3 and N=4, respectively, at $B \geq 7$ T with the effective value $R_W = 1.5$. However, the positions of the transitions are slightly higher in the energy scale. One of the possible reasons for this discrepancy is the change of the effective confining potential (the parabolic confinement frequency) at high magnetic field. It appears from our calculations that there is a visible suppression of the electron-electron interaction in the observed experimental data. Indeed, the experimental evidence [6] is that the QD has a high degree of a circular symmetry. We can conclude that for this device the confining potential dominates in the electron dynamics.

The origin of magic values of the orbital momenta which occur after the onset of the MDD phase for $N = 4$ ($L_z = 6, 10, 14...$) can be understood from the symmetry properties of the total wave function of a few-electron QD. The classical minimum-energy configuration of a few point charges in a parabolic potential is highly symmetric [14]. Such configurations takes place if N point charges create equidistant nodes on the ring, with the angle $\alpha = 2\pi/N$. It is reasonable to suppose that a quantum ground state is localized around the equilibrium classical state (see below). However, we have to take into account that the total wave function should be antisymmetric. Note that the rotation on the angle α is equivalent to a cyclic permutation of the particle coordinates on the ring. Since the spatial part of the total wave function Φ_r is the eigenstate of L_z operator, one can require the fulfillment of the following symmetry condition

$$R_z(\alpha)\Phi_r = \exp(i\frac{2\pi}{N}L_z)\Phi_r \equiv P_p\Phi_r = \varepsilon\Phi_r \tag{4}$$

Here $\varepsilon = \pm$ is the parity of the permutation for N point charges located on the ring. If the permutation is odd, then $\exp(i\frac{2\pi}{N}L_z)\Phi_r = -\Phi_r$, and the total orbital momentum takes values $L_z = N(2k+1)/2$ ($k = 1, 2...$). If the permutation is even, one has $\exp(i\frac{2\pi}{N}L_z)\Phi_r = \Phi_r$. As a result, the lowest quantum states carry the magic orbital momenta $L_z = Nk$ ($k = 1, 2...$) (see also [15]). We recall, however, that the classical configuration ($R_W \gg 1$) can contain a few rings (see Fig.2). In this case the magic numbers can be determined by the number of electrons in the external or in the internal rings starting from the orbital momentum $L_z^{(MDD)}$ of the MDD phase. For nonzero magnetic field this number must obey also the condition Eq.(6) (see below). Although for different strengths R_W the HF results provide different sequence of the total orbital momentum for the ground states, we found that these sequences are subject to this rule at a nonzero magnetic field.

The total energy for the Hamiltonian (1) can be written in the following form

$$E_N = \Big\langle \sum_{i=1}^{N} \Big[\frac{p_x^2 + p_y^2}{2m} + \frac{m}{2}(\omega_0^2 + \omega_L^2) r^2 + \omega_L \ell_z \Big]_i \Big\rangle + \langle V_C + g^* \mu_B B S_z \rangle, \tag{5}$$

where $\omega_L = \omega_c/2$ and $\langle ... \rangle$ means a mean field value. For a nonzero magnetic field we have a sequence of different energies for a fixed value of the total orbital momentum L_z. In virtue of Hellman-Feynman theorem, we obtain for the lowest equilibrium state for a fixed value of the the total orbital momentum L_z at a given magnetic field

$$\frac{\partial E_N}{\partial \omega_L}\Big|_{L_z = L} = 0 \quad \Longrightarrow \quad L_z = \Big\langle \sum_{i=1}^{N} \ell_z \Big\rangle = -m\,\omega_L \Big\langle \sum_{i=1}^{N} r_i^2 \Big\rangle = -\omega_L \Im \tag{6}$$

Since the moment of inertia $\Im \sim R_W^{2/3}$, one obtains an estimation for the optimal sequence of values of the ground state orbital momentum L_z at different values of the magnetic field (ω_L) and for different values of the Wigner parameter R_W ($\Im$) (see Fig.2).

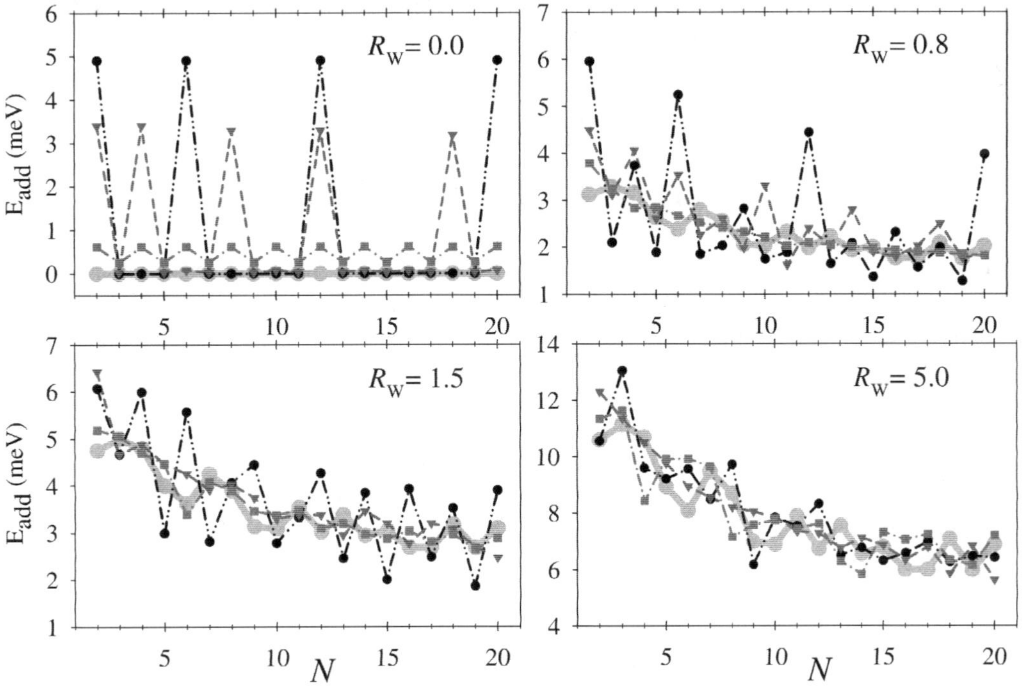

Figure 3. The addition energy E_{add} as a function of the magnetic field B and the strength parameter R_W. The values of E_{add} at different values of the magnetic field: $-\cdot\cdot-$ line connects the maxima ● for $B = 0$ T; $--$ line connects the maxima $\vee$ for $B = 2$ T; $-\cdot-$ line connects the maxima ■ for $B = 15$ T; a thick line displays the classical limit.

While the electron-electron interaction is important for the explanation of spin and orbital ground-state oscillations of QDs in the magnetic field (cf [11]), for small number of electrons the confinement energy may become prevalent over the Coulomb energy [12], indeed. To investigate the evolution of the shell structure, we calculate the addition energy for the QD up to N=20 electrons at different values of the strength parameter R_W and the magnetic field B (see Fig.3). At $R_W = 0$ (noninteracting case) one observes a strong shell structure effects for $B = 0$ and 2 T. For $B = 0$ T the magic numbers (including spin) turn out to be the usual sequence of the two-dimensional isotropic oscillator, that is 2, 6, 12, 20 [5, 6]. For $B = 2$ T we find new shells as if the confining potential were be a deformed harmonic oscillator without a magnetic field. The magic numbers are 2,4, 8, 12, 18. In this case shell structure occurs whenever the ratio of the two eigenmodes of the Hamiltonian (1) (without the Zeeman term) $\Omega_{\pm} = \Omega \pm \omega_c/2$ is a rational number with a small numerator and denominator [13]. With the increase of the interaction strength $R_W = 0.8$ the shell structure survives only without the magnetic field. The magnetic field compresses the energy gaps between energy levels forming the Hall liquid. At relatively large $R_W = 5$ one observes the onset of the classical limit (Thomson model) [14]. We recall that the classical limit can be reached also at a high magnetic field but a relatively small R_W. In the classical limit, the total energy for one-ring equilibrium configuration is

$$E_N^{cl} = (3/8)(2R_W S_N)^{2/3} N\hbar\omega_0, \quad S_N = \sum_{j=1}^{N-1} \frac{1}{\sin \frac{\pi}{N} j}. \tag{7}$$

While the quantum shell structure diminishes, one observes the onset of the *classical shell structure* (see Fig.3) due to an optimal packing of electrons in various two-dimensional rings with $R_W \gg 1$. In the classical limit one has, for example, the following sequences of rings: $N = 6 \Rightarrow (1,5)$,...,$N = 10 \Rightarrow (2,8)$,...,$N = 20 \Rightarrow (1,7,12)$ [14].

4. Summary

We developed an efficient, numerically stable HF procedure and compare our results with available experimental and theoretical results [8]. We found that, in order to reproduce the experimental data, the screening of the effective electron-electron interaction in the HF approach should be decreased with the increase of the magnetic field. Based on our procedure, we have shown that the effective interaction gives rise to dynamical symmetries of N-electron QDs ($N \leq 20$). These symmetries manifest themselves as near-degeneracies in the quantum spectrum and produce maxima in the addition spectra for specific electron numbers at specific values of the magnetic field. Varying the effective strength parameter R_W and the strength of the magnetic field we discuss the evolution of the localized states, formed in the ground state, from $N = 2$ to $N = 20$ electrons. The increase of the magnetic field strength and/or the effective electron-electron interaction strength diminishes the quantum shell structure and gives rise to the *classical shell structure* produced by an optimal packing of electrons in two-dimensional QDs. We propose a transparent recipe to determine the ground state orbital momenta at arbitrary values of the magnetic field and the strength of the effective Coulomb interaction.

Acknowledgments

This work was partly supported by Grant No. FIS2008-00781/FIS (Spain) and RFBR Grants No. 08-02-00118 (Russia).

References

[1] Kouwenhoven L P, Austing D G and Tarucha S 2001 *Rep. Prog. Phys.* **64** 701
[2] Reimann S M and Manninen M 2002 *Rev. Mod. Phys.* **74** 1283
[3] Yannouleas C and Landman U 2007 *Rep. Prog. Phys.* **70** 2067

[4] Nazmitdinov R G 2009 *Physics of Particles and Nuclei* **40** 71
[5] Tarucha S *et al* 1996 *Phys. Rev. Lett.* **77** 3613
[6] Nishi Y *et al* 2006 *Phys. Rev. B* **74** 033306
[7] Wigner E P 1934 *Phys. Rev. B* **46** 1002
[8] Maksym P A *et al* 2009 *Phys. Rev. B* **79** 115314
[9] Puente A, Serra Ll and Nazmitdinov R G 2004 *Phys. Rev. B* **69** 125315
[10] Simonović N S and Nazmitdinov R G 2008 *Phys. Rev. A* **78** 032115
[11] Maksym P A and Chakraborty T 1990 *Phys. Rev. Lett.* **65** 108; 1992 *Phys. Rev. B* **45** 1947; Wagner M, Merkt U and Chaplik A V 1992 *Phys. Rev. B* **45** 1951; Hawrylak P 1993 *Phys. Rev. Lett.* **71** 3347 Dineykhan M and Nazmitdinov R G 1997 *Phys. Rev. B* **55** 13707; 1999 *J. Phys.: Condens. Matter* **11** L83
[12] Macucci M, Hess K and Iafrate G J 1993 *Phys. Rev. B* **48** 17354; 1995 *J. Appl. Phys.* **77** 3267; Wojs A *et al* 1996 *Phys. Rev. B* **54** 13707
[13] Heiss W D and Nazmitdinov R G 1996 *Phys. Lett. A* **222**, 309; 1997 *Phys. Rev. B* **55** 16310; 1998 *Pis'ma v ZhETF* **68** 870 (*JETP Lett.* **68** 915)
[14] Thomson J J 1904 *Phil. Mag.* **7** 237; Partoens B and Peeters F M 1997 *J. Phys.: Condens. Matter* **9** 5383
[15] Ruan W Y *et al* 1995 *Phys.Rev. B* **51** 7942

International Conference on Theoretical Physics Dubna-Nano 2010 IOP Publishing
Journal of Physics: Conference Series **248** (2010) 012018 doi:10.1088/1742-6596/248/1/012018

Effective Theory of Electron-Hole Exchange in Semiconductor Quantum Dots

E S Kadantsev and P Hawrylak

Quantum Theory Group, Institute for Microstructural Sciences, National Research Council, Ottawa, Canada K1A 0R6

E-mail: ekadants@babylon.phy.nrc.ca

Abstract. Semiconductor self-assembled quantum dots (QDs) are expected to become building blocks of novel optoelectronic devices capable of generating single photons and entangled photon pairs (EPPs) on demand. The major barrier to generation of EPPs is the so-called "bright exciton splitting" originating from the long-range electron-hole exchange interaction. A $\mathbf{k} \cdot \mathbf{p}$ effective mass theory of electron-hole exchange in semiconductor quantum dots is presented (*Phys. Rev. B* **81**, 045311). The matrix element responsible for the "bright" exciton splitting in the effective exciton Hamiltonian is identified and analyzed. An excitonic fine structure for a model quantum dot with quasi-two-dimensional anisotropic harmonic oscillator (2DLAHO) confining potential is considered as a function of the shape and external fields.

1. Introduction

One of the promising applications of semiconductor quantum dots (QDs) is the generation of entangled photon pairs (EPPs) on demand [1, 2, 3, 4]. Polarization entangled photon pairs can be efficiently generated using the biexciton cascade process (BCP) [5] in which a biexciton radiatively decays into the ground state via two different indistinguishable paths involving two intermediate dipole-active ("bright") exciton states. The major barrier to EPPs generation in BCP is the splitting of the intermediate "bright" exciton levels which distinguishes the paths of radiative decay and, as a result, destroys the entanglement. The splitting of the two bright exciton states is controlled by the long range electron-hole exchange (LRE) interaction and depends on the dot's shape, size, and chemical composition. External magnetic [6, 2, 4] and electric fields have been applied in the hope to manipulate the exciton fine structure [7, 8, 9].

Investigation of the electron-hole exchange interaction in QDs, particularly its LRE part, should help in the development of the EPP generation schemes. The LRE electron-hole interaction in bulk semiconductors was studied almost 40 years ago [10, 11, 12] . Recent advances in single QD spectroscopies motivated re-examination of the electron-hole exchange in semiconductor systems with spatial confinement of charge carriers within the framework of envelope function approximation [13, 14, 15, 16, 17, 18, 19] and, more recently, from the point of view of atomistic empirical pseudopotential and tight-binding approximation [20, 21, 22].

Within the envelope function approximation, Takagahara [13] have shown that if the electron-hole pair envelope contains only $Y_{00}(\theta, \phi)$ angular momentum component, then the long-range part of the electron-hole exchange vanishes. In Ref. [17], Takahagara derived effective eight-band excitonic Hamiltonian which takes into account electron-hole exchange interaction. Within the envelope function formalism of Refs. [13, 17], the LRE is a dipole-dipole interaction. It was demonstrated numerically in Ref. [17] that LRE vanishes for the ground state "bright" exciton doublet in QDs with rotationally symmetric 3D confining potential. A similar to that of Ref. [17] representation of the electron-hole exchange was discussed by Maialle *et. al.* [14] in connection with the exciton spin dynamics in quantum wells.

Within the empirical atomistic approach it has been demonstrated that the physical origin of the long-range exchange interaction in QDs might be different from that of bulk [20] due to the loss of local orthogonality on the unit cell scale between the electron and hole quasi-particle orbitals. It also has been shown that the "bright" exciton splitting can have either nonzero (dots of C_{2v} point group) or zero (dots of D_{2h} point group) magnitude in the case of "shape-symmetric" dots [22] and the importance of the dot's true point group symmetry underlined by both the macroscopic shape and the zinc-blende crystal structure has been realized. Recently, Dalessi and Dupertuis extended [23] the effective $\mathbf{k}\cdot\mathbf{p}$ approach to account for the true point group symmetry of the semiconductor nanostructures which might enable one to *qualitatively predict* excitonic fine structure based on the envelop theories.

In this work, we will re-examine electron-hole exchange within the framework of the envelope approximation. We derive an effective "four band" excitonic Hamiltonian which includes effects of the LRE interaction. The elements of the effective Hamiltonian are expressed explicitly in terms of the envelope functions. The "microscopic parts" of single-particle orbitals are integrated out and enter the effective Hamiltonian as numerical parameters. The number of bands is truncated to two conduction and two valence bands for the simplicity of the interpretation. We present explicit expression for the matrix element responsible for the "bright" exciton splitting and, therefore, establish a new sufficient condition for the LRE quenching.

Effective units of length and energy are used throughout unless otherwise specified. The lengths are measured in the effective Bohrs $a_{eB} = \epsilon\hbar^2/(m^* e^2)$, where ϵ is dielectric constant and m^* is the conduction band effective mass. The energies are measured in effective Hartrees, $1\,\mathrm{H} = m^* e^4/(\epsilon\hbar)^2 = 2\,\mathrm{Ry}^*$. For example, using material parameters for GaAs, $a_{eB} = 97.9$ Å and $1\,\mathrm{H} = 11.86$ meV.

2. Theory of exciton fine structure in envelope function approximation

The excitonic states are obtained by diagonalizing the many-body Hamiltonian. Let $c^\dagger$ (c) denote electron creation (annihilation) operator, respectively. The many-body Hamiltonian in a semiconductor is

$$\hat{H} = \hat{H}_e + \hat{H}_h + \hat{H}_{int}, \quad \hat{H}_e = \sum_i \varepsilon_i c_i^\dagger c_i, \quad \hat{H}_h = \sum_\alpha \varepsilon_\alpha c_\alpha^\dagger c_\alpha,$$

$$\hat{H}_{int} = \frac{1}{2}\sum_{ijkl} c_i^\dagger c_j^\dagger c_k c_l \langle ij|V|kl\rangle + \frac{1}{2}\sum_{\alpha\beta\gamma\delta} c_\alpha^\dagger c_\beta^\dagger c_\gamma c_\delta \langle \alpha\beta|V|\gamma\delta\rangle + \sum_{i\beta\gamma l} c_i^\dagger c_\beta^\dagger c_\gamma c_l \left(\langle i\beta|V|\gamma l\rangle - \langle i\beta|V|l\gamma\rangle\right),$$

$$V_{ABCD} = \langle AB|V|CD\rangle = \int\!\!\int \frac{\phi_A^\dagger(\mathbf{x}_1)\phi_B^\dagger(\mathbf{x}_2)\phi_C(\mathbf{x}_2)\phi_D(\mathbf{x}_1)}{\epsilon(\mathbf{r}_1,\mathbf{r}_2)|\mathbf{r}_1-\mathbf{r}_2|} d\mathbf{x}_1 d\mathbf{x}_2 \tag{1}$$

The Hamiltonian (1) is obtained from the full many-body Hamiltonian by assuming that the quasi-particle orbitals belong to two distinct manifolds (conduction/valence band) and retaining the interactions within the conduction and valence bands (the first and the second terms in $\hat{H}_{int}$) as well as the interband direct $\langle i\beta|V|\gamma l\rangle$ and exchange $\langle i\beta|V|l\gamma\rangle$ terms. The indices $\{i,j,k,l\}$ and $\{\alpha,\beta,\gamma,\delta\}$ run over the valence and conduction band states, and $\varepsilon_i(\varepsilon_\alpha)$ denote the quasi-particle energies of conduction (valence) states. V_{ABCD} is a "screened" Coulomb-like matrix element involving "two-dimensional" integration over the combined space-spin variable $\mathbf{x} = \{\mathbf{r}, s\}$.

To obtain the excited states, the Hamiltonian (1) is diagonalized in the basis of all electron-hole pairs $|X'\rangle = c_{c\sigma'}^\dagger c_{v\tau'}|g.s.\rangle$, $|X\rangle = c_{c\sigma}^\dagger c_{v\tau}|g.s.\rangle$, where $|g.s.\rangle$ is a single Slater determinant with fully occupied valence and empty conduction bands ($c^\dagger_{v\tau}|g.s.\rangle = c_{c\sigma}|g.s.\rangle = 0$) and $v\tau$ ($v\tau'$) and $c\sigma$ ($c\sigma'$) is a composite index describing valence and conduction band states, respectively. The matrix element of (1) is

$$\langle X|\hat{H}|X'\rangle = \langle g.s.|c^\dagger_{v\tau}c_{c\sigma}\hat{H}c^\dagger_{c\sigma'}c_{v\tau'}|g.s.\rangle = \delta_{c\sigma,c\sigma'}\delta_{v\tau,v\tau'}\left(\varepsilon_{c\sigma} - \varepsilon_{v\tau} + E_{g.s.}\right)$$
$$-\langle c\sigma v\tau'|V|v\tau c\sigma'\rangle + \langle c\sigma v\tau'|V|c\sigma' v\tau\rangle - \Sigma(v\tau', v\tau)\delta_{c\sigma,c\sigma'} + \Sigma(c\sigma, c\sigma')\delta_{v\tau,v\tau'}, \quad (2)$$

where $E_{g.s.} = \langle g.s.|\hat{H}|g.s.\rangle$ is the Hartree-Fock energy of the ground state and $\Sigma(A,B) = \sum_\alpha \left(\langle A\alpha|V|\alpha B\rangle - \langle A\alpha|V|B\alpha\rangle\right)$ is the so-called self-energy.

In this work, the quasi-particle states are computed in the effective mass approximation (EMA). Within the EMA, each quasi-particle state is uniquely specified by the band label of the periodic part $u(\mathbf{r})$ of the Bloch eigenstate at $\mathbf{k} = 0$ ($vj_z, vj'_z, c\sigma, c\sigma'$) and by the index of the envelop function $F(\mathbf{r})$ (q, r, s, p), for example, $\phi^q_{vj_z}(\mathbf{r}) = F^q(\mathbf{r})u_{vj_z}(\mathbf{r})$. By "expanding" the composite indices as $v\tau \to \{vj_z, q\}$, $v\tau' \to \{vj'_z, r\}$, $c\sigma \to \{c\sigma, s\}$, $c\sigma' \to \{c\sigma', p\}$, neglecting the contribution from the self-energies, and "measuring" the excitations with respect to the ground state ($E_V = 0$), we arrive at the following expression for the matrix element (2)

$$\langle g.s.|c^\dagger_{vj_zq}c_{c\sigma s}\hat{H}c^\dagger_{c\sigma'p}c_{vj'_zr}|g.s.\rangle = \delta_{vj_zq,vj'_zr}\delta_{c\sigma s,c\sigma'p}(\varepsilon_{c\sigma s} - \varepsilon_{vj_zq}) - \langle c\sigma s, vj'_zr|V|vj_zq, c\sigma'p\rangle +$$
$$+\langle c\sigma s, vj'_zr|V|c\sigma'p, vj_zq\rangle \quad (3)$$

The effective Hamiltonian which involves only envelope functions is obtained by integrating out "microscopic" degrees of freedom (periodic parts of Bloch functions). The derivation follows that of Ref. [17]. First, we postulate the form of the "microscopic" functions. The conduction (valence) band "microscopic" functions are of s (p) symmetry with spin (hole angular momentum) projections $\pm 1/2$ ($\pm 3/2$)

$$u_{c1/2}(\mathbf{r}) = \langle \mathbf{r}|c1/2\rangle = \chi_s(r)Y_{00}(\hat{r})|\alpha\rangle, \quad u_{c-1/2}(\mathbf{r}) = \langle \mathbf{r}|c-1/2\rangle = \chi_s(r)Y_{00}(\hat{r})|\beta\rangle,$$
$$u_{v-3/2}(\mathbf{r}) = \langle \mathbf{r}|v-3/2\rangle = \chi_p(r)Y_{11}(\hat{r})|\alpha\rangle, u_{v+3/2}(\mathbf{r}) = \langle \mathbf{r}|v+3/2\rangle = \chi_p(r)Y_{1-1}(\hat{r})|\beta\rangle. \quad (4)$$

The Coulomb attraction matrix element is approximated as

$$\langle c\sigma s, vj'_zr|V|vj_zq, c\sigma'p\rangle = \frac{1}{\epsilon}\delta_{c\sigma c\sigma'}\delta_{vj_zvj'_z}\int\int d\mathbf{r}_1 d\mathbf{r}_2 \frac{F^{s\dagger}_{c\sigma}(\mathbf{r}_1)F^p_{c\sigma'}(\mathbf{r}_1)F^{r\dagger}_{vj'_z}(\mathbf{r}_2)F^q_{vj_z}(\mathbf{r}_2)}{|\mathbf{r}_1 - \mathbf{r}_2|}, \quad (5)$$

where ϵ is a static dielectric function.

The exchange matrix element can be thought of as a Coulomb-like interaction of two "transition densities", where each "transition density" is a product of the electron and hole quasi-particle orbitals. First, we decompose the exchange integral into the short-range V^E_{SR} (exchange within Wigner-Seitz cell) and long-range (exchange between the two Wigner-Seitz cells) V^E_{LR} contributions. The "effective" expression for the exchange matrix elements in terms of the envelop functions is obtained in a way which follows "standard" $\mathbf{k}\cdot\mathbf{p}$ derivation, the only difference is the nature of the microscopic integrals to be parameterized. One postulates the "slowly varying nature" of the envelop functions and factors them out of the integral expression. As a result of this procedure, we obtain the following "master" expression for the exchange matrix element

$$\langle c\sigma s, vj'_zr|V|c\sigma'p, vj_zq\rangle = V^E_{SR}(c\sigma s, vj'_zr; c\sigma'p, vj_zq) + V^E_{LR}(c\sigma s, vj'_zr; c\sigma'p, vj_zq),$$
$$V^E_{SR}(c\sigma s, vj'_zr; c\sigma'p, vj_zq) = E_{SR}(H^{int}_{SR})^{c\sigma,vj_z}_{c\sigma',vj'_z}\int d\mathbf{r}F^{s\dagger}_{c\sigma}(\mathbf{r})F^q_{vj_z}(\mathbf{r})F^p_{c\sigma'}(\mathbf{r})F^{r\dagger}_{vj'_z}(\mathbf{r}),$$
$$V^E_{LR}(c\sigma s, vj'_zr; c\sigma'p, vj_zq) = -\frac{4\pi}{3}\mu^2\left((\mathbf{d}^0_{c\sigma'vj'_z})^\dagger\cdot\mathbf{d}^0_{c\sigma vj_z}\right)\int d\mathbf{r}F^{s\dagger}_{c\sigma}(\mathbf{r})F^q_{vj_z}(\mathbf{r})F^p_{c\sigma'}(\mathbf{r})F^{r\dagger}_{vj'_z}(\mathbf{r})$$
$$-\mu^2\sum_{\gamma,\delta=1}^{3}(\mathbf{d}^0_{c\sigma vj_z})_\gamma(\mathbf{d}^0_{c\sigma'vj'_z})^\dagger_\delta\int\int\left(\frac{\partial^2 F^{s\dagger}_{c\sigma}(\mathbf{r}_1)F^q_{vj_z}(\mathbf{r}_1)}{\partial r_1^\gamma\partial r_1^\delta}\right)\frac{F^p_{c\sigma'}(\mathbf{r}_2)F^{r\dagger}_{vj'_z}(\mathbf{r}_2)}{|\mathbf{r}_1 - \mathbf{r}_2|}d\mathbf{r}_1 d\mathbf{r}_2 \quad (6)$$

where E_{SR} and μ^2 are two numerical constants parameterizing short- and long-range exchange interaction, respectively, $(H_{SR}^{int})_{c\sigma',vj'_z}^{c\sigma,vj_z}$ is an element of short-range exchange "microscopic" matrix

$$\mathbf{H}_{\mathbf{SR}}^{\mathbf{int}} = \begin{pmatrix} & & |c\sigma',vj'_z\rangle & |c\sigma',vj'_z\rangle & |c\sigma',vj'_z\rangle & |c\sigma',vj'_z\rangle \\ & & |\frac{1}{2},-\frac{3}{2}\rangle & |\frac{1}{2},\frac{3}{2}\rangle & |-\frac{1}{2},-\frac{3}{2}\rangle & |-\frac{1}{2},+\frac{3}{2}\rangle \\ |c\sigma,vj_z\rangle & |\frac{1}{2},-\frac{3}{2}\rangle & 1 & 0 & 0 & 0 \\ |c\sigma,vj_z\rangle & |\frac{1}{2},+\frac{3}{2}\rangle & 0 & 0 & 0 & 0 \\ |c\sigma,vj_z\rangle & |-\frac{1}{2},-\frac{3}{2}\rangle & 0 & 0 & 0 & 0 \\ |c\sigma,vj_z\rangle & |-\frac{1}{2},+\frac{3}{2}\rangle & 0 & 0 & 0 & 1 \end{pmatrix}, \tag{7}$$

and $\mathbf{d}^0_{c\sigma vj_z}$ is "microscopic" dipole

$$\begin{array}{llccc} & & (\mathbf{d}^0_{c\sigma vj_z})_x & (\mathbf{d}^0_{c\sigma vj_z})_y & (\mathbf{d}^0_{c\sigma vj_z})_z \\ |c\sigma,vj_z\rangle & |\frac{1}{2},-\frac{3}{2}\rangle & -1 & -i & 0 \\ |c\sigma,vj_z\rangle & |\frac{1}{2},+\frac{3}{2}\rangle & 0 & 0 & 0 \\ |c\sigma,vj_z\rangle & |-\frac{1}{2},-\frac{3}{2}\rangle & 0 & 0 & 0 \\ |c\sigma,vj_z\rangle & |-\frac{1}{2},+\frac{3}{2}\rangle & 1 & -i & 0. \end{array} \tag{8}$$

The numerical parameters E_{SR} and μ^2 can be determined as to reproduce the excitonic fine structure in bulk semiconductors and, therefore, implicitly contain the screening effects.

Taking into account (7) and (8), a "block" Hamiltonian corresponding to the exchange interaction between two electron-hole pairs $F_c^p F_v^r$ and $F_c^s F_v^q$ can be presented in the form

$$\begin{pmatrix} & & |c\sigma',vj'_z\rangle & |c\sigma',vj'_z\rangle & |c\sigma',vj'_z\rangle & |c\sigma',vj'_z\rangle \\ & & |\frac{1}{2},-\frac{3}{2}\rangle & |\frac{1}{2},+\frac{3}{2}\rangle & |-\frac{1}{2},-\frac{3}{2}\rangle & |-\frac{1}{2},+\frac{3}{2}\rangle \\ |c\sigma,vj_z\rangle & |\frac{1}{2},-\frac{3}{2}\rangle & \delta_0^{SRE,L}+\delta_0^{LRE,L}+\delta_0^{LRE,N} & 0 & 0 & \delta_{12}^{LRE,N} \\ |c\sigma,vj_z\rangle & |\frac{1}{2},+\frac{3}{2}\rangle & 0 & 0 & 0 & 0 \\ |c\sigma,vj_z\rangle & |-\frac{1}{2},-\frac{3}{2}\rangle & 0 & 0 & 0 & 0 \\ |c\sigma,vj_z\rangle & |-\frac{1}{2},+\frac{3}{2}\rangle & \delta_{21}^{LRE,N} & 0 & 0 & \delta_0^{SRE,L}+\delta_0^{LRE,L}+\delta_0^{LRE,N} \end{pmatrix}, \tag{9}$$

where we separated different contributions to the exchange based on their origin (SRE or LRE) and integral type ("local" and "nonlocal"). The contributions are

$$\begin{aligned} \delta_0^{SRE,L} &= E_{SR}S, \quad \delta_0^{LRE,L} = -\frac{8\pi\mu^2}{3}S, \quad S = \int d\mathbf{r}F_c^{s\dagger}(\mathbf{r})F_v^q(\mathbf{r})F_c^p(\mathbf{r})F_v^{r\dagger}(\mathbf{r}), \\ \delta_0^{LRE,N} &= -\mu^2\left(R_{xx}+R_{yy}\right), \quad \delta_{12}^{LRE,N} = \mu^2\left(R_{xx}+2iR_{xy}-R_{yy}\right), \quad \delta_{21}^{LRE,N} = \mu^2\left(R_{xx}-2iR_{xy}-R_{yy}\right), \\ R_{\delta\gamma} &= \int\int\left\{\left(\frac{\partial^2}{\partial r_1^\delta \partial r_1^\gamma}\right)F_c^{s\dagger}F_v^q\right\}\frac{F_c^p(\mathbf{r}_2)F_v^{r\dagger}(\mathbf{r}_2)}{|\mathbf{r}_1-\mathbf{r}_2|}d\mathbf{r}_1 d\mathbf{r}_2. \end{aligned} \tag{10}$$

With regard to the above exchange expressions we note the following. (a) The short-range exchange causes splitting between the "bright" ($|\frac{1}{2},-\frac{3}{2}\rangle, |-\frac{1}{2},+\frac{3}{2}\rangle$) and "dark" doublets ($|\frac{1}{2},+\frac{3}{2}\rangle, |-\frac{1}{2},-\frac{3}{2}\rangle$) by moving "bright" doublet up in energy by $\delta_0^{SRE,L}$. SRE does not split the "bright" doublet. The integral describing SRE is "local" in nature and involves overlap between two electron-hole pairs. (b) Long-range exchange contains expressions of two types: a "local" term $\delta_0^{LRE,L}$ which arises from $\mu^2\left((\mathbf{d}^0_{c\tau'v\sigma'})^\dagger\cdot\mathbf{d}^0_{c\tau v\sigma}\right)$ and "nonlocal" terms $\delta_0^{LRE,N}$, $\delta_{12}^{LRE,N}$, and $\delta_{21}^{LRE,N}$ which involve differentiation operators applied to the electron-hole pair envelopes. The "nonlocal" terms describe dipole-dipole interaction between the electron-hole transition densities. (c) The splitting between "bright" and "dark" doublets is affected by LRE through

Figure 1. Exciton fine structure as a function of anisotropy of the confining potential.

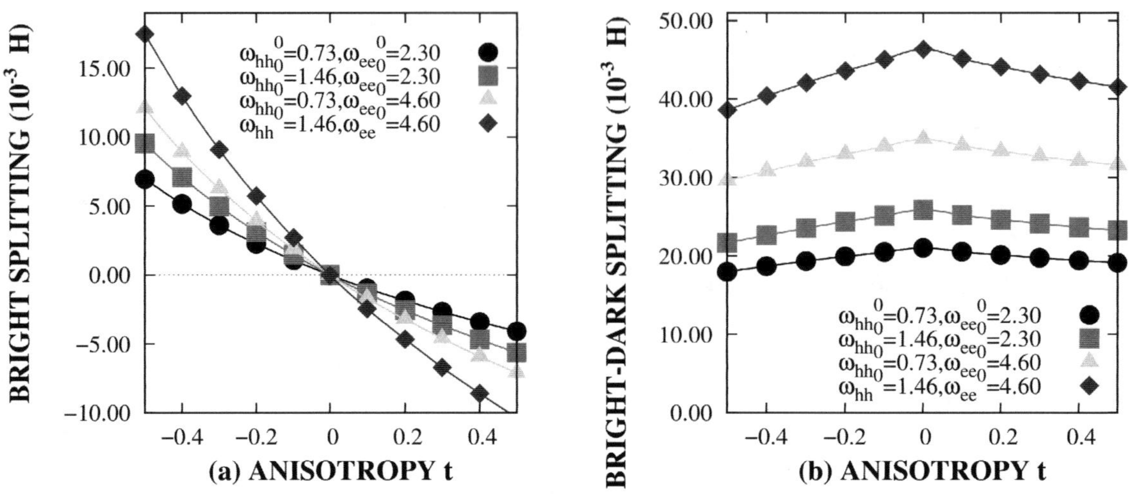

the "local" LRE term $\delta_0^{LRE,L}$ and "nonlocal" LRE term $\delta_0^{LRE,N}$. (d) The LRE, in general, splits the "bright" doublet by coupling the electron-hole pairs with "completely opposite" z-projections of the angular momentum, for example, $|-\frac{1}{2},\frac{3}{2}\rangle$ and $|\frac{1}{2},-\frac{3}{2}\rangle$. The terms which are responsible for the "bright" exciton splitting are "nonlocal" expressions $\delta_{12}^{LRE,N}$ and $\delta_{21}^{LRE,N}$. (e) The splitting within the "dark" doublet is not described in our model. This stems from the fact that the microscopic parts were chosen to be "pure" spinors and the two-by-two spin-density matrix corresponding to "dark" states ($|-\frac{1}{2},-\frac{3}{2}\rangle$ and $|\frac{1}{2},\frac{3}{2}\rangle$) vanishes identically.

The splitting of "bright" exciton levels vanishes provided that $\delta_{12}^{LRE,N}$ and $\delta_{21}^{LRE,N}$ vanish which constitutes a condition for LRE quenching within our model. In the next section, we demonstrate numerically and analytically that this condition is fulfilled in the case of isotropic 2D HO confining potential.

3. Exciton fine structure of quasi-two-dimensional anisotropic harmonic oscillator (2DLAHO) quantum dot

In this section, we will investigate excitonic fine-structure for a model quantum dot in which the confining potential is of 2D-like anisotropic harmonic oscillator type. We note that HO oscillator spectrum has been observed in self-assembled quantum dots [24].

Within 2DLAHO model, the Hamiltonian for holes (electrons) is

$$\hat{H}_{h(e)} = -\frac{1}{2M_{||,h(e)}}\left(\nabla_x^2+\nabla_y^2\right)+\frac{1}{2}M_{||,h(e)}\left(\omega_{x,h(e)}^2x^2+\omega_{y,h(e)}^2y^2\right)-\frac{1}{2M_{\perp,h(e)}}\nabla_z^2+\frac{1}{2}M_{\perp,h(e)}\omega_{z,h(e)}^2z^2,$$
$$\omega_{x,h(e)}=\omega_{h(e)}^0(1+t),\quad \omega_{y,h(e)}=\frac{\omega_{h(e)}^0}{1+t},\quad \omega_{x,h(e)}\omega_{y,h(e)}=(\omega_{h(e)}^0)^2=const,\quad \omega_{z,h(e)}\gg\omega_{h(e)}^0, \tag{11}$$

where $M_{||,h(e)}$ and $M_{\perp,h(e)}$ are effective masses, $\omega_\gamma(\gamma=x,y,z)$ denotes confinement frequency in x, y, and z directions, respectively, $\omega_{h(e)}^0$ determines the typical energy scale of the confining potential. The energies and lengths are expressed in the effective units.

Figure 2.

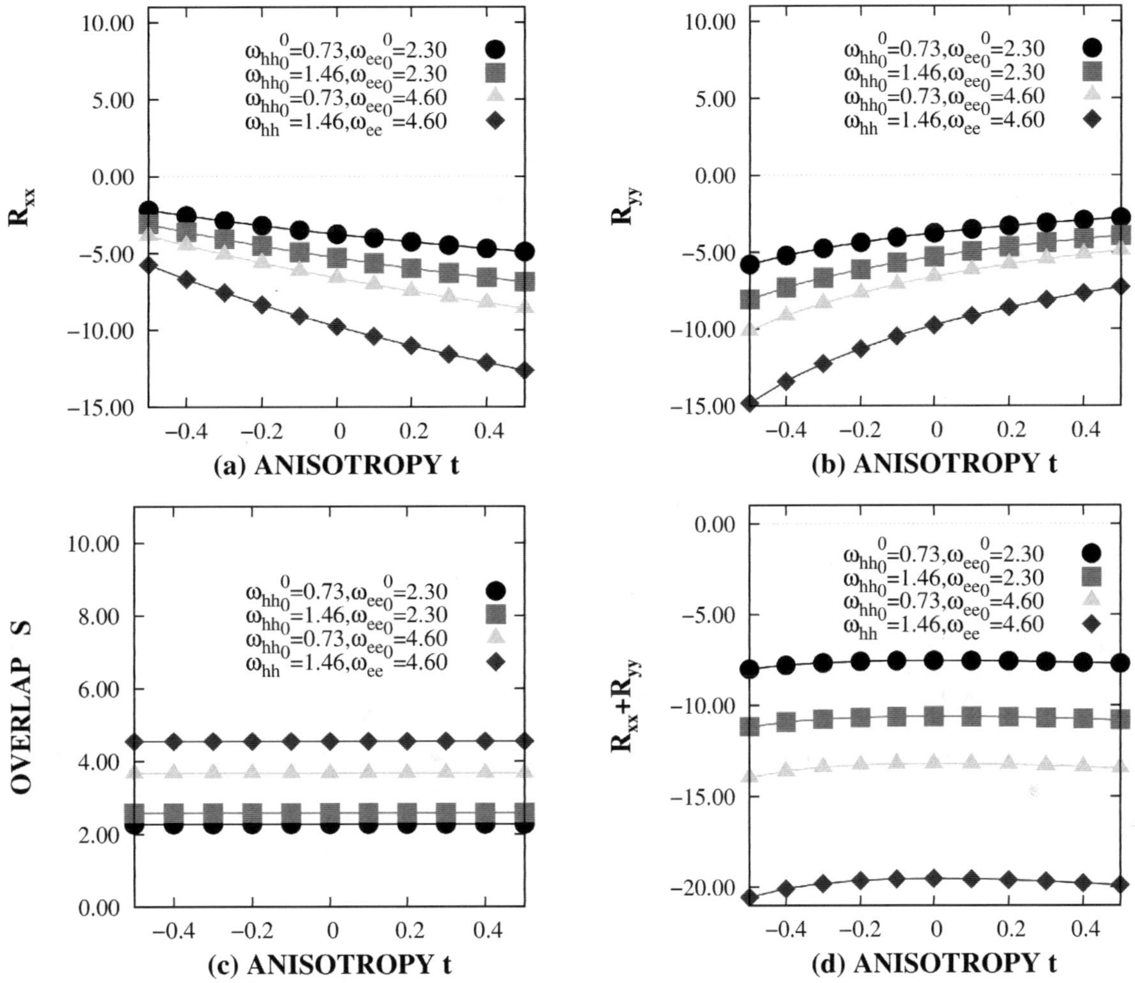

The excitonic fine structure is studied as a function of anisotropy t. For $t < 0$, $\omega_x < \omega^0$ and $\omega_y > \omega^0$ and the confinement along x is weaker than along y. For $t > 0$, $\omega_x > \omega^0$ and $\omega_y < \omega^0$. In the case $t = 0$, the confining potential for holes and electrons is rotationally symmetric.

In Ref. [25], an explicit expression for matrix elements $\delta_{12}^{LRE,N}$ and $\delta_{21}^{LRE,N}$ was derived in the case of $s_e s_h$ envelop and it was demonstrated that $\delta_{12}^{LRE,N}$ and $\delta_{21}^{LRE,N}$ vanish provided that the "confinement" of *electron-hole pair envelope* is identical in the x and y directions ($\alpha_x = \alpha_y$). Figure 1a shows "bright" exciton doublet splitting as a function of lateral anisotropy t as computed from the effective theory for four different combinations of the confinement frequencies. The splitting energy is defined as $E_X^y - E_X^x$, where E_X^x and E_X^y are ground state "bright" exciton energy levels polarized along the x and y directions, respectively. The "bright" exciton splitting decreases as the magnitude of the anisotropy $|t|$ decreases and vanishes at $|t| = 0$. The passing through $t = 0$ is accompanied by the change of polarization of the "bright" ground state. For $t < 0$ $(\omega_x < \omega_y)$, $E_X^y > E_X^x$ and the "bright" ground state is dipole active along the x-axis, whereas for $t > 0$ $(\omega_x < \omega_y)$ the "bright" ground state is dipole active along the y-axis. In our model, the "bright" ground state is always dipole-active along the axis of weaker confinement. Figure 1b shows the splitting between the "bright" ground state and the "dark" doublet.

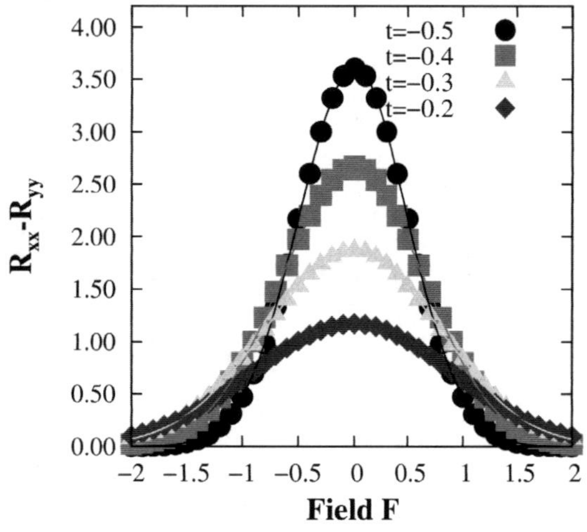

Figure 3. "Nonlocal" LRE matrix element $R_{xx} - R_{yy}$ which determines the "bright" exciton splitting as a function of lateral electric field F.

The Figure 2 shows R_{xx} (a) and R_{yy} (b) "nonlocal" contributions to the LRE exchange as a function of anisotropy for several confinement frequencies. We can see that R_{xx} increases and R_{yy} decreases in magnitude as t goes from -0.5 to 0.5. Due to this trend, the difference $R_{xx} - R_{yy}$ which determines the "bright" exciton splitting decreases as a function of $|t|$. At $t = 0$, $R_{yy} = R_{xx}$ and the exchange matrix element which couple two "bright" exciton levels vanishes. The "polarization" of the "bright" ground state is determined by the sign of $R_{xx} - R_{yy}$.

Figure 2c shows the overlap S between $s_e s_h$ pairs which, within the envelop approximation, determines the SRE ($\delta_0^{SRE,L}$) as well as the "contact" contribution from the LRE ($\delta_0^{LRE,L}$). The electron-hole pair overlap S is insensitive to the anisotropy t due to the imposed "area conservation" condition $\omega_{x,h(e)}\omega_{y,h(e)} = (\omega^0_{h(e)})^2 = const$. Figure 2d shows $R_{xx} + R_{yy}$ which determines the "nonlocal" LRE contribution to the "bright"-"dark" splitting. We find that $\delta_0^{SRE,L} > 0$ and its effect is to move the "bright" doublet up in energy. The "local" ($\delta_0^{LRE,L}$) and "nonlocal" LRE ($\delta_0^{LRE,N}$) act against each other.

4. Application of external fields

In the previous section, we have demonstrated that the magnitude of the "bright" exciton splitting can be controlled through the shape of the confining potential. It is, therefore, of great interest to examine the effects of external fields on the excitonic fine structure. For example, QDs can be placed between Schottky gates for the application of vertical and lateral electric fields. Recently, it has been demonstrated [9] that by applying an in-plane electric field it is possible to fine-tune photon cascades originating from the recombination of multiexciton complexes in QDs.

Within our model, the application of lateral electric field $\vec{F} = F\mathbf{e}_x + F\mathbf{e}_y$ of magnitude $|F|$ displaces the "origin" of electron and hole envelopes in xy-plane from $(0, 0)$ to (x_0^e, y_0^e) and (x_0^h, y_0^h), respectively. The single-particle energy levels are rigidly shifted by the Stark shift whereas the spacing between the single-particle energy levels of HO is not affected.

Evaluating the integrals which control "bright" exciton splitting $R_{xx}(F), R_{yy}(F), R_{xy}(F)$ as a function of field F for $s_e s_h$ electron-hole envelope, we obtain

$$R_{xx}(F) = R_{xx}(0) \exp\left(-\frac{2\alpha_{xc}\alpha_{xv}}{\alpha_x}\Delta x_0^2 - \frac{2\alpha_{yc}\alpha_{yv}}{\alpha_y}\Delta y_0^2\right),$$

$$R_{yy}(F) = R_{yy}(0) \exp\left(-\frac{2\alpha_{xc}\alpha_{xv}}{\alpha_x}\Delta x_0^2 - \frac{2\alpha_{yc}\alpha_{yv}}{\alpha_y}\Delta y_0^2\right), \quad R_{xy}(F) = 0, \tag{12}$$

where $R_{xx}(0)$ and $R_{yy}(0)$ are matrix elements in the absence of the electric field ($F = 0$). Field dependence of $R_{xx}(F) - R_{yy}(F)$ which determines the "bright" exciton splitting is shown on Figure 3 for four different initial values of the lateral anisotropy.

At $F = 0$, the "bright" exciton splitting attains maximum value which is determined by the initial shape anisotropy of the confining potential. The larger the magnitude of the anisotropy, the larger is the initial ($F = 0$) "bright" exciton splitting. As F increases in magnitude, the electron and hole envelopes are pulled out in the opposite direction and the magnitude of the splitting is reduced. Since the separation between the envelopes depends on F^2, the splitting is independent of the sign of the field F.

Acknowledgments

The authors acknowledge stimulating discussions with Drs. M. Zielinski, M. Korkusinski, R. Williams, and D. Dalacu. E. K. thanks the Government of Canada for establishing the NRC-NSERC-BDC Nanotechnology Partnership.

[1] Shields A J 2007 *Nature Photonics* **1** 215
[2] Stevenson R M, Young R J, Atkinson P, Cooper K, Ritchie D A and Shields A J 2006 *Nature* **439** 179
[3] Akopian N, Lindner N H, Poem E, Berlatzky Y, Avron J, Gershoni D, Gerardot B D and Petroff P M 2006 *Phys. Rev. Lett.* **96** 130501
[4] Greilich A, Schwab M, Berstermann T, Auer T, Oulton R, Yakovlev D R, Bayer M, Stavarache V, Reuter D and Wieck A 2006 *Phys. Rev. B* **73** 045323
[5] Benson O, Santori C, Pelton M and Yamamoto Y 2000 *Phys. Rev. Lett.* **84** 2513
[6] Bayer M, Ortner G, Stern O, Kuther A, Gorbunov A A, Forchel A, Hawrylak P, Fafard S, Hinzer K, Reinecke T L, Walck S N, Reithmaier J P, Klopf F and Schafer F 2002 *Phys. Rev. B* **65** 195315
[7] Kowalik K, Krebs O, Lemaitre A, Laurent S, Senellart P, Voisin P and Gaj J A 2005 *Appl. Phys. Lett.* **86** 041907
[8] Vogel M, Ulrich S M, Hafenbrak R, Michler P, Wang L, Rastelli A and Schmidt O G 2007 *Appl. Phys. Lett.* **91** 051904
[9] Reimer M E, Korkusinski M, Dalacu D, Lefebvre J, Lapointe J, Poole P J, Aers G C, McKinnon W R, Hawrylak P and Williams R L 2008 *Phys. Rev. B* **78** 195301
[10] Knox R S 1963 *Theory of excitons* (*Solid State Phys. Suppl.* vol 5) (Berlin: Academic Press)
[11] Denisov M M and Makarov V P 1973 *Phys. Stat. Sol. (b)* **56** 9
[12] Bir G L and Pikus G E 1975 *Symmetry and Strain-Induced Effects in Semiconductors* (New York: Wiley)
[13] Takagahara T 1993 *Phys. Rev. B* **47** 4569
[14] Maialle M Z, de Andrada e Silva E A and Sham L J 1993 *Phys. Rev. B* **47** 15776
[15] Efros A L, Rosen M, Kuno M, Nirmal M, Norris D J and Bawendi M 1996 *Phys. Rev. B* **54** 4843
[16] Gupalov S V, Ivchenko E L and Kavokin A V 1998 *J. of Exp. and Theor. Phys.* **86** 388
[17] Takagahara T 2000 *Phys. Rev. B* **62** 16840
[18] Gupalov S V and Ivchenko E L 2000 *Phys. Solid State* **42** 1976
[19] Glazov M M, Ivchenko E L, Besombes L, Leger Y, Maingault L and Mariette H 2007 *Phys. Rev. B* **75** 205313
[20] Franceschetti A, Wang L W, Fu H and Zunger A 1998 *Phys. Rev. B* **58** R13367
[21] Goupalov S V and Ivchenko E L 2001 *Phys. Solid State* **43** 1867
[22] He L, Gong M, Li C F, Guo G C and Zunger A 2008 *Phys. Rev. Lett.* **101** 157405
[23] Dalessi S and Dupertuis M A 2010 *Phys. Rev. B* **81** 125106
[24] Raymond S, Studenikin S, Sachrajda A, Wasilewski Z, Cheng S J, Sheng W, Hawrylak P, Babinski A, Potemski M, Ortner G and Bayer M 2004 *Phys. Rev. Lett.* **92**
[25] Kadantsev E S and Hawrylak P 2010 *Phys. Rev. B* **81** 045311

Journal of Physics: Conference Series **248** (2010) 012019 doi:10.1088/1742-6596/248/1/012019

Local Franck-Condon factors in suspended carbon nanotube quantum dots

Fabio Cavaliere[1], Eros Mariani[2], Renaud Leturcq[3,4], Christoph Stampfer[3,5] and Maura Sassetti[1]

[1] Dipartimento di Fisica, Università di Genova and CNR-SPIN, Via Dodecaneso 33, 16146 Genova, Italy

[2] Centre for Graphene Science, School of Physics, University of Exeter, Stocker Road, Exeter, EX4 4QL, United Kingdom

[3] Laboratory for Solid State Physics, ETH Zurich, 8093 Zurich, Switzerland

[4] IEMN CNRS-UMR 8520, ISEN, Avenue Poincaré, BP 60069, 59652 Villeneuve d'Ascq Cedex, France

[5] JARA-FIT and II. Institute of Physics, RWTH Aachen University, 52074 Aachen, Germany

E-mail: cavalier@fisica.unige.it

Abstract. We investigate the effects of the coupling between quantized vibrational modes (vibrons) and electronic degrees of freedom in suspended carbon nanotube quantum dots. The elastic deformations couple both to the total dot charge and to its spatial density fluctuations. The latter, neglected in the Anderson-Holstein model, have profound consequences if the size of the vibron and of the dot do not coincide, as is generically the case in experimental devices. When the vibron is smaller than the quantum dot, spatially varying local Franck-Condon factors emerge with striking effects on the transport properties of the system. The theoretical results are supported by our experimental observations.

1. Introduction

The intriguing transport properties of single-wall carbon nanotubes (CNTs) attracted considerable interest in recent years [1]. Impurities or geometrical defects along a CNT may behave as tunnel barriers leading to the formation of a CNT quantum dot connected to the external CNT parts acting as source and drain contacts [2, 3, 4]. Quantum dots formed in CNTs display a Coulomb diamond pattern showing signatures of the fourfold degeneracy of electronic states in graphene (out of which the CNT are made) [5]. On top of their peculiar electronic structures, suspended CNTs possess a rich variety of vibrational degrees of freedom, ranging from stretching and breathing modes to soft transverse bending ones [6]. Electrons tunnelling in suspended CNT quantum dots have been shown to excite stretching vibrons, leading to the appearance of *vibrational sidebands* in the nonlinear conductance spectra [7, 8, 9, 10].

Most commonly, the transport properties of suspended CNT dots are described in terms of the Anderson-Holstein (AH) model [11, 12], in which the vibrational motion couples to the *total* excess charge of the dot. The AH model leads to the *Franck-Condon* (FC) *blockade* [13] recently observed in experiments [9], with *position-independent* FC factors [13, 14, 15, 16].

In general, however, vibrons couple both to the total charge and to the spatial modulations of the charge density in the dot [17, 18, 19]. The scenario obtained considering the latter can differ

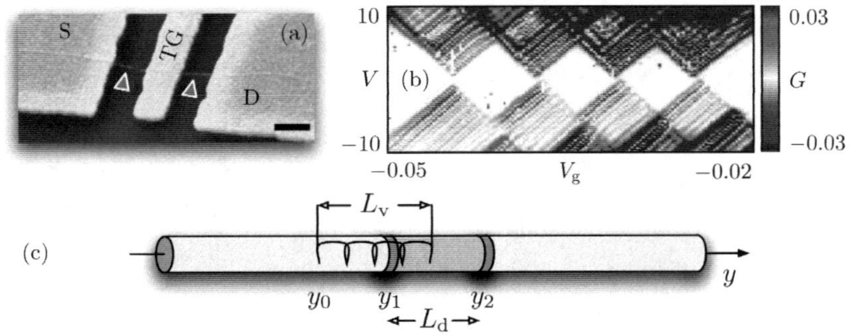

Figure 1. (a) SEM image of the suspended CNT device (scale bar: 200 nm): red marks denote its position relative to source (S) and drain (D) contacts and the suspended top gate (TG). (b) Colour plot of the differential conductance G (units e^2/h) as a function of top gate V_g (units V) and bias V (units mV) voltages. (c) Scheme of the CNT dot (orange) connected to CNT contacts (yellow) via tunneling barriers (red) at $y_{1,2}$. The vibron is shown as a spring (see text).

significantly from that of the conventional AH model. Indeed, if the vibron length L_v is smaller than the dot size L_d, *position-dependent* FC factors arise [19], with dramatic consequences on the transport properties. Indeed, if the vibron locates asymmetrically along the dot *only* the electrons tunneling across the barrier *next to the vibron* may excite the vibrational motion. As a result, a *selective suppression* of the family of conductance traces corresponding to tunneling through the barrier farther from the vibron occurs. Only when $L_v \gg L_d$ the effects of the density fluctuations are less relevant and one recovers the standard AH model. In this paper, the transport properties of a suspended CNT dot will be explored, with a particular emphasis on the regime $L_v < L_d$. The theory will be compared with the results of our transport experiment performed on a suspended CNT, which also exhibits the suppression of a family of conductance traces. The CNT is suspended between source and drain contacts, to which a bias voltage V is applied, see Fig. 1a. Here, a quantum dot is formed below an electrically isolated suspended top gate [9], and its excess number of electrons may be tuned by means of the top-gate voltage V_g. Figure 1b shows a color map of the differential conductance G as a function of V_g and V. Besides the almost perfect fourfold symmetry of the Coulomb blockade diamonds, one observes a rich spectrum of excited states related to the excitation of stretching vibrons [9, 6], with a level spacing of about 0.8 meV corresponding to a vibron size $L_v \approx 60$ nm [9]. The dot length, estimated from the size of Coulomb diamonds, is $L_d \approx 240$ nm $> L_v$. Two main features are present: (*i*) a striking suppression of conductance traces with negative slope, and (*ii*) a regular alternating pattern of positive (PDC, red) and negative differential conductance (NDC, blue). By taking into account charge density fluctuations in the regime $L_v < L_d$, we show how the point (*i*) can be addressed as a result of asymmetric FC factors stemming from vibronic excitations located near the barrier on the source side [1]. We point out that the usual AH model with symmetric FC factors *is not* able to reproduce such a suppression of conductance traces. In parallel, we show that asymmetries between the tunneling rates on different barriers and orbital states in the dot yield a *dynamical trapping* of electronic states which can finally be responsible for the repeated PDC/NDC pattern of point (*ii*) [20, 21, 22].

[1] According to the bias scheme of the device [9], conductance traces with negative slope correspond to tunneling events through the drain barrier.

2. Position-dependent electron-vibron coupling

We illustrate the mechanism giving rise to *position-dependent* FC factors studying the simplest model for the dot, namely a single spinless electronic state, coupled to the lowest vibrational stretching mode. The extension of this model to the case of a CNT-based quantum dot will be presented in Sec. 3. Tunnel barriers at $y_1 = 0$ and $y_2 = L_{\rm d}$ define the dot region while the vibron occupies the CNT region $y_0 \leq y \leq y_0 + L_{\rm v}$, see Fig. 1c. The dot is modeled as a Luttinger liquid (LL) with open boundary conditions (OBC) [23], with the electronic field operator $\psi(y)$ satisfying $\psi(y_1) = \psi(y_2) = 0$. The field operator is decomposed as [23] $\psi(y) \approx e^{ik_{\rm F}y}\psi_{\rm R}(y) + e^{-ik_{\rm F}y}\psi_{\rm L}(y)$ with $k_{\rm F}$ is the Fermi momentum and $\psi_{\rm R/L}(y)$ the field operators for right (R) and left (L) movers. Due to OBC, $\psi_{\rm L}(y) = -\psi_{\rm R}(-y)$, allowing a description in terms of right movers, with

$$\psi_{\rm R}(y) = \frac{1}{\sqrt{2\pi\xi}} e^{i\vartheta} e^{i\pi y N/L_{\rm d}} e^{i\phi(y)} \tag{1}$$

the bosonized form of their field operator. Here, N is the dot excess charge zero mode, ξ a short length cutoff and $[\theta, N] = i$. Furthermore,

$$\phi(y) = \sum_{q>0} \frac{1}{\sqrt{q}} e^{-\xi\pi q/2L_{\rm d}} \left[e^{-i\pi q y/L_{\rm d}} b_q^\dagger + \text{h.c.}\right] \tag{2}$$

is the plasmonic field with $b_q^\dagger$ creating the q–th plasmon mode with frequency $\omega_q = \pi v_{\rm c} q/L_{\rm d}$ and group velocity v_c. In the following, we will assume $v_c \simeq v_{\rm F}$ due to gate-induced screening [24]. The dot and vibron Hamiltonians ($H_{\rm d}$, $H_{\rm v}$) read ($\hbar = 1$)

$$H_{\rm d} = E_{\rm c} N^2 + \sum_q \omega_q b_q^\dagger b_q = H_{\rm d}^{\rm (N)} + H_{\rm d}^{\rm (pl)} \quad ; \quad H_{\rm v} = \frac{p_0^2}{2M} + \frac{M\omega_0^2}{2} x_0^2 , \tag{3}$$

with $E_{\rm c}$ the charging energy, M the vibron mass, $\omega_0 = \pi v_{\rm s}/L_{\rm v}$ its frequency and $v_{\rm s}$ the stretching mode velocity [24]. In a CNT, $v_{\rm s} < v_{\rm c}$ which in experiments implies $\omega_0 < \omega_1$ [9, 8]. The vibron coordinate x_0 represents the amplitude of the distortion field $u(y) = \sqrt{2} x_0 \sin\left[\pi(y - y_0)/L_{\rm v}\right]$, while p_0 is the conjugate momentum. The electron-vibron coupling is described by

$$H_{\rm d-v} = c \int_{\max[0,y_0]}^{\min[L_{\rm d},y_0+L_{\rm v}]} {\rm d}y\, \rho(y) \partial_y u(y) \tag{4}$$

with c the deformation potential coupling constant [24, 25, 26] and $\rho(y)$ the electron density of the dot. In the spirit of LL we approximate $\rho(y) \approx \rho_{\rm R}(y) + \rho_{\rm R}(-y)$ where $\rho_{\rm R}(y) = (N/2L_{\rm d}) + (1/2\pi)\partial_y\phi(y) = \rho_0/2 + \delta\rho(y)$ is the electron density of right movers [23] with $\rho_0 = N/L_{\rm d}$ the average density and $\delta\rho(y) = (1/2\pi)\partial_y\phi(y)$ the density fluctuations. This approximation neglects terms $e^{2ik_{\rm F}y}\psi_{\rm R}^\dagger(y)\psi_{\rm L}(y) + \text{h.c.}$ with spatial oscillations of short-wavelength $k_{\rm F}^{-1}$ and is valid if $L_{\rm v} \gg k_{\rm F}^{-1}$, which implies $N \gg L_{\rm d}/\pi L_{\rm v}$, a condition satisfied in our experiment.

Introducing generalized plasmonic coordinates $x_q = \left(b_q + b_q^\dagger\right)/\sqrt{2\omega_q}$ the coupling in Eq. (4) can be decomposed into a contribution $H_{\rm d-v}^{\rm (N)} = \sqrt{M} c_0 x_0 N$ due to the zero mode and $H_{\rm d-v}^{\rm (pl)} = x_0\sqrt{M}\sum_{q=1}^{\infty} c_q x_q$ due to the plasmonic density fluctuations. Here we have introduced

$$c_0 = \frac{\omega_0^{\frac{3}{2}}\delta}{\pi}\lambda_{\rm m} \sin\left[\pi\left(1 + \frac{y_0}{L_{\rm d}}\right)\right]\theta\left(\frac{y_0}{L_{\rm d}}\right)\theta\left(1 + \frac{y_0}{L_{\rm d}}\right) \tag{5}$$

$$c_q = \frac{2\omega_0^{\frac{3}{2}}\sqrt{\omega_1}\lambda_{\rm m} q^2\delta^2}{\pi(1 - q^2\delta^2)}\left\{\sin\left[\pi\frac{y_0}{L_{\rm d}}\right] + \sin\left[\pi q\delta\left(1 + \frac{y_0}{L_{\rm d}}\right)\right]\right\} \tag{6}$$

where $\delta = L_{\rm v}/L_{\rm d}$ and $\lambda_{\rm m} = c/(v_{\rm s}\sqrt{M\omega_0})$. Note that, for simplicity, here we have quoted only the expressions for $\delta \leq 1$ and $y_0 + L_{\rm v} \leq L_{\rm d}$. The total Hamiltonian $H_{\rm dot-vib} = H_{\rm d} + H_{\rm v} + H_{\rm d-v}$ is quadratic in the vibron and plasmon generalized coordinates: introducing the two vectors $X_\mu = \left(\sqrt{M}x_0, x_1, x_2, \ldots\right)$ and $P_\mu = \left(p_0/\sqrt{M}, p_1, p_2, \ldots\right)$ it has the form $\sum_{\mu=0}^{\infty}\left[P_\mu^2 + X_\mu^2\right]/2 + X_0\sum_{\mu=1}^{\infty} c_\mu X_\mu + c_0 X_0 N$ and is thus easily diagonalized [27, 16]. We first proceed diagonalizing $H_{\rm d-v}^{(\rm pl)}$ and introduce the new normal modes $\bar{X}_\mu = (\bar{x}_0, \bar{x}_1, \bar{x}_2, \ldots)$, $\bar{P}_\mu = (\bar{p}_0, \bar{p}_1, \bar{p}_2, \ldots)$ with $X_\mu = \sum_{\nu=0}^{\infty} k_{\mu\nu}\bar{X}_\nu$ and $P_\mu = \sum_{\nu=0}^{\infty} k_{\mu\nu}\bar{P}_\nu$. The diagonalizing matrix elements $k_{\mu\nu} = c_\mu\left(\Omega_\nu^2 - \omega_\mu^2\right)^{-1} k_{0\nu}$ for $\mu \geq 1$ depend on $k_{0\nu} = \left[1 + \sum_{q=1}^{\infty} c_q^2/\left(\Omega_\nu^2 - \omega_q^2\right)^2\right]^{-1/2}$, with the new eigenfrequencies Ω_ν given by the self-consistent equation $\Omega_\nu^2 = \omega_0^2 + \sum_{q=1}^{\infty} c_q^2/\left(\Omega_\nu^2 - \omega_q^2\right)$. The new modes with $\nu \geq 1$ represent dressed plasmons with energy $\Omega_\nu > \omega_\nu$, while the term $\nu = 0$ with $\Omega_0 < \omega_0$ represents the vibron excitation dressed by a plasmonic cloud. In order to eliminate the remaining term $H_{\rm d-v}^{(\rm N)} = c_0 N \sum_{\nu\geq 0} k_{0\nu}\bar{x}_\nu$ we employ a Lang-Firsov transformation [23], generated by $\mathcal{U} = \exp\left[-iN\sum_{\nu\geq 1}\eta_\nu\bar{p}_\nu\right]$ with $\eta_\nu = c_0 k_{0\nu}/\Omega_\nu^2$. Finally, we obtain $H_{\rm dot-vib} = E_{\rm c}^* N^2 + \sum_{\nu=0}^{\infty}\Omega_\nu\left(a_\nu^\dagger a_\nu + 1/2\right)$, where $a_\nu^\dagger$ creates the ν–th eigenmode and $E_{\rm c}^* = E_{\rm c} - \sum_{\nu\geq 0}\left(c_0 k_{0\nu}/\sqrt{2}\Omega_\nu\right)^2$ contains a polaronic shift. Dot-vibron states are labeled by $|N, l\rangle$, with l the occupation number of the dressed-vibron states.

The above transformations affect the electronic field as well. Crucially, due to the mixing of vibrons and plasmons induced by $H_{\rm d-v}$, vibronic operators appear in the electronic field. Since we are interested into electronic tunneling at energies *smaller* than the plasmon excitations, we restrict to the Hilbert space of the dressed-vibron excitations with $\nu = 0$. The truncated field operator after the above transformations reads

$$\psi_{\rm R}(y) = \frac{e^{i\vartheta}}{\sqrt{2\pi\xi}} e^{i\pi y N/L_{\rm d}} e^{-[\lambda_N + \lambda_-(y)]\left(a_0^\dagger - a_0\right)} e^{i\lambda_+(y)\left(a_0^\dagger + a_0\right)} . \tag{7}$$

$$\lambda_N = c_0/\sqrt{2\Omega_0^3} \quad ; \quad \lambda_\pm(y) = \sqrt{k_{00}\frac{\omega_1}{\Omega_0}}\sum_{q=1}^{\infty}\frac{c_q \sin\left(\frac{\pi q y}{L_{\rm d}} + \frac{\pi}{4} \pm \frac{\pi}{4}\right)}{\Omega_0^2 - \omega_q^2} . \tag{8}$$

The transformed field operator Eq. (7) removes an electron from the system, with the simultaneous excitation of the vibron mode.

In the sequential regime, tunneling rates can be evaluated via the golden rule. For the localized injection of an electron at y along the nanotube, with a scattering $l \to l'$ of the vibron state, one obtains $\Gamma(y) \propto X_{l,l'}(y)$, with $X_{l,l'}(y) = 2\pi\xi\left|\langle N+1, l'|\psi_{\rm R}^\dagger(y)|N, l\rangle\right|^2$. From Eq. (7),

$$X_{ll'}(y) = e^{-\lambda^2(y)}\left[\lambda^2(y)\right]^{l_> - l_<}\frac{l_<!}{l_>!}\left\{L_{l_<}^{l_> - l_<}[\lambda^2(y)]\right\}^2 , \tag{9}$$

where $\lambda^2(y) = [\lambda_N + \lambda_-(y)]^2 + \lambda_+^2(y)$, $l_< = \min(l, l')$, $l_> = \max(l, l')$ and $L_a^b(x)$ are the generalized Laguerre polynomials.

This constitutes our main result: Eq. (9) represents a *position-dependent* Franck-Condon factor, due to the position-dependent effective electron-vibron coupling $\lambda(y)$, in sharp contrast with the predictions of the AH model. The latter fails in cases when $\max\{\lambda_\pm(y)\} \gg \lambda_N$ so that the position dependence of the coupling can no longer be neglected. Two regimes can be identified according to the vibron being smaller than the dot ($L_{\rm v} < L_{\rm d}$) or larger ($L_{\rm v} > L_{\rm d}$).

For a *small vibron* ($L_{\rm v} < L_{\rm d}$), the spatial modulations of the electron density $\delta\rho(y)$ can never be neglected. In particular, if the vibron is located *within* the dot ($0 \leq y_0 \leq L_{\rm d} - L_{\rm v}$) we

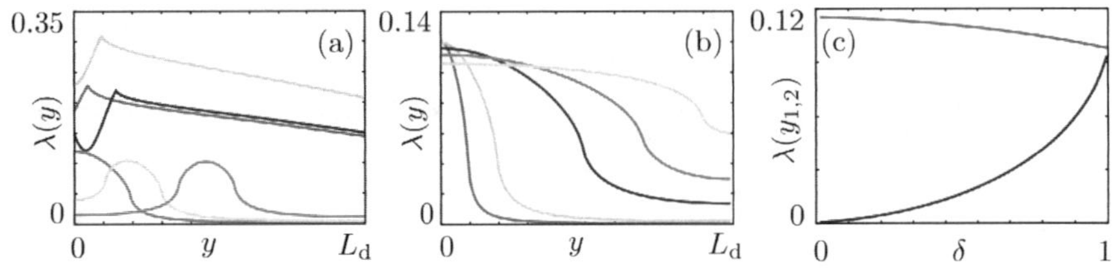

Figure 2. (a) Effective coupling $\lambda(y)$ for $\delta = 0.2$ and different positions of the center of the vibron $y_c = y_0 + L_v/2$: (red) $y_c = -L_v/4$; (green) $y_c = 0$; (blue) $y_c = L_v/4$; (purple) $y_c = L_v/2$; (cyan) $y_c = L_d/4$; (orange) $y_c = L_d/2$. (b) Plot of $\lambda(y)$ for $y_0 = 0$ and different vibron sizes: (red) $\delta = 0.1$; (green) $\delta = 0.2$; (blue) $\delta = 0.5$; (purple) $\delta = 0.7$; (cyan) $\delta = 0.9$. (c) Effective coupling at the tunneling barriers $\lambda(y_\alpha)$ as a function of δ for $y_0 = 0$: (red) $\alpha = 1$; (blue) $\alpha = 2$. In all panels, $v_c/v_s = 32$ and $\lambda_m = 3$ (assuming a CNT waist $\approx$1 nm) [24].

find $\lambda_N \propto c_0 \to 0$, see Eq. (5), and density fluctuations provide the *only* mechanism through which electrons and vibrons couple via $\lambda_\pm(y)$. When the vibron crosses either tunneling barrier ($y_0 < y_i < y_0 + L_v$ for $i = 1, 2$) on the other hand, both λ_N and $\lambda_\pm(y)$ are of the same order. Fig. 2a shows $\lambda(y)$ for $\delta = 0.2$ and different vibron positions. When the latter sits entirely within the dot, the coupling is sizeable *only* in the proximity of the vibron. With the vibron across the dot ($y_0 < y_1 < y_0 + L_v$), $\lambda(y)$ increases due to the contribution of λ_N but is nonetheless peaked in the vibron region. The above results imply a localized suppression of the tunneling conductance which would allow to detect the vibron location and extension, in contrast with the AH model, which would predict a *constant* suppression in the whole dot region.

Figure 2b shows $\lambda(y)$ for a vibron sitting next to the left tunneling barrier case ($y_0 = y_1 = 0$) and different values of δ: as L_v increases the spatial dependence of $\lambda(y)$ becomes less dramatic. The values of the effective coupling at the two tunneling barriers $\lambda_{1,2} = \lambda(y_{1,2})$ are particularly important in order to describe lateral tunneling in CNT-dots embedded into longer CNTs, as in the experiment discussed here. In Fig. 2c, $\lambda_{1,2}$ are shown as a function of δ for $y_0 = y_1 = 0$: when $\delta \to 0$ the small value of λ_2 implies that the excitation of vibron modes is inhibited for tunneling on the right barrier. As discussed in the next section, this lies at the origin of the strong suppression of a family of conductance traces observed in our experiment.

For $\delta = 1$ and a symmetric vibron $y_0 = (L_d - L_v)/2$ one has $\lambda_1 = \lambda_2$. In this symmetric situation the lateral tunneling may be formally treated within an *effective* AH model with coupling constant $\lambda_{AH} = \lambda_{1,2}$. Notice that the physical mechanism of electron-vibron coupling is however different from the AH model, since here the coupling to the total charge vanishes, $\lambda_N = 0$.

In the case of a *large vibron* ($L_v > L_d$), with $\delta \gg 1$ and the dot embedded into the vibron, the term $\partial_y u(y)$ in Eq. (4) is essentially constant over the integration domain (the dot region). As a result, the contribution due to density fluctuations $\delta\rho(y)$ averages to zero yielding $\lambda_\pm(y) = 0$ and the only residual coupling due to the total charge N. As a result, one obtains $\lambda_N \approx \lambda_m \cos(\pi y_0/L_v)$, a position-*independent* coupling, and the usual AH model is recovered. The maximum value of the effective coupling ≈ 0.24, obtained for $\delta < 1$ depends strongly on the ratio v_s/v_c and on λ_m. In all the calculations above, we have adopted usual parameters for graphene and CNTs. However, several effects may lead to a renormalization of the above quantities. As an example, the coupling to the radial breathing mode has been shown to reduce the charge velocity [17, 28], thus increasing v_s/v_c and allowing to reach $\lambda_1 > 1$ while still $\lambda_2 \ll 1$. Also the value of the deformation potential coupling parameter c has been experimentally found larger than expected [29], which would lead to a higher estimate for λ_m.

3. Transport properties of a vibrating CNT

The above model can be easily extended to the case of a CNT with two valley $\eta = \pm 1$ and two spin channels $\sigma = \pm 1$ with $N_{\eta\sigma}$ excess electrons per channel. It is convenient to introduce total (+) and relative (-) charge (c) and spin (s) modes $N_{c+} = \sum_{\eta,\sigma=\pm1} N_{\eta,\sigma}$, $N_{c-} = \sum_{\eta,\sigma=\pm1} \eta N_{\eta,\sigma}$, $N_{s+} = \sum_{\eta,\sigma=\pm1} \sigma N_{\eta,\sigma}$ and $N_{s-} = \sum_{\eta,\sigma=\pm1} \sigma\eta N_{\eta,\sigma}$. The vibron and the electrons couple only through the c+ mode and therefore the diagonalization of their coupled dynamics proceeds in analogy with the previous section. Neglecting the high energy sector of plasmons, the model Hamiltonian is $\mathcal{H} = \mathcal{H}_0 + \mathcal{H}_{\text{shift}} + \mathcal{H}_t$ where

$$\begin{aligned}
\mathcal{H}_0 &= \frac{E_c}{8}(N_{c+} - N_g)^2 + \frac{\varepsilon}{8}\left(N_{c-}^2 + N_{s+}^2 + N_{s-}^2\right) + \Omega_0 \bar{a}_0^\dagger \bar{a}_0 \,; \\
\mathcal{H}_{\text{shift}} &= \frac{\Delta\varepsilon}{2}(N_{c+} - N_{c-}) \,; \\
\mathcal{H}_t &= \sum_{\alpha=1,2}\sum_{\eta,\sigma} \tau_{\alpha,\eta} \psi^\dagger_{R,\eta,\sigma}(y_\alpha)\Phi^{(\alpha)}_{R,\eta,\sigma}(y_\alpha) + \text{h.c.}\,.
\end{aligned}$$

Here, $N_g \propto V_g$ is the number of charges induced by the gate voltage, $\varepsilon = \pi v_F/L_d$ and E_c is the charging energy renormalized by the polaronic shift. The term $\mathcal{H}_{\text{shift}}$ describes an energy mismatch of the CNT subbands with $\eta = \pm 1$ [30, 31]. Finally, the tunneling hamiltonian $\mathcal{H}_t$ connects the dot and the leads. Here, $\psi_{R,\eta,\sigma}(y)$ is the right movers field operator of the four-branches model. Due to the coupling of the vibron to the $c+$ mode dressed vibron operators appear in $\psi_{R,\eta,\sigma}(y)$, in which we further neglect the high-energy dynamics of the collective charge and spin modes in analogy to Eq. (7). Furthermore, $\Phi^{(\alpha)}_{R,\eta,\sigma}(y)$ is the field operator of right movers in lead α and $\tau_{\alpha,\eta}$ the tunneling amplitude for the level η with spin σ. Leads, modeled as LLs [4] with four branches and OBC at the tunneling barriers $y = y_\alpha$, are kept at electrochemical potentials whose difference is proportional to the bias voltage V.

To describe the steady-state occupation probability of the dot states in the sequential regime, we employ a rate equation [13, 32]. Tunneling rates into the branch η on barrier α read

$$\Gamma^{(\text{in})}_{\alpha,\eta} = \Gamma_0 \left|\frac{\tau_{\alpha,\eta}}{\tau_{2,+1}}\right|^2 X_{ll'}(y_\alpha) f\left(\Delta E + (-1)^{\alpha+1}\frac{eV}{2}\right), \tag{10}$$

where $\Gamma_0 = 2\pi\nu|\tau_{2,+1}|^2/\xi^2$, ν is the density of states assumed equal on both leads and $f(E)$ is the Fermi function at temperature T with ΔE the energy difference between initial and final dot states. They contain the *barrier-dependent* FC factors discussed in the previous section [2]. It is useful to introduce the parameter $A = |\tau_{1,\eta}|^2/|\tau_{2,\eta}|^2$, describing the asymmetry between between source and drain tunnel barriers, and $\gamma = |\tau_{\alpha,-1}|^2/|\tau_{\alpha,+1}|^2$.

Sequential tunneling is justified for $k_B T \gg \Gamma_0$. Additionally, the above treatment neglects coherences among vibron states, which is appropriate if $\Omega_0 \gg \Gamma_0$. In our experiment both conditions are satisfied, as we can estimate $k_B T \approx 90\mu$eV (for $T \approx 1$ K), $\Gamma_0 \approx 1$ μeV (from current-voltage characteristics) and $\Omega_0 \approx 0.8$ meV (from the separation of vibron subbands). We also extract $E_c \approx 4.5$ meV from the size of the Coulomb diamonds [9].

Restricting for simplicity to $N_g \approx N_g^* = 1/2 + 4K + 3\varepsilon/2E_c$ where the resonance between $N_{c+} = 4K$ with $N_{\eta,\sigma} = K\ \forall\ \eta,\sigma$ and $N_{c+} = 4K+1$ with K a positive integer occurs, the rate equation is numerically solved and the differential conductance $G = \partial I/\partial V$ is obtained.

As already discussed in the introduction, the nonlinear transport spectra exhibit two main features: (i) a striking suppression of conductance traces with negative slope and (ii) a regular pattern of alternating PDC and NDC. Concerning (i), the suppression of traces with negative slope can *only* be reproduced by our model if *asymmetric* FC factors are employed in the

[2] For tunneling out, a similar expression holds.

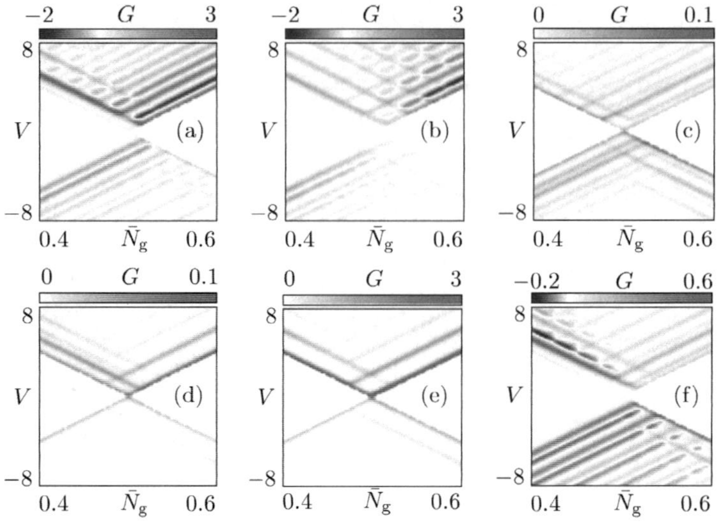

Figure 3. Colour maps of the numerical differential conductance G (units 0.01 e^2/h) as a function of $\bar{N}_g = N_g - N_g^*$ (see text) and V (units meV). (a) Plot for $A = 1/40$, $\gamma = 40$, $\lambda_1^2 = 2$, $\lambda_2^2 = 0.2$ and $\Delta\varepsilon = 0.48$ meV; (b) as in (a) but $\lambda_1^2 = \lambda_2^2 = 2$ (AH model); (c) as in (a) but for $A = \gamma = 1$; (d) as in (a) but for $\gamma = 1$; (e) as in (a) but for $\Delta\varepsilon = 0$; (f) as in (a) but for $A = 1$. In all panels, $\Omega_0 = 0.8$ meV, $k_B T = 50$ μeV, $E_c = 4.5$ meV and $\Gamma_0 = 0.8$ μeV.

calculation. A simple asymmetry of the tunneling barriers $A \neq 1$ is not able to reproduce such a feature, regardless the asymmetry strength. Concerning (*ii*), the occurrence of NDC requires $\gamma > 1$, $\Delta\varepsilon \approx \Omega_0/2$ and $A < 1$.

Given the bias setup [9], traces with negative slope correspond to transitions on the right tunnel barrier: therefore we assume $\lambda_2 \ll \lambda_1$ with $\lambda_2 \approx 0$, corresponding to a vibron *smaller* than the dot, located next to the left tunneling barrier. Figure 3a shows G as calculated with parameters chosen with all the above prescriptions. The results of our model are in excellent agreement with the experimental findings, see Fig. 1b. In particular, we note an almost perfect suppression of the conductance traces with negative slope due to the smallness of λ_2. In sharp contrast, results obtained within the standard AH model ($\lambda_1 = \lambda_2$) but same other parameters, shown in Fig. 3b, display no suppression of the conductance traces with negative slope, despite the strong asymmetry of the tunneling barriers.

Figures 3c-f illustrate the role of the other parameters of the system in creating NDC, for fixed $\lambda_1 \neq \lambda_2$ as in Fig. 3a. For a symmetrical setup $A = \gamma = 1$, no NDC can be achieved, see Fig 3c. Choosing $\gamma > 1$ ($\gamma < 1$) results in a bottleneck for the tunneling events involving the state $\eta = +1$ ($\eta = -1$). This leads to a dynamical trapping of one of the orbital states, which acquires a high occupation probability with a corresponding decrease in the current [20, 21, 22]. No qualitative difference is present between $\gamma < 1$ and $\gamma > 1$. Tuning $\gamma = 1$ instead suppresses NDC, as shown in Fig. 3d, since no dynamical trapping can be achieved.

A regular pattern of repeating NDC and PDC traces is obtained if $\Delta\varepsilon \approx \Omega_0/2$. In this case, the vibrational subbands for $\eta = +1$ are interleaved with those of $\eta = -1$, resulting in an alternating dynamical trapping and un-trapping of the dot states. With completely degenerate states $\Delta\varepsilon = 0$, as shown in Fig. 3e, trapping and un-trapping compete and the net result is the disappearance of NDC. We note that the smallness of the $\Delta\varepsilon$ required for observing NDC is not sufficient to spoil appreciably the fourfold symmetry of the Coulomb blockade diamonds observed in Fig. 1b. Tuning the left/right tunnel barriers to $A < 1$ yields NDC lines with the *same* positive slope at any voltage. With $A = 1$ one has NDC with positive slope for $V < 0$ and suppressed NDC with *negative* slope for $V > 0$ (see Fig. 3f) in contrast with the experiment.

4. Conclusions

A microscopic model describing the coupling between electrons and vibrons in a suspended CNT-quantum dot has been presented. In contrast with the usual AH model, vibrons couple both to the total dot charge and to the *spatial fluctuations* of the electronic density. As a result, a *position-dependent* effective coupling $\lambda(y)$ and related position-dependent FC factors arise when the vibron is smaller than the dot, while the conventional results of the AH model are recovered for vibrons *larger* than the dot itself. Small vibrons located asymmetrically with respect to the dot center result in a large effective coupling on the tunnel barrier closer to the vibron. This implies a strong suppression of one family of conductance traces in the transport characteristics of the system. Such a suppression is indeed observed in our experiment and cannot be explained with the usual AH model. The alternating NDC/PDC traces observed in the experiment have been addressed in terms of asymmetric tunnel couplings on the tunneling barriers and between different orbital state of the CNT dot.

Acknowledgments

Financial support by INFM-CNR via Seed Project PLASE001 and “Progetto giovani”, and by the EU-FP7 via ITN-2008-234970 NANOCTM are gratefully acknowledged.

References

[1] Charlier J -C, Blase X, and Roche S 2007 *Rev. Mod. Phys.* **79** 677
[2] Postma H W, Teepen T, Yao Z, Grifoni M, and Dekker C 2001 *Science* **293** 76
[3] Nazarov Yu V and Glazman L I 2003 *Phys. Rev. Lett.* **91** 126804
[4] Kleimann T, Cavaliere F, Sassetti M, and Kramer B 2002 *Phys. Rev. B* **66** 165311
[5] Cobden D H and Nygård J 2002 *Phys. Rev. Lett.* **89** 046803
[6] Mariani E and von Oppen F 2009 *Phys. Rev. B* **80** 155411
[7] LeRoy B J, Lemay S G, Kong J, and Dekker C 2004 *Nature* **432** 371
[8] Sapmaz S *et al.* 2006 *Phys. Rev. Lett.* **96** 026801
[9] Leturcq R *et al.* 2009 *Nature Phys.* **5** 327
[10] Hüttel A K *et al.* 2009 *Phys. Rev. Lett.* **102** 225501
[11] Zazunov A, Feinberg D, and Martin T 2006 *Phys. Rev. B* **73** 115405
[12] Shen X Y, Dong B, Lei X L, and Horing N J M 2007 *Phys. Rev. B* **76** 115308
[13] Koch J and von Oppen F 2005 *Phys. Rev. Lett* **94**, 206804
[14] Franck J 1926 *Transactions of the Faraday Society* **21** 536; Condon E 1926 *Phys. Rev.* **28** 1182
[15] Braig S and Flensberg K 2003 *Phys. Rev. B* **68** 205324
[16] Mitra A, Aleiner I, and Millis A J 2004 *Phys. Rev. B* **69** 245302
[17] Izumida W and Grifoni M 2005 *New J. Phys.* **7** 244
[18] Flensberg K 2008 *New J. Phys.* **10** 059801
[19] Cavaliere F, Mariani E, Leturcq E, Stampfer C, and Sassetti M 2010 *Phys. Rev. B* **81** 201303(R)
[20] Ciorga M *et al.* 2002 *Appl. Phys. Lett.* **80** 2177
[21] Cavaliere *et al.* 2004 *Phys. Rev. Lett.* **93** 036803
[22] Cavaliere F, Braggio A, Sassetti M, and Kramer B 2004 *Phys. Rev. B* **70** 125323
[23] Fabrizio M and Gogolin A 1995 *Phys. Rev. B* **51** 17827
[24] Tipical CNT parameters: $c \approx 30$ eV, $v_s \approx 2.4 \cdot 10^4$ m/s, $v_F = 8 \cdot 10^5$ m/s and $\rho_0 \approx 6.7 \cdot 10^{-7}$ Kg/m^2 [25].
[25] Dresselhaus M S and Eklund P C 2000 *Adv. Phys.* **49** 705
[26] Suzuura H and Ando T 2002 *Phys. Rev. B* **65** 235412
[27] Ullersma P 1966 *Physica* **32** 27
[28] De Martino A and Egger R 2003 *Phys. Rev. B* **67** 235418
[29] Bolotin K *et al.* 2008 *Phys. Rev. Lett.* **101** 096802
[30] Egger R 1999 *Phys. Rev. Lett.* **83** 5547
[31] Yoshioka H and Okamura Y 2002 *J. Phys. Soc. Jpn.* **71** 2512
[32] Merlo M, Haupt F, Cavaliere F, and Sassetti M 2008 *New J. Phys.* **10** 023008
[33] Haupt F, Cavaliere F, Fazio R, and Sassetti M 2006 *Phys. Rev. B* **74** 205328

Journal of Physics: Conference Series **248** (2010) 012020 doi:10.1088/1742-6596/248/1/012020

The effects of polarization fields on exciton binding energy in GaN quantum dots

A Asgari[1,a,b], **S. Asadzadeh**[a]
[a] Photonics Group, Research Institute for Applied Physics, University of Tabriz Tabriz 51665-163, Iran
[b] School of Electrical, Electronic and Computer Engineering, The University of Western Australia Crawley, WA 6009, Australia

E-mail: asgari@tabrizu.ac.ir

Abstract. In this paper the exciton binding energy in wurtzite GaN quantum dot is investigated by considering the influences of polarization in details. We find that the strong built-in electric field gives rise to an obvious modification of conduction band profile of QDs and leads to remarkable electron- hole spatial separation. This effect has a significant influence on exciton states and optical properties of the QDs. The relationship between exciton states and QD size and QD confinement potential is studied. Our results represent the decrease of exciton binding energy by considering piezoelectric polarization.

1. Introduction

Semiconductor quantum dots (QDs) are systems of paramount interest in nanoscience and nanotechnology [1]. They are the natural evolution of band-gap/wave-function engineering in semiconductors; however, in contrast to quantum wells and wires, they exhibit a discrete i.e., atomic like energy spectrum and, more important, their optical response is dominated by few-particle/exciton effects. Moreover, several QD-based technological applications have been proposed, such as lasers, charge-storage devices, fluorescent biological markers, and all-optical quantum information processors [2-6]. On the other hand, the wide-band-gap III-nitrides based semiconductor heterostructures AlGaN/GaN have recently attracted much attention due to conspicuous device applications in electronics and optoelectronics, such as high brightness blue/green light-emitting diodes (LED)s and laser diodes [7-10]. The physical interest in these systems is strongly related to their excitonic properties, which have attracted attention both with respect to the single-exciton problem and the exciton-exciton interaction. Also, it is found that the electronic and optical properties of wurtzite (WZ) III-nitride heterostructures are affected by the strong built-in electric field induced by the piezoelectricity and spontaneous polarizations. The magnitude of the strong built-in electric field is estimated to be in the order of MV/cm [11–15]. In this paper the exciton binding energy in wurtzite GaN quantum dot is investigated by considering the influences of polarization fields in details using finite difference method and also a variational procedure.

[1] Corresponding Author: asgari@tabrizu.ac.ir, Tel: +98-411-3393005, Fax: +98-411-3347050.

2. Model and Discussions

In order to investigate the effects of the piezoelectricity and spontaneous polarization on exciton states and optical properties of WZ QD structures, let us now consider the undoped WZ-GaN cylindrical quantum dot in AlGaN media (Fig.1). For the quantum dot the radius is R and the height is L.

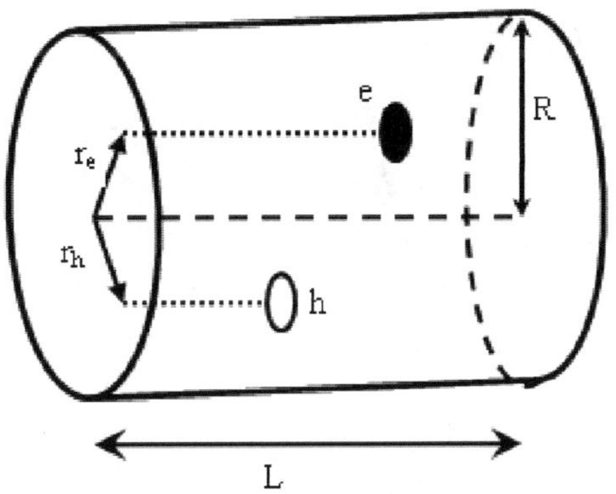

Figure 1. The proposed cylindrical shaped GaN QD within a AlGaN layer.

In order to study the electronic structures of QD, different methods have been experienced. The single band method is used in this study. In the frame work of the envelope function, and the effective mass theory, the Hamiltonian can be written as:

$$H = \frac{-\hbar^2}{2}\nabla\frac{1}{m^*(x,y,z)}\nabla + V(x,y,z) \quad (1),$$

Where

$$m^*(x,y,z) = \begin{cases} m^*_{GaN} & in\ QD \\ m^*_{AlGaN} & else \end{cases} \quad (2),$$

$$V(x,y,z) = \begin{cases} 0 & in\ QD \\ \Delta E_c\ or\ \Delta E_v & else \end{cases} \quad (3).$$

In which m^*_{GaN} and m^*_{AlGaN} are the effective mass of electrons in GaN and AlGaN respectively, ΔE_c, *and* ΔE_v are the conduction band and valance band discontinuity, respectively, and [16]:

$$\Delta E_C = 0.7(x\times 6.13 + (1-x)\times 3.42 - x(1-x) - Eg_0)\ \text{eV} \quad (4).$$

Where Eg_0 is the GaN band gap, and x notifies Al- molar fraction in the barrier. As the system is under build in electric field effects, the Hamiltonian should be modified as:

$$H = \frac{-\hbar^2}{2}\nabla\frac{1}{m^*(x,y,z)}\nabla + V(x,y,z) + e\vec{F}_d.\vec{r} \quad (5).$$

Where F_d demonstrates the built in electric field. It should be mentioned that III- nitrides in the wurtzite phase, have a strong spontaneous macroscopic polarizations and large piezoelectric coefficients. The abrupt variation of the polarization at interfaces, gives rise to a large polarization sheet charges which, in turn, creates a noticeable built-in electric field [17]. Therefore, the optical properties of wurtzite AlGaN/GaN QDs are affected by the 3D confinement of electrons and holes, and the strong built-in electric field in the QDs region, which in turn, causes simulation of these systems, extremely challenging task.

The Built in electric field which applied in the equations is [18]:

$$F_d = \frac{L_{br}(P_{tot}^{br} - P_{tot}^{d})}{\varepsilon_0(L_d \varepsilon_{br} + L_{br}\varepsilon_d)} \quad (6).$$

Where $\varepsilon_{br(d)}$, is the relative dielectric constant of the barrier (dot), $P_{tot}^{br/d}$ is the total polarization and $L_{br/d}$ is the width of the barrier and height of the dot.

$$P_{tot}^{br/d} = P_{piezo}^{br/d} + P_{sp}^{br/d} \quad (7).$$

The Piezoelectric polarization has two parts, one induced by the lattice Mismatch (ms), and the other caused by the thermal strain (ts): $P_{piezo}^{br/d} = P_{ms}^{br/d} + P_{ts}^{br/d}$, where $P_{ms}^{br/d} = 2(e_{31} - e_{33}\frac{c_{13}}{c_{33}})(\frac{a-a_0}{a})$ and $P_{ts}^{d} = -3.2\times10^{-4} c/m^2$ [18], e_{31} and e_{33} are the piezoelectric coefficients, 'a' is the lattice constant of $Al_xGa_{1-x}N$ and is equal to $a = (0.077\ x + 3.189\ )\mathring{A}$, and c_{13} and c_{33} are elastic constant. (All other material parameters can be found in [19]). The spontaneous polarization for $Al_xGa_{1-x}N$ is x dependent, and considered as: $P_{sp} = (-0.052x - 0.029)$. To solve the Schrödinger equation, we used finite difference numerical methods.

The exciton Hamiltonian in this system and in cylindrical coordinate can be defined as [20]:

$$\begin{aligned} H = &-\frac{\hbar^2}{2m_e^*}\left\{\frac{\partial^2}{\partial r_e^2} + \frac{1}{r_e}\frac{\partial}{\partial r_e} + \frac{r_e^2 - r_h^2 + r^2}{r_e r}\frac{\partial^2}{\partial r_e \partial r}\right\} \\ &-\frac{\hbar^2}{2m_h^2}\left\{\frac{\partial^2}{\partial r_h^2} + \frac{1}{r_h}\frac{\partial}{\partial r_h} + \frac{r_h^2 - r_e^2 + r^2}{r_h r}\frac{\partial^2}{\partial r_h \partial r}\right\} - \frac{\hbar^2}{2\mu}\left\{\frac{\partial^2}{\partial r^2} + \frac{1}{r}\frac{\partial}{\partial r}\right\} - \frac{\hbar^2}{2m_e^*}\frac{\partial^2}{\partial z_e^2} \\ &-\frac{\hbar^2}{2m_h^*}\frac{\partial^2}{\partial z_h^2} + V_e(r_e, z_e) + V_h(r_h, z_h) - \frac{e^2}{\varepsilon\sqrt{r^2 + (z_e - z_h)^2}} \end{aligned} \quad (8).$$

Where $r = |r_e - r_h| = \sqrt{r_e^2 + r_h^2 - 2r_e r_h \cos(\varphi_e - \varphi_h)}$ and $\mu = m_e^* m_h^* / m_e^* + m_h^*$. The Schrödinger equation base on this Hamiltonian is solved by variational method using the wave function of:

$$\psi(r_e, r_h, r, z_e, z_h) = f(r_e) f(r_h) g(z_e) g(z_h) e^{-\alpha\sqrt{r^2 + (z_e - z_h)^2}} \quad (9).$$

Where f *and* g is the electron and hole wavefunctions, respectively, and α is the variational parameter. The electron and hole wavefunction can be find solving the Schrödinger equation based on Hamiltonian represented on (5).

To find the exciton energy one should solve the Schrödinger equation based on Hamiltonian represented on (8) and find the $\langle\psi|H|\psi\rangle$. Where $\langle\psi|H|\psi\rangle$ includes 8 sentences (5 for kinetics and 3 for potential) as:

$$k1=\int_{-\infty}^{\infty}\int_{-\infty}^{\infty}\int_{0}^{\infty}\int_{0}^{\infty}\int_{0}^{\infty}\frac{\hbar^2}{2m_h}f^2(r_e)f^2(r_h)g^2(z_e)g^2(z_h)e^{-2\alpha\sqrt{r^2+(z_e-z_h)^2}}\left[\left(\frac{f'(r_e)}{f(r_e)}\right)^2+\frac{\alpha\left(r_e^2-r_h^2+r^2\right)}{r_e\sqrt{r^2+(z_e-z_h)^2}}\frac{f'(r_e)}{f(r_e)}\right]r_e dr_e r_h dr_h r dr dz_e dz_h \quad (10),$$

$$k_2=\int_{-\infty}^{\infty}\int_{-\infty}^{\infty}\int_{0}^{\infty}\int_{0}^{\infty}\int_{0}^{\infty}\frac{\hbar^2}{2m_h}f^2(r_e)f^2(r_h)g^2(z_e)g^2(z_h)e^{-2\alpha\sqrt{r^2+(z_e-z_h)^2}}\left[\left(\frac{f'(r_h)}{f(r_h)}\right)^2+\frac{\alpha\left(r_h^2-r_e^2+r^2\right)}{r_e\sqrt{r^2+(z_e-z_h)^2}}\frac{f'(r_h)}{f(r_h)}\right]r_e dr_e r_h dr_h r dr dz_e dz_h \quad (11),$$

$$k_3=\int_{-\infty}^{\infty}\int_{-\infty}^{\infty}\int_{0}^{\infty}\int_{0}^{\infty}\int_{0}^{\infty}\frac{\hbar^2}{2\mu}f^2(r_e)f^2(r_h)g^2(z_e)g^2(z_h)e^{-2\alpha\sqrt{r^2+(z_e-z_h)^2}}\left[\frac{\alpha^2 r^2}{r^2+(z_e-z_h)^2}\right]r_e dr_e r_h dr_h r dr dz_e dz_h \quad (12),$$

$$k_4=\int_{-\infty}^{\infty}\int_{-\infty}^{\infty}\int_{0}^{\infty}\int_{0}^{\infty}\int_{0}^{\infty}\frac{\hbar^2}{2m_e}f^2(r_e)f^2(r_h)g^2(z_e)g^2(z_h)e^{-2\alpha\sqrt{r^2+(z_e-z_h)^2}}\left[\left(\frac{g'(z_e)}{g(z_e)}\right)^2-\frac{2\alpha(z_e-z_h)}{r_e\sqrt{r^2+(z_e-z_h)^2}}\frac{g'(z_e)}{g(z_e)}+\frac{\alpha^2(z_e-z_h)^2}{r^2+(z_e-z_h)^2}\right]r_e dr_e r_h dr_h r dr dz_e dz_h \quad (13),$$

$$k_5=\int_{-\infty}^{\infty}\int_{-\infty}^{\infty}\int_{0}^{\infty}\int_{0}^{\infty}\int_{0}^{\infty}\frac{\hbar^2}{2m_h}f^2(r_e)f^2(r_h)g^2(z_e)g^2(z_h)e^{-2\alpha\sqrt{r^2+(z_e-z_h)^2}}\left[\left(\frac{g'(z_h)}{g(z_h)}\right)^2+\frac{2\alpha(z_e-z_h)}{r_e\sqrt{r^2+(z_e-z_h)^2}}\frac{g'(z_h)}{g(z_h)}+\frac{\alpha^2(z_e-z_h)^2}{r^2+(z_e-z_h)^2}\right]r_e dr_e r_h dr_h r dr dz_e dz_h \quad (14),$$

$$p_1=\int_{-\infty}^{\infty}\int_{-\infty}^{\infty}\int_{0}^{\infty}\int_{0}^{\infty}\int_{0}^{\infty}V_e(r_e,z_e)f^2(r_e)f^2(r_h)g^2(z_e)g^2(z_h)e^{-2\alpha\sqrt{r^2+(z_e-z_h)^2}}r_e dr_e r_h dr_h r dr dz_e dz_h \quad (15),$$

$$p_2=\int_{-\infty}^{\infty}\int_{-\infty}^{\infty}\int_{0}^{\infty}\int_{0}^{\infty}\int_{0}^{\infty}V_h(r_h,z_h)f^2(r_e)f^2(r_h)g^2(z_e)g^2(z_h)e^{-2\alpha\sqrt{r^2+(z_e-z_h)^2}}r_e dr_e r_h dr_h r dr dz_e dz_h \quad (16),$$

$$p_3 = \int_{-\infty}^{\infty}\int_{-\infty}^{\infty}\int_{0}^{\infty}\int_{0}^{\infty}\int_{0}^{\infty} \frac{-e^2}{4\pi\varepsilon\varepsilon_0\sqrt{r^2+(z_e-z_h)^2}} f^2(r_e)f^2(r_h)g^2(z_e)g^2(z_h)e^{-2\alpha\sqrt{r^2+(z_e-z_h)^2}} r_e dr_e r_h dr_h r dr dz_e dz_h \quad (17).$$

By minimizing the equation $E_{ex}(\alpha)=\frac{\langle\psi|H|\psi\rangle}{\langle\psi|\psi\rangle}$ regards α: and using $E_b = E_e + E_h - E_{ex}$, the binding energy of exciton can be obtained.
The electron and hole energy as function of QD radius (Fig 2) and as function of QD height (Fig. 3) are calculated.

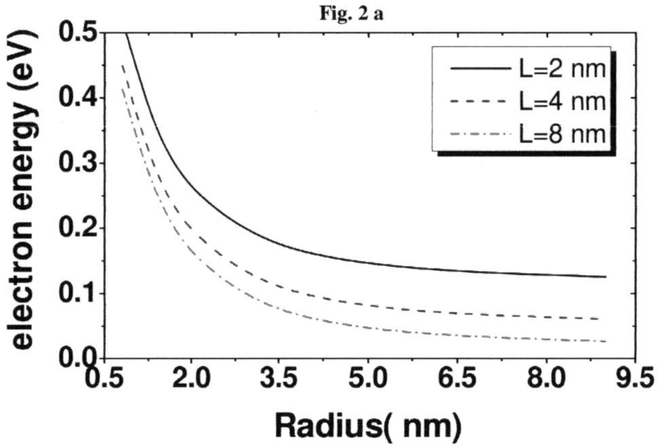

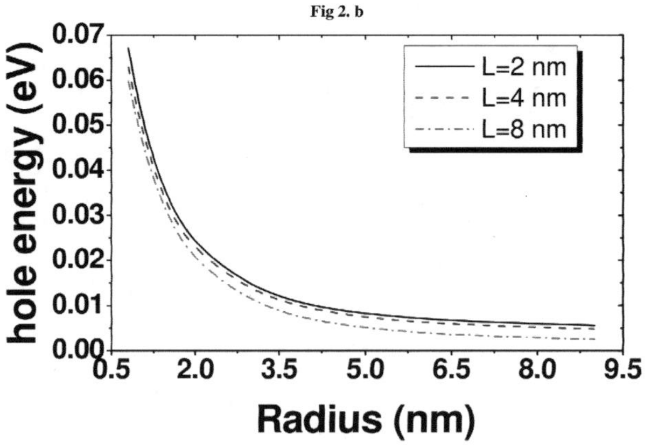

Figure 2. Electron (a) and hole (b) ground state energies as function of QD radius for different QD height and barrier Al content of x=0.3.

As shown in Fig. 2, with increasing the QD radius the electron and hole ground state energy decreases and almost saturate for large QDs.

Fig. 3 shows the difference between the electron and hole energy with and without the polarization effects. With including the polarization effects, a Stark shift happens.

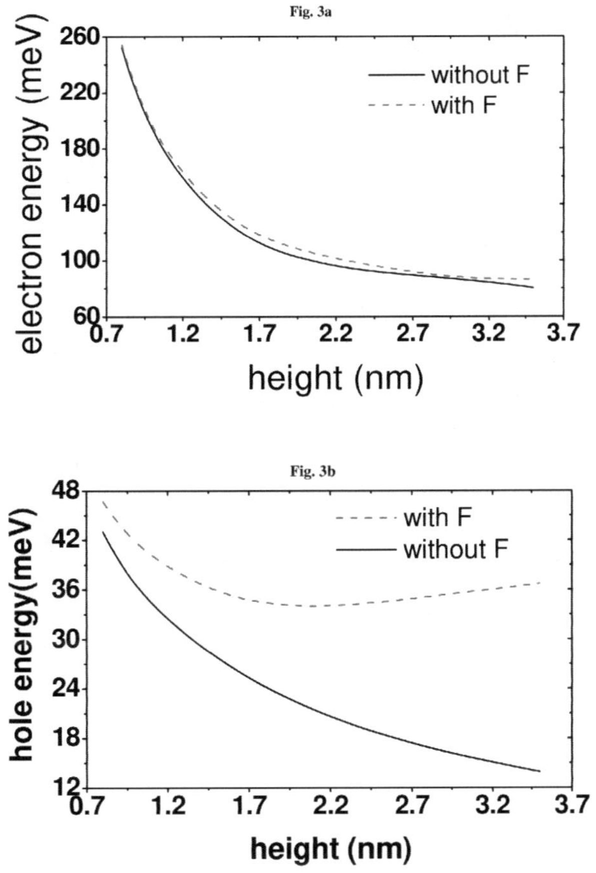

Figure 3. Electron (a) and hole (b) ground state energies as function of QD height with and without polarization effects (x=0.12, R=4 nm).

Using these results, one can find the exciton binding energies. Exciton binding energy is calculated as function of QD radius (Fig. 4), QD height (Fig. 5), and also QD confinement potential, Al mole fraction, (Fig. 6). As indicated in Fig. 4, the maximum binding energy for exciton is happened for QDs with radius of about 1 nm.

Our calculation indicates that the decreasing of QD height increase the exciton binding energy. Also, the effects of polarization fields are much for large QDs. the exciton binding energy is calculated for different values of surrounding potential that depends on aluminium concentration (x). Our results show exciton binding energy increases with the increase of aluminium concentration (x).

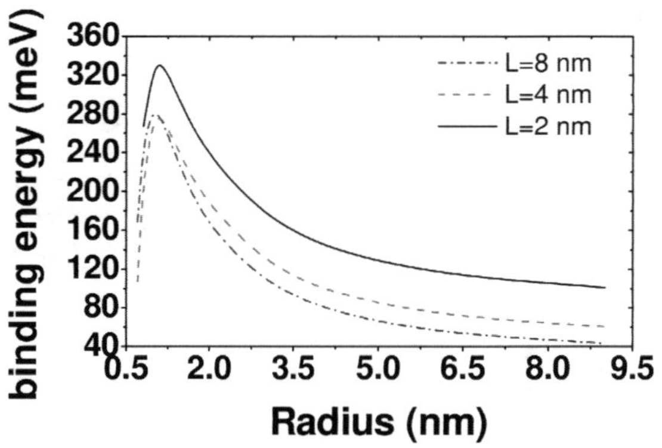

Figure 4. Exciton binding energy as function of QD readies and for different QD height (x=0.3) including the polarization fields.

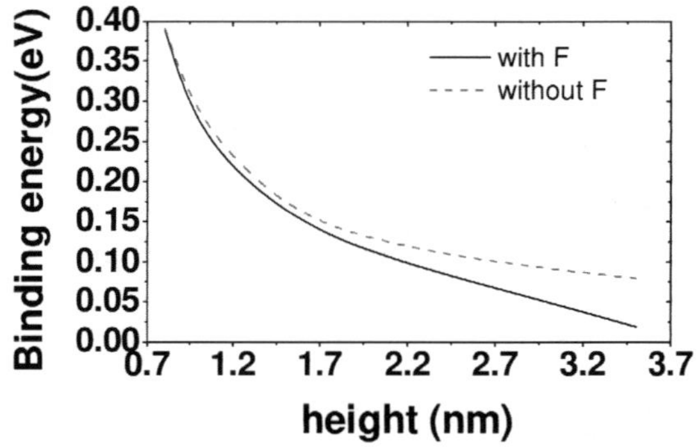

Figure 5. Comparison of exciton binding energy by considering polarization and without it for cylindrical QD with radius R=4nm and x=0.12

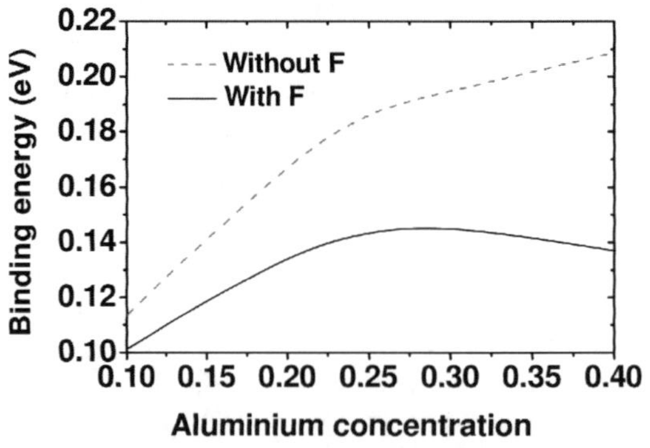

Figure 5. Comparison of exciton binding energy by considering polarization and without it for cylindrical QD with radius R=4nm and height L=2nm.

Journal of Physics: Conference Series **248** (2010) 012020 doi:10.1088/1742-6596/248/1/012020

3. Conclusion

In summery the effects of polarization field in GaN quantum dot, on binding energy of excitons have been studied. Our calculations results show that the strong built-in electric field gives rise to an obvious modification of conduction band and valance band profile of QDs and leads to remarkable electron- hole spatial separation. This effect has a significant influence on exciton states and optical properties of the QDs. The relationship between exciton states and height, radius and confinement potential of QDs is studied.

4. References

[1] L. Jacak, P. Hawrylak, and A. Wojs, 1998, Quantum Dots, Springer, Berlin.

[2] H. Saito, K. Nishi, and S. Sigou, 2001, Appl. Phys. Lett. 78, 267.

[3] J. J. Finley, M. Skalitz, M. Arzberger, A. Zrenner, G. Bohm, and G. Abstreiter, 1998, Appl. Phys. Lett. 73, 2618.

[4] T. Lundstrom, W. Schoenfeld, H. Lee, and P. M. Petroff, 1999, Science 286, 2312.

[5] M. Bruchez, Jr., M. Moronne, P. Gin, S. Weiss, and A. P. Alivisatos, 1998, Science 281, 2013.

[6] E. Biolatti, R. C. Iotti, P. Zanardi, and F. Rossi, 2000, Phys. Rev. Lett. 85, 5647.

[7] Y.K. Kuo, J.Y. Chang, M.C. Tsai, S.H. Yen, 2009, Appl. Phys. Lett. 95 011116.

[8] Y. Yang, X.A. Cao, C.H. Yan, 2009, Appl. Phys. Lett. 94, 041117.

[9] J. Coraux, V. Favre-Nicolin, H. Renevier, M.G. Proietti, B. Amstatt, E. Bellet-Amalric, B. Daudin, 2007, J. Appl. Phys. 101 056106.

[10] O. Weidemann, P.K. Kandaswamy, E. Monroy, G. Jegert, M. Stutzmann, M. Eickhoff, 2009, Appl. Phys. Lett. 94, 113108.

[11] S.F. Chichibu, M. Sugiyama, T. Kuroda, A. Tackeuchi, T. Sota, S.P. Denbaars, S. Nakamura, Y. Ishida, H. Okumura, 2001, Appl. Phys. Lett. 79, 3600.

[12] P. Ruterana, S. Kret, A. Vivet, G. Maciejewski, P. Dluzewski, 2002, J. Appl. Phys. 91, 8979.

[13] R.A. Taylora, J.W. Robinsona, J.H. Ricea, A. Jarjoura, J.D. Smitha, R.A. Oliverb, G.A.D. Briggsb, M.J. Kappersc, C.J. Humphreysc, Y. Arakawad, 2004, Physica E 21, 285.

[14] J.J. Shi, Z.Z. Gan, 2003, J. Appl. Phys. 94, 407.

[15] J.J. Shi, C.X. Xia, S.Y. Wei, Z.X. Liu, 2005, J. Appl. Phys. 97, 083705.

[16] O. Ambacher, R. Dimitrov, M. Stutzmann, B. E. Foutz, 1999, J. Appl. Phys. 85, 3222.

[17] O. Mayrock, H. J. Wünsche, F. Henneberger, 2000, Phys. Rev. B 62, 16870.

[18] S. De Rinaldis, I. D Amico, E. Biolatti, R. Rinaldi, R. Cinagolani, F. Rossi, 2002, Phys. Rev. B 65, 081309R.

[19] R. Cingolani A. Botchkarev, H. Tang, H. Morkoç, G. Traetta, G. Coli, M. Lomascolo, A. Di Carlo, F. Della Sala, and P. Lugli , 2000, Phys. Rev. B. 61, 2711.

[20] Safwan, S. A., Hekmat, M. H., and . El-Meshad, N.A., 2007, FIZIKA A 16 , 1.

Journal of Physics: Conference Series **248** (2010) 012021 doi:10.1088/1742-6596/248/1/012021

An orbital entanglement in two-electron quantum dots in a magnetic field

A Chizhov[1,3] **and R G Nazmitdinov** [1,2,3]

[1] BLTP, Joint Institute for Nuclear Research, 141980 Dubna, Russia
[2] Departament de Física, Universitat de les Illes Balears, 07122 Palma de Mallorca, Spain
[3] Dubna International University, 141980 Dubna, Russia

E-mail: chizhov@theor.jinr.ru, rashid@theor.jinr.ru

Abstract. We consider a simple model for two-electron quantum dots (QDs) in the perpendicular magnetic field. With the aid of this model we study the degree of the orbital entanglement as a function of the magnetic field strength and the QD shape. We found that the circular QD exhibits no entanglement in the magnetic field. Although the entanglement increases with the increase of the QD deformation at a fixed value of the magnetic field, it approaches the constant value at the fixed deformation with the increase of the magnetic field.

1. Introduction

Nowadays there is an enormous experimental and theoretical activity focused on the study of properties of few-electron quantum dots (QDs) [1, 2, 3]. This interest is motivated by: i) possible applications of QDs in nanoelectronics and ii) due to a rapid development of the field of quantum computing. It is expected that QDs could lead to novel device applications in fields such as quantum cryptography and information storage. It is also widely believed that the entangled states of the electrons confined in a QD may give a natural realization of a quantum bit or "qubit" [4]. However, an entanglement being one of the most subtle and intriguing phenomena in nature is not yet well understood [5]. The questions how to efficiently produce and control it, for example, in QDs are among fundamental as well as technological problems.

In general, in finite fermionic systems the entanglement is associated with a correlated dynamics of electrons. The correlated motion of electrons in QDs and, therefore, the entanglement could be controlled by externally applied electromagnetic fields or by varying the parameters of QDs. The simplest QD with the essential features of more complex systems contains two electrons. When two electron move in the external field created by the confining potential and by a homogeneous constant magnetic field, the spin part of the total wave function evolves from the singlet to the triplet state (see [3] and references therein). At the same time the spatial part changes in accordance with the antisymmetric nature of the total wave function. The spin singlet wave function is always maximally entangled. In this case one can study how the external field and parameters of the system would influence the spatial function and, therefore, the *spatial* entanglement which may have important effects in various applications.

Recently we developed an analytical approach to study the degree of entanglement in a model of two-coupled harmonic oscillators as a function of time [6]. Note, that in condensed matter physics this model is used as a starting point for analysis of electronic properties of QDs in a

Journal of Physics: Conference Series **248** (2010) 012021 doi:10.1088/1742-6596/248/1/012021

perpendicular magnetic field [3]. In this communication we report the results of a detailed study of the effect of the magnetic field and shape parameters of a two-electron QD on the degree of the orbital entanglement.

2. The Model

In theoretical studies of QDs the underlying lattice of the semiconductor material is taken into account by using an effective mass for the conduction electrons. For vertical QDs, the resulting confining potential is, to a good approximation, harmonic [1]. In particular, for the typical voltage ~ 1 V applied to the gate, the confining potential is some eV deep which is large compared to the few meV of the confining frequency [7, 8]. Hence, the electron wave function is localized close to the minimum of the well which always can be approximated by a parabolic potential. Indeed, this approximation enables one to reproduce quite well experimental ground state transitions under the perpendicular magnetic field, observed in recent experiments with few-electron QDs (see details in [3, 9, 10]).

We consider a system of two electrons whose motion is restricted to the xy plane by a parabolic potential. The system is also subject to an external magnetic field applied in the vertical direction (z). The full Hamiltonian thus reads

$$H = \sum_{i=1}^{2} h_i + V(\mathbf{r}_1, \mathbf{r}_2) + H_{\rm spin} = \sum_{i=1}^{2} \left[\frac{1}{2m^*} \left(\mathbf{p} - \frac{e}{c}\mathbf{A} \right)^2 + \frac{1}{2} m^* \left(\omega_x^2 x^2 + \omega_y^2 y^2 \right) \right]_i + V(\mathbf{r}_1, \mathbf{r}_2) + H_{\rm spin} , \tag{1}$$

where $H_{\rm spin} = g^* \mu_B (\mathbf{s}_1 + \mathbf{s}_2) \cdot \mathbf{B}$ describes the Zeeman energy and $\mu_B = |e|\hbar/2m_e c$ is the Bohr magneton. Here m^* and g^* are the effective electron mass and g-factor, respectively, and we use the planar polar coordinates ($r^2 = x^2 + y^2$). The confining potential is approximated with a 2D harmonic oscillator with two different confining frequencies ω_x and ω_y, which are not equal in a general case. Although the three-dimensional nature of QDs is important for analysis of experimental data [3], one is able to reproduce the experimental results with the effective two-dimensional electron-electron interaction [11, 9].

For the perpendicular magnetic field ($\mathbf{B} \parallel z$) we choose a gauge described by the vector $\mathbf{A} = [\mathbf{B} \times \mathbf{r}]/2 = \frac{1}{2} B(-y, x, 0)$. In the following we assume $g^* = 0$ (the parabolic quantum well with $g^* = 0$ can be grown [12]) and neglect the contribution of the Zeeman term. In this communication we consider the evolution of the entanglement in the ground state of the QD due to the applied magnetic field and shape parameters. To proceed further, we neglect the interaction between electrons $V(\mathbf{r}_1, \mathbf{r}_2) = 0$ in order to shed light on the *spatial* entanglement of the single-electron dynamics due to the magnetic field. In this case the spatial part of the total wave function is factorized. In particular, for the singlet state $\Psi(\mathbf{r}_1, \mathbf{r}_2, \mathbf{s}_1, \mathbf{s}_2) = \psi(\mathbf{r}_1)\psi(\mathbf{r}_2)\xi(\mathbf{s}_1, \mathbf{s}_2)$ ($\xi(\mathbf{s}_1, \mathbf{s}_2)$ is the spin wave function), there is a spin entanglement, while the spatial motion of each electron is independent. The spatial dynamics for each electron is described by the Hamiltonian

$$h = \frac{{p_x}^2 + {p_y}^2}{2m^*} + \frac{m^*}{2} \left(\omega_1^2 x^2 + \omega_2^2 y^2 \right) - \omega_L l_z \tag{2}$$

In the chosen gauge the magnetic field contributions are taken into account by dint of an effective parabolic confinement with two effective frequencies $\omega_1^2 = \omega_x^2 + \omega_L^2$, $\omega_2^2 = \omega_y^2 + \omega_L^2$, and with a magnetic rotation $\omega_L l_z$. Here $\omega_L = |e|B/2m^*c$ is the Larmor frequency and $l_z = xp_y - yp_x$ is an orbital momentum. The Hamiltonian (2) can be recast in the form of two coupled oscillators

$$h = \hbar\omega_1 \hat{c}_1^\dagger \hat{c}_1 + \hbar\omega_2 \hat{c}_2^\dagger \hat{c}_2 + i\hbar g_1 \left(\hat{c}_1^\dagger \hat{c}_2 - \hat{c}_2^\dagger \hat{c}_1 \right) - i\hbar g_2 \left(\hat{c}_1^\dagger \hat{c}_2^\dagger - \hat{c}_2 \hat{c}_1 \right) , \tag{3}$$

where $\hat{c}_{1,2}$ ($\hat{c}^\dagger_{1,2}$) are the annihilation (creation) boson operators of the fields with the energies $\hbar\omega_{1,2}$. The interplay between these fields is governed by dint of the coupling constants

$$g_1 = \frac{\omega_L}{2}\frac{\omega_1+\omega_2}{\sqrt{\omega_1\omega_2}}, \quad g_2 = \frac{\omega_L}{2}\frac{\omega_1-\omega_2}{\sqrt{\omega_1\omega_2}}. \tag{4}$$

By means of the Bogoliubov transformation $\hat{a}_k = \sum_{m=1}^2 \left(A^k_m \hat{c}_m + B^k_m \hat{c}^\dagger_m\right)$, the Hamiltonian (3) is reduced to the diagonal form

$$h = \sum_{k=1}^{2} \hbar\Omega_k \left(\hat{a}^\dagger_k \hat{a}_k + 1/2\right) \tag{5}$$

with the two eigenmodes

$$\begin{aligned} \Omega_k^2 &= [\eta_+ + 2(g_1^2 - g_2^2) + (-1)^{k+1}\Delta]/2\,, \\ \Delta &= [\eta_-^2 + 4\eta_+(g_1^2 - g_2^2) + 8\omega_1\omega_2(g_1^2+g_2^2)]^{1/2}\,, \end{aligned} \tag{6}$$

where $\eta_\pm = \omega_1^2 \pm \omega_2^2$. The coefficients A^k_m and B^k_m are defined in [13].

3. Entanglement of the ground state

Here, we focus upon the eigenstate of the Hamiltonian (5) with $\hat{n}_1 = \langle \hat{a}^\dagger_1 \hat{a}_1\rangle = 0$, $\hat{n}_2 = \langle \hat{a}^\dagger_2 \hat{a}_2\rangle = 0$, which is the ground state of the QD. In this case, the quantum state of the fields $\hat{c}_1$ and $\hat{c}_1$ governed by the Hamiltonian (3) can be described by a Gaussian Wigner function of the form (see details in [6])

$$W^E(\xi) = (4\pi^2\sqrt{\det V})^{-1} \exp\left\{-\frac{1}{2}\zeta^T \mathbf{V}^{-1}\zeta\right\} \tag{7}$$

Here, $\zeta^T = (q_1, p_1, q_2, p_2)$ is a transposed four-vector whose elements are the quadrature-component variables defined as $c_i = (q_i + ip_i)/\sqrt{2}$. The variance matrix

$$\mathbf{V} = \begin{pmatrix} \mathbf{X}_1 & \mathbf{Y} \\ \mathbf{Y}^T & \mathbf{X}_2 \end{pmatrix} \tag{8}$$

is determined by the matrices $\mathbf{X}_i$ and $\mathbf{Y}$ which forms are given in [6]. The Gaussian form of the Wigner function enables one to calculate analytically the measure of entanglement between the fields in the form of the logarithmic negativity via the symplectic spectrum of the partial transpose of the variance matrix. As a result, the logarithmic negativity

$$E = -\frac{1}{2}\log_2(4\mathcal{A})\,, \tag{9}$$

where

$$\mathcal{A} = \mathcal{B} - \sqrt{\mathcal{B}^2 - \det\mathbf{V}}\,, \quad \mathcal{B} = \frac{\det\mathbf{X}_1 + \det\mathbf{X}_2}{2} - \det\mathbf{Y}\,, \tag{10}$$

determines the degree of entanglement for $E > 0$. For $E \leq 0$ the composite state is separable according to the Peres-Horodecki criterion [14].

In Fig.1 one observes that a circular QD ($\omega_x = \omega_y$) does not manifest the orbital entanglement. It means that there appears no correlations between coordinate and momentum distributions of an electron, corresponding to the mutually orthogonal projections, i.e., the distributions of x (y)-values and p_y (p_x)-values are independent. The deviation from the circular shape $\omega_y/\omega_x \neq 1$

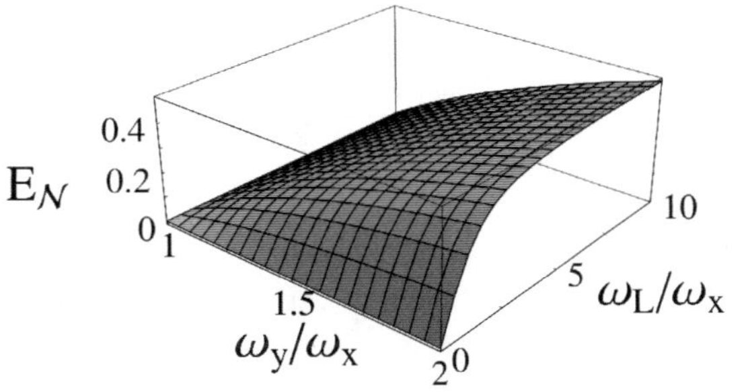

Figure 1. The logarithmic negativity E as function of the magnetic field (the Larmor frequency) and the deformation of the system.

(the onset of deformation) at nonzero value of the magnetic field ω_L/ω_x yields the correlations between the coordinate and the momentum of the mutually orthogonal projections. Moreover, as can be seen in Fig.1, the entanglement increases with the increase of the deformation. The increase of the magnetic field leads to the saturation effect in the entanglement for a fixed QD deformation. For example, at $\omega_y/\omega_x = 2$ the entanglement reaches the value $E \sim 0.4$ at $\omega_L/\omega_x \approx 2.5$ which remains almost unchanged at higher values of the magnetic field.

4. Summary

We consider a simple model for two-electron two-dimensional QDs with the anisotropic parabolic confinement. The considered model demonstrates that the orbital entanglement can be controlled in time by appropriate choice of the magnetic field strength and the choice of the shape parameters of the systems. The obtained results show that even the simple system exhibits rich dynamics from the point of view of the information theory and possible physical applications. Further analysis is required for better understanding of the orbital entanglement in two-electron QDs in the magnetic field.

Acknowledgments

This work was partly supported by RFBR Grant No. 08-02-00118 (Russia) and by Grant No. FIS2008-00781/FIS (Spain).

References

[1] Kouwenhoven L P, Austing D G, and Tarucha S 2001 *Rep. Prog. Phys.* **64** 701
[2] Yannouleas C and Landman U 2007 *Rep. Prog. Phys.* **70** 2067
[3] Nazmitdinov R G 2009 *Physics of Particles and Nuclei* **40** 71
[4] Hanson R *et al* 2007 *Rev. Mod. Phys* **79** 1217
[5] Amico L, Fazio R, Osterloh A and Vedral V 2008 *Rev. Mod. Phys* **80** 517
[6] Chizhov A V and Nazmitdinov R G 2008 *Phys. Rev. A* **78** 0641302
[7] Bimberg D, Grundmann M and Ledentsov N N 1998 *Quantum Dot Heterostructures* (New York: John Willey & Sons)
[8] Chakraborty T 1999 *Quantum Dots* (Amsterdam: Elsevier)
[9] Maksym P A *et al* 2009 *Phys. Rev. B* **79** 115314
[10] Puente A, Pons M and Nazmitdinov R, see the present proceedings
[11] Simonović N S and Nazmitdinov R G 2008 *Phys. Rev. A* **78** 032115
[12] Ellenberger C *et al* 2006 *Phys. Rev. Lett.* **96** 126806
[13] Chizhov A V and Nazmitdinov R G 1990 *Int. J. Mod. Phys. B* **4**, 2335
[14] Peres A 1996 *Phys. Rev. Lett.* **77**, 1413; Horodecki P, 1997 *Phys. Lett. A* **232**, 333

Journal of Physics: Conference Series **248** (2010) 012022 doi:10.1088/1742-6596/248/1/012022

The Scattering of Hot Electrons by Phonons in AlGaN/GaN Quantum Dot

A Asgari[1,a,b], **S. Babanejad**[a]
[a] Photonics Group, Research Institute for Applied Physics, University of Tabriz Tabriz 51665-163, Iran
[b] School of Electrical, Electronic and Computer Engineering, The University of Western Australia Crawley, WA 6009, Australia

E-mail: asgari@tabrizu.ac.ir

Abstract. A detailed study of the energy and momentum relaxation together with the determination of the dominant scattering mechanisms in the wide band gap III-N semiconductor materials are crucial for obtaining a comprehensive understanding of the carrier dynamics, especially at high operating electric fields. In this work we study a AlGaN/GaN quantum dots. The energy loss rate of quantum dot are calculated as a function of the external controllable parameters electrons density, quantum dot size and lattice temperature for the interaction of the charge carriers with bulk acoustic (deformation potential and piezoelectric scattering) and optical phonons.

1. Introduction

Nitride semiconductors have recently attracted considerable interest in view of their potentialities in semiconductor electronics [1] and optoelectronics. On the one hand, their wide direct band gap makes them suitable for the implementation devices emitting in the blue/UV spectral region [2–7]. On the other hand, strong quantum confinement effects would enable nitride-based devices to operate at room temperature. In fact, single-photon sources [8] and low-threshold lasers [9] have been recently demonstrated in GaN quantum dots and microdisks, respectively.

The electrons in a bulk or quantum structures become energetic (or hot) when a high electric field is applied, and the thermal equilibrium between the electron gas and the lattice is destroyed. The electrons relax by losing their excess energy to the lattice. This process follows several steps. First, the electrons having enough energy emit longitudinal optical (LO) phonons, Then, the emitted LO phonons decay into acoustic phonons and the excess energy drains out to the lattice. Finally, the electron gas and the lattice reach equilibrium separately at different temperatures, T and TL, respectively, coupled only weakly through electron-phonon interaction. In the steady state the input power is equal to the power loss to the lattice. At high electron temperatures the dominant mechanism for energy relaxation is emitting LO phonons. However, there are several parameters affecting the power loss of hot electrons: the electron-phonon coupling, electronic screening, hot-phonon and finite-thickness effects. Throughout the years, there has been a large amount of theoretical effort to explain the experimental results for the power loss of electrons in semiconductor bulk and lower dimensional structures [9–11]. In order to understand the power loss mechanism of hot electrons in GaN quantum

[1] Corresponding Author: asgari@tabrizu.ac.ir, Tel: +98-411-3393005, Fax: +98-411-3347050.

dots (microdisks), we made a detailed theoretical investigation. In this paper, the energy loss rate of quantum disk are calculated as a function of the external controllable parameters such as electrons density, n_s, disk width, L, lattice temperature, TL, and electron temperature, Te, for the interaction of the charge carriers with the acoustic (deformation potential and piezoelectric scattering) and optical phonons.

2. Model and Discussions

To study the properties of the hot electron scattering by phonon in GaN quantum dot, a GaN disk shaped structure, with radius (R~1) µm much larger than the its height ($L = 2-20\ nm$), has been assumed. The disk is sandwiched between two $Al_xGa_{1-x}N$ barriers, and surrounded by vacuum in the inplane directions (Fig. 1).

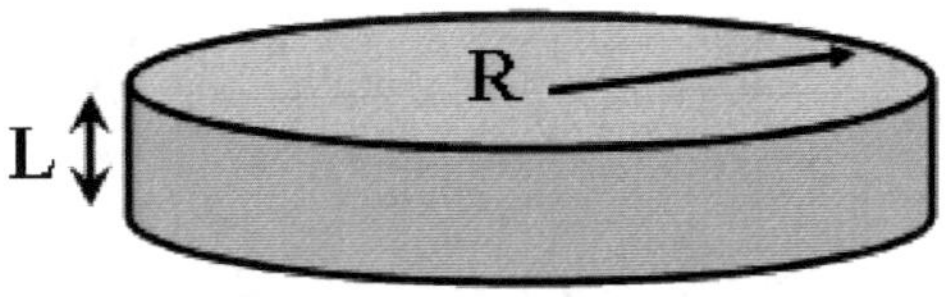

Figure 1. The proposed cylindrical shaped GaN QD within a AlGaN layer.

The single- and few-particle properties are modelled within the envelope-function and effective mass approximations. As a first step, we compute the single-particle states of electrons and holes. The condition R>>L, warrants the separation between the inplane (x, y) and vertical degrees of freedom (z). The z component of the single-particle eigenstates can thus be obtained by solving the following one-dimensional Schrödinger equation:

$$-\frac{\hbar^2}{2}\frac{\partial}{\partial z}\left(\frac{1}{m(z)}\frac{\partial \psi_\alpha}{\partial z}\right)+V_\alpha(z)\psi_\alpha = E_\alpha\psi_\alpha \qquad (1),$$

where a =e, h. The potential $V_\alpha(z)$accounts for the band offset and for the effect of the built-in electric field:

$$F_d = \frac{L_{br}(P_{tot}^{br} - P_{tot}^{d})}{\varepsilon_0(L_d\varepsilon_{br} + L_{br}\varepsilon_d)} \qquad (2).$$

Where $\varepsilon_{br(d)}$ is the relative dielectric constant of the barrier (dot), $P_{tot}^{br/d}$ is the total polarization and $L_{br/d}$ is the width of the barrier and height of the dot.

$$P_{tot}^{br/d} = P_{piezo}^{br/d} + P_{sp}^{br/d} \qquad (3).$$

The Piezoelectric polarization includes: one part induced by the lattice mismatch (ms), and the other caused by thermal strain (ts): $P_{piezo}^{br/d} = P_{ms}^{br/d} + P_{ts}^{br/d}$, where $P_{ms}^{br/d} = 2(e_{31} - e_{33}\frac{c_{13}}{c_{33}})(\frac{a-a_0}{a})$ and $P_{ts}^{d} = -3.2\times10^{-4}\,c/m^2$ [12]. e_{31} and e_{33} are the piezoelectric coefficients, c_{13} and c_{33} are elastic constants, and 'a' is the lattice constant of $Al_xGa_{1-x}N$ and is $a = (0.077\ x + 3.189)\ \mathring{A}$. (All other material parameters can be found in [13]).

The spontaneous polarization for $Al_xGa_{1-x}N$ is Al molar fraction dependent and is given by: $P_{sp} = (-0.052x - 0.029)$.

We employed the effective electron-temperature model, which assumes T > TL. Taking TL = 0, the power loss per electron is

$$P = \int E \frac{df(E)}{dt} N(E)\, d(E), \qquad (4).$$

Where E is electron energy, f(E) is electron distribution function, and N(E) is the density of states. The functionality of electron distribution function as time can be represented by:

$$\frac{df(E)}{dt} = \int_q \int_{k'} \{W(k,k') f(k')[1 - f(k)] - W(k,k') f(k)[1 - f(k')]\} \times \left[\frac{V}{8\pi^3}\right]^2 dk'\, dq \qquad (5),$$

and W(k,k') and W(k',k) are the absorption and emission scattering rate respectively and denoted as:

$$W(k,k') = \frac{2\pi}{\hbar} |M(k',k,q)|^2\, \delta(E_{k'} - E_k - \hbar\omega), \qquad (6),$$

$$W(k',k) = \frac{2\pi}{\hbar} |M(k,k',q)|^2\, \delta(E_{k'} - E_k + \hbar\omega), \qquad (7),$$

and

$$M(k',k,q) = \langle k' | H^a_{e-ph} | k + q \rangle \qquad (8),$$

$$M(k,k',q) = \langle k' | H^e_{e-ph} | k - q \rangle \qquad (9),$$

Here, $|k\rangle$, and $|k'\rangle$ is the electron wavefunction which can be obtained from Schrödinger equation. $H^{a(e)}{}_{e-ph}$ is the Hamiltonian for phonon absorption (emission) [14-15].

Now, using numerical methods one can solve the equation (4) and find the phonons energy loss rates (ELR). For this purpose, we first solve it for infinite GaAs and GaN quantum wells and compared the results in fig.2. As shown in this figure, the optical phonon ELR for GaN is higher than GaAs, but the Acoustic phonon ELR behaviour is different.

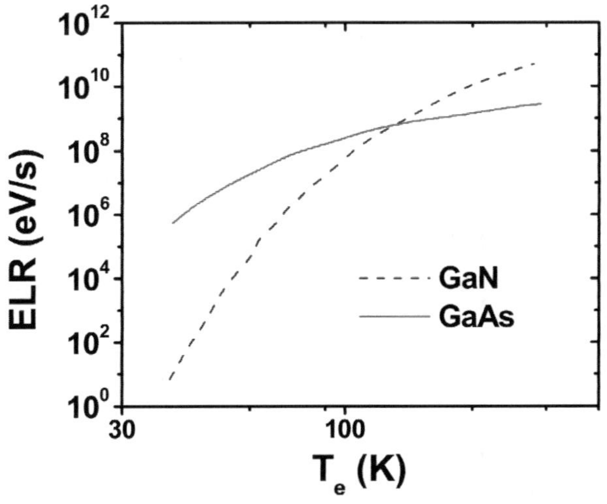

Figure 2. The comparison of optical phonon energy loss rate (ELR) in GaN and GaAs infinite quantum well with well width of 20 nm and TL=20K.

Fig. 3 shows the optical phonon ELR for GaN microdisks as function of microdisk thickness for different electron temperate. As evident from the figure, with increasing the disk size the ELR decreases; and also it is hardly depends on the electron temperature and increases with its increasing.

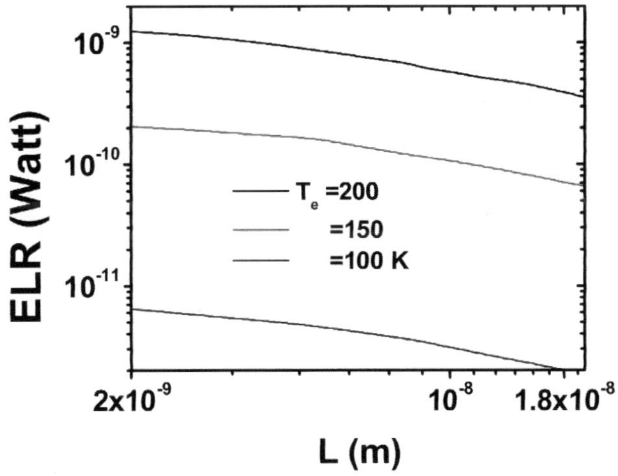

Figure 3. The optical phonon energy loss rate (ELR) in GaN microdisks as function of disk thickness, for different electron temperature and Tl=4.2 K.

In Fig. 4, we calculated the different part of acoustic phonon ELR. As depicted in the figure, for TL=4.2 and Te=10K, the piezoelectric ELR is much higher than the deformation potential. Also it is found that the thicker disks have low acoustic phonon ELR. This behaviour is same for optical phonon ELR (Fig. 5). As shown in Fig. 5, for higher temperature the ELR difference between two disks is much higher than at low temperatures.

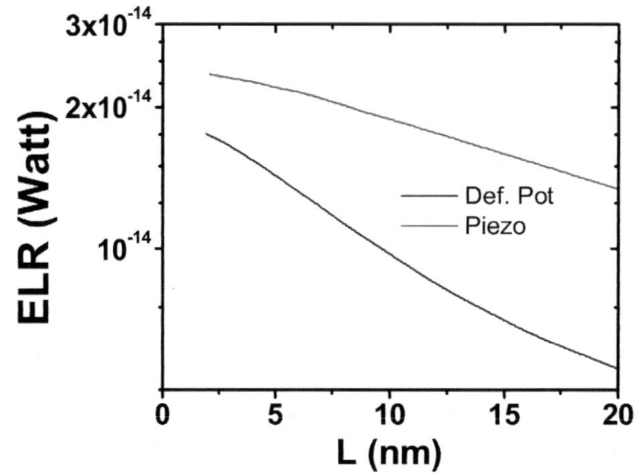

Figure 4. The Acoustic phonon energy loss rate (ELR) in GaN microdisks as function of disk thickness. TL=4.2 K. and Te=10K.

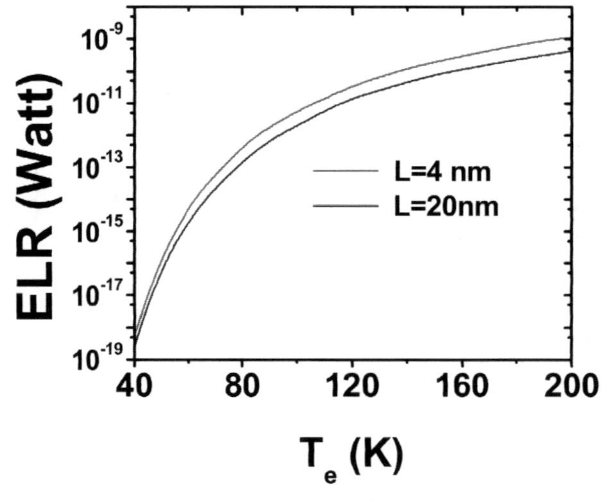

Figure 5. The optical phonon energy loss rate (ELR) in GaN microdisks as function of electron temperature for different disk thickness at TL=4.2 K.

The different part of acoustic phonon ELR (piezoelectric and deformation potential) are calculated for different electron temperature. Our calculation results show that for very low electron temperature (Te<12 K), the piezoelectric is dominant scattering mechanism, but for Te>12K, the deformation potential is dominant. In Fig. 6, the piezoelectric, the deformation potential and the total acoustic phonon ELR is plotted as function of electron temperature.

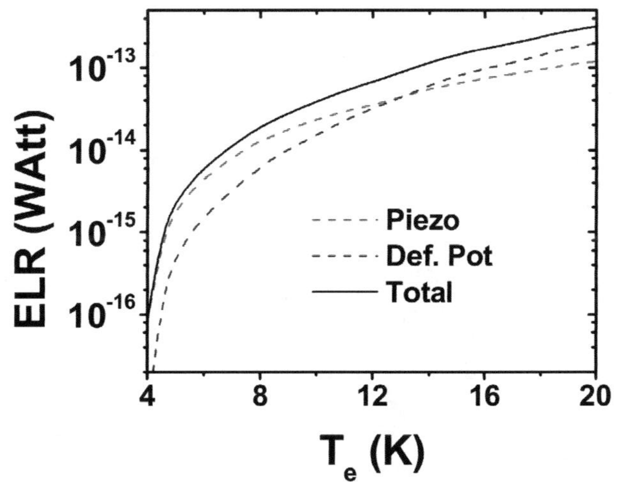

Figure 6. The acoustic phonon energy loss rate (ELR) in GaN microdisks as function of electron temperature at TL=4.2 K. The disk thickness is 4nm. The solid line is the sum of piezoelectric (blue dashed) and deformation potential (red dashed) energy loss rate.

Finally, the total phonon ELR is shown in Fig. 7 for the GaN microdisks with thickness of 4 nm and TL=4.2 K. It is found that the acoustic phonon is dominant scattering mechanism at temperature lower that ~100K, and the optical phonon is dominant scattering mechanism for higher temperature. It should bentioned that the electron density used in calculation is $1\times10^{17}\,m^{-2}$.

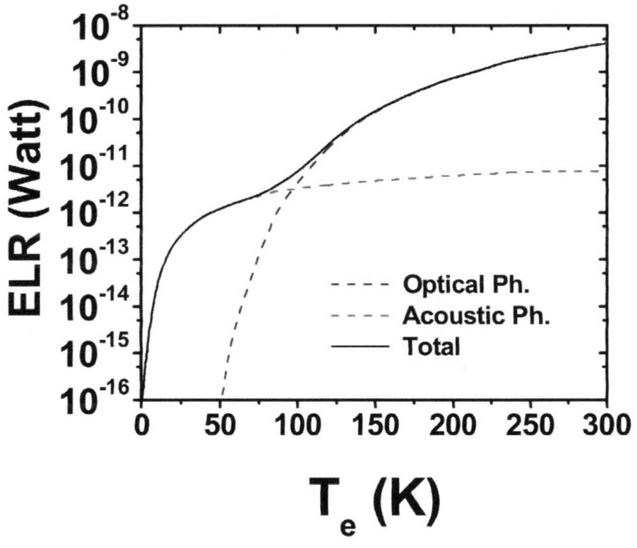

Figure 7. The phonon energy loss rate (ELR) in GaN microdisks (R=1um, L=4nm) as function of electron temperature at Tl=4.2 K. The solid line is the sum of optical (blue dashed) and acoustic (red dashed) phonons energy loss rate.

3. Conclusion

We have studied the energy loss rate by electrons via acoustic (deformation potential and piezoelectric mechanism) and optical phonons with using an accurate numerical solution for calculation, in GaN quantum dots (microdisks). It is found that for electron temperature higher than 100K, the optical phonons scattering is dominant and for lower than 70K, the acoustic phonons is overcome. Also, our calculation results show that for very low electron temperature (Te<12 K), the piezoelectric ELR is much higher than the deformation potential, but for Te>12K, the deformation potential ELR is higher than the piezoelectric.

4. References

[1] A. Asgari, M. Kalafi, L. Faraone, 2004, J. Appl. Phys 95, 1185.
[2] S. Nakamura, S. Pearton, G. Fasol, 2000, The Blue Laser Diode, Springer, New York.
[3] S.N. Mohammad, A.A. Salvador, H. Morkoc, Proc. IEEE 83, 1306 (1995)
[4] R. Davis, 1991, Proc. IEEE 79, 702.
[5] R.A. Mair, K.C. Zeng, J.Y. Lin, H.X. Jiang, B. Zhang, L. Dai, A. Botchkarev, W. Kim, H. Morkoc, M.A. Khan, 1998, Appl. Phys. Lett. 72, 1530.
[6] S. Chang, N.B. Rex, R.K. Chang, G. Chong, L.J. Guido, 1999, Appl. Phys. Lett. 75, 166.
[7] S.L. McCall, A.F.J. Levi, R.E. Slusher, S.J. Pearton, R.A. Logan, 1992, Appl. Phys. Lett. 60, 289.
[8] S. Kako, C. Santori, K. Hoshino, S. Gotzinger, Y. Yamamoto, Y. Arakawa, 2006, Nature materials 5, 887.
[9] E. Conwell, 1963, High Field Transport in Solids, Academic Press, New York.
[10] B. K. Ridley, 1982, J. Phys. C: Solid State Phys. 15, 5899.
[11] S. Das Sarma, J. K. Jain, and R. Jalabert, 1988, Phys. Rev. B 37, 1228.
[12] S. De Rinaldis, I. D Amico, E. Biolatti, R. Rinaldi, R. Cinagolani, F. Rossi, 2002, Phys. Rev B, 65, 081309R.
[13] R. Cingolani, A. Botchkarev, H. Tang, H. Morkoç G. Traetta, G. Coli, M. Lomascolo A. Di Carlo, F. Della Sala, and P. Lugli, 2000, Phys Rev B 61, 2711.
[14] A. J. Vickers, 1992, phys. Rev. B, vol. 46, 13313.
[15] B. K. Ridley, 1991, Rep. Prog. Phys. 54, 169.

Journal of Physics: Conference Series **248** (2010) 012023 doi:10.1088/1742-6596/248/1/012023

Dynamics of Na clusters in/on insulating substrates

M Bär[1], B Faber[1], P-G Reinhard[1], [2] P M Dinh,[2] E Suraud,[2] P Wopperer

[1] Institut für Theoretische Physik, Universität Erlangen, Staudtstrasse 7, D-91058 Erlangen, Germany
[2] Laboratoire de Physique Théorique, UMR CNRS – Université Paul Sabatier, 118, route de Narbonne F-31062 Toulouse Cédex, France

E-mail: paul-gerhard.reinhard@physik.uni-erlangen.de

Abstract. We apply a hierarchical model for Na clusters in/on insulating substrate (Ar or MgO) to explore three different regimes of cluster dynamics. The effect of the interface is scrutinized for optical response of Na_8 embedded in Ar matrix and deposited on MgO(001) surface. The different mechanical properties of Ar and MgO surface are explored in studying the collision with a Na_6 cluster at different initial energies. The stabilization of highly charged Na clusters through a substrate is discussed for strong laser ionization of Na_8 in Ar matrix and on MgO surface.

1. Introduction

The impressive progress of laser physics has motivated research on the dynamics of clusters exposed to strong perturbations, see e.g. [1]. Practically, a great deal of cluster research is done for clusters in contact with a substrate, deposited or embedded, see e.g. [2–4]. Most of these studies are concerned with the details of structure and chemical bonding. Highly excited dynamics of embedded or deposited clusters is still at the edge of present days possibilities, in particular from the theoretical side. The challenge is to describe large substrate systems together with the quantum dynamical details of an electronically active, but small, subsystem. For such situations, hierarchical approaches have been developed called quantum-mechanical molecular-mechanical modeling (QM/MM) [5–8]. These have been recently implemented for metal clusters in contact with electronically inert substrates [9–11]. We will report here a few selected applications of the QM/MM modeling for Na cluster with Ar or MgO substrate to cluster dynamics: the effect of the interface on optical response, the dynamics of cluster deposition, and (hindered) Coulomb explosion of a cluster which is highly charged by a strong laser pulse.

2. Formal framework

The hierarchical QM/MM model has been described in detail elsewhere, for MgO substrate in [11, 12] and for Ar substrate in [9, 13, 14], for a recent review see [15]. Here we give a brief summary.

The constituents of the Na cluster and their degrees of freedom are: $\varphi_n(\vec{r},t)$, $n = 1...N_{\rm el}$ for the valence electrons of the Na cluster, and the positions of the Na^+ ions. The Na cluster is treated in by time-dependent local-density approximation with molecular dynamics (TDLDA-MD), the valence electrons quantum-mechanically and the ions classically [16, 17]. The MgO

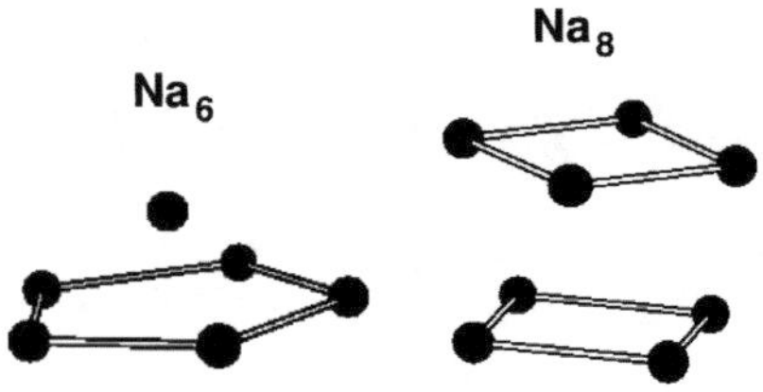

Figure 1. Spatial structure of the free Na_6 and Na_8 clusters.

substrate is composed of two species, Mg^{2+} cations and O^{2-} anions with the degrees of freedom: $\vec{R}_{i^{(c)}}$, $i^{(c)} = 1...M$ for the positions of the O cores, $\vec{R}_{i^{(v)}}$, $i^{(v)} = 1...M$ for the centers of the O valence clouds, $\vec{R}_{i^{(k)}}$, $i^{(k)} = 1...M$ for the positions of the Mg^{2+} ions. The cations are electrically inert and are modeled as charged point particles. The anions are easily polarizable which is accounted for by associating to it two constituents : A valence electron distribution and the complementing core. All parts of the MgO substrate are described as classical degrees of freedom. The difference $\mathbf{R}^{(c)} - \mathbf{R}^{(v)}$ describes the electrical dipole moment of an O^{2-} anion. The Mg and O ions are arranged in crystalline order corresponding to bulk MgO. The dynamical degrees of freedom for Mg and O are taken into account in an active cell of the MgO(001) surface region underneath the Na cluster. The active cell is embedded in an "outer region" of MgO material where only ions are kept fixed, while oxygen dipoles remain fully dynamical. Beyond that region, only the Madelung potential is considered. The effect of the outer region on the active part is given by a time-independent shell-model potential taken over from [18]. The Ar substrate is composed of only once species, neutral Ar atoms, each of which being characterized by its position and dipole moment (exactly as O^{2-} anions).

The atomic dipoles allow to explicitly treat the dynamical polarizability of the atoms through polarization potentials [19], which are the dominant agents at long range. The dipole parameters are adjusted to reproduce the dynamical polarizability of the material at low frequencies. The short-range repulsion between the constituents is obtained by adding especially constructed local core potentials. In case of Ar, the Van-der-Waals interaction is also accounted for in an approximate way [13].

The modeling for the case of Ar substrate was developed and tested first in [20] and its TDLDA extension in [9]. The QM/MM model for MgO substrate was developed first for noble metal ad-atoms and clusters [21–24] and has been extended to alkaline clusters in [11].

3. Results and discussion

3.1. A word on cluster structure

The ionic structures of free Na_6 and Na_8 are illustrated in figure 1. Na_6 has fivefold symmetry, consisting out of a ring of 5 ions plus one ion at the symmetry axis on top of the ring. Na_8 is a highly symmetric configuration consisting out of two rings of 4 ions each twisted relative to each other by 45° to minimize Coulomb energy.

Various substrates are considered. The MgO(001) substrate consists of six layers each containing 784 Mg^{2+} ions and 784 O^{2-} ions. The ions in the lowest layer are fixed to prevent it from relaxing and forming an artificial second surface. The active cell consists of three layers, each containing square arrangements of 242 Mg^{2+} cations and 242 O^{2-} anions [11, 12]. The Ar(001) substrate is built from layers of 8×8 Ar atoms. These plaquettes are copied periodically in lateral direction to simulate a planar surface. Six layers are used if not explicitly quoted differently. Embedding in Ar matrix is simulated by using a large Ar cluster at whose center a cavity is carved by removing 13 Ar atoms [13]. We will consider here Ar_{164} and Ar_{434} surroundings.

3.2. Optical response

Figure 2 shows the dipole strengths for Na_8 in various configurations. The right panel

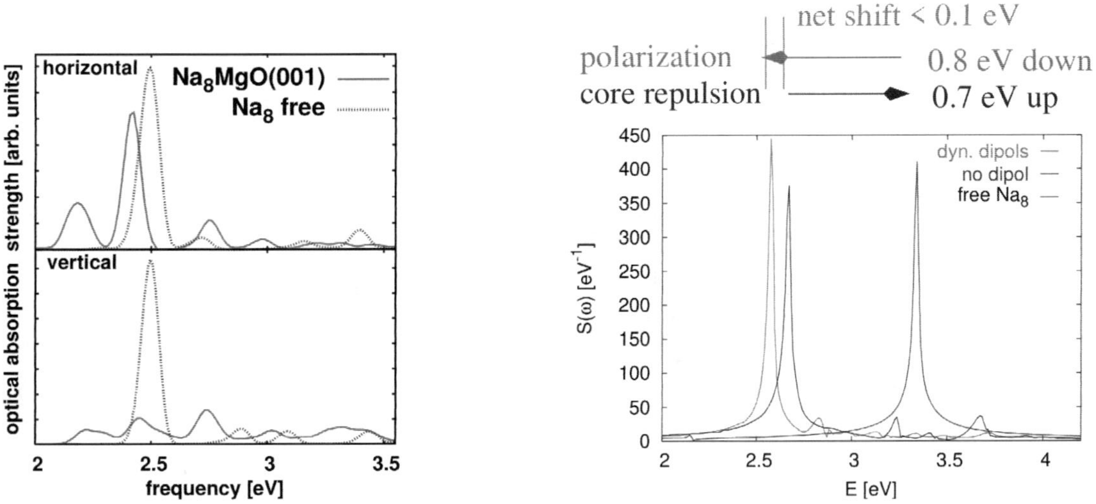

Figure 2. Optical absorption spectra of Na_8 in different environments. Left: Comparison of free Na_8 with Na_8@MgO(001) for modes vertical to the MgO surface (lower) and parallel to it (upper). Right: Comparison of free Na_8 (black) with Na_8 embedded in an Ar_{164} matrix. The full calculation is shown with blue line and is denoted "dyn. dipol". The case "no dipol" (red curve) shows a calculation where the polarization potentials had been switched off.

demonstrates the effect of embedding in an Ar matrix for the mode along z-direction (symmetry axis). All spectra exhibit a clean Mie plasmon peak with only little strength left at other places. The surprising fact is that the average resonance position changes only very little by the embedding (compare blue with black line). We have also computed the average peak frequency by switching off the polarization potentials, thus leaving only the effect of short range repulsion (red thin line) which causes a strong blue shift. This large shift is almost perfectly counterweighted by the attractive effect of dynamical polarization of the medium. The balance can change a little bit when considering substrates of heavier rare-gases with their larger polarizabilities [25].

The left panel of figure 2 shows the effect of deposition on MgO. The modes horizontal to the surface (upper left panel) show only moderate changes. There is a slight red shift of the average resonance peak and a spectral fragmentation because one $1ph$ state is coming close to the resonance. The net red shift emerges from the fact that there is only a small core repulsion in horizontal direction (coming from surface corrugation). The vertical mode is heavily perturbed by fragmentation over several $1ph$ states. This is caused by symmetry breaking which enables coupling of the negative-parity dipole mode to the swamp of $1ph$ modes with even parity [26].

3.3. Dynamics of cluster deposition

Figure 3 illustrates the deposition dynamics of a Na_6 cluster at moderate collisional energies for Ar(100) and MgO(100) surface. A broader range of energies and configurations for Ar substrates is discussed in [15, 27]. The Na_6 center-of-mass starts from 15 a_0 above the surface facing the ring side towards the surface. The initial boost gives it a kinetic energy $E^0_{\rm kin}$ of 0.1 Ry (left) or 0.01 Ry (right) per Na ion. The low energy case (middle and right panels) show fast stopping and quick capture of the Na cluster. The other aspects of the process develop very differently for

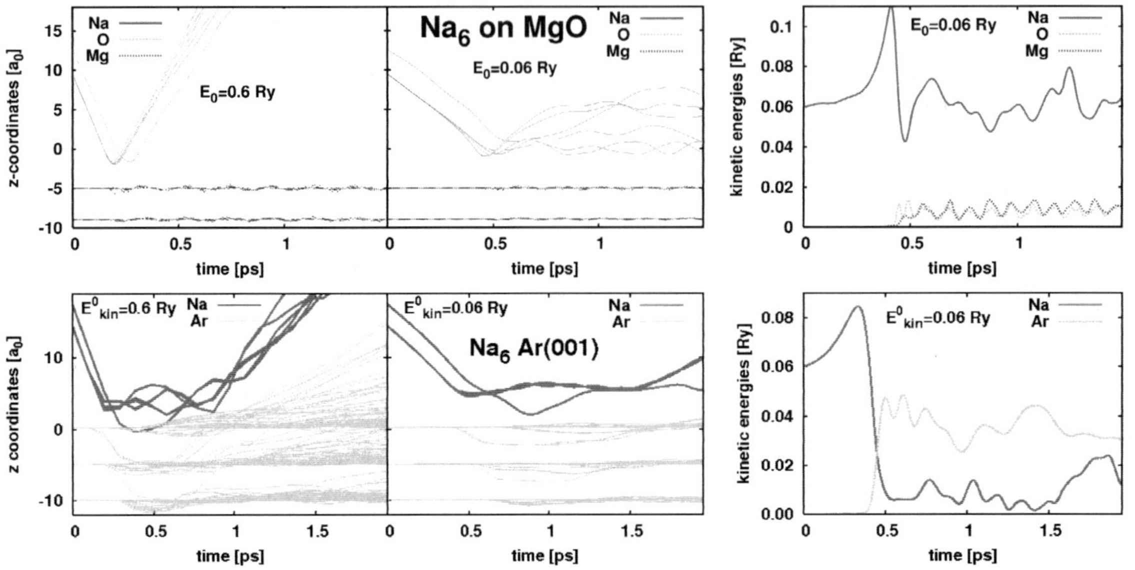

Figure 3. Dynamics of deposition of a Na_6 cluster on MgO(001) substrate (upper panels) or Ar(001) (lower panels). Left: time-evolution of z-coordinates for an impinging kinetic energy of 0.6 Ry. Middle: time-evolution of z-coordinates for an much smaller impinging kinetic energy of 0.06 Ry. Right: time evolution of kinetic energies of cluster ions and substrate atoms for a kinetic energy of 0.06 Ry.

the two different substrates. For Ar(100), stopping is achieved by the substrate which is highly excited and leaves little energy to the cluster just sufficient to allow gentle oscillations of the top ion through the ring forth and back. The Mg(100) surface, however, is hardly excited at all and the sticking of Na_6 is achieved by conversion of the initial center-of-mass energy into huge internal cluster excitation. The complementing information on kinetic energies (right panels) confirms this: the soft substrate Ar(100) absorbs nearly all impact energy while the rather hard MgO(100) substrate remains almost unaffected. Both cases agree in that the main energy transfer proceeds at a very short time scale, within a few ten fs. Further thermal equilibration takes much longer at a scale of ns, far beyond simulation time [28]. The high energy cases (left panels) show both a reflection of the cluster, but different in detail: MgO(001) leads to hard reflection with little energy loss while the Ar(001) surface absorbs a great deal of incoming energy and is substantially stirred up.

The dissipative effect of the Ar substrate depends much on the number of active layers. There are experiments which consider only a few layers of Ar deposited on a metal carrier [29] which delivers a mechanically inert support. On the other hand, the Ar layers behave as a soft mechanical stopper in a deposition process while being chemically inert [30]. Figure 4 shows the effect of varying the number of layers (2, 4 and 8). The pattern of the z-coordinates (left panels) indicate already a significant change from almost "hard" reflection to dissipative attachment. The kinetic energies (middle panels) corroborate this, showing increasing energy transfer to the substrate with increasing number of layers. The right panel contains a global analysis of absorbed energy. There is a strong increase of absorption for the first few layers which then levels off to an asymptotic value at around 6 layers.

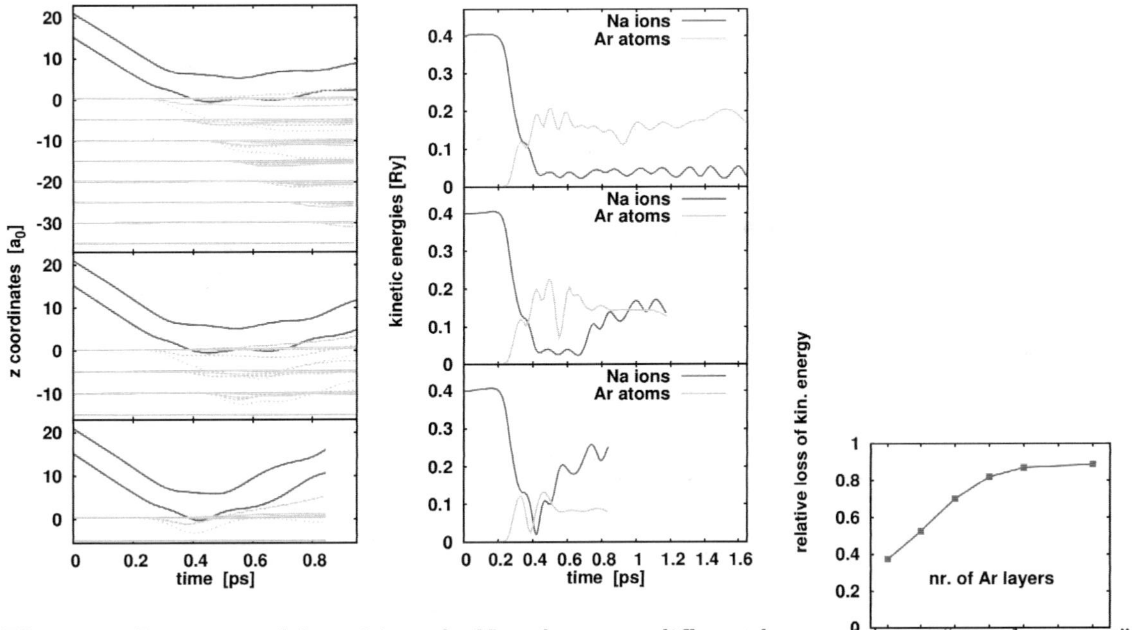

Figure 4. Dynamics of deposition of a Na_8 cluster on different layers of Ar on “metal support”. Left: time-evolution of z-coordinates for an impinging kinetic energy of 0.4 Ry for 2 layers (lower), 4 layers (middle) and 8 layers (upper). Middle: time evolution of kinetic energies of cluster ions and substrate atoms for the cases as shown in the left panels. Right: Relative loss of cluster kinetic energy ($E_{\mathrm{kin,out}}/E_{\mathrm{kin,in}}$) as a function of the number of layers.

3.4. Dynamics after strong laser ionization

Laser irradiation leads to strong ionization of the cluster which can exceed the limits of Coulomb-stability of the cluster. This has been much studied in the past under the notion “appearance size”, for a review see [31]. For Na clusters, the smallest Na cluster with charge state $Q = 2^+$ has been observed for Na_{10}^{++} in experiments with laser pulses [32]. Free Na_8 becomes unstable for ionization $Q \geq 2^+$. The interesting question is how a substrate modifies the charge stability.

Figure 5 shows results of laser excitation of Na_8 embedded in Ar_{434}. The short and intense laser pulse produces very quickly a net ionization stage of $Q = 2^+$ (lower), 3^+ (middle), or 4^+ and so deposits a large amount of energy in the Na_8 cluster. This fast initial electron emission proceeds much similar as for the free cluster. The finite Ar environment does not impose any hindrance for the escaping electrons. The Coulomb pressure generated by the almost instantaneous ionization starts to drive the cluster ions apart. The dominant effect is radial expansion, visualization in the figure in terms of radial coordinates $r = \sqrt{x^2 + y^2 + z^2}$. The cavity inside Ar is spacious. Thus we see for the initial 200 fs a fast expansion/explosion of the cluster almost as in the free case. The expansion is stopped abruptly if the cluster encounters the repulsive Ar core. The stopping radius increases with the charge state Q. After stopping, the cluster turns over into oscillations for a while. The Ar atoms take up the momentum from the stopped ions which then propagates like a sound wave through the Ar medium. The perturbation $Q = 4^+$ is strong enough to produce finally an emission of Ar atoms from the outmost shell. The sound wave distributes the energy very quickly over all shells exciting them to more or less hefty fluctuations. The Ar amplitudes increase with increasing charge state Q which is related to the initial amount of Coulomb energy $\propto Q^2$. In spite of the strong energy inflow, the Ar matrix stays basically intact and represents an efficient cage for the highly charged Na cluster. Charge states 2^+ and 3^+ are finally stabilized. Only transient stabilization is achieved for $Q = 4^+$. At

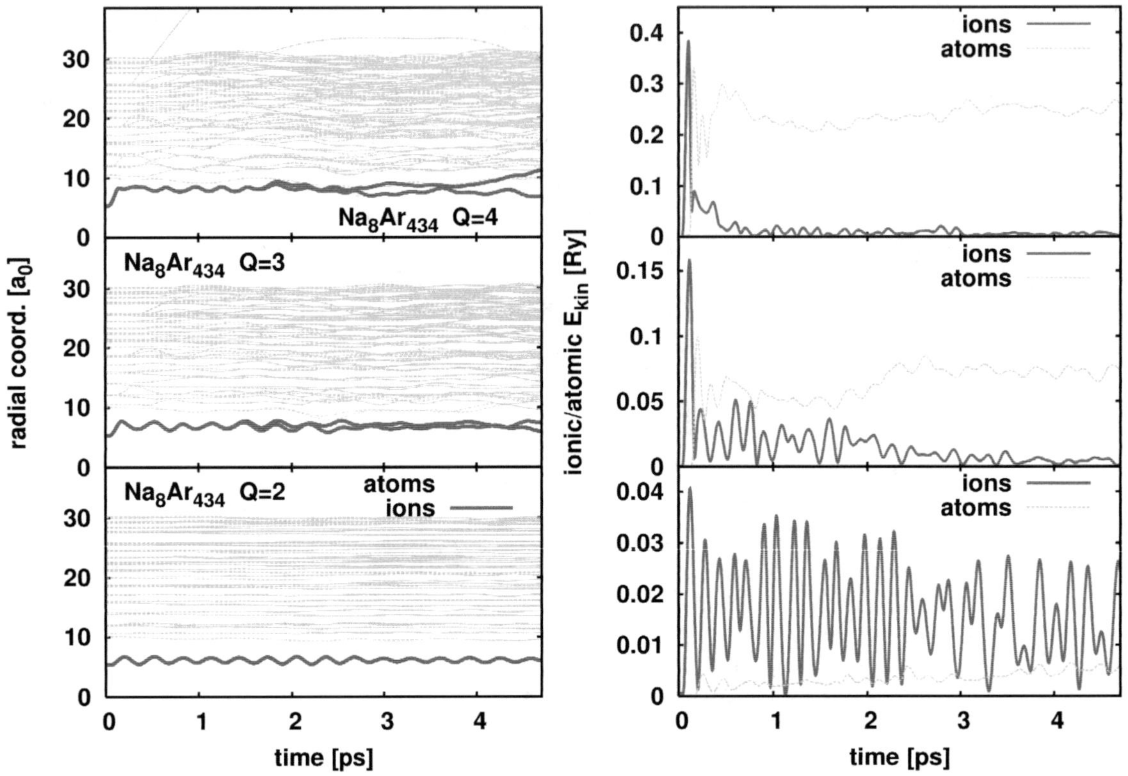

Figure 5. Dynamics of a Na_8 cluster embedded in an Ar_{434} and excited by a short laser pulse to initial charges states $Q = 2^+$ (lower), $Q = 3^+$ (middle), and $Q = 4^+$ (upper). The laser pulse had frequency $\omega = 0.14$ Ry, pulse length $T_{\rm pulse} = 50$ fs, polarization along the z axis (symmetry axis of Na_8) and intensity tuned to obtain the wanted ionization stages Q. Left: time-evolution of radial-coordinates of cluster ions (red thick curve) and substrate atoms (green thin dashes). Right: time evolution of kinetic energies of cluster ions and substrate atoms.

about 4 ps, the cluster expansion revives and starts a Coulomb driven diffusion through the medium (visible when looking for longer time spans [33]).

The kinetic energies shown on the right panels allow to read off a few time scales. The initial Coulomb explosion lasts only for about 200 fs. The distribution of energy over the Ar shells is done within 0.5-0.7 ps for $Q = 3^+$ and even faster for $Q = 4^+$. It propagates at about the speed of the sound wave which is about 30 a_0/ps, close to the velocity of sound in pure Ar. The more gentle stopping for $Q = 2^+$ does not transfer so much energy at the first encounter. The motion turns to stable oscillations of embedded Na_8^{++} where energy transport between the various sub-systems proceeds on much longer time scales [28].

The stabilization by a cage for embedded clusters is plausible. One wonders what the effect of the substrate may be for deposited clusters. This is illustrated in figure 6 for the case of Na_8@MgO(001) laser-ionized to charge state $Q = 3^+$. The upper panel shows for comparison free Na_8 under the same conditions . The z-coordinates (upper left panel) signal a clear Coulomb explosion. The trajectories (upper right panel) indicate a strictly radial expansion of all the 8 ions. The case of the deposited cluster (lower panels) behaves much differently. The cluster does not breakup nor release single ions. The system sticks together in spite of the huge charge state, of the dramatic perturbation and of much open space above the surface. Here, the glue

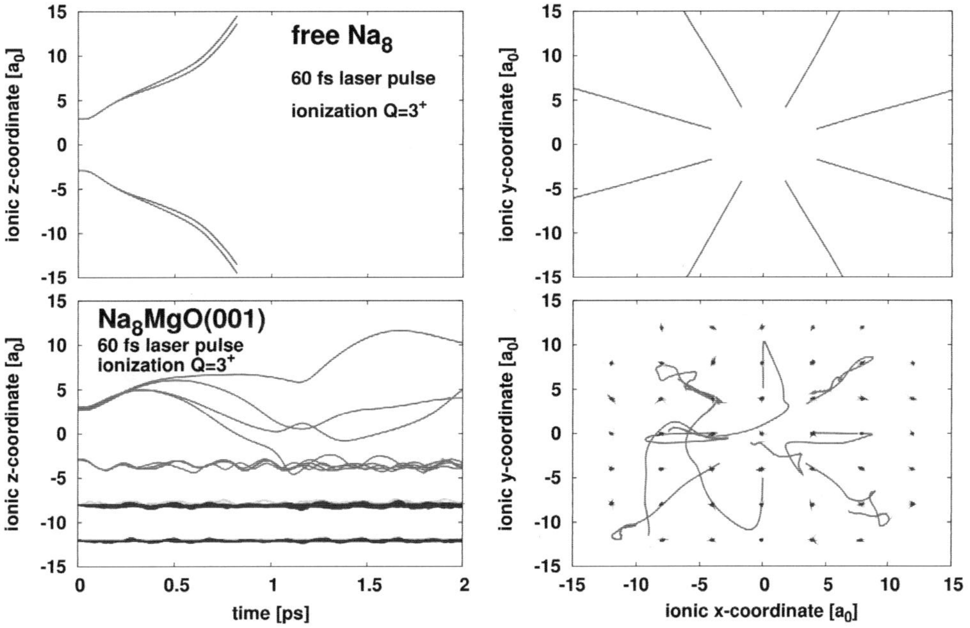

Figure 6. Dynamics of a free Na_8 cluster (upper) and a Na_8 cluster (lower) deposited on MgO(001) surface and excited by a short laser pulse to initial charges state $Q = 3^+$. The laser pulse had frequency $\omega = 0.14$ Ry, pulse length $T_{\rm pulse} = 60$ fs, and intensity tuned to obtain ionization $Q = 3^+$. Left: time-evolution of z-coordinates of cluster ions (red) and substrate atoms (Mg green, O blue). Right: trajectories of Na ions and substrate atoms in the x-z-plane. atoms. The z-direction is vertical to the surface and x horizontal.

is delivered by the strong and attractive polarization interaction with the substrate. The final state is not a compact cluster. It rather looks like an almost unperturbed lower ring of 4 ions still tied to the surface while the outer 4 ions remain only loosely connected and undergo large spatial fluctuations.

4. Conclusions and outlook

We have applied a recently developed hierarchical modeling for Na clusters in contact with Ar or MgO substrate to three different examples from cluster dynamics with the aim to explore the effect of the interface.

The optical response of metal clusters produces a dominant peak, the Mie plasmon resonance. Its average position is almost unaffected by the interface. This is result of a subtle balance between core repulsion and attractive dynamical polarizability of the material. However, the detailed fragmentation of the spectrum over $1ph$ states near the resonance is volatile and can change with change of surroundings. In particular, the reflection symmetry breaking through a substrate surface dissolves the mode vertical to the surface into a very broad distribution.

The collision of a cluster with Ar(001) or MgO(001) surface shows at first glance similar trends, reflection at high initial energies and cluster attachment for low energies. However, the mechanisms work out differently in detail. The soft Ar substrate acts as an efficient energy absorber producing a soft landing in case of attachment and a strongly dissipative "reflection" which, in fact, almost destroys the impact spot of the surface. The hard material MgO(001), on the other hand, takes up very little energy from the colliding cluster. This produces almost clean reflection but leaves a large amount of internal cluster excitation in case of attachment.

Laser excitation was tuned to produce quickly highly charged clusters and so to allow a study of Coulomb driven cluster expansion (or explosion). We find that embedding in Ar matrix as well as deposition on MgO surface both stabilize high charge states of the cluster which would Coulomb-explode immediately for a free cluster. The mechanisms, however, are different. Embedding in a rare-gas matrix confines the cluster by inertia of the surrounding atoms. Deposition on a polarizable, insulating substrate stabilizes the high charges by a strong and attractive polarization interaction.

Acknowledgments

The authors thank T. Moskaleva and N. Rösch for their invaluable help with benchmark data during the development of the QM/MM model. This work was supported by the Deutsche Forschungsgemeinschaft (RE 322/10-1).

References

[1] Fennel T, Meiwes-Broer K H, Tiggesbäumker J, Reinhard P G, Dinh P M and Suraud E 2010 *Rev. Mod. Phys.* **82** 1793
[2] Meiwes-Broer K H, ed 2000 *Metal clusters at surfaces* (Berlin: Springer)
[3] Meiwes-Broer K H, ed 2006 *Clusters at Surfaces: Electronic Properties and Magnetism* vol 82 Applied Phys. A, special issue
[4] Meiwes-Broer K H and Berndt R, eds 2007 *Atomic Clustes at Surfaces and in Thin Films* vol 45 Eur. Phys. J. D, topical issue
[5] Gresh N, Parisel O and Giessner-Prettre C 1999 *THEOCHEM* **458** 27
[6] Rösch N, Nasluzov V A, Neyman K M, Pacchioni G and Vayssilov G N 2004 in J Leszczynski, ed, *Computational Material Science* (Amsterdam: Elsevier) Theoretical and Computational Chemistry Series, Vol. 15 p 367
[7] Moret M E, Tapavicza E, Guidoni L, Röhrig U, Sulpizi M, Tavernelli I and Rothlisberger U 2005 *Chimia* **59** 493
[8] Bucher D, Raugei S, Guidoni L, Peraro M D, Rothlisberger U, Carloni P and Klein M L 2006 *Biophysical Chemistry* **124** 292
[9] Gervais B, Giglio E, Jaquet E, Ipatov A, Reinhard P G and Suraud E 2004 *J. Chem. Phys.* **121** 8466
[10] Fehrer F, Reinhard P G, Suraud E, Giglio E, Gervais B and Ipatov A 2006 *Appl. Phys. A* **82** 151
[11] Bär M, Moskaleva L V, Winkler M, Reinhard P G, Rösch N and Suraud E 2007 *Euro. Phys. J. D* **45** 507
[12] Baer M 2006 *Metallcluster in Kontakt mit Edelgaassubstraten* Ph.D. thesis Universität Erlangen/Nürnberg
[13] Fehrer F, Reinhard P G, Suraud E, Giglio E, Gervais B and Ipatov A 2005 *Appl. Phys. A* **82** 151
[14] Fehrer F, Mundt M, Reinhard P G and Suraud E 2005 *Ann. Phys. (Leipzig)* **14** 411
[15] Dinh P M, Reinhard P G and Suraud E 2009 *Phys. Rep.* **485** 43
[16] Reinhard P G and Suraud E 2003 *Introduction to Cluster Dynamics* (New York: Wiley)
[17] Calvayrac F, Reinhard P G, Suraud E and Ullrich C A 2000 *Phys. Rep.* **337** 493
[18] Nasluzov A, Neyman K, Birkenheuer U and Rösch N 2001 *J. Chem. Phys.* **115** 17
[19] Dick B G and Overhauser A W 1958 *Phys. Rev.* **112** 90
[20] Duplàa F and Spiegelmann F 1996 *J. Chem. Phys.* **105** 1492
[21] Matveev A V, Neyman K M, Pacchioni G and Rösch N 1999 *Chem. Phys. Lett.* **299** 603
[22] Ferrari A M, Xiao C, Neyman K M, Pacchioni G and Rösch N 1999 *Phys. Chem. Chem. Phys.* **1** 4655
[23] Inntam C, Moskaleva L V, Neyman K M, Nasluzov V A and Rösch N 2006 *Appl. Phys. A* **82** 181
[24] Rösch N, Nasluzov V A, Neyman K M, Pacchioni G and Vayssilov G N 2004 in J Leszczynski, ed, *Computational Material Science* (Amsterdam: Elsevier) Theoretical and Computational Chemistry Series, Vol. 15 p 367
[25] Fehrer F, Dinh P M, Suraud E and Reinhard P G 2007 *Phys. Rev. B* **75** 235418
[26] Montag B and Reinhard P G 1995 *Phys. Rev. B* **51** 14686
[27] Dinh P M, Fehrer F, Reinhard P G and Suraud E 2007 *Euro. Phys. J. D* **45** 415
[28] Fehrer F, Reinhard P G and Suraud E 2005 *Appl. Phys. A* **82** 145
[29] Irawan T, Boecker D, Ghaleh F, v Issendorf B and Hövel H 2005 *Appl. Phys. A* **82** 81
[30] Lau J T, Wurth W, Ehrke H U and Achleitner A 2003 *Low Temperature Physics* **29**(3) 223–227
[31] Näher U, Björnholm S, Frauendorf S, Garcias F and Guet C 1997 *Phys. Rep.* **285** 245
[32] Schmitt C, Ellert C, Schmidt M and Haberland H 1997 *Z. Phys. D* **42** 145
[33] Fehrer F, Dinh P M, Baer M, P-G, Reinhard and Suraud E 2007 *Euro. Phys. J. D* **45** 447

Journal of Physics: Conference Series **248** (2010) 012024 doi:10.1088/1742-6596/248/1/012024

The self-interaction correction in the time domain

P M Dinh[1], J Messud[1], P-G Reinhard[2] and E Suraud[1]

[1]Laboratoire Physique Théorique, IRSAMC, CNRS and UPS, Université de Toulouse, 118 Route de Narbonne, 31062 Toulouse cedex, France

[2]Institut für Theoretische Physik, Universität Erlangen, Staudstrasse 7, D91058 Erlangen, Germany

E-mail: `suraud@irsamc.ups-tlse.fr`

Abstract. We provide a brief outline of the Self Interaction Correction (SIC) used in connection with the Local Density Approximation (LDA) of Density Functional Theory (DFT), pointing out especially the implications for Time Dependent theories (TDDFT). We discuss how the self-interaction problem can be resolved by a clean formulation of the SIC theory including in an explicit manner the ortho-normalization of Kohn Sham orbitals. We finally show a few examples of applications of the theory.

1. Introduction

The dynamics of finite Fermion systems under strong perturbations has focused numerous investigations, in many domains of physics, ranging from electronic dynamics, especially in clusters [1] and plasma physics [2] to nuclear collisions [3]. From the theoretical point of view the problem is still far from being solved. Several difficulties remain, both at the formal and practical levels. The formulation of well founded theories for the SIC thus remains a key issue. The practical goal of our investigations is specifically to study the dynamics of clusters and molecules subject to an irradiation by a laser or a charged projectile. We aim at providing a fully dynamical description of such processes including electron and ion dynamics, as well as possible response of the environment when present (as e.g. in the case of solvated molecules [4] or embedded/deposited clusters [5, 6]). We furthermore require a high degree of accuracy especially at low energy, as we aim, in particular, to analyze phenomena close to ionization thresholds. Density Functional Theory (DFT) [7], extended in the time domain (Time Dependent DFT = TDDFT) [8], appears here as a most promising theory for addressing such questions. DFT provides a major tool of investigation in numerous physical and chemical systems since several decades, both for static and dynamical problems. It now constitutes a well founded theoretical basis on which to develop elaborate approaches. It, nevertheless, still raises numerous formal difficulties. We shall address here the specific well known pathology of the self-interaction error which is non-ignorable in the dynamical description of irradiation processes which usually involve electronic emission.

The simplest approximation to DFT is the well known Local Density Approximation (LDA). It provides a simple ansatz to account for exchange and correlation on the basis of a Fermi Gas approximation. It constitutes an extremely useful tool in DFT because of its simplicity and robustness. The LDA approximation furthermore provides a most sound basis for many theoretical developments. But it suffers from a self-interaction error: one electron interacts with the mean-field, itself a functional of the total density (hence including this one electron itself). Each electron thus spuriously interacts with itself. This pathology, typical of Hartree mean-field, is explicitly cured by the exchange term in the Hartree-Fock approximation. The approximate LDA exchange nevertheless does not suffice to restore a proper cancellation, which leads to a spurious residual self interaction. This defect can unfortunately have sizable consequences, especially on ionization properties (Ionization Potential (IP) in particular), which is catastrophic for the description of irradiation processes. The aim of this contribution is to discuss strategies to cure this defect by a Self-Interaction Correction (SIC) in a truly dynamical context compulsory for a proper description of irradiation dynamics.

The paper is organized as follows. After a brief introduction to DFT and LDA we illustrate the self-interaction problem on a typical example. We then introduce standard SIC approaches, explain their limitations and the need for improved formulations. We finally present a recently introduced clean formulation of SIC and illustrate the improvement it brings on typical examples.

2. From DFT to LDA

There exist many well documented presentations of DFT, LDA and TDDFT. We refer the reader to these various references for more details [7, 8]restrict the present discussion to a brief introduction of DFT and LDA as basis for the further discussion. The considerations are restricted to electrons here, ions being treated as classical particles all over. We work in the Kohn-Sham [9] scheme of DFT in which the system is represented by a set $\{\psi_\alpha, \alpha = 1, ...\}$ of single-electron wave functions. From them one constructs the one-body density, $\rho = \sum_\alpha |\psi_\alpha|^2 = \sum_\alpha \rho_\alpha$, which is the key ingredient of DFT according to Hohenberg and Kohn's theorem [10]. The resulting theory involves a single-electron Hamiltonian, $h[\rho] = h_0 + U_{\rm H} + U_{\rm xc}$, where $h_0 = -\frac{\hbar^2}{2m}\Delta$ is the kinetic term and $U_{\rm H}$ the Hartree (direct) Coulomb interaction. The exchange correlation term, $U_{\rm xc}$, is a functional of the density $\rho(\mathbf{r})$, which has to be properly chosen.

The Local Density Approximation (LDA) provides the simplest approximation for $U_{\rm xc}$ and consists in assuming the system as infinite and homogeneous to compute this term, locally. This amounts to replace $U_{\rm xc}[\rho^\infty]$ by $U_{\rm xc}[\rho(\mathbf{r})]$, where $\rho(\mathbf{r})$ is the local density and ρ^∞ the density of the corresponding infinite system with homogeneous density $\rho(\mathbf{r})$. The exchange correlation term thus becomes a simple function of the local density $\rho(\mathbf{r})$. We denote by $U_{\rm LDA}$ the corresponding one-body potential ($U_{\rm LDA} = U_{\rm H} + U_{\rm xc}$). By construction, LDA should be valid only for systems in which the density $\rho(\mathbf{r})$ varies sufficiently slowly. In fact, it performs usually much better than expected. It can thus be successfully applied to situations with non-vanishing density gradients and it thus constitutes a sound basis for the construction of more elaborate approximation schemes.

3. The SIC problem

In spite of its many successes, LDA suffers from the self interaction problem since the approximate exchange term does not cancel exactly the self interaction contribution in the direct term. This is a key issue in electronic systems in which Coulomb exchange provides a very large fraction (typically 3/4) of the total binding of a system. A poor treatment of the self interaction then necessarily leads to measurable pathologies. The simplest example is provided by the IP which sensitively depends on the self interaction. Indeed the LDA approximation exhibits a wrong asymptotic behavior of the electronic potential : a departing electron does not feel the correct potential due to electrons remaining in the system. This causes an incorrect value of the IP when estimated from the Highest Occupied Molecular Orbital (HOMO), which reflects a violation of Koopmann's theorem [11]. Although acceptable in a purely static context, this defect becomes dramatic in a truly dynamical situation in which one wants to follow ionization "on the fly".

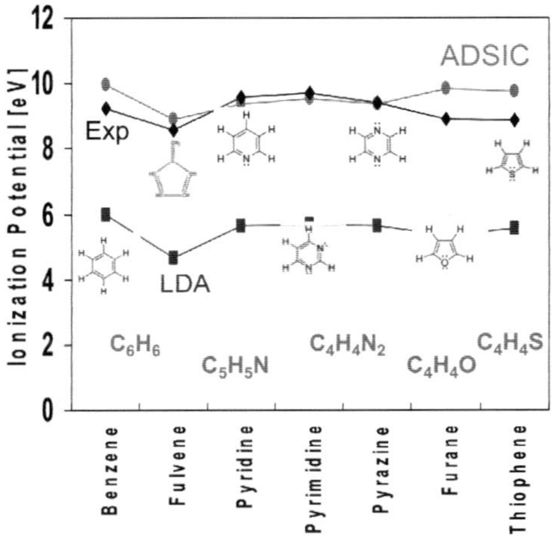

Figure 1. Ionization Potentials of carbon cycles : experimental, LDA and ADSIC values. See text for details. Computations are from [12].

Figure 1 shows IP's in the case of a sequence of carbon cycles, in which carbon atoms are replaced by nitrogen or oxygen atoms. In the whole sequence of test cases the effect is the same: a strong underestimation of the IP, with a relative error of order 50 %. As will be clear below this defect reflects the absence of SIC in LDA which thus badly fails.

4. Self Interaction Correction (SIC) and Optimized Effective Potential (OEP)

4.1. Basics of SIC-OEP

The basic SIC method was introduced in the late 1970's and consists in subtracting "by hand" the spurious self interaction [13].This amounts to define a total SIC energy $E_{\rm SIC}$ and the corresponding one-body Hamiltonian, obtained from variation of $E_{\rm SIC}$ with respect to ψ^*_α, as

$$E_{\rm SIC} = E_{\rm LDA}[\rho] - \sum_{\beta=1}^{N} E_{\rm LDA}[|\psi_\beta|^2], \tag{1}$$

$$h_{{\rm SIC},\alpha} = h_{\rm LDA} - U_{\rm LDA}[|\psi_\alpha|^2] = h_0 + U_{{\rm SIC},\alpha}, \tag{2}$$

where $U_{{\rm SIC},\alpha} = U_{\rm LDA}[\rho] - U_{\rm LDA}[\rho_\alpha]$. A problem is here that the resulting one-body Hamiltonian does depend on the orbital on which it acts which, in turn, has inconvenient consequences. The orbital dependence a priori breaks the ortho-normality of the ψ_α, which may severely plague any time dependent SIC calculations, if formulated that way. This ortho-normality problem is well known since the early days of SIC methods and several strategies have been developed to circumvent it, in particular in the stationary case.

A standard way out is to construct an approximate single one-body Hamiltonian common to all orbitals. This approach relies on a general method to do so, applicable to any mean-field theory (such as Hartree Fock in particular) and which is known as the Optimized Effective Potential (OEP) [14],having been introduced much earlier than SIC. The idea beyond is remarkably simple and elegant. It first consists in parameterizing the single-particle orbitals $\varphi_i^{V_0}$ by a common (optimal) potential V_0 (yet to be constructed). We introduce purposely a new notation in order to properly distinguish these orbitals from the original ψ_α wave functions, which by construction belong to a more general Hamiltonian. One then writes down the total energy (Hartree-Fock, SIC, ...) using these parameterized wave functions as single electron orbitals. There remains to construct the potential V_0 itself. This is achieved by optimizing the total energy with respect to the potential V_0, which first provides V_0. From V_0 one then obtains the associated wave functions $\varphi_i^{V_0}$. The problem has then been solved, in a variational manner,

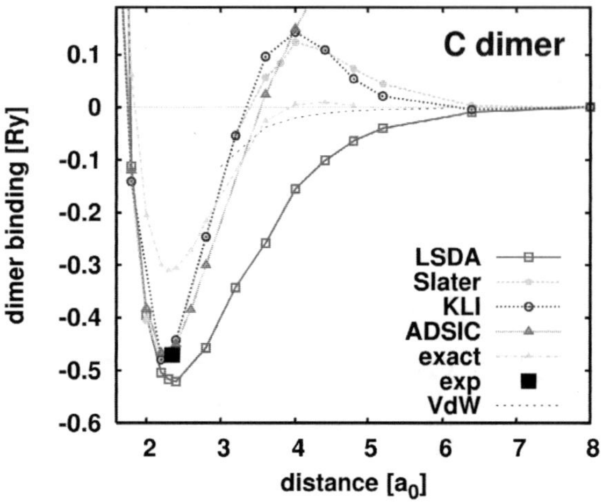

Figure 2. Potential energy surface of the dimer C_2 calculated with various DFT schemes, with or without SIC. The full square indicates the experimental values of the bond length and the binding energy. The LSDA (LDA including spin) presents an acceptable qualitative behavior, but misses the ground state energy and strongly misses the vibration frequency. All SIC calculations perfectly reproduce ground state (bond length, ground state energy, vibration frequency), but badly miss the shape of the potential energy surface at intermediate distances.

but in a restricted Hilbert space, "parameterized" by V_0, which by construction is a potential common to all orbitals.

In the SIC case one can formally summarize the SIC-OEP procedure by :

$$h_{\text{OEP}}\,|{\varphi_i}^{V_0}) = (h_0 + V_0)\,|{\varphi_i}^{V_0}) = \varepsilon_i|{\varphi_i}^{V_0}), \tag{3}$$

$$E_{\text{OEP}} = E_{\text{SIC}}[\{{\varphi_i}^{V_0}\}], \tag{4}$$

$$\frac{\delta E_{\text{OEP}}}{\delta V_0} = 0 \rightarrow V_0 \rightarrow {\varphi_i}^{V_0}. \tag{5}$$

One can furthermore separate V_0 into 3 components (associated to various degrees of approximations) : $V_0 = V_S + V_K + V_C$. The first term V_S corresponds to the simplest Slater approximation [16], the complementing term V_K associated to V_S provides the well known KLI approximation $(V_S + V_K)$ [17] and finally V_C is a correction usually assumed small. One can derive explicit expressions for these various terms, but most calculations have been performed at Slater or KLI levels, the full OEP-SIC equations being in most cases extremely demanding to solve in practical cases [15]. We write down for illustration the (simple) Slater term :

$$V_S = \sum_i \frac{|{\varphi_i}^{V_0}|^2}{\rho} U_{\text{LDA},i} \quad , \tag{6}$$

where $U_{\text{LDA},i} = U_{\text{LDA}}\left[|{\varphi_i}^{V_0}|^2\right] = U_{\text{LDA}}[\rho_i]$ which nicely shows the average (over single electron orbitals) character of the produced OEP. The Slater approximation can be further simplified to provide the Average Density SIC (ADSIC) [18] by making the replacement $\rho_i \rightarrow \rho/N$ (where N is the total number of electrons in the system). This particularly simple ADSIC approximation has been successfully applied to metal clusters [18] and organic systems [12]. This is illustrated in figure 1 where it can be seen that ADSIC provides a very good reproduction of carbon cycle IP's: it remarkably solves the IP problem and restores Koopmann's theorem.

4.2. Limitations

In spite of their successes, neither ADSIC nor the more involved Slater or KLI approximations are fully satisfying. The potential energy surface of C_2, as shown in figure 2, demonstrates one of the problems. ADSIC is totally off at large distances while Slater and KLI show a strange bump at intermediate distances. This bump spoils dissociation properties, a key issue when considering

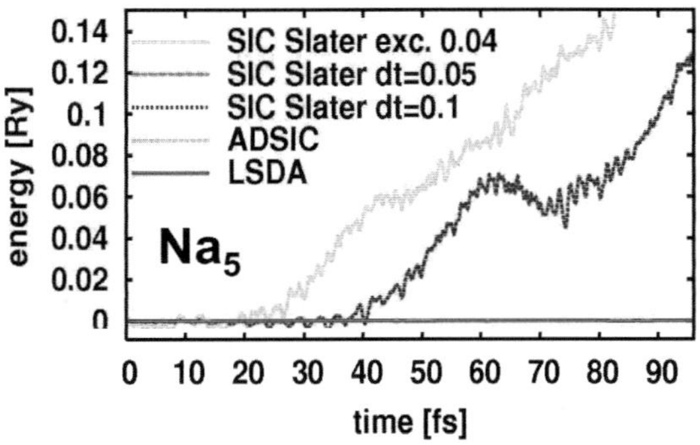

Figure 3. Time evolution of the total energy of ground state Na_5 cluster in various approximations as indicated. Two time steps have been used in the Slater case and led to results exactly superimposing each other.

irradiation processes. It reflects a "conflict" at the side of single-electron wave functions, namely the fact that in this transitional range single electron orbitals have to find a compromise between localization (asymptotic configuration) and delocalization (molecular state).

The case of the dynamical evolution of Na_5 presented in figure 3 is even more dramatic. The total energy is plotted as a function of time and found to be badly violated in a standard Slater calculation. This defect can be proven analytically. It is then found that energy and zero force theorem are explicitly violated in time, in Slater and KLI approximations [19].

These pathologies call for an improvement of SIC approaches, both in its full and OEP approximate versions. This is what will be discussed in the following section.

5. SIC revisited

5.1. Time Dependent SIC

As noted above, the basic primary defect of the original SIC formulation is the lack of a proper ortho-normalization of single-electron orbitals. A simple cure is to enforce ortho-normalization within the variational principle itself, by introducing a set of Lagrange parameters $\lambda_{\beta\alpha}$. The variational principle then reads

$$\delta_{\psi_\alpha^*}(E_{\rm SIC} - \sum_{\beta,\gamma}(\psi_\beta|\psi_\gamma)\lambda_{\gamma\beta}) = 0 \tag{7}$$

which leads to the set of coupled equations

$$h_{\rm SIC,\alpha}|\psi_\alpha) = \sum_\beta \lambda_{\beta\alpha}|\psi_\beta), \tag{8}$$

$$0 = (\psi_\beta|U_{\rm LDA}[\rho_\beta] - U_{\rm LDA}[\rho_\alpha]|\psi_\alpha). \tag{9}$$

The second equation, called "symmetry condition", stems from the ortho-normality constraint and becomes a crucial ingredient of the theory. Thanks to the variational formulation, there necessarily exists a solution to this set of equations. The enforcement of ortho-normalization in SIC was already proposed some times ago in static calculations [20, 21]. Very recently, the formalism was extended to the time domain [22, 23]. We briefly outline it below.

Starting point is the quantum action including explicit ortho-normalization and on which variational principle is applied

$$0 = \delta \int_{t_0}^{t} dt' \Big(E_{\rm SIC} - \sum_\alpha(\psi_\alpha|\mathrm{i}\hbar\partial_t|\psi_\alpha) - \sum_{\beta,\gamma}(\psi_\beta|\psi_\gamma)\lambda_{\gamma\beta}\Big) \quad . \tag{10}$$

This leads to

$$(\hat{h}_\alpha - \mathrm{i}\hbar\partial_t)|\psi_\alpha) = \sum_\beta |\psi_\beta)\lambda_{\beta\alpha} \quad , \tag{11}$$

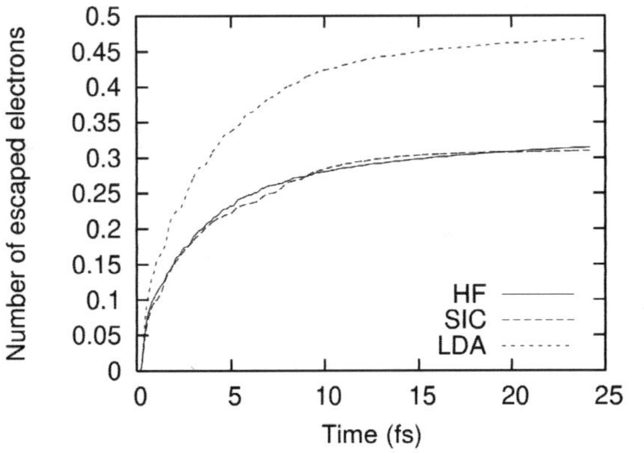

Figure 4. Ionization as a function of time of a simple model dimer (1D calculation) in LDA, TDHF and TDSIC with explicit ortho-normalization. Calculations have been performed at exchange only level making TDHF exact. From [24].

to be complemented again by the symmetry condition (9). The difficult point is now to propagate single electron orbitals. Indeed the Lagrange multipliers imply a non-trivial propagation scheme as they are explicitly time dependent. A way out is to introduce two sets of single electron wave functions. The idea behind the double set strategy is to exploit a degree of freedom which is usually neglected in DFT. One can perform any unitary transformation among the occupied single-electron orbitals without changing the total Slater determinant. In practice, we introduce two sets of single-electron orbitals linked by such an unitary transformation. The first set $\{\psi_\alpha\}$ (as introduced above) serves to fulfill the symmetry condition (9). It turns out that it provides usually well localized wave functions. The second set $\{\varphi_i\}$ serves to diagonalize the SIC Hamiltonian, Eq. (8). Then, the single electron equations of motion become

$$(\hat{h}_{\rm SIC} - {\rm i}\hbar\partial_t)|\varphi_i) = 0 \qquad (12)$$

which can be solved with standard time propagation

$$|\varphi(t)) = e^{-{\rm i}\int_{t_0}^{t} dt' \hat{h}_{\rm SIC}/\hbar}|\varphi(t_0)) \qquad (13)$$

which resolves the propagation puzzle. In practice, we propagate the two sets of wave functions simultaneously, the $\{\varphi_i\}$ are determined from time evolution (13) and the unitary transformation to the $\{\psi_\alpha\}$ is determined at each time step to fulfill the symmetry condition (9). Note also that the SIC Hamiltonian is constructed from the $\{\psi_\alpha\}$ orbitals.

The capability of this new Time Dependent SIC (TDSIC) formulation is illustrated in figure 4, in which we compare a TDSIC calculation with Time Dependent Hartree Fock (TDHF) and LDA. Calculations have been performed in the exchange only approximation such that TDHF becomes exact in that case and thus provides a simple benchmark. As can be seen from the figure TDSIC fully matches TDHF leading to exactly the same ionization after an initial perturbation. In turn, the LDA calculation leads to a much larger ionization due to the significantly higher IP in that case. Note however that the TDSIC is numerically expensive and this calls for further approximations.

5.2. Generalized OEP and Slater

The double set strategy also provides a means to reformulate the OEP approach so that one can re-derive Slater-like and KLI-like approximations in a much better justified manner and thus with a larger range of applicability. One intuitively sees that the $\{\varphi_i\}$ are probably structurally close to the SIC-OEP ${\varphi_i}^{V_0}$ wave functions. This actually provides the basis for the development of the so called the Generalized Slater (GSlat) approximation [24] which can be formulated both

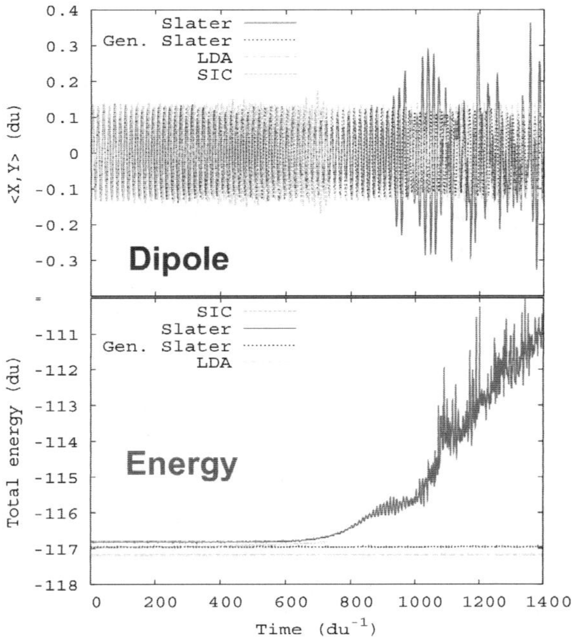

Figure 5. Dipole (upper panel) and total energy (lower panel) of a 6 electron quantum dot following an initial perturbation plotted as a function of time in various approximations. Note the obvious defect of Slater, cured at GSlat level on this time span, and properly cured by TDSIC. Adapted from [25].

in a static and a dynamical context [25]. Because of the localization of the $\{\psi_\alpha\}$ wave functions, one can demonstrate that the usual Slater approximation becomes more accurate, provided the potential is expressed in terms of the $\{\psi_\alpha\}$. The OEP problem, in turn, remains solved at the level of the ${\varphi_i}^{V_0}$ wave functions. This formally reads

$$h_{\rm OEP}\,|{\varphi_i}^{V_0}) \;=\; \varepsilon_i|{\varphi_i}^{V_0}), \tag{14}$$

$$V_0 \;\simeq\; V_{\rm GSlat} = \sum_\alpha \frac{|\psi_\alpha|^2}{\rho} U_{\rm LDA}[|\psi_\alpha|^2], \tag{15}$$

with both sets connected by the proper unitary transform. In its stationary version, GSlat provides results extremely close to the full SIC ones. For example, it restores dissociation properties of C_2 (figure 2). It also provides improved values of polarizabilities as compared to Slater [26]. Time-Dependent GSlat also constitutes a better approximation than the mere Slater. Still, the formal shortcomings of Slater (energy and ZFT violations) are not strictly cured formally. The onset of the instability is merely postponed. This has, nevertheless, the interesting consequence that TDGSlat can be used for spectral analysis in the linear regime, while Slater cannot. The point is illustrated in figure 5 where total energy and dipole momentum of a small quantum dot (containing 6 electrons) are plotted as a function of time for various schemes.

6. Conclusions

We have discussed in this paper the self-interaction problem in finite electronic systems, especially from a dynamical point of view. We have proposed an improved scheme for solving several of the original SIC pathologies. The self interaction problem is a well known defect of LDA and stems from the fact that exchange is treated in an approximate manner in LDA. In spite of the successes of LDA it is in particular unable to properly describe ionization properties because of this self interaction problem. This has dramatic consequences on the description of dynamical scenarios in irradiation processes. The SIC method is supposed to cure this problem.

But it also suffers from pathologies linked in particular to the loss of ortho-normalization of wave functions by construction. The OEP method, in conjunction with SIC, allows to derive a formally clean approximate theory to full SIC, but produces extremely involved equations. The simplest approximations, Slater and KLI, although being used frequently, still suffer from severe defects, both formally and practically. We have thus proposed to revisit the SIC problem by reformulating the theory with explicit ortho-normalization. We have, furthermore, developed a two set strategy which is practically compulsory in the time domain, but which also exhibits interesting properties in the static case. It allows, for example, to derive an extension of the Slater approximation, the Generalized Slater approximation, which provides an interesting a wider range of applicability in statics and in the linear domain of dynamics. Nevertheless, the ultimate approach still remains the full TDSIC and it would be extremely valuable to optimize it further from a numerical point of view in order to apply it to large realistic systems in a truly dynamical context.

Acknowledgments

The authors thank the Deutsche Forschungsgemeinschaft (RE 322/10-1) and the Agence Nationale de la Recherche (ANR-06-BLAN-0319-02) for financial support during the realization of this work.

[1] Calvayrac F *et al.* 2000 *Phys. Rep.* **337** 493; Reinhard P-G and Suraud E *Introduction to Cluster Dynamics* 2003 (Wiley, New York); Fennel Th *et al.*, 2010 *Rev. Mod. Phys.* **82** 1793
[2] Ichimaru S, Iyetomi H and Tanaka S 1987 *Phys. Rep.* **149** 91
[3] Bertsch G F, Das Gupta A 1988 *Phys. Rep.* **160** 190; Abe Y *et al.* 1996 *Phys. Rep.* **275** 49
[4] Ndongmouo-Taffoti U F *et al.* 2010 *Eur. Phys. J. D* **58** 131
[5] Meiwes-Broer K H (Ed.) 2000 *Metal Clusters at Surfaces: Structure, Quantum Properties, Physical Chemistry* (Springer, Berlin,)
[6] Dinh P M, Reinhard P-G and Suraud E 2009 *Phys. Rep.* **485** 43
[7] Dreizler R M and Gross E K U 1990 *Density Functional Theory: An Approach to the Quantum Many-Body Problem* (Springer-Verlag, Berlin); Parr R G and Yang W 1989 *Density-Functional Theory of Atoms and Molecules* (Oxford University Press, Oxford); Kohn W 1999 *Rev. Mod. Phys.* **71** 1253
[8] Marques M A, Ullrich C A, Nogueira F, Rubio A, Burke K and Gross E K U 2006 *Time Dependent Density Functional Theory* Lect. Notes in Physics **706** (Springer, Berlin)
[9] Kohn W and Sham L J 1965 *Phys. Rev.* **140** 1133
[10] Hohenberg P and Kohn W 1964 *Phys. Rev.* **136** 864
[11] Koopmans T 1934 *Physica* **1** 104
[12] Ciofini I, Adamo C and Chermette H 2005 *J. Phys. Chem.* **123** 121102
[13] Perdew J P 1979 *Chem. Phys. Lett.* **64** 127; Kümmel S and Kronik L 2008 *Rev. Mod. Phys.* **80** 3
[14] Sharp R T and Horton G K 1953 *Phys. Rev.* **90** 317; Talman J D and Shadwick W F 1976 *Phys. Rev. A* **14** 36
[15] Körzdörfer T, Kümmel S and Mundt M 2008 *J. Chem. Phys.* **129** 014110
[16] Slater J C 1951 *Phys. Rev.* **81** 385
[17] Krieger J B, Li Y and Iafrate G J 1990 *Phys. Lett. A* **146** 256;
[18] Legrand C, Suraud E and Reinhard P-G 2002 *J. Phys. B* **35** 1115
[19] Mundt M, Kümmel S, van Leeuwen R and Reinhard P-G 2007 *Phys. Rev. A* **75** 050501
[20] Pederson M, Heaton R A and Lin C C 1984 *J. Chem. Phys.* **80** 1972
[21] Goedecker S and Umrigar C J 1997 *Phys. Rev. A* **55** 1765
[22] Messud J, Dinh P M, Reinhard P-G and Suraud E 2008 *Phys. Rev. Lett.* **101** 096404
[23] Messud J, Dinh P M, Reinhard P-G and Suraud E 2009 *Ann. Phys. (NY)* **324** 955
[24] Messud J, Dinh P M, Reinhard R-G and Suraud E 2008 *Chem. Phys. Lett.* **461** 316
[25] Messud J, Dinh P M, Reinhard R-G and Suraud E 2009 *Phys. Rev. A* **80** 044503
[26] Messud J, Wang Z P, Dinh P M, Reinhard R-G and Suraud E 2009 *Chem. Phys. Lett.* **479** 300

Nanoparticles dynamics on a surface: fractal pattern formation and fragmentation

Veronika V. Dick, Ilia A. Solov'yov, Andrey V. Solov'yov

Frankfurt Institute for Advanced Studies, Goethe University, Ruth-Moufang-Str. 1, 60438 Frankfurt am Main, Germany

E-mail: solovyov@fias.uni-frankfurt.de

Abstract. In this paper we review our recent results on the formation and the post-growth relaxation processes of nanofractals on surface. For this study we developed a method which describes the internal dynamics of particles in a fractal and accounts for their diffusion and detachment. We demonstrate that these kinetic processes determine the final shape of the islands on surface after post-growth relaxation. We consider different scenarios of fractal relaxation and analyze the time evolution of the island's morphology.

1. Introduction

The process of formation and post-growth relaxation of deposited structures is still not well understood (see e.g. [1–6] and references therein). The understanding of formation and post-growth relaxation processes would allow one to explore the self-organization processes of particles on the surface in obtaining patterns with predictable properties.

The conventional technique of thin-film growth by deposition of atoms [1, 2], small atomic clusters [1, 3] and molecules [2, 7, 8] on surfaces gives a possibility to construct materials with pre-defined properties. Recent experiments show that patterns with different morphology can be formed in the course of cluster deposition process on a surface [1,3,9]. Among other possible shapes, droplet-like and fractalic islands have been observed in various systems [1, 3, 9]. It was shown that the island morphology depends on various factors, such as the temperature [1,3,4,10,11], particle size [12], particle deposition rate [10,13,14], substrate roughness [15,16], concentration of impurities in the system [3,4,17] and interparticle interaction energies [3,10].

The formation of fractals provides a natural framework for studying disordered structures on surface because fractals are generally observed in far-from-equilibrium growth regime. During the last years fractal shapes have been observed for a variety of systems. For example, fractals consisting of Ag [4, 11, 17], Au [18], Fe-N [19] clusters and C_{60} molecules [20, 21] have been fabricated on different surfaces with the use of the cluster deposition technique [2, 7]. The self-organization dynamics of particles on a surface was also studied theoretically [5, 6, 16, 22]. An efficient theoretical tools for describing particle dynamics on a surface is the diffusion limited aggregation (DLA) method [23], which was used for the study of the fractal formation process on a surface [10, 13, 16, 22].

In the present paper we investigate further the process of fractal formation and also the post-growth fractal stability on a surface. For this we have developed a method describing the internal dynamics of particles in a fractal and accounting for their diffusion and detachment.

Journal of Physics: Conference Series **248** (2010) 012025 doi:10.1088/1742-6596/248/1/012025

We demonstrate that these kinetic processes determine the shape of the islands on surface being formed in the course of island formation and the following post-growth relaxation. We consider different scenarios of fractal relaxation and analyze the time evolution of the island's morphology. The results of our calculations are compared with experimental measurements of the post-growth relaxation of silver cluster fractals on a graphite substrate [4, 11, 17].

2. Theoretical methods

In this section we discuss the theoretical methods used for studying the dynamics of particles on a surface. Computations were performed with the use of the MBN Explorer computer package, which is developed for structure optimization, simulation of dynamics and growth processes in various nanosystems [5,6,24–33]. Below we describe the computational method used in our work and explain how the internal dynamics of particles in a fractal has been accounted for.

To study the diffusion of particles on surface we used a version of the kinetic Monte Carlo (KMC) method [34–36]. The idea of the KMC method is as follows. The time-evolution of a molecular system is modeled stepwise in time. With a certain probability, at each step of the simulation, the system undergoes a structural transformation. The new configuration of the system is then used as the starting point for the next simulation step. The transformation of the system is governed by the kinetic rates, input into the KMC method. Note, that at each simulation step the system can be transformed into one of several states. Thus, in the KMC method, the probability for the system to attain a certain configuration is proportional to the corresponding kinetic rate. Due to its probabilistic nature, the KMC method allows to study dynamical processes on time scales significantly exceeding the time scales of the conventional molecular dynamics simulation. Therefore, this method is ideal for studying dynamical processes involving large ensembles of particles, in which the details of single particle dynamics are not so important, and it can be parameterized by a few relatively simple kinetic rates.

For our studies we use a modification of the conventional KMC method. For each particle on surface we first determine the number of its possible diffusion directions. Thereby, a particle can either diffuse freely on surface, or diffuse along the periphery of the already pre-formed structure on surface (see Fig. 1). A diffusion direction of the particle is chosen randomly in such a manner that all possible diffusion directions are equally probable. For the chosen diffusion direction the probability for particle diffusion is then calculated, and the particle is translated in this direction with the calculated probability. To speed up the calculation, we simulate particle dynamics on a two-dimensional hexagonal grid, on which a particle has up to six neighbors, as illustrated in Fig. 1. The size of a single grid cell in this case is defined by the particle radius.

To simulate the growth process the following procedure has been adopted. At every step of the simulation new particles are deposited on the surface according to the deposition rate and occupy some of the free cells in the grid. Simultaneously, the already deposited free particles diffuse on the surface with the rate

$$\Gamma = \frac{1}{\Delta t} = \nu_1 \exp\left[-\frac{E_a}{kT}\right], \tag{1}$$

where E_a is the activation energy, ν_1 is the attempt escape rate, T is the temperature of the system, k is the Boltzmann constant, and Δt is the simulation time step, which defines the characteristic time for particle diffusion on surface. As follows from Eq. (1) the values of E_a and T define the characteristic time scale of the fractal growth. In this paper we focus on the study of silver cluster fractals on a graphite substrate, and therefore, assume E_a fixed.

The simulation time step Δt is related to the particle diffusion coefficient D on surface as follows [6]

$$\Delta t = \frac{d_0^2}{4D}, \tag{2}$$

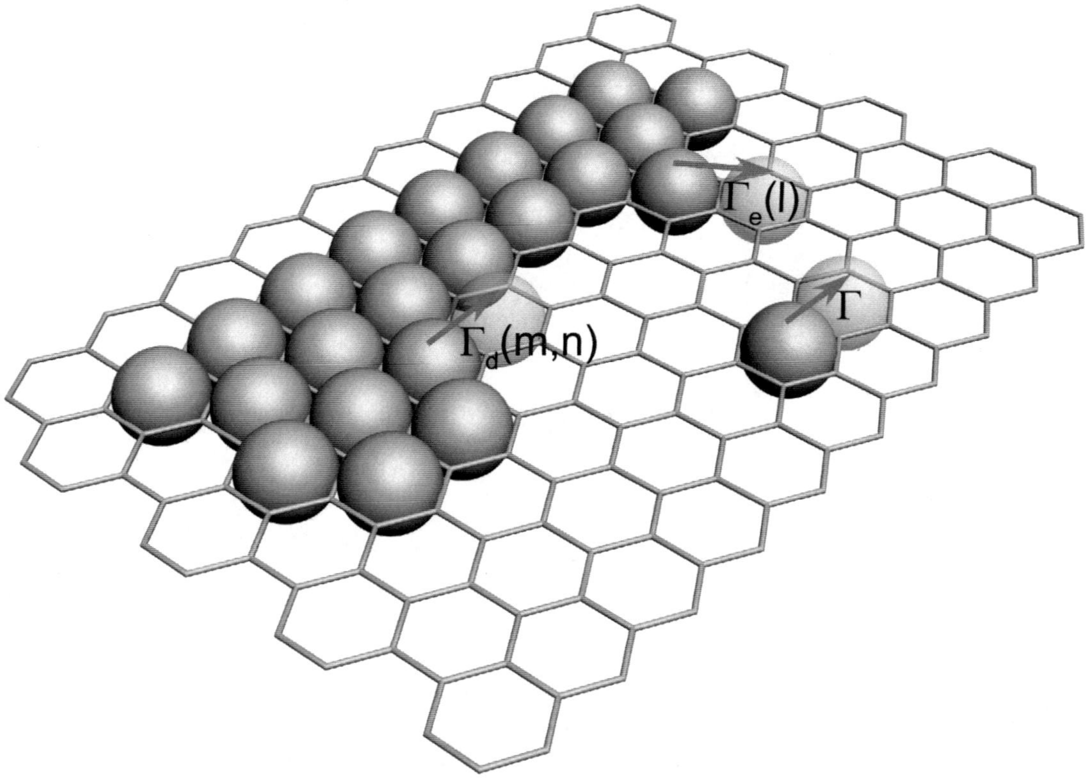

Figure 1. Arrangement of deposited particles on the hexagonal grid. The important processes which govern pattern formation on surface are indicated by arrows: Γ is the free particle diffusion rate, $\Gamma_d(m,n)$ is the diffusion rate along the periphery of an island, and $\Gamma_e(l)$ is the particle detachment rate from an island. The diffusion along the periphery depends on the number of broken bonds (m) and the number of maintained neighboring bonds (n). The particle detachment rate depends on the number of broken bonds (l). In the depicted example $m = 3$, $n = 1$, $l = 2$.

where d_0 is the particle diameter. The experimental value of the diffusion coefficient of an Ag_{500} cluster on graphite at room temperature is equal to $2 \cdot 10^{-7}$ cm^2/s [17]. Substituting this value in Eq. (2), one obtains $\Delta t = 78$ ns.

The relaxation of an island on surface is driven by diffusion of particles along the island's periphery and particle detachment from the island. Both processes are schematically illustrated in Fig. 1. The diffusion and the detachment rates depend on the activation energy and the particle-particle interaction. In the Arrhenius approximation, the diffusion rate of a particle along the periphery of the fractal reads as:

$$\Gamma_d(m,n) = \nu_2 \exp\left[-\frac{mE_b}{kT} - \frac{n\Delta\epsilon}{kT} - \frac{E_a}{kT}\right], \quad (3)$$

where m is the number of bonds that are broken due to the particle motion, $E_b > 0$ is the binding energy barrier of two particles, n is the number of maintained neighboring bonds between two particles and $\Delta\epsilon \leq E_b$ is the diffusion energy barrier [11, 22], ν_2 is the attempt escape rate. Equation (3) describes the probability of a particle to overcome a potential energy barrier, which for a particle diffusing along the fractal periphery is parameterized by the energies E_b, $\Delta\epsilon$, and E_a. Parameters E_b and $\Delta\epsilon$ define the potential energy barrier for particle diffusion along the fractal periphery, while E_a characterizes the time scale of particle diffusion [6].

Note that Eq. (3) does not account for the bonds which may be created in the system when a particle diffuses. This feature of Eq. (3) is easy to understand. Indeed, in our model the particle diffusion process is considered stepwise, i.e., at each step of the computation a particle is displaced with a certain probability on the distance equal to its diameter in a random direction (see Fig. 1). But prior the particle is displaced to its new location there is no information about the newly created bonds in the system. Therefore only those bonds which the particle forms in its neighborhood prior the displacement influence its diffusion dynamics.

The evaporation (detachment) rate of a particle from the island in the Arrhenius approximation is given by

$$\Gamma_e(l) = \nu_3 \exp\left[-\frac{lE_b}{kT} - \frac{\Delta\mu}{kT} - \frac{E_a}{kT}\right], \tag{4}$$

where l is the number of bonds broken after particle detachment from the island, $\Delta\mu$ is the chemical potential of particle detachment [3,11,22,37], ν_3 is the attempt escape rate of a particle from its equilibrium position. For the simulations of the fractal fragmentation presented below we assume:

$$\nu_2 \simeq \nu_3 = \nu. \tag{5}$$

This approximation is valid when the characteristic attempt escape rate of a particle leading to its diffusion or detachment are close. From Eqs. (1)-(4) one derives that the probability of different kinetic processes in the system depends on the values of E_b, $\Delta\epsilon$, $\Delta\mu$, which are called below as the kinetic parameters. For the convenience, in this paper we define all the kinetic parameters in the units of kT (1 $kT = 0.026$ eV) at room temperature (300 K).

3. Results

3.1. Fractal growth

Using the methods described in the pervious section we consider particle dynamics on a 500×500 nm^2 substrate with periodic boundary conditions. According to the periodic boundary conditions every particle which reaches a boundary of the substrate is removed to the opposite boundary. This method allows to keep the surface density of deposited particles constant when no new particles are added into the system. Periodic boundary conditions provide a useful tool for studying processes on a small surface fragment with particle surface density as in experiment. Thus, the periodic boundary conditions allow to study formation of a single (or several fractals), as if it was surrounded by many similar fractals on a macroscopically large substrate.

To illustrate the process of particle selforganization on surface we consider two different scenarios shown in Fig. 2. We deposit the particles on surface with the rate $F = 1.026 \times 10^9$ particle/cm^{2}s being close to the experimental value [4]. The diameter of a particle in the simulations is taken 2.5 nm, i.e. as in the Ag_{500} cluster [4,11,17]. It is known from experiment that particles nucleate more efficiently in the vicinity of point defects [16]. Therefore, to speed up the nucleation process we have introduced five immovable point defects on surface that become the fractal growth centers as illustrated in Figs. 2a-b.

The variation of the kinetic parameters E_b, $\Delta\epsilon$ and $\Delta\mu$ results in the formation of different patterns on surface. Figure 2 shows the morphology of fractals on surface during the particle deposition process simulated for $E_b = 30$ kT, $\Delta\epsilon = 10$ kT, $\Delta\mu = 100$ kT (plot a), and for $E_b = 60$ kT, $\Delta\epsilon = 20$ kT, $\Delta\mu = 100$ kT (plot b) after $t = 50$ s of simulation. The comparison of the fractal morphologies shown in Fig. 2a and Fig. 2b shows that the thickness of fractal branches

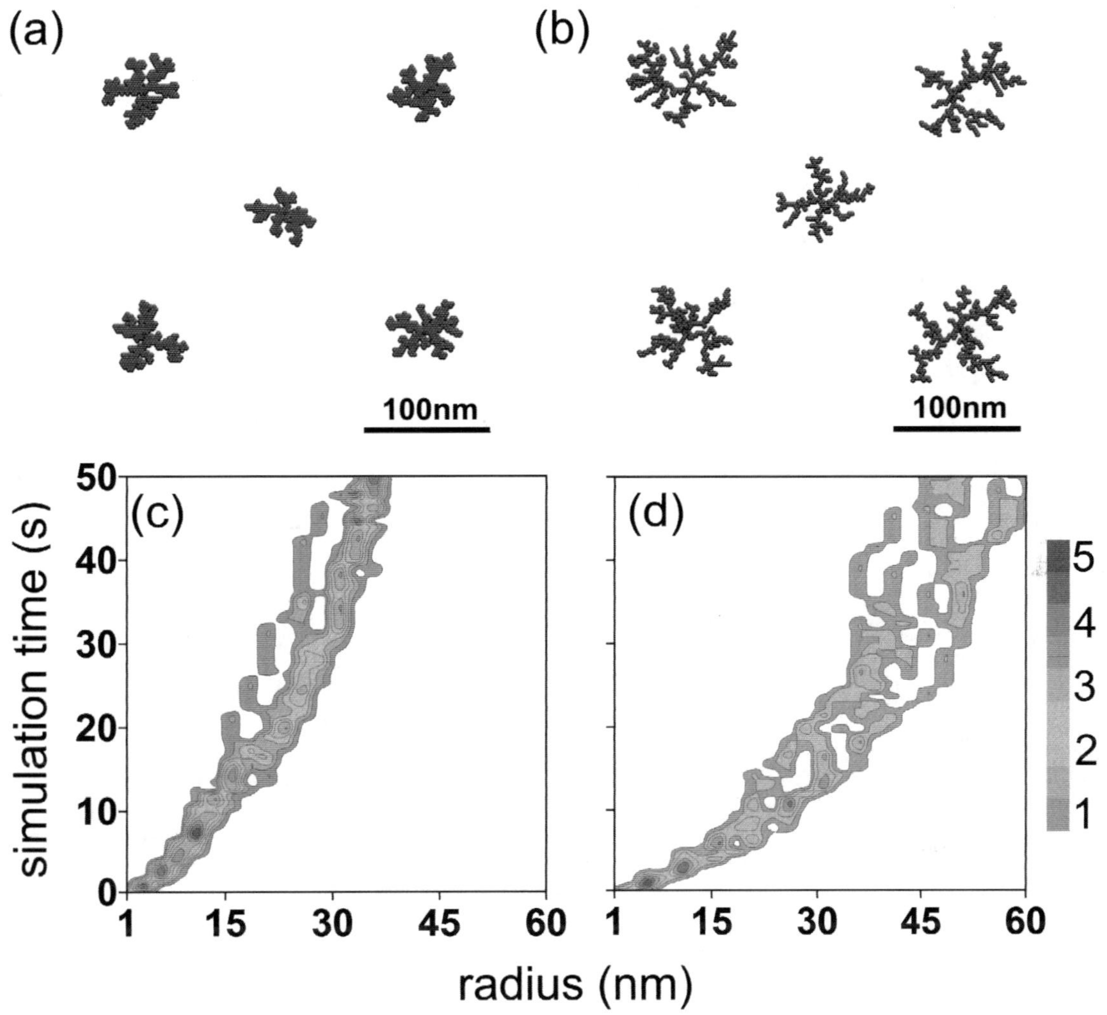

Figure 2. Formation of fractals on surface during the particle deposition process simulated for $E_b = 30\ kT$, $\Delta\epsilon = 10\ kT$, $\Delta\mu = 100\ kT$ (a) and for $E_b = 60\ kT$, $\Delta\epsilon = 20\ kT$, $\Delta\mu = 100\ kT$ (b). Plots (a) and (b) show the morphology of the formed structures after $t = 50$ s of simulation. Contour plots (c) and (d) show the time evolution of the size distribution calculated for the emerging fractals on surface at room temperature for the kinetic parameters introduced in (a) and (b) respectively.

increases at smaller values of kinetic parameters. This observation can be easily understood. The morphology of fractals is mainly governed by the time needed for a newly deposited particle to reach the growth region, and by the time needed for a particle to find an optimum position within the fractal (rearrangement time). The later process is governed by the parameter $\Delta\epsilon$. For $\Delta\epsilon = 10\ kT$ (see Fig. 2a) the rearrangement time becomes shorter than for $\Delta\epsilon = 20\ kT$, as seen from Eq. (3), leading to an increase of the fractal branch thickness. If $\Delta\epsilon = 20\ kT$ (see Fig. 2b), the particles within the fractal need much time to hop, and therefore, the fractals with less optimized particle structure, i.e. with thinner branches, are formed. Note that the thickness of fractal branches is related to the cluster mobility on surface, which is also determined by the

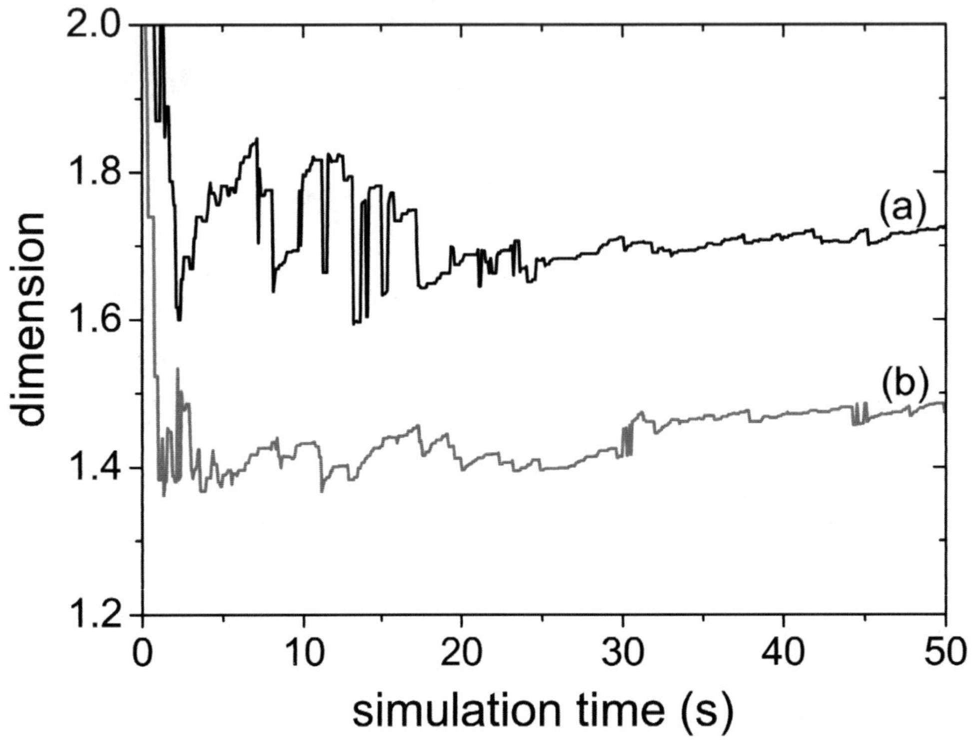

Figure 3. Time evolution of the average fractal dimension calculated using Eq. (7) for the structures shown in Fig. 2. Curves (a) and (b) illustrate the time evolution of the fractal dimension for the fractalic islands shown in Fig. 2a and Fig. 2b respectively.

size of the deposited clusters. This fact has been demonstrated experimentally for antimony clusters consisting of 90, 150, 500 and 1400 atoms [38].

The thickness of the fractal branches influences the compactness of the structures formed on surface. Figures 2c-d show the time evolution of the size distribution calculated for the fractals shown in Figs. 2a-b respectively. It is clear from Figs. 2c-d that the average radius of a fractal grows nonlinearly with time. At each time frame the distribution of fractal sizes for the fractals with thicker branches (see Fig. 2c) has a smaller width than that for the distribution of the fractals with thinner branches (see Fig. 2d). This happens because in the case of fractals with thicker branches the peripherial particles have enough time to rearrange and form a more compact structure, while in the case of fractals with thinner branches the inner dynamics of particles in the fractal is almost frozen. The fact that fractals with thinner branches grow almost without any rearrangements leads to the formation of larger size structures, as compared to the more compact fractals with thicker branches during the same simulation time interval.

An important characteristic of a structure on surface is the fractal dimension d_f. The Hausdorff fractal dimension is generally defined as [39, 40]:

$$d_f = \lim_{l \to 0} \frac{\log[N(l)]}{\log[1/l]}. \tag{6}$$

Here $N(l)$ is the number of self-similar structures of linear size l needed to cover the whole structure. In practice, the fractal dimension is usually calculated by the box-counting method or using the mass-formula for dimension introduced in [41]:

$$N = \rho \left(\frac{2R}{d_0} \right)^{d_f}, \tag{7}$$

where N is the number of particles in the system, R is the radius of the minimal circumscribed circle of the structure, d_0 is the diameter of a particle and ρ is the ratio of the covered surface to the entire surface area (packing density). In the case of the hexagonal grid $\rho = \pi/2\sqrt{3}$.

The time evolution of the average fractal dimension calculated using Eq. (7) for the structures shown in Fig. 2 is presented in Fig. 3. Here the averaging is performed over the five structures which are formed during the growth process. Figure 3 shows that the fractal dimension calculated for fractals with thicker branches is higher than the fractal dimension calculated for the structures with thinner branches. This happens because the morphology of a fractal with thicker branches is closer to a compact droplet for which the fractalic dimension is equal to 2.

Figure 3 illustrates that the fractal dimension for the structures with thicker branches experiences more significant fluctuations in the beginning of the simulation. This happens, because the structure of small fractals with thicker branches experiences more significant rearrangements due to the formation of compact droplets in the core of the fractal (see Fig. 2a). This results in an increase of the fractal dimensionality. When the number of particles attached to the fractal core reaches a critical value, the time needed for the particle rearrangement becomes larger than the characteristic period for a new particle attachment and the fractal branches are getting formed. This fixes the fractal dimension at a certain value.

3.2. Fractals fragmentation

The structures shown in Figs. 2a-b experience further transformation after the particle deposition has finished. Depending on the kinetic parameters the time after which a noticeable change in the fractal morphology can be detected is different. To illustrate the difference in the post-growth relaxation processes we considered fragmentation of a single fractal by varying the kinetic parameters of our model. The initial structure of the fractal used to study the fragmentation process is different from the structures presented in Fig. 2, and is shown in Fig. 4a. This fractal was grown with the method described above and has a diameter of ~ 638 nm, which is close to the diameter of the experimentally grown structures discussed in [4, 11, 17].

We present several scenarios of fractal relaxation. The rate of fractal decay depends on chosen parameters of our model, which define the morphology of fragments emerging during the relaxation of a fractal. The snapshots of the structures which emerge on the surface after different time intervals are shown in the Fig. 4 and give an illustrative example of three different fragmentation scenarios of a fractal. Figure 4b corresponds to the "explosion" of a fractal and is the fastest fragmentation scenario. In this case the interaction energy E_b between particles is weak and the probability of a particle to evaporate from the fractal is of the same order of magnitude as the probability of a particle to diffuse along the fractal periphery. This fragmentation scenario can be realized in experiment if the temperature of the system is rapidly elevated after the fractal formation. Figures 4c and 4d illustrate the fragmentation into a group of compact and non-compact islands respectively. The possibility of compact droplets formation depends on the ratio between E_b and $\Delta\epsilon$ as follows from Eqs.(3)-(4). At higher value of the barrier energy $\Delta\epsilon$ the diffusion rate of a particle along the periphery of an island is comparable with the rate of particle detachment, leading to the formation of highly disordered droplets. These fragmentation scenarios shown in Figs. 4c-d are expected to occur in experiments in which impurities are added to the system, causing the fractal to break apart [4].

The important quantity characterizing the fractal fragmentation process is the number of fragments at a given time. The minimum size of a fragment corresponding to a single particle is equal to one. The time evolution of the number of fragments calculated for different sets of kinetic parameters is shown in Fig. 5a-b. Lines 1 and 2 in Fig. 5a show the time evolution of

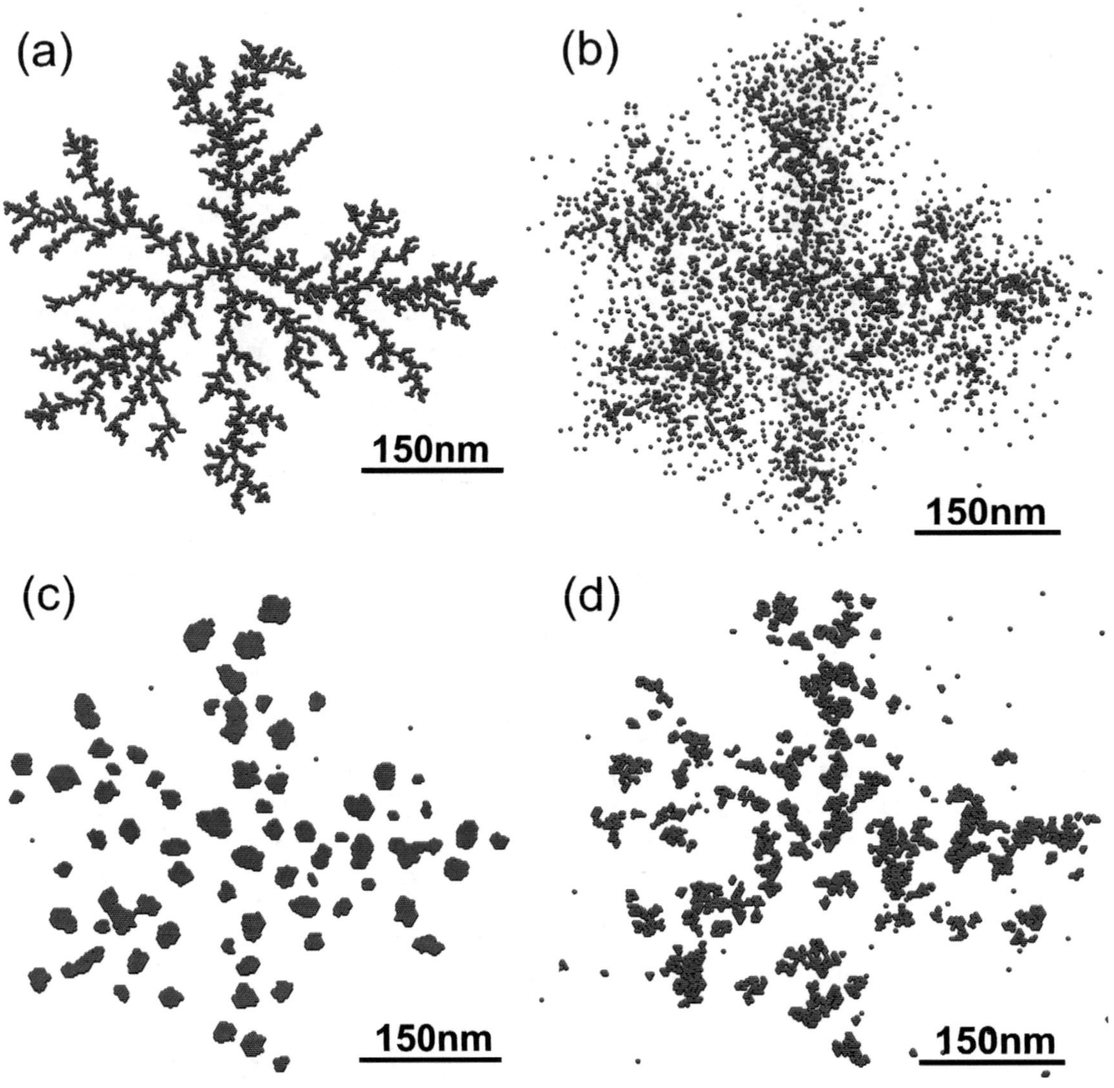

Figure 4. Evolution of the fractal structure during the post-growth relaxation process. Plot (a) shows the initial structure of a fractal. Plots (b)-(d) show the snapshots of the system at different fragmentation stages: (b) rapid fragmentation of a fractal after $t = 0.02$ s with $E_b = 1\ kT$, $\Delta\epsilon = 0.2E_b$, $\Delta\mu = 2\ kT$; (c) the diffusion of particles along the branch of a fractal is a more preferable process, fractal fragments into a group of compact islands after $t = 0.05$ s with $E_b = 4\ kT$, $\Delta\epsilon = 0.1E_b$, $\Delta\mu = 2\ kT$; (d) slow diffusion of particles along the branch causes the fractal to fragment into a group of dispersed islands after $t = 0.05$ s with $E_b = 4\ kT$, $\Delta\epsilon = 0.8E_b$, $\Delta\mu = 2\ kT$.

the number of fractal fragments in the case of relatively weak binding energy between particles being equal to 1 kT and 2 kT respectively. The number of fragments rapidly approaches the asymptotic value, equal to the total number of particles in the fractal.

Figure 6 shows the distributions of island sizes calculated after 0.78 s of simulation for the fractal shown in Fig. 4a. The calculations were performed on a $2.5 \times 2.5\ \mu s^2$ substrate with

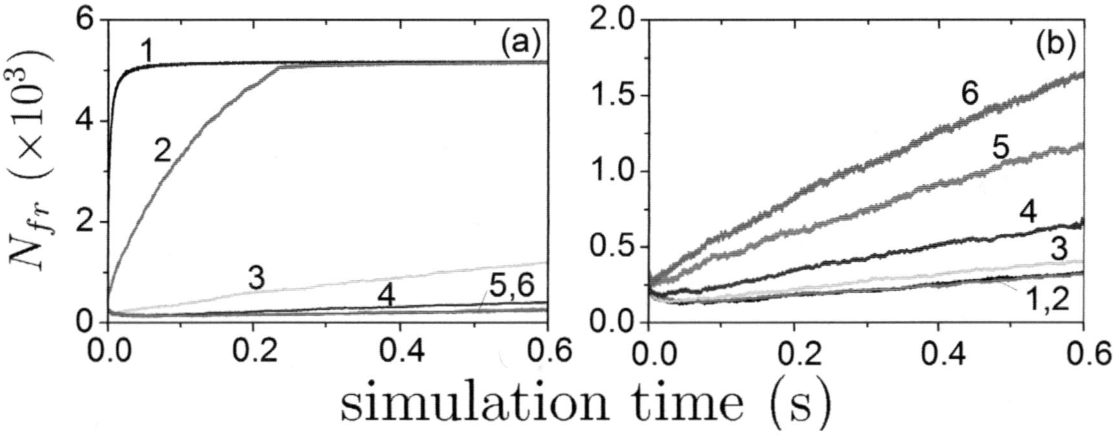

Figure 5. Time evolution of the number of fragments N_{fr} calculated for the fractal structure shown in Fig. 4a on a substrate 2.5×2.5 μm^2 with periodic boundary conditions. The fractal fragmentation in this example is considered for $\Delta\mu = 2$ kT for different values of the binding energy E_b and the barrier energy $\Delta\epsilon$. Plots (a) show the results of calculation obtained for $\Delta\epsilon = 0.2E_b$ by varying the binding energies between two particles. The lines 1-6 correspond to $E_b = (1,\ 2,\ 3,\ 4,\ 5,\ 6)$ kT respectively. Plots (b) represent the results obtained for $E_b = 4$ kT with varying $\Delta\epsilon$. The lines 1-6 correspond to $\Delta\epsilon = (0,\ 0.1,\ 0.2,\ 0.5,\ 0.8,\ 1)E_b$ respectively.

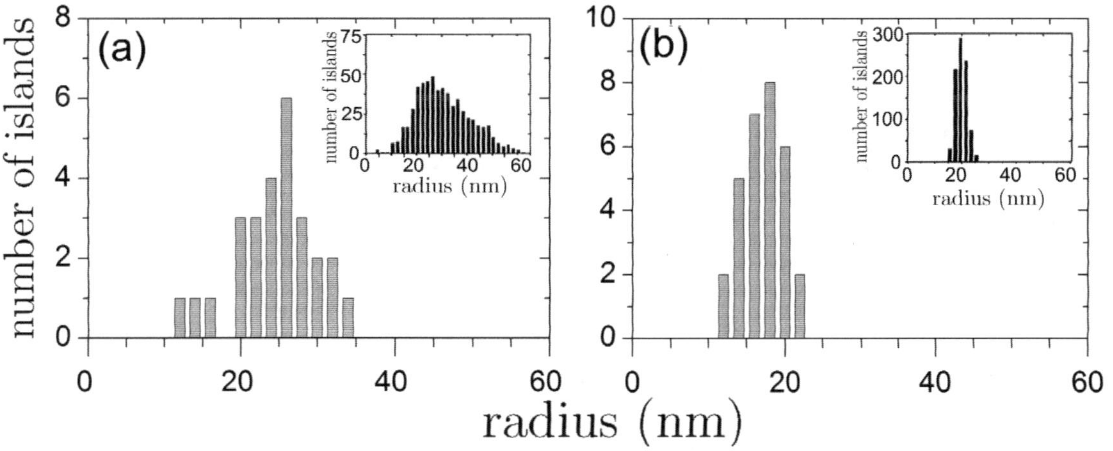

Figure 6. Distributions of island sizes after the fractal fragmentation obtained for different sets of the kinetic parameters: (a) $E_b = 4$ kT, $\Delta\epsilon = 0.1$ E_b, $\Delta\mu = 2$ kT; (b) $E_b = 3$ kT, $\Delta\epsilon = 0.2$ E_b, $\Delta\mu = 10$ kT. Insets show the results of the experimental measurements: (a) for perturbed silver fractals after annealing, and (b) after adding of oxide impurities to silver clusters [4].

periodic boundary conditions for the kinetic parameters $E_b = 4\ kT$, $\Delta\epsilon = 0.1E_b$, $\Delta\mu = 2\ kT$. In this case the diffusion of particles along the fractal periphery is the dominating process. The increased rate of particle peripheral diffusion leads to the higher probability of island rearrangement and to the formation of a broad distribution of island sizes, as seen in Fig. 6a. The inset in Fig. 6a shows the results of experimental observations for silver fractals after annealing at 600 K. The size distribution of the silver cluster islands has a broad width with the most probable radius of silver islands ~ 25 nm. The same value follows from the theoretical calculation.

Figure 6b shows the distribution of island sizes calculated for $E_b = 3\ kT$, $\Delta\epsilon = 0.6\ kT$, $\Delta\mu = 10\ kT$. The result of the numerical calculation is compared with the experimental observation on silver fractals grown from oxidized silver nanoparticles [4] shown in the inset. The most abundant radius of silver cluster island in this case is 18 nm, being in good agreement with the results of our calculations as seen from Fig. 6b.

Note that in Fig. 6 the width and the position of the maximum in the theoretical distributions are close to experiment while the absolute value in the distributions differs significantly. This happens because in our calculation we consider a single fractal, while the experimental measurements deal with many fractals on a surface. A similar shape of the calculated distributions and the measured ones shown in Fig. 6 allows one to conclude that for $N_{fractals}$ fractals on surface the number of islands formed after fragmentation is equal to

$$N_{tot}\left(R\right) = N_{fractals} \cdot \langle N_{isl}\left(R\right)\rangle, \tag{8}$$

where $\langle N_{isl}\left(R\right)\rangle$ is the average number of islands formed after fragmentation of a single fractal.

4. Conclusion

We have performed a theoretical analysis of the growth and post-growth processes occurring in nanofractals on a surface using the method, which describes the internal dynamics of particles in a fractal and accounts for their diffusion and detachment. We demonstrated that these kinetic processes are responsible for the morphology of the islands created on a surface.

The developed method includes three kinetic parameters, which are determined by interatomic interactions in the system. The kinetic parameters can be also obtained from experiment and are specific for different types of substrates and deposited materials. We have analyzed the fractal stability on surface by varying the kinetic parameters over a wide range of values and revealed three principally different fragmentation scenarios of a fractal.

5. Acknowledgments

This work was supported by the European EXCELL project. The possibility to perform complex computer simulations at the Frankfurt Center for Scientific Computing is also gratefully acknowledged.

References

[1] Meiwes-Broer K H (ed) 2000 *Metal Cluster at Surfaces: Structure, Quantum Properties, Physical Chemistry* (Berlin: Springer-Verlag)
[2] Drexler K E 1992 *Nanosystem* (USA: Willey)
[3] Bréchignac C, Houdy P and Lahmani M (eds) 2007 *Nanometerials and Nanochemistry* (Cambridge University Press)
[4] Lando A, Kébaïli N, Cahuzac P, Masson A and Bréchignac C 2006 *Phys. Rev. Lett.* **97** 133402–(1–4)
[5] Dick V V, Solov'yov I A and Solov'yov A V 2009 *AIP Conf. Proc.* **1197** 76–88
[6] Dick V, Solov'yov I A and Solov'yov A V 2010 *http://arxiv.org/abs/1001.3992*
[7] Rubahn H G 2004 *Basics of Nanotechnology* (Denmark: Weinheim)
[8] Solovyeva V, Keller K and Huth M 2009 *Thin Solid Films* **517** 6671–6676
[9] Jensen P 1999 *Rev. Mod. Phys.* **71** 1695–1735

[10] Barabási A L and Stanley H E (eds) 1995 *Fractal Concepts in Surface Growth* (Cambridge: Cambridge University Press)
[11] Bréchignac C, Cahuzac P, Carlier F, Colliex C, de Frutos M, Kébaïli N, Roux J L, Masson A and Yoon B 2003 *Eur. Phys. J. D* **24** 265–268
[12] Carlier F, Benrezzak S, Cahuzac P, Kébaïli N, Masson A, Srivasta A K, Colliex C, de Frutos M and Bréchignac C 2006 *Nanoletters* **6** 1875–1879
[13] Bardotti L, Jensen P, Hoareau A, Treilleux M and Cabaud B 1995 *Phys. Rev. Lett.* **74** 4694–4697
[14] Scott S A and Brown S A 2006 *Eur. Phys. J. D* **39** 433–438
[15] Kébaïli N, Benrezzak S, Cahuzac P, Masson A and Bréchignac C 2009 *Eur. Phys. J. D* **52** 115–118
[16] Schmidt M, Kébaïli N, Lando A, Benrezzak S, Baraton L, Cahuzac P, Masson A and Bréchignac C 2008 *Phys. Rev. B* **77** 205420–(1–4)
[17] Lando A, Kébaïli N, Cahuzac P, Colliex C, Couillard M, Masson A, Schmidt M and Bréchignac C 2007 *Eur. Phys. J. D* **43** 151–154
[18] Liu H and Reinke P 2007 *Surf. Sci.* **601** 3149–3157
[19] Li W, Zheng X and Fei W 2009 *Vacuum* **83** 949–952
[20] Liu H and Reinke P 2006 *J. Chem. Phys.* **124** 164707–(1–5)
[21] Böttcher A, Weis P, Jester S S, Löffler D, Bihlmeier A, Klopper W and Kappes M M 2005 *Phys. Chem. Chem. Phys.* **7** 2816–2820
[22] Irisawa T, Uwaha M and Saito Y 1995 *Europhys. Lett.* **30(3)** 139–144
[23] Witten Jr T A and Sander L M 1981 *Phys. Rev. Lett.* **47** 1400–1403
[24] Solov'yov I A, Mathew M, Solov'yov A V and Greiner W 2008 *Phys. Rev. E* **78** 051601–(1–13)
[25] Semenikhina V V, Lyalin A G, Greiner W and Solov'yov A V 2008 *J. Exp. Theor. Phys.* **106** 678–689
[26] Solov'yov I A, Geng J, Solov'yov A V and Johnson B F G 2009 *Chem. Phys. Lett.* **472** 166–170
[27] Geng J, Solov'yov I A, Zhou W, Solov'yov A V and Johnson B F G 2009 *J. Phys. Chem. C* **113** 6390–6397
[28] Koshelev A, Shutovich A, Solov'yov I A, Solov'yov A V and Greiner W 2003 *Proceedings of International Workshop "From Atomic to Nano-scale", Old Dominion University* 184–194
[29] Solov'yov I A, Solov'yov A V and Greiner W 2003 *Int. J. Mod. Phys. E* **13** 697–736
[30] Obolensky O I, Solov'yov I A, Solov'yov A V and Greiner W 2005 *Comp. Lett.* **1** 313–318
[31] Solov'yov I A, Solov'yov A V, Greiner W, Koshelev A and Shutovich A 2003 *Phys. Rev. Lett.* **90** 053401–(1–4)
[32] Solov'yov I A, Geng J, Solov'yov A V and Johnson B F G 2009 *AIP Conf. Proc.* **1197** 89–102
[33] Geng J, Solov'yov I A, Reid D G, Skelton P, Wheatley A E H, Solov'yov A V and Johnson B F G 2010 *Phys. Rev. B* **81** 214114–(1–13)
[34] Levi A C and Kotrla M 1997 *J. Phys.: Condens. Matter* **9** 299–344
[35] Albano E 1996 *Heter. Chem. Rev.* **3** 389–418
[36] Voter A F 2005 *Radiation Effects in Solids* (Springer, NATO Publishing Unit, Dordrecht, The Netherlands) chap Introduction to the Kinetic Monte Carlo Method
[37] Frenkel J (ed) 1955 *Kinetic Theory of Liquids* (Dover)
[38] Bréchignac C, Cahuzac P, Carlier F, Colliex C, de Frutos M, Kébaïli N, Roux J L, Masson A and Yoon B 2001 *Eur. Phys. J. D* **16** 265–269
[39] Hausdorff F 1919 *Math. Annal.* **79** 157–179
[40] Martinez F, Cabrerizo-Vilchez M A and Hidalgo-Alvarez R 2001 *Physica A* **298** 387–399
[41] Feder J 1988 *Fractals* (New York: Plenum Press) ISBN 0-306-42851-2

20-nanogold $Au_{20}(T_d)$ cluster and its hollow cage isomers: structural and energetic properties

E S Kryachko[1,2] and F Remacle[2]

[1]Bogolyubov Institute for Theoretical Physics, Kiev-143, 03680 Ukraine
[2]Department of Chemistry, Bat. B6c, University of Liège, B-4000 Liège, Belgium

E-mail: [1]eugene.kryachko@ulg.ac.be; [2]fremacle@ulg.ac.be

Abstract. The present work discusses the capability of 20-nanogold low-energy clusters to encage atomic and molecular species and investigates hollow cages of Au_{20}, their structures and stability, and their void reactivity. We begin with performing a systematic computational search of hollow cages on the potential energy surface of doubly anionic Au_{20} which, according to the experimental abundance spectra of Au_{20}^{2-}, has an approximately degenerate ground state. Since the computed second electron affinity of the neutral tetrahedral ground-state cluster $Au_{20}(T_d)$ agrees well with the experimental value $EA_2^{expt}(Au_{20})$, this ground state is thus partly occupied by $[Au_{20}(T_d)]^{2-}$ whose original neutral T_d-symmetry is broken. Determining the other, yet unknown isomer and applying 'reverse' charge state mappings $Z = -2 \Rightarrow Z = 0, \pm 1$, we identify stable and low-energy Au_{20} hollow cages which are further studied from different angles and compared with $Au_{20}^Z(T_d)$. The void reactivity of these 20-nanogold hollow cages is the key theme - it is suggested that, together with the global characteristics, such as the ionization potential and electron affinity, the molecular electrostatic potential and HOMO-LUMO patterns are actually tools that may shed a light on the general features of voids of these golden fullerenes and their capability to encage H and Li. The confinement character of the studied golden fullerenes is compared with the classical examples, C_{60} in particular.

1. Introduction

The discovery in the 1980s of the buckyball C_{60} and larger fullerenes [1] enables to form, by trapping guest atoms, ions, and molecules into their nanosized voids (nanovoids), so called 'endohedral' or @-fullerenes which manifest a variety of remarkable features. This discovery raised the question of existence of similar fullerene-like structures or hollow cages for other chemical elements. Among the latter, gold is a particular element – definitely 'noblesse oblige' – because of its rather unique properties which are mostly dictated by the strong relativistic effects [2].

The first fullerene-type cluster of gold – a 'golden fullerene' [3] - is the smallest hollow icosahedral cage $Au_{12}(I_h)$ that was predicted in 2002 [4] and, as demonstrated computationally [5-7], is only stable endohedrally, i. e. as $M@Au_{12}$ with a metal atom M = W, V, Nb, and Ta. The cage Au_{14} is similar: it is bound as the endohedral golden fullerene $M@Au_{14}$ and unbound in the gas phase [8]. Larger hollow cages $Au_{N=16-18}$ in the anionic charge state have recently been observed in photoelectron spectroscopy experiments [9]. Their voids are characterized by a diameter that determines a spatial confinement of ca. 5.5 Å [9], i. e., these voids are large enough to accommodate some guest atoms. The concept of golden fullerenes has recently been extended to N > 18 and already reached 55 [10] and 72 [11]. Among them is, for instance, the ground-state and highly stable icosahedral golden fullerene $Au_{32}(I_h)$

[3] whose HOMO-LUMO gap $\Delta := \varepsilon_{LUMO} - \varepsilon_{HOMO}$, defined as the difference of the energy eigenvalues of the HOMO (highest occupied molecular orbital) and of the LUMO (lowest unoccupied molecular orbital), falls into the interval of 1.5 – 2.5 eV, being thus close to $\Delta \in$ (1.57,1.80 eV) of C_{60} [12].

Generally speaking, an arbitrary 3D molecular structure is either space-filled, compact, without any void or possesses some voids (emptiness) that result in a hollow cage shape [9]. The space-filled structures, such as for example, the magic gold cluster $Au_{20}(T_d)$ [13][19] (see also the current work [14] and references therein) are expected to be more energetically stable, at least in the neutral charge state (see particularly [8,13,15-20]).

2. Au_{20} fullerenes

It is generally anticipated that golden fullerenes inherit many exceptional features of the classical fullerenes C_n - the foremost one is the chemical reactivity which is broadly defined as a capability of a given molecular species to form chemical bonding patterns while interacting or contacting with other atoms, ions, or molecules. This is by large the reason why hollow cage gold nanoparticles (NPs) have recently attracted a great deal of attention in nanoscience. Since 20-nanogold fullerenes are definitely not the ground-state structures in the neutral charge state Z = 0 and since a tetrahedral space-filled shape T_d is even preserved in the charge states $Z = \pm 1$ [13,14], the question rises whether and how they can be detected experimentally and also if they can be synthesized? In general, low-energy molecular isomers are hard to identify experimentally, though the experimental identification of low-energy gold isomers has recently been reported in the series of works [21-25], e. g.: (**i**) Ref. [22] where four isomers of Au_{20}^- lying higher the ground-state within an energy interval of ~0.58 eV were observed; (**ii**) Ref. [21] that identifies four isomers of Au_7 placed by $\leq$ 0.74 eV above the ground state. This experimental search has to be conducted in parallel with the theoretical research on golden cage isomers, by a direct analogy with a similar research of low-energy carbon fullerene's isomers [26].

What are the experiments that have been performed for 20-nanogold? Only few such experiments have been conducted [13,22,27-36]. In 1992, using UV photoelectron spectroscopy, Taylor et al. [36] reported the (first) electron affinity of Au_{20}, $EA_1^{expt}(Au_{20}) = 2.732 \pm 0.050$ eV. In 2003, Li et al. [13] measured the experimental photoelectron spectroscopy spectrum of Au_{20} and obtained $EA_1^{expt}(Au_{20}) = 2.745 \pm 0.015$ eV. They also provided a strong computational evidence of the existence of the ground-state neutral $Au_{20}(T_d)$ cluster that possesses a remarkably large HOMO-LUMO gap $\Delta(Au_{20})$ = 1.77 eV, thus exceeding $\Delta^{expt}(C_{60})$ = 1.57 eV by 0.2 eV. The collision induced dissociation experiments were also conducted in order to study the fragmentation channels of singly and multiply charged Au_{20}^+, Au_{20}^{2+}, and Au_{20}^{3+} clusters [28-30].

In parallel, the experimental abundance spectra of Au_N^{2-} for $12 \leq N \leq 30$, which were produced by laser vaporization of gold metal, storage of the anions Au_n^- in a Penning trap, their mass selection, and a final electron attachment, definitely confirmed that the doubly charged negative cluster Au_{20}^{2-} exists and is stable [31-35]. Hence, Au_{20} has a positive second electron affinity $EA_2^{expt}(Au_{20})$ estimated, within the liquid drop model, as equal to ~ 0.5 eV. Moreover, as demonstrated in Figure 2 of [32], the ground state on the dianionic potential energy surface (PES) of 20-nanogold is quasi doubly degenerate, implying that there exist two isomers of Au_{20}^{2-} which share it. Since the ZPE-corrected $EA_2(Au_{20}(T_d))$ is computed equal to 0.525 eV for B3LYP and to 0.433 eV for PW91 [14] (see the next Section), it is quite close to the experimental value which suggests that the $Au_{20}^{2-}(T_d)$ lies at the very bottom of the PES. Besides, the Coulomb repulsion of the negative excess charges of $Au_{20}^{2-}(T_d)$ makes it, via breaking a T_d symmetry, a hollow cage with an external diameter of ~1.05 nm and a void one of ~4 - 5 Å. In contrast, the fullerene dianion C_{60}^{-2} is metastable in the ground state with the lifetime of ≈ 20 s (T = 0 K) with respect to autodetachment [37]. This feature therefore makes the chemical reactivities of the classical and golden fullerenes somewhat different.

The other(s) ground-state dianionic isomer of 20-nanogold has not yet been reported. Its identification partly initiated this work. In fact, the goal of this work is twofold: (**α**) to conduct a systematic search on the dianionic PES of 20-nanogold to determine the unknown bottom-energy structure [say, cage I^{-2}, since we define the neutral ground-state cluster 0 := $Au_{20}(T_d)$ and correspondingly, 0^{-2}] and, in addition, one novel low-energy hollow Au_{20}^{-2} cage II^{-2}; and (**β**) via sequentially mapping of the above PES to the PESs corresponding to Z = -1, 0, and +1, to show that in the neutral and cationic charge states cage II splits into two, II^Z and III^Z (Z = 0, +1), and to analyze different facets of the void reactivity of these new cages in different charge states. In particular, we investigate whether they are capable to encapsulate some atom or ion and show that some of the cages can indeed be doped, e. g. with a Li atom. Hence, these charge-state mappings suggest that cage I could be experimentally detected.

2.1. Computational methodology

All computations of gold clusters were carried out with the density functional (DF) potentials B3LYP, BP86, and PW91PW91 (PW91 for short) in conjunction with the energy-consistent 19-($5s^2 5p^6 5d^{10} 6s$) valence electron relativistic effective core potential of the Los Alamos double-zeta type LANL2DZ. The GAUSSIAN 03 package of quantum chemical programs [38] was used throughout the present work. All geometrical optimizations were carried out with the keywords "Tight" and "Int = UltraFine". The harmonic vibrational frequencies and concomitant zero-point energies (ZPE) were calculated in order to locate the energy-minimum structures and to distinguish them from the saddles. The basis set comprised of the standard Pople's basis set 6-311++G(d,p) is additionally used for the H and Li atoms.

2.2. Structure and stability of 20-nanogold cages

The structures of three novel cages I, II, and III of Au_{20} in the charge state Z = 0, characterized by external diameters of about 0.8-1.2 nm which are slightly larger that, 0.7 nm, of C_{60}, are shown in Figure 1. Regarding their voids, these cages are essentially hollow in the 0-charge state with cage or void diameters of ca. 4.5 – 6.0 Å. Structurally, cage I resembles a helix. Cage II exhibits a bilayer motif and is formed by two planar $Au_{10}(D_{2h})$ clusters which are twisted relative to each other by a dihedral angle of 33.1° and bonded to each other by means of 20 additional metallic bonds. Due to its quite specific bilayer shape, cage II could be synthesized.

All these cages are definitely energy minima on the 20-gold PESs for Z = 0, ±1, and -2 with real harmonic vibrational frequencies only, which are distributed, say for Z = 0, within 11 – 160, 12 – 118, and 15 – 154 cm^{-1}, respectively. As neutral, they are closed-shell clusters which, as reported in Table, are highly stable with respect to the atomization channel $Au_{20}^{I,II,III} \Rightarrow 20$ Au. This stability is corroborated by high absolute values, of ~33 – 34 eV, of the corresponding energies, $-\Delta E_f$, and enthalpies, $-\Delta H_f$, of formation, despite a large entropy effect. The latter increases the Gibbs free energies, ΔF_f^{298}, of formation at T = 298 K and correspondingly lowers the cages' stabilities by 6-7 eV, but they still remain strongly bound. Since II is formed by two bonded planar $Au_{10}(D_{2h})$ clusters, its corresponding energy of formation amounts to -2.34 eV with B3LYP DFT and to -4.34 eV with the PW91, and is hence strongly DF dependent. Cage II is less stable than I in the neutral charge state by approximately -43.15 – (-43.20) = 0.05 (- PW91; 0.11 – BP86, 0.37 – B3LYP) eV. On the other hand, in the same charge state, cage III is more stable compared to I by 0.11-0.16 eV. Nevertheless, within the numerical accuracy, they are located very close to each other.

As suggested above, on the dianionic PES, $Au_{20}^{-2}(T_d)$ shares the very bottom of the dianionic PES with I^{-2}, though actually, in a strict computational sense, the latter lies slightly below $Au_{20}^{-2}(T_d)$ by ~0.05 eV. Relative to the asymptote comprising of 18 Au + 2 Au^-, I^{-2} is essentially stable that is manifested by ΔE_f = -34.32 eV and a Gibbs free energy ΔF_f^{298} = -27.56 eV at the B3LYP computational level. Notice that in the dianionic charge state, II and III merge to one another and are less stable than I by 0.39 eV. Via the mapping $Z = -2 \Rightarrow Z = -1$, I approaches $Au_{20}^{-1}(T_d)$ from above by 0.59 – 0.77 eV, which only differ by 0.01 – 0.19 eV from the energy range reported in recent diffraction experiments [22], and by construction, II^{-1} and III^{-1} are identical. The neutral cages I, II, and III are placed energetically within 1.40-1.85 eV relative to the ground-state cluster $Au_{20}^{Z=0}(T_d)$. On the other hand, on the cationic sheet of the 20-gold PES, III becomes closer to $Au_{20}^{+1}(T_d)$ by 0.66 eV. Both energy gaps can likely be achievable in experiments.

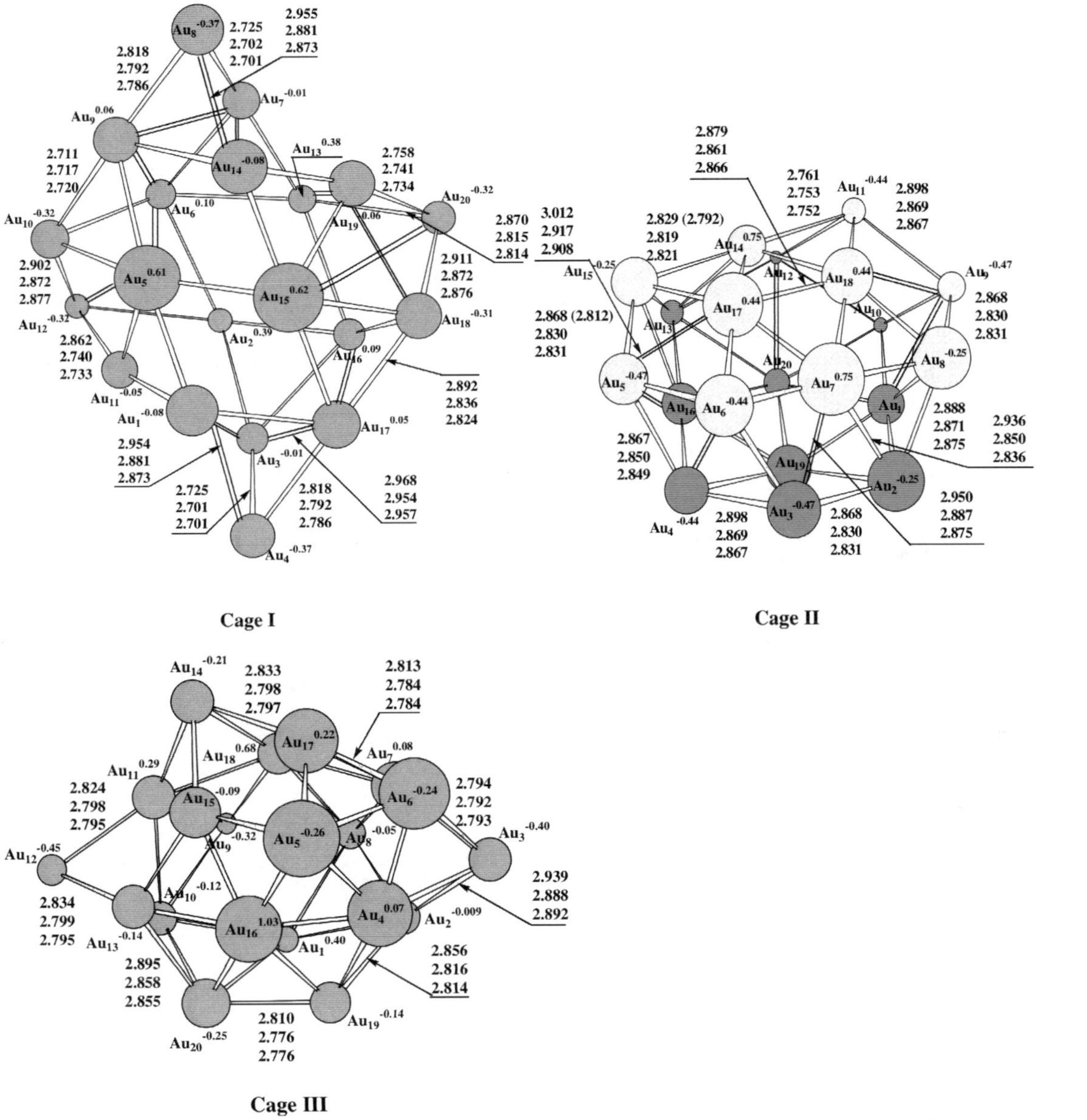

Figure 1. The golden hollow cage clusters I, II and III in the neutral charge state Z = 0. Cage I resembles a helix with the void diameter of ~5.9 Å. Its external diameter is ~1.2 nm. The void diameter of the cage II is approximately equal to 0.84 nm, i. e. it slightly exceeds that of $Au_{20}(T_d)$. As readily seen in this Figure 1, II is composed of the two planar $Au_{10}(D_{2h})$ clusters colored correspondingly in yellow (light in b/w) and brown (correspondingly dark). The angle between the longest diagonals Au_5-Au_9 and Au_3-Au_{12}, each of which connects the two vertex gold atoms mostly separated from each other, amounts to 33.1°. Cage III is characterized by a void diameter, R(Au_9-Au_{16}), of 5.5 Å. Its external diameter associated with R(Au_3-Au_{12}) amounts to 0.98 nm. Selected bond lengths are given in Å (reading from top to bottom: B3LYP, BP86, and PW91 DF potentials). The B3LYP Mulliken charges of some gold atoms are indicated by superscripts.

Table 1. Selected properties (in eV) of the cages I, II, and III in the neutral charge state. ΔE_f is defined as the ZPE-corrected energy of formation of the cage, taken with respect to twenty non-interacting gold atoms. ΔH_f determines the corresponding heat of atomization and ΔF_f^{298} is the corresponding free Gibbs energy at T = 298 K. ΔE is the relative ZPE-corrected energy of the cage with respect to the ground-state cluster $Au_{20}{}^{Z}(T_d)$. IE_1 is the first ionization energy and EA_1 the adiabatic electron affinity.

Selected	Cage I			Cage II			Cage III		
properties	B3LYP	BP86	PW91	B3LYP	BP86	PW91	B3LYP	BP86	PW91
$-\Delta E_f$	33.94	41.73	43.20	33.62	41.62	43.15	34.05	41.89	43.37
$-\Delta H_f$	33.95	41.76	43.24	33.62	41.64	43.41	34.07	41.92	43.41
$-\Delta F_f^{298}$	27.15	34.82	36.28	26.84	34.71	36.44	27.20	34.92	36.38
ΔE	1.48	1.56	1.60	1.85	1.67	1.65	1.42	1.40	1.43
$-\varepsilon_{HOMO}$	5.71	5.48	5.36	5.50	5.32	5.20	5.81	5.64	5.52
$-\varepsilon_{LUMO}$	4.52	5.08	4.97	4.37	4.93	4.79	4.21	4.82	4.71
Δ	1.19	0.40	0.39	1.13	0.40	0.42	1.59	0.82	0.81
IE_1	6.692	6.852	6.730	6.539	6.722	6.602	6.647	6.862	6.744
EA_1	3.543	3.748	3.630	3.776	3.628	3.502	3.340	3.359	3.280

2.3. One-electron properties

The HOMO and LUMO eigenvalues and the corresponding HOMO-LUMO gap Δ, the first adiabatic electron affinities (EA_1) and ionization potentials (IE_1) are the key features that determine chemical reactivity. All these features of cages I, II, and III are summarized in Table 1. The HOMOs are placed ca. 0.4-0.9 eV higher the HOMO of $Au_{20}(T_d)$. Cages I and II feature rather narrow gaps $\Delta \approx 0.4$ eV (BP86 and PW91), compared to $Au_{20}(T_d)$. The EA_1 of the studied cages are significantly large, falling within 3.3 - 3.8 eV, thus exceeding the $EA_1^{B3LYP}(Au_{20}(T_d)) = 2.606$ eV (notice that $EA_1(C_{60}) = 2.57$ eV) by ca. 0.5-1.2 eV. The IE_1 of I, II, and III are lower - within 6.6-6.7 eV - compared to IE_1^{PW91} of $Au_{20}(T_d) = 7.34$ eV and $IE_1^{B3LYP}(Au_{20}(T_d)) = 7.398$ eV. Altogether, the above distinctions between the studied hollow cages I, II, and III and the space-filled cluster $Au_{20}(T_d)$ allow us to conclude that the reactivity of these cages is drastically different from that of $Au_{20}(T_d)$. Yet another indicator of the reactivity of a given molecule or cluster is the chemical hardness η that is approximately defined as a half of the difference between the IE_1 and EA_1, viz., $\eta \approx (IE_1 - EA_1)/2$. Within this reactivity scale, II with lowest hardness, $\eta^{B3LYP}(II) = 1.38$ eV, is more reactive than both I and III, with I being more reactive than cage III : $\eta^{B3LYP}(II) < \eta^{B3LYP}(I) = 1.57 \text{ eV} < \eta^{B3LYP}(III) = 1.65$ eV. All cages are more reactive than $Au_{20}(T_d)$ which is characterized by $\eta^{B3LYP}(Au_{20}(T_d)) = 2.40$ eV.

2.4. Quantifying void reactivity of Au_{20} cages

When molecules interact with one another, chemical bonding patterns arise between their atoms. Typically, these patterns are spread over the outer space of each interacting partner. In this sense, cages are different – by definition, they possess atoms which are assumed to be capable to form chemical bonds with molecules from the outer space as well as from the inner one (inside the cage, in voids), i.e. with those which are encapsulated or confined within the cage. This hence implies a bifunctionality of the chemical reactivity - the outer or exo-reactivity and the inner, void, or endo-reactivity - of some atoms which compose cages and thus suggests potential routes to control both kinds of chemical reactivities. The features, such as the ionization energies and electron affinities that indicate how a given cage overall reorganizes either upon removal or upon addition of an electron, which are examined in the previous Subsection are those which are usually invoked to characterize chemical reactivity. They are, however, the global characteristics of a cage that cannot be partitioned into those which may be solely ascribed either to the outer or to the inner void.

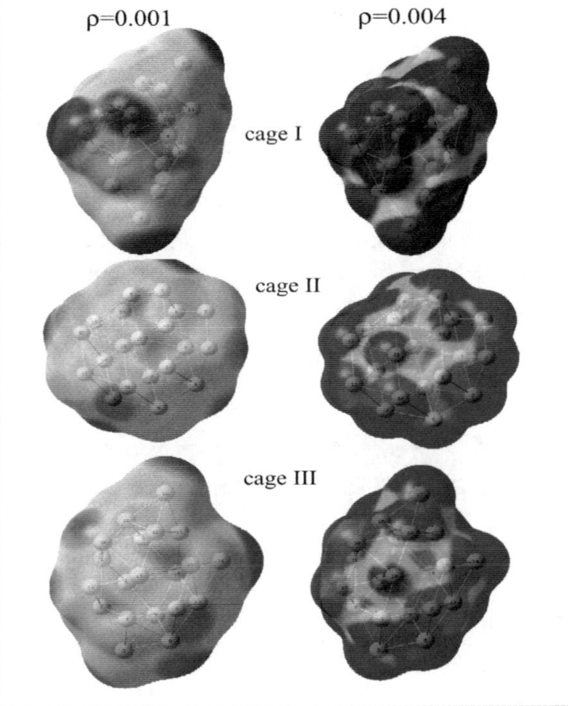

Figure 2. The MEPs of cage I, II, and III mapped from -0.016 (red) to +0.016 (blue) $|e|/(4\pi\varepsilon_0 a_0)$ onto 0.001 $|e|\cdot Å^{-3}$ isosurface of the one-electron density $\rho(\mathbf{r})$ and from -0.01 to +0.01 $|e|/(4\pi\varepsilon_0 a_0)$ onto 0.004 (the right column, correspondingly, for cages I, II, and III respectively) $|e|\cdot Å^{-3}$ isosurface of the one-electron density $\rho(\mathbf{r})$. $\rho(\mathbf{r})$ is computed at the B3LYP computational level.

The void regions of cages I, II, and III are in fact spatially confined areas which can accommodate some heteroatom M or heteroatoms (dopant). A remarkable feature of these cages is that they all have only one-atom-layer that separates the void from the outer surface which may thus facilitate the direct control of the outer reactivity from the inner one. A typical point of view that often prevails is that the void reactivity of a cage is a direct consequence of the spatial confinement, i.e., the dopant feels the cage's boundary, and therefore, the size of dopant plays a decisive role for a stable encapsulation. Put in other words, if the size of the void is of the same order of magnitude or comparable with that of the dopant, one may anticipate that doping is stable. Obviously, doping influences the outer reactivity. It is well known that the patterns of the molecular orbitals, particularly of the HOMO and LUMO, and the concept of the molecular electrostatic potential (MEP) [39] are crucial for understanding of both facets of chemical reactivity. For example, the buckyball fullerene C_{60} exhibits two different behaviors of the MEP [40]. It is positive in the entire void region, where C_{60} is capable to encage atoms, ions, and some molecules, and reaches the outer central regions of the pentagon rings. On the contrary, it is negative in the outer region and the most negative at the midpoints of the bonds linked two hexagonal rings.

The MEP patterns of the studied golden fullerenes I, II, and III are plotted in Figure 2. These MEPs look quite different from that of C_{60}. First: the outer MEP is nonnegative, as demonstrated by blue converging to green regions. The void MEP of cage I can be both, either negative - as shown in red, particularly in the neighborhood of the gold atoms Au_1 and Au_5 - or positive, that is in blue, as in a small and rather deep (due to its color) 'pocket' around Au_{17}. By analogy with C_{60}, the neighborhood of the latter atom is therefore attractive to accommodate an atom. The MEP patterns of cage II are positive in its outer region and admit both signs in the void. Two symmetric MEP regions in the neighborhood of the gold atoms Au_{12}, Au_{14}, Au_{20}, and Au_3, Au_7, and Au_{19} are negative, whereas the other two, around Au_{13}, Au_{17}, and Au_{10}, Au_{18}, are positive, implying thus two different sites for encaging. In addition, there exists a 'pocket' with a slightly negative charge in the center of the MEP of cage III and the surface mapped on the $\rho(\mathbf{r}) = 0.004$ isocontour is essentially positive. To summarize: the MEPs of the studied cages I, II and III imply that their voids are capable, in general, to confine neutral, as well as both, positively and negatively, charged atomic and molecular guests.

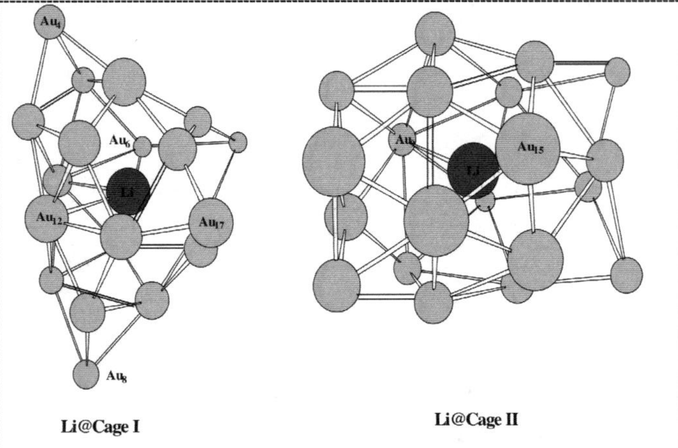

Figure 3. The endohedral golden fullerenes Li@cage I and Li@cage II in the neutral charge state. The geometry of Li@cage I is determined by the following B3LYP distances $R(Au_4\text{-}Au_8) = 9.789$, $R(Au_4\text{-}Au_{12}) = 7.282$, and $R(Au_8\text{-}Au_{17}) = 8.055$ Å. The Li atom forms bonds with Au_6 and Au_{12} of the cage I with the equal bond lengths $R(Li\text{-}Au_6) = R(Li\text{-}Au_{12}) = 2.745$ Å. The latter are comparable with those of Li@cage II which amount to $R(Li\text{-}Au_{15}) = 2.811$ and $R(Li\text{-}Au_9) = 2.832$ Å.

It is rather natural to consider the very small doping atoms H and Li. The atomic radii of H amounts to ~0.25 – ~0.53 Å, thus implying its easiness to be accommodated within the voids of the studied Au_{20} cages I, II, and III. It is however not the case because geometric arguments are in general insufficient for characterizing encaging that is close to a typical molecular process involving charge transfer: the computational scenarios of the initial endo-bondings of H within these cages end with the final exo-bondings to the Au-Au bonds and corresponding binding energies of -2.1 – -2.3 eV. In contrast, we found that $Au_{20}(T_d)$, while interacting with H, undergoes rather considerable extension of its void that becomes sufficiently large to encage the latter by the three planar 9-coordinated inner Au^i atoms and to form the centered triangle $Au_{1,2,3}^{i\,+0.09}$-$H^{-0.33}$, where H carries a negative Mulliken charge, with the binding energy equal to -0.53 eV. Note that latter is slightly higher than the energy calculated in [41] by means of the PW91 DF and a plane-wave basis set. One may suggest that the emergence of the confinement propensity of the 3D space-filled and compact cluster $Au_{20}(T_d)$ is the result of two main factors: the softness of its void Au-Au bonds whose stretching modes fall around ~100 cm^{-1}, compared to C_{60}, and a lower value of its electron affinity compared to the above cages.

The MEP patterns of cages I and II and their differences with the MEP of C_{60} are corroborated by encaging Li into the former and forming the endohedral golden fullerenes Li@cage I and Li@cage II which are demonstrated in Figure 3. In both endohedral systems, Li forms rather strong covalent Au-Li bonds with bond lengths of ca. 2.75 – 2.83 Å that are reflected in the large, by the absolute value, binding energies of 3.78 and 3.67 eV, respectively. The latter exceed by factor of ~ 2.7 the absolute value of the binding energy of the endohedral fullerene $Li@C_{60}$ and is simply explained by two facts: first, $IE_1(\text{cages}) > IE_1(Li) = 5.62$ eV and second, $EA_1(\text{cages}) > EA_1(C_{60}) > EA_1(Li) = 0.42$ eV, implying that the studied hollow golden cages have a stronger ionizing power on Li than the buckyball C_{60}. It is also worth mentioning that the endo-binding energy of $Li@Au_{20}(T_d)$ amounts to -2.49 eV, and hence it is weaker compared to that of cages I and II, though the latter are less stable than $Au_{20}(T_d)$.

Acknowledgments

This work was partially supported by the AIP 'Clusters and Nanowires' Project of the Belgian Federal Government, the EC FET proactive NanoICT Project 'MOLOC', and the F.N.R.S. (Belgium) FRFC Project 2.4565.06.

References

[1] Kroto H W, Heath J R, O'Brien S C, Curl R F and Smalley R E 1985 *Nature (London)* **318** 162
[2] Pyykkö P 2008 *Chem. Soc. Rev.* **37** 1967
[3] Johansson M P, Sundholm D and Vaara J 2004 *Angew. Chem. Int. Ed.* **43** 2678
[4] Pyykkö P and Runeberg N 2002 *Angew. Chem. Int. Ed.* **41** 2174
[5] Li X, Kiran B, Li J, Zhai H J and Wang L S 2002 *Angew. Chem. Int. Ed.* **41** 4786
[6] Yoon B, Koskinen P, Huber B, Kostko O, v Issendorff B, Häkkinen H, Moseler M and Landman U 2007 *Chem. Phys. Chem.* **8** 157
[7] Sun Q, Wang Q, Jena P and Kawazoe Y 2008 *ACS Nano* **2** 341
[8] Gao Y, Bulusu S and Zeng X C 2005 *J. Am. Chem. Soc.* **127** 156801
[9] Bulusu S, Li X, Wang L-S and Zeng X C 2006 *Proc. Natl. Acad. Sci. U.S.A.* **103** 8326
[10] Schmid G 2008 *Chem. Soc. Rev.* **37** 1909
[11] Karttunen A J, Linnolahti M, Pakkanen T A and Pyykkö P 2008 *Chem. Commun.* 465
[12] Taylor J, Guo H and Wang J 2001 *Phys. Rev. B* **63** 121104(R)
[13] Li J, Li X, Zhai H-J and Wang L-S 2003 *Science* **299** 864
[14] Kryachko E S and Remacle F 2007 *Int. J. Quantum Chem.* **107** 2922
[15] Apra E, Ferrando R and Fortunelli A 2006 *Phys. Rev. B* **73** 205414
[16] Fernández E M, Soler J M, Garzón I L and Balbás L C 2004 *Phys. Rev. B* **70** 165403
[17] de Bas B S, Ford M J and Cortie M B 2004 *J. Mol. Struct. (Theochem)* **686** 193
[18] Fernández E M, Soler J M and Balbás L C 2006 *Phys. Rev. B* **73** 235433
[19] Wang J, Bai J, Jellinek J and Zeng X C 2007 *J. Am. Chem. Soc.* **129** 4110
[20] Krishnamurty S, Shafai G S, Kanhere D G, de Bas B S and Ford M J 2007 *J. Phys. Chem. A* **111** 10769
[21] Gruene P, Rayner D M, Redlich B, van der Meer A F G, Lyon J T, Meijer G and Fielicke A 2008 *Science* **321** 674
[22] Xing X, Yoon B, Landman U and Parks J H 2006 *Phys. Rev. B* **74** 165423
[23] Huang W, Bulusu S, Pal R, Zeng X C and Wang L-S 2009 *ACS Nano* **3** 1225
[24] Häkkinen H, Moseler M, Kostko O, Morgner N, Hoffmann M A and von Issendorff B 2004 *Phys. Rev. Lett.* **93** 093401
[25] Filsinger F, Erlekam U, v. Helden G, Küpper J and Meijer G 2008 *Phys. Rev. Lett.* **100** 133003
[26] Killblane C, Gao Y, Shao N and Zeng X C 2009 *J. Phys. Chem. A* **113** 8839
[27] The NIST, http://webbook.nist.gov, reports only two of them dealing with the first electron affinity
[28] Becker St, Dietrich G, Hasse H-U, Klisch N, Kluge H-J, Kreisle D, Krückeberg St, Lindinger M, Lützenkirchen K, Schweikhard L, Weidele H and Ziegler J 1994 *Z. Phys. D* **30** 341
[29] Ziegler J, Dietrich G, Kruckeberg S, Lutzenkirchen K, Schweikhard L and Walther C 1998 *Hyperfine Inter.* **115** 171
[30] Vogel M, Hansen K, Herlert A and Schweikhard L 2001 *Eur. Phys. J. D* **16** 73
[31] Herlert A, Krückeberg S, Schweikhard L, Vogel M and Walther C 1999 *Phys. Scr.* **T80** 200
[32] Schweikhard L, Herlert A and Vogel M 1999 *Phil. Mag. B* **79** 1343
[33] Yannouleas C and Landman U 2000 *Phys. Rev. B* **61** R10587
[34] Herlert A, Schweikhard L and Vogel M 2001 *Eur. Phys. J. D* **16** 65
[35] Herlert A and Schweikhard L 2003 *Int. J. Mass. Spectrometry* **229** 19
[36] Taylor K J, Pettiette-Hall C L, Cheshnovsky O and Smalley R E 1992 *J. Chem. Phys.* **96** 3319
[37] Drew A and Cederbaum L S 2000 *Phys. Rev. A* **63** 012501
[38] Frisch M J et al. 2004 *GAUSSIAN 03* (Revision C.02) (Wallington, CT: Gaussian, Inc.)
[39] Politzer P and Truhlar D G (Eds) 1981 *Chemical Applications of Atomic and Molecular Electrostatic Potentials* (New York: Plenum)
[40] Mauser H, Hirsch A, van Eikema Hommes N J R and Clark T 1997 *J. Mol. Model.* **3** 415
[41] Molina L M and Hammer B 2005 *J. Catalysis* **233** 399

Dynamics of magnetic nano-particle assembly

V N Kondratyev [1 2]
[1] Physics Department, Taras Shevchenko National University, Pr. Acad. Glushkova 2, Kiev, UA-03022 Ukraine
[2] Bogolubov Laboratory of Theoretical Physics, JINR, 141980, Dubna, Russia

E-mail: vkondra@univ.kiev.ua

Abstract. Ferromagnetically coupled nano-particle assembly is analyzed accounting for inter- and intra- particle electronic structures within the randomly jumping interacting moments model including quantum fluctuations due to the discrete levels and disorder. At the magnetic jump anomalies caused by quantization the magnetic state equation and phase diagram are found to indicate an existence of spinodal regions and critical points. Arrays of magnetized nano-particles with multiple magnetic response anomalies are predicted to display some specific features. In a case of weak coupling such arrays exhibit the well-separated instability regions surrounding the anomaly positions. With increasing coupling we observe further structure modification, plausibly, of bifurcation type. At strong coupling the dynamical instability region become wide while the stable regime arises as a narrow islands at small disorders. It is shown that exploring correlations of magnetic noise amplitudes represents convenient analytical tool for quantitative definition, description and study of supermagnetism, as well as self-organized criticality.

1. Introduction

The assemblies of magnetic nano-particles provide an opportunity to develop new materials with characteristics beyond traditional solids and bring plausible benefits in *'figures of merits'* for advanced technology and therapy. In particular, several laboratory tasks, e.g., injection, sample preparation, manipulation, reaction control, detection, separation etc, can be integrated in a single nano-magnet assembly, i.e., 'lab on a chip' system, cf. [1, 2] and refs. therein. In this contribution we consider analytical tools employed to specify, quantify and analyze such devices in respect with magnetodynamics.

At sufficiently high density of nano-magnets the assembly structure changes from superpara- (SPM) to superferro-magnetism (SFM) [1-4] inducing, thereby, jerky magnetodynamics with sharp discontinuities in the array magnetization process. The electronic spatial quantization is well-known to represent specific nano-particle feature bringing the sharp step-like discontinuities of the magnetic response in varying fields due to the Zeeman splitting [4, 5]. Familiar examples are related to the singlet-triplet transitions (cf., e.g., [4] and refs. therein) in semiconductor quantum dots and carbon nanotubes in magnetic field. We compare in this contribution different cases and demonstrate that this particular feature makes the system of nano-particles significantly different from traditional ensemble of magnetic elements with constant spins. In particular, we demonstrate that an effect of multiple discontinuity anomalies can bring in SFM structure some new phases related to self-organized (SO) criticality.

2. Superferromagnetism with disorder

Recent theoretical studies [3, 4] of realistic supercrystalline objects show, that the relationship between inter-particle magnetic interaction and the disorder substantially determines the magnetic structure and dynamics. Then an occurrence of SO criticality represents, perhaps, one of the most interesting phenomenon. At such a regime magnetic induction of nano-particle arrays displays erratic stochastic discontinuities with rather wide distribution of jump amplitudes.

2.1. Mean-field approach to ensemble of interacting discrete moments at a disorder

At a modeling of the nano-particle assembly magnetism we employ very general form for namo-particle magnetic moment $m=\mu\sum_n \nu_n\theta(b-b_n)$ with moment unit μ and step function $\theta(x)$ depending on local magnetic field *b*. The set of quantities $\{\nu,b\}_n \equiv \{\nu_n,b_n\}$ determines the value of magnetic moment directed along the field vector (for more details see [4] and refs. therein). The conditions $\nu_n \neq 0$ simulate the discrete anomalies of moment jumps. The field independent spin ½ corresponds to $\{\nu,b\}_0=\{-g/2,-\infty\}$, $\{\nu,b\}_1=\{g,0\}$ with the g-factor g and $\nu_n=0$ at $n\neq 0,1$. The particles with a singlet-triplet transition at a field B_{st} are associated with the non-zero parameters $\{\nu,b\}_{-1}=\{-g,-\infty\}$; $\{\nu,b\}_0=\{g,-B_{st}\}$; $\{\nu,b\}_1=\{g,B_{st}\}$. In addition, we consider the particle array with 5 such jumps.

Magnetisation of a system composed from Π elements with moments m_i can be represented as $P=\sum_i m_i/V=\langle m\rangle/V_D$ with total volume V and an area occupied by *i*th particle $V_D=V/\Pi$. For SFM with ferromagnetic coupling of a strength J between the nearest neighbor (nn) particles the respective Ising term $-J\sum_{ij}m_i m_j$ contributes to the Hamiltonian. Here the sum runs over the nn elements. Besides, inhomogeneity and disorder in the form of defects, grain boundaries, impurities lead to random crystalline anisotropy and varying interaction strengths in the super-crystalline heterostructure. Such effects can be accounted for by the random fields h_i. Considering also dynamical components of h_i allows improve inexactness of the model description with nn interaction. The central limit theorem suggests, thereby, that the random fields obey the Gaussian distribution $W(h)=\exp\{-h^2/R^2\}/R\sqrt{\pi}$ of a width *R*, which we call the disorder. The total Hamiltonian H of an array in a field H can be expressed through an interaction of the particle magnetic moment m_i with local fields b_i

$$b_i=H(t)+JV_D\sum_{j\subset nn}P_j+h_i,\ \mathrm{H}=-\sum_i m_i b_i \qquad (1)$$

We refer for this model as randomly jumping interacting moments (RJIM) model [3, 4].

2.2. SFM structure within the mean-field approximation

The nonequilibrium system corresponding to the Hamiltonian Eq. (1) can be analysed by employing the mean-field approach implying an equal inter-particle interaction strength with a coupling constant $J_{ij}=J/\Pi$. The local field in Eq. (1) is then simplified to the form $b_i^{mf}=H(t)+JP+h_i$ with an averaged over a sample magnetization *P*, see above. The random fields can be viewed, therefore, as mean-field fluctuations (cf, e.g., [3, 4, 6]). In the thermodynamic limit $\Pi\to\infty$ we calculate the magnetic state equation (MSE)

$$P=\int dhW(h)m(H+JP+h) \qquad (2)$$

and the magnetic susceptibility $\chi=[\chi_{\mathrm{NI}}^{-1}-J]^{-1}$ with $\chi_{NI}=\Sigma_i W(b-b_i)$ representing the susceptibility of an array without inter-particle interaction (i.e. J=0). Figure 1 shows the dependence of magnetisation *P* for such non-interacting particle assembly on the disorder *R* and magnetic field *b*. As is seen magnetic jump anomalies induced by quantization result in discontinuities in MSE. At increasing disorder such stepwise behaviour is smeared out.

It is worthy to notice here, that each particle jump anomaly brings two MSE discontinuities, see Fig. 1B. Therefore, Fig. 1C shows part of 6 jumps from the total 10 discontinuities on magnetization curve.

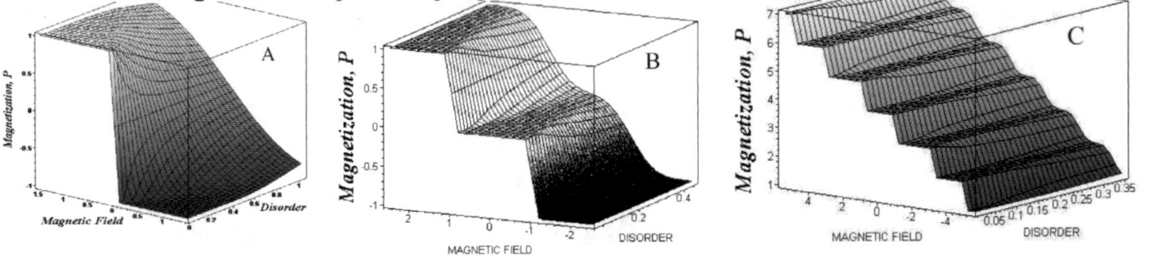

Figure 1. Magnetic state equation, MSE, of an assembly of nano-particles with constant spin (panel A), singlet-triplet transition (panel B) and 5 jump anomalies (panel C).

2.3. Magnetic phase diagram of SFM

Ferromagnetic inter-particle coupling brings the hysteresis loops in magnetization curves at the external field strengths corresponding to discontinuity anomalies. Such a behaviour can be understood in terms of avalanche propargation. When the local field b_i of some i-th lattice element passes through the value b_n the moment changes step-wise. Due to the ferromagnetic interaction a jumping moment can cause some of the nearest neighbors to jump, which may in turn trigger some of their neighbors, and so on, generating, thereby, magnetic avalanche, cf., [3, 4]. As a consequence some sharp stepwise discontinuity arises on magnetization curves. Within the mean-field approach MSE is given by the solution of an equation $P=P[H+JP]$, cf., Eq. (2). Therefore, the value $J\chi_{NI}$ represents an average number of induced jumps per single jumping moment and the negatively defined susceptibility yields adiabatic spinodal region for SFM, cf. Eq. (2). Since the number of induced moment jumps exceeds 1 the system favours to evolve in an avalanche spanning almost entire sample with a macro-magnetization discontinuity. Evidently, the relation $\chi_{NI} \geq J^{-1}$ corresponds to the instability condition. Such spinodal regions are located insight of contour lines on $\{B,R\}$-plane as is indicated in Fig. 2. The upper and lower field-lines meet at the critical point. As is seen in Fig. 2B and Fig. 2C for the case of multiple jumps in particle magnetization the array with weak coupling experience well isolated spinodal regions. The respective magnetodynamics is, therefore, quite similar to the case of a single jump. However, for strong coupling the region of instability is very spread so that we observe a confined isolated stability domain at small disorders.

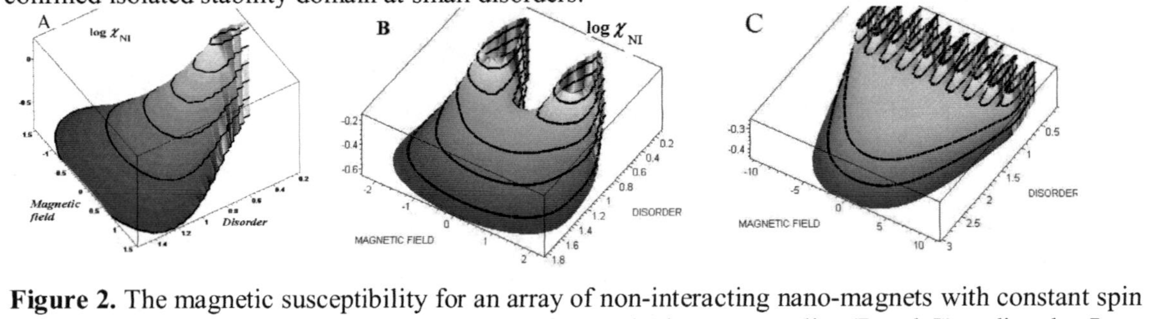

Figure 2. The magnetic susceptibility for an array of non-interacting nano-magnets with constant spin (panel A), singlet-triplet transition at B_{st}=1 (Panel B) and 5 jump anomalies (Panel C) at disorder R.

2.4. SFM structure in magnetodynamics

As is seen above from the mean-field analysis the SFM objects display a spinodal regions at small disorder and approaches stable behaviour at large values R. Such a structure implies pronounced hysteresis loop at the instability region. Such a loop vanishes to a nearly smooth magnetization curve at increasing disorder. The intermediate regime matches a vicinity of SO criticality measured for as a difference between the number of induced jumps and 1, $d = J\chi_{NI} - 1$. Since at such conditions the mean number of induced jumps shows an extremum approaching 1, the mean linear size l_b of the biggest avalanche S_b can be estimated as $l_b^{mf} \approx (1+d)/2$ of the total linear size.

Since within the mean-field treatment an average number of the moments induced to jump by a single jumping moment is site independent for the noise size distribution one obtains [4]

$$D_{mf}(S) \sim S^{-3/2} \exp\{-Sd^2/2\} \qquad (3)$$

We perform the simulations for SFM of simple cubic lattice of a size $(30)^3$. The cumulative size distribution $C(S)=\sum_{N\geq S} D(N)$ allows one to reduce statistical errors in the data analysis. Figure 3A represents distributions *C(S)* integrated over the magnetization reversal hysteresis loop and averaged over 100 events of, e.g., array samples. One sees a clear transition from the `U' shape distribution at small disorders to an abrupt exponential suppression of large size avalanches at large disorders. At transitional values *R* the distribution shows behaviour close to the power law dependence $C \sim S^{-\tau}$ with an exponent $\tau \approx 0.85$ corresponding to 1.85 for *D(S)* and being slightly different from the mean-field estimate.

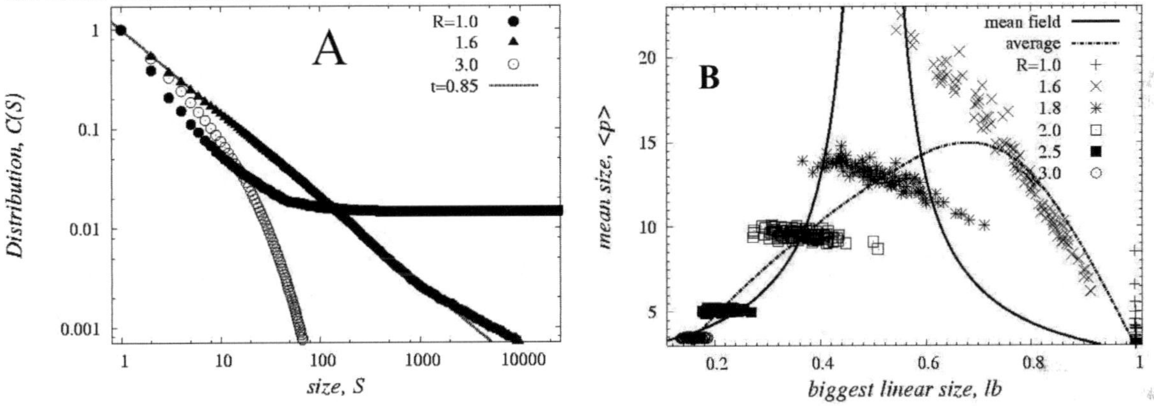

Figure 3. Panel A - Cumulative avalanche size distributions are compared to the power law with an exponent $\tau = 0.85$. Results of the RJIM model for $(30)^3$ simple cubic lattice are shown for disorders R=1 - solid circles, 1.6 - solid triangles, 3 - open circles. Panel B - Mean avalanche size versus the linear size of the biggest avalanche in units of array length. Results of RJIM model at various disorders are shown by particles while dashed-dotted line joins respective average values for each disorder. Solid line displays the prediction of the mean-field approximation in the thermodynamic limit.

Making use of the analytical form Eq. (3) we analyze some analytical tools which might be employed in order to specify and analyze SFM systems. Similarly to methods of high energy physics (cf, e.g., [7,8] and refs. therein) the correlations of avalanche size distribution might provide the criticality signals and a tool specifying and quantifying magnetic structure. For certain *i*th (re)magnetization event we define the mean noise signal

$$\langle p\rangle = \sum_S S\, D(S) / \sum_S D(S) = (\Pi - S_b)/(N_{tot} - 1) \qquad (4)$$

where the sum runs over the avalanche sizes *S* excluding the biggest one, the quantity N_{tot} gives the total number of noisy jumps, i.e. avalanches. Substituting Eq. (3) into Eq. (4) the mean avalanche size, i.e. mean value for magnetic emission signals, is evaluated to be $\langle p\rangle_{mf} \sim |d|^{-1} + const(d)$, and diverges at critical conditions, i.e. $d \to 0$, in the thermodynamic limit $\Pi \to \infty$.

If the system undergoes a critical behaviour being a precursor of SO criticality in some particular (re)magnetization events, strong correlations will appear in magnetic noise. For instance, in case of magnetodynamics we can study correlations between the strongest signal (i.e., the largest avalanche S_b) and the mean signal value $\langle p \rangle$ for remaining avalanches in this particular event, e.g., from numerical model or experimental data. These correlations are similar to the Campi scatter plots [7,8]. In Fig. 3B we plot the mean avalanche size versus the length of largest avalanche for certain event. The data in Fig. 3B were obtained from RJIM simulations assuming various disorders as partially presented in Fig. 3A. In Fig. 3B we can clearly distinguish two branches corresponding to under-critical, i.e., large size of the biggest jump amplitude and small mean value, and overcritical, i.e. small size of the biggest avalanche and small average values. The right branch consists mainly of events with small disorders, while the left branch originates from events having large R. The set of two branches meet in the critical region. The results of the mean field approximation in the thermodynamic limit are in a reasonable agreement with numerical data for overcritical disorders. At sub-critical conditions the mean field approach reproduces only qualitatively the RJIM model simulations.

3. Conclusions

We considered the magnetic structure of nano-particle assemblies accounting for inter- and intra-particle properties. Spatial quantization due to intra-particle confinement brings sharp step-vise anomalies for magnetic moments in external field. As was shown in [3,4] such moment jumps in conjunction with ferromagnetic inter-particle coupling induce jerky magnetodynamics with sharp steps in the array (de)magnetization process. Making use of the RJIM model [3,4] we account for quantum effects due to the particle discrete levels, inter-particle coupling and disorder. MSE and phase diagramm of such system are demonstrated to exhibit spinodal regions on $\{R,B\}$-plane. In a case of weak coupling these regions of instability are well separated and surround the positions of the particle magnetic anomalies. With increasing coupling we observe further structure modification, plausibly, of bifurcation type. At strong coupling the instability region become wide while the stable regimes are reduced to narrow islands at small disorders. We consider analytical tools exploring correlations of magnetic noise amplitudes and demonstrate an application for quantitative definition, description and study of SFM structure and origin, as well as self-organized criticality. Further analysis based on numerical simulations within RJIM model and implications of proposed tools will shed a light on the origin and properties of this transition.

Acknowledgments
The author thanks JINR (Dubna, Russia) for the warm hospitality and the financial support.

References

[1] Bedanta S and Kleemann W 2009 *J. Phys. D: Appl. Phys.* **42** 013001
[2] Kondratyev V N and Lutz H O 1998 *Phys. Rev. Lett.* **81** 4508; 1999 *Eur. Phys. J.* D **9** 483
[3] Kondratyev V N 2002 *Phys. Rev. Lett.* **88** 221101; 2002 *JAERI-Research* 2001–057
[4] Kondratyev V N 2006 *Phys. Lett.* A **354** 217; 2008 *J. Phys.CS* **129** 012013; 2010 *J. Phys.CS* **200** 012093; 2010 *Phys.Stat. Sol.* B **247** 685
[5] Kondratyev V N 2002 *J. Nucl. Sci. Technol.* **1** Suppl. 2. 550; 2002 *J. Nucl. Radiochem. Sci.* **3** 205
[6] Kondratyev V N 1993 *Phys. Lett.* A **179** 209; 1994 *Phys. Lett.* A **190** 465; 1996 *Z. Phys.* B **99** 473
[7] Kondratyev V N 1997 *AIP Conf. Proc.* **416** 447; 2008 *J. Phys.CS* **129** 012020; 2009 *Bull. RAS Physics* **73** 491
[8] Kondratyev V N, Lutz H O and Ayik S 1997 *J. Chem. Phys.* **106** 7766

Journal of Physics: Conference Series **248** (2010) 012028 doi:10.1088/1742-6596/248/1/012028

Magnetic flux structures in various shaped composite structures with d- and s-wave superconductors (d-dots)

Masaru Kato[1, 5], Tomio Koyama[2,5], Masahiko Machida[3,5], Shuichi Kawamata[4], and Takekazu Ishida[4,5]

[1]Department of Mathematical Sciences, Osaka Prefecture University, 1-1, Gakuencho, Nakaku, Sakai, Osaka 599-8531, Japan

[2]IMR, Tohoku University, 2-1-1 Katahira Aoba-ku, Sendai 980-8577, Japan

[3] CCSE, JAEA, 6-9-3 Higashi-Ueno Taito-ku, Tokyo 110-0015, Japan

[4] Department of Physics and Electronics, Osaka Prefecture University, 1-1, Gakuencho, Nakaku, Sakai, Osaka 599-8531, Japan

[5] CREST, JST, 5, Sanbancho, Chiyoda-ku, Tokyo 102-0075, Japan

kato@ms.osakafu-u.ac.jp

Abstract. Utilizing phenomenological two-components Ginzburg-Landau equation, we investigate the vortex structures in nano-sized composite structures of d- and s-wave superconductors (d-dot), which shows spontaneous half-quantum magnetic fluxes. Especially, Effects of disorders in junctions between d- and s-wave superconductors are investigated. Holes in junction at the corners do not affect the magnetic flux patterns. Although the large hole at the corner changes the position of the flux. A wide disorder along the junction affects the magnetic structures much. At the edges of the non-disordered junction, there appear fractional fluxes.

1. Introduction

Nano-structured superconductors show much different property compared to bulk superconductors. Also, superconducting devices, especially superconducting qubit, have attracted much attention. Previously, we showed another candidate for such superconducting devices [1-17].

Our idea is based on the symmetry of high-Tc superconductors. Since the discovery of the high-Tc superconductors, the pairing symmetry, as well as its origin, has been controversial, but phase-sensitive experiment is crucial for determining the symmetry of the Cooper pairs. One such phase-sensitive experiment is a corner junction experiment [18, 19]. In this experiment, a square-shaped high-Tc superconductor is connected with a conventional s-wave superconductor with two Josephson junctions around a corner. Each junction is parallel to either the x- or the y-axis of the high-Tc superconductor. The critical current between the s- and the d-wave superconductors is zero under a zero external magnetic field because of the sign change of superconducting order parameter between Cooper pairs that go through the junction along the x direction and other Cooper pairs that go through another junction along the y direction. This is in apparent contrast with a corner junction between both s-wave superconductors for which the critical current is largest under zero external field. Therefore, from this experiment, the symmetry of the Cooper pairs in high-Tc superconductors was determined as d-wave, especially $d_{x^2-y^2}$. In addition, this experiment showed that the critical current became maximum when the total magnetic flux in the corner junction was $\Phi_0/2$. Here, $\Phi_0 = hc/2e$ is the flux quantum. This means the state with a half-flux quantum magnetic flux is stable and that in the superconducting states of both superconductors, the stable states have a spontaneous magnetic flux at the corner without an external field. Experiments showing this property have been done. Especially, Hilgenkamp et al. showed in a zigzag junction between Nb and YBCO that a spontaneous magnetic flux appears at every corner under a zero external field [20]. This spontaneous magnetic flux is useful and can be used as a spin or a bit by itself.

In contrast to the previous approach for using the spontaneous half-flux quantum, we considered nano-sized d-wave superconductor embedded in a s-wave matrix [1–17]. We call it d-dot. We want to consider the whole d-dot system as one element, not as individual half-quantum fluxes. The single half-quantum flux is useful because there are two opposite directions that the flux can take. Also, our d-dot has two degenerate states if the spontaneous magnetic fluxes appear. This is because the state with spontaneous magnetic fluxes under zero external field breaks time-reversal symmetry. This property is independent of the shape, so a d-dot in any shape always has two degenerate stable states. Therefore, the d-dot as a whole can be considered as a single element with two level states. Experiments to establish these d-dots are now in progress [21].

In this study we investigate the magnetic flux structures for various shaped d-dot Especially we discuss the effect of disorders in the junctions. For this investigation, we use two-component Ginzburg-Landau equation and the finite element method, which are explained in next section.

2. Model

For analyzing the spontaneous half-quantum magnetic fluxes, we must take into account the anisotropy of d-wave superconductivity. Therefore we use following two-component Ginzburg-Landau free energy for d-wave superconductivity and s-wave superconductivity. For d-wave superconductors [1,6,7,11,13,14,16,17],

$$\begin{aligned}
\mathrm{F}_d\left(\Delta_s,\Delta_d,\boldsymbol{A}\right)= \int_\Omega \Bigg(& \frac{3\alpha}{8}\lambda_d\left[|\Delta_d|^2-\frac{4\ln T_{cd}/T}{3\alpha}\right]^2+\alpha\lambda_d\left[|\Delta_s|^4+2\frac{\alpha_s}{\alpha}|\Delta_s|^2\right] \\
&+\alpha\lambda_d\left[2|\Delta_s|^2\left(\Delta_d\right(^2+\frac{1}{2}\left(\Delta_s^{*2}\Delta_d^2+\Delta_s^2\Delta_d^{*2}\right)\right] \\
&+\frac{1}{4}\alpha\lambda_d v_F^2\left[|\boldsymbol{\Pi}\Delta_d^*|^2+2|\boldsymbol{\Pi}\Delta_s|^2+\left(\Pi_x^*\Delta_s\Pi_x\Delta_d^*-\Pi_y^*\Delta_s\Pi_y\Delta_d^*+\mathrm{H.c.}\right)\right] \\
&+\frac{1}{8\pi}|\boldsymbol{h}-\boldsymbol{H}|^2+\frac{1}{8\pi}\left(\operatorname{div}\boldsymbol{A}\right)^2\Bigg)\,d\Omega
\end{aligned} \quad . \tag{1}$$

And for s-wave superconductors,

$$\begin{aligned}
\mathrm{F}_s(\Delta_s,\Delta_d,\boldsymbol{A}) = \int_\Omega \Bigg(& \frac{\alpha}{2}\lambda_s\left[|\Delta_s|^2 - \frac{\ln T_{cs}/T}{\alpha}\right]^2 + \frac{\alpha\lambda_s}{2}\left[\frac{3}{8}|\Delta_d|^4 + \frac{\alpha_d}{\alpha}|\Delta_d|^2\right] \\
& + \frac{\alpha\lambda_s}{2}\left[2|\Delta_s|^2|\Delta_d|^2 + \frac{1}{2}\left(\Delta_s^{*2}\Delta_d^2 + \Delta_s^2\Delta_d^{*2}\right)\right] \\
& + \frac{1}{8}\alpha\lambda_s v_F^2\left[|\boldsymbol{\Pi}\Delta_d^*|^2 + 2|\boldsymbol{\Pi}\Delta_s|^2 + \left(\Pi_x^*\Delta_s\Pi_x\Delta_d^* - \Pi_y^*\Delta_s\Pi_y\Delta_d^* + \mathrm{H.c.}\right)\right] \\
& + \frac{1}{8\pi}|\boldsymbol{h}-\boldsymbol{H}|^2 + \frac{1}{8\pi}(\operatorname{div}\boldsymbol{A})^2\Bigg)d\Omega
\end{aligned} \quad . \tag{2}$$

Here Δ_d and Δ_s are d- and s-wave superconducting order parameters, respectively. $\Pi = \frac{\hbar}{i}\nabla - \frac{e}{c}\boldsymbol{A}$ is the gauge-invariant derivative and $\alpha = \frac{7\varsigma(3)}{8(\pi T)^2}$. λ_d and λ_s are the strengths of the coupling constants for the d- and the s-wave interaction channels, respectively. We assume repulsive interaction for s-wave (d-wave) channel for d-wave superconductor (s-wave superconductor), respectively. T_{cd} and T_{cs} are transition temperatures for d- and s-wave superconductors, respectively. $\boldsymbol{h} = \nabla\times\boldsymbol{A}$ is a microscopic magnetic flux density and $\boldsymbol{H}$ is an external magnetic field.

In these free energies, an anisotropy of the d-wave super- conductivity appears in the coupling term of the gradient of both order parameters, where the two terms that contain the gradients along the x and the y directions have different signs.

We also use alternative model, in which we consider the junction region explicitly. In the d-wave superconducting region, we only consider d-wave superconducting order parameter and in s-wave SC region, we only consider the s-wave SC order parameter and the free energy for each region is given as,

$$\mathrm{F}_d(\Delta_d,\boldsymbol{A}) = \int_\Omega\left(\frac{3\alpha}{8}\lambda_d\left[|\Delta_d|^2 - \frac{4\ln T_{cd}/T}{3\alpha}\right]^2 + \frac{1}{4}\alpha\lambda_d v_F^2|\boldsymbol{\Pi}\Delta_d^*|^2 + \frac{1}{8\pi}|\boldsymbol{h}-\boldsymbol{H}|^2 + \frac{1}{8\pi}(\operatorname{div}\boldsymbol{A})^2\right)d\Omega\,, \tag{3}$$

$$\mathrm{F}_s(\Delta_s,\boldsymbol{A}) = \int_\Omega\left(\frac{\alpha}{2}\lambda_s\left[|\Delta_s| - \frac{\ln T_{cs}/T}{\alpha}\right]^2 + \frac{1}{4}\alpha\lambda_s v_F^2|\boldsymbol{\Pi}\Delta_s^*|^2 + \frac{1}{8\pi}|\boldsymbol{h}-\boldsymbol{H}|^2 + \frac{1}{8\pi}(\operatorname{div}\boldsymbol{A})^2\right)d\Omega\,. \tag{4}$$

And in the junction region, we consider coupling between both order parameters as,

$$\begin{aligned}
\mathrm{F}_M(\Delta_s,\Delta_d,\boldsymbol{A}) = \int_\Omega\Bigg(& |\lambda_d|\frac{3b}{2}\left[|\Delta_d|^2 + \frac{a_d}{3b}\right]^2 + b\left[|\Delta_s|^2 + \frac{a_s}{2b}\right]^2|\lambda_d| + b|\lambda_d|\left[2|\Delta_s|^2|\Delta_d|^2 + \frac{1}{2}\left(\Delta_s^{*2}\Delta_d^2 + \Delta_s^2\Delta_d^{*2}\right)\right. \\
& \left. + \frac{1}{4}v_F^2\left\{|\boldsymbol{\Pi}\Delta_d^*|^2 + 2|\boldsymbol{\Pi}\Delta_s^*|^2 + \left(\Pi_x^*\Delta_s\Pi_x\Delta_d^* - \Pi_y^*\Delta_s\Pi_y\Delta_d^* + \mathrm{H.c.}\right)\right\}\right] + \frac{1}{8\pi}|\boldsymbol{h}-\boldsymbol{H}|^2 + \frac{1}{8\pi}(\operatorname{div}\boldsymbol{A})^2\Bigg)d\Omega
\end{aligned} \quad . \tag{5}$$

In order to minimize these free energies, we use the finite-element method [12–15] because we want to investigate variously shaped d-dots, especially disordered junction case in this study. We show an example of element partition in Fig. 1 for the alternative model. In this example the junction region is divided into smaller elements because the spatial variation of order parameters are rapid.

In the finite element method, we expand the order parameters and the vector potential in each element by using the area coordinates, which are defined for *e*-th element as

$$N_i^e(x,y) = \frac{1}{2S_e}\left(a_i^e + b_i^e x + c_i^e y\right) \quad (i=1,2,3)\ . \tag{6}$$

Here $a_i^e = x_j^e y_k^e - x_k^e y_j^e$, $b_i^e = y_j^e - y_k^e$, and $c_i^e = x_k^e - x_j^e$, where (i,j,k) is a cyclic permutation of $(1,2,3)$ and (x_i^e, y_i^e) is the coordinate of i-th vertex of the e-th element. The order parameters and vector potential are given as,

$$\Delta_d(x,y) = \sum_e \sum_{i=1}^{3} \Delta_{di}^e N_i^e(x,y), \tag{7}$$

$$\Delta_s(x,y) = \sum_e \sum_{i=1}^{3} \Delta_{si}^e N_i^e(x,y), \tag{8}$$

$$\boldsymbol{A}(x,y) = \sum_e \sum_{i=1}^{3} \boldsymbol{A}_i^e N_i^e(x,y), \tag{9}$$

where Δ_{di}^e and Δ_{si}^e are the values of the d-wave and the s-wave order parameters at the i-th vertex in e-th element, respectively. And $\boldsymbol{A}_i^e$ is the value of the vector potential at the i-th vertex in e-th element. Inserting these expression into previous free energy expression and minimizing it about those values, Δ_{di}^e, Δ_{si}^e and $\boldsymbol{A}_i^e$, we get the GL equations in the finite element method.

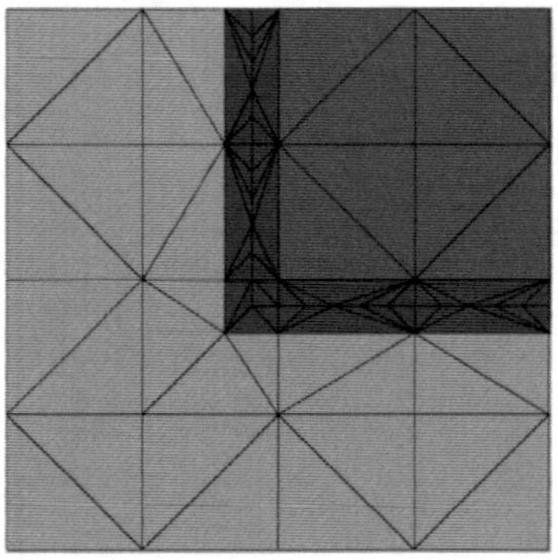

Figure 1. Example of finite elements partition.

3. Numerical Results

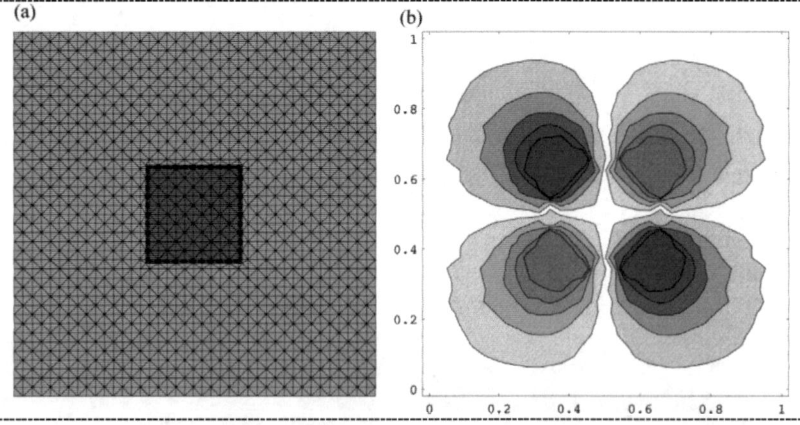

Figure 2. A square shaped d-dot. Configuration (a) and distribution of magnetic field (b) are shown.

First we show the magnetic field distribution for regular squared d-dot under zero external field in figure 1. There appear four spontaneous half-quantum magnetic fluxes at the corners of square-shaped d-wave superconducting region. They are aligned anriferromagnetically, because the repulsive interaction between magnetic fluxes. Also, in figure 2, we show the order parameter structures of s- and d- wave superconductivity. From this figure, interference of two order parameters in the junction region appears clearly.

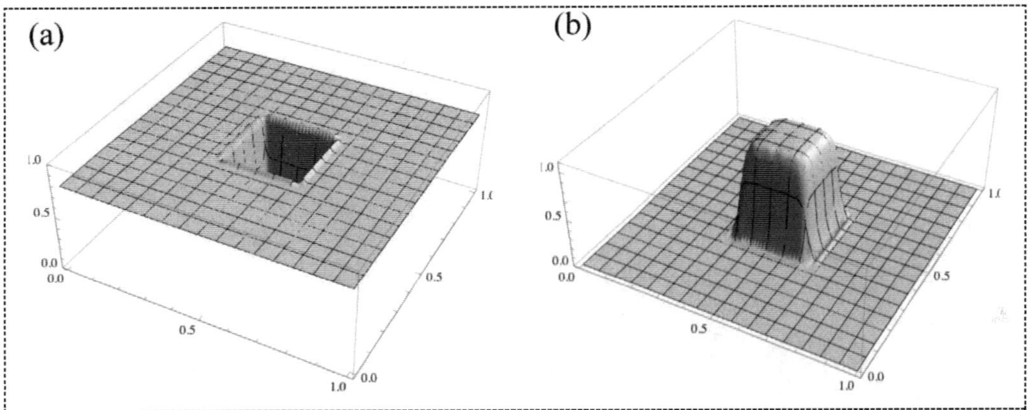

Figure 3. A square shaped d-dot. Oder parameter structure for s-wave(a) and d-wave (b) are shown.

Next we consider the disorders in the junctions. In figure4, we show the magnetic structures for d-dot with small disorder at the corners. As shown in figure 4 (a), the junction is broken at the corners. But the magnetic structures remain almost same as the perfect one (figure 2). This is because the spontaneous current between two superconductors flows around this hole. So small disorder at the corner does not affect the spontaneous current much.

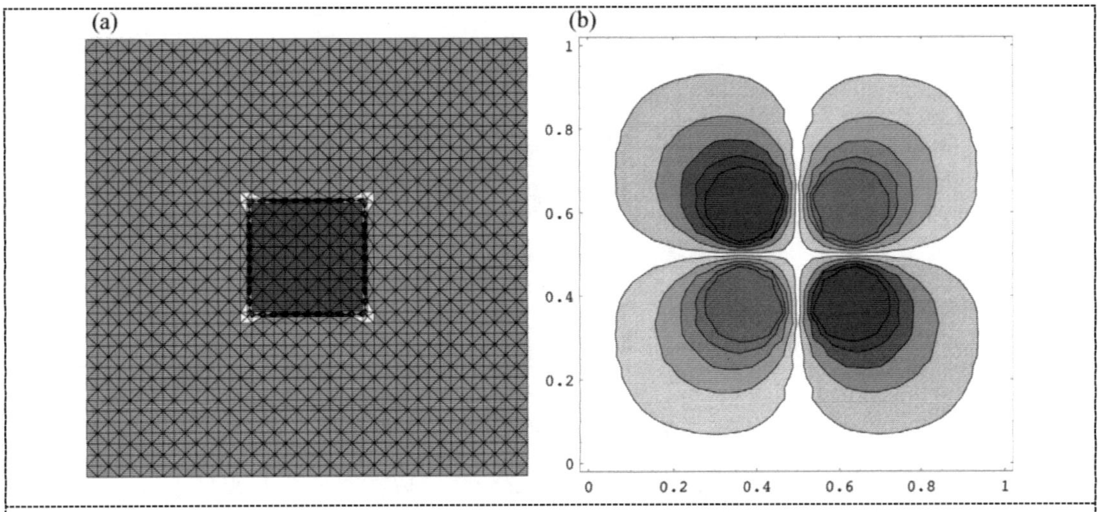

Figure 4. A square shaped d-dot with disorder of junction at corners. Configuration (a) and distribution of magnetic field (b) are shown.

When the defect in the junction becomes large, then situation becomes different. As shown in figure 5, distribution of magnetic field is different from those in figures 2 and 4. In this case, the magnetic fluxes appear inside of the d-wave superconducting region. In previous cases, the phase of the order parameter changes around a corner by π, because of the phase change of d-wave order parameter. But in this case, the phase of the s-wave order parameter fixed to be constant, and the phases of d-wave order parameter fixed at both edges around a corner to be continuously connected the s-wave order parameter. Then the phases of the d-wave order parameter at these edges are different by π and in the d-wave region between these two edges the phase of d-wave order parameter varies gradually as shown in figure 6. So inside of d-wave SC region, current circulate around a point, which is apart from the corner. And half-quantum magnetic flux appears at this point. There is another solution where the phase change occurs around the corner, but it costs much kinetic energy because the current should flow large area around the large hole. Also because of this large hole, d-wave order parameter is free from the outside s-wave order parameter and its phase changes freely along this hole. In previous case, around the corner the phase of d-wave order parameter closely related to those of s-wave order parameter through the junction.

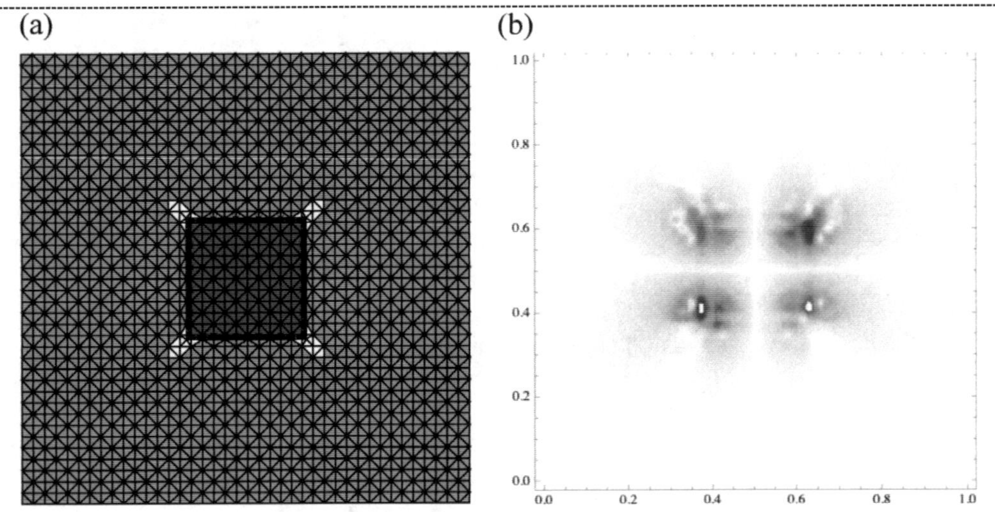

Figure 5. A square shaped d-dot with large disorders of junction at corners. Configuration (a) and distribution of magnetic field (b) are shown.

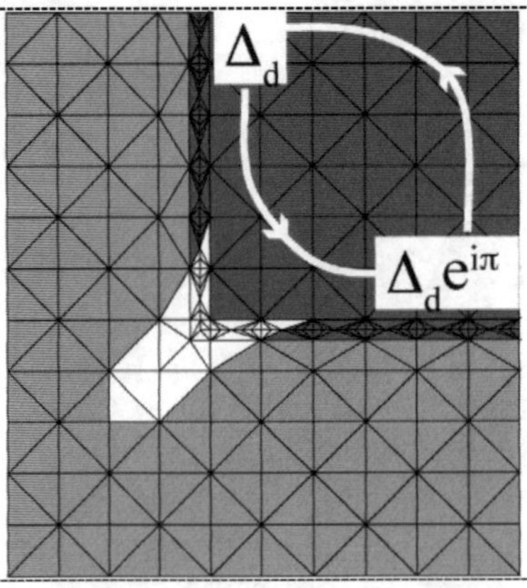

Figure 6. Phase change of d-wave superconducting order parameter at the corner junction with a large hole.

Next we consider the case in which junction is broken in large area. In figure 7 we show an example where junction is broken over three corners. In this case, magnetic flux appears at the corner with good junction and the edges of the junction. There is no magnetic flux at the upper-right corner. Total magnetic flux is zero, because total phase change of order parameter around the d-wave region should be zero, otherwise the kinetic energy costs much. For other largely damaged junction cases, similar flux structures appear. Such magnetic flux structure were observed experimentally. Therefore avoiding disorder along the junction is crucial for forming clear magnetic flux structure.

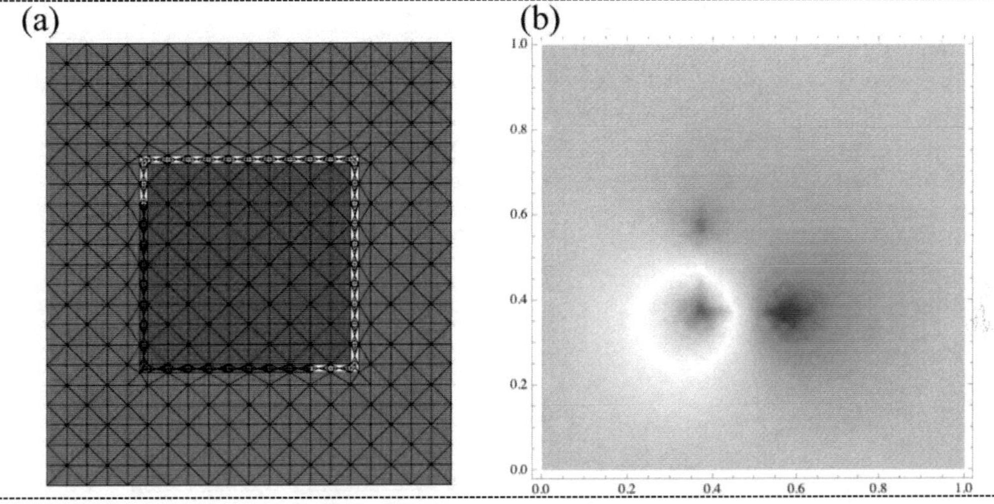

Figure 7. A square shaped d-dot with broken junction in large area. Configuration (a) and distribution of magnetic field (b) are shown.

4. Summary and Conclusions

In this study we consider the magnetic flux structures in various shaped composite structures with d- and s-wave superconductors (d-dot). Especially we consider the effect of the broken junctions. Effect of disorders or holes at the corner junction depends on their size. For smaller holes, magnetic flux pattern remains as that of perfect case, but for larger holes, centers of spontaneous magnetic fluxes shift to inside of d-wave superconducting region. But overall structures remain same for both cases. Larger disorder along the junction is harmful to the magnetic flux structures.

Experimental realization of our d-dot system is a future problem.

References

[1] Kato M, Ako M, Machida M, Koyama T and Ishida T 2004 Physica C, **412-414** 352
[2] Koyama T, Machida M, Kato M and Ishida T 2004 Physica C **412-414** 358
[3] Machida M, Koyama T, Kato M and Ishida T 2004 Physica C **412-414** 367
[4] Ako M, Machida M, Koyama T, Ishida T and Kato M 2004 Physica C, **412-414** 544
[5] Fujii M, Yoshikawa H, Miki S, Kawamata S, Satoh K, Yotsuya T, Kato M, Machida M, Koyama T, Terashima T, Tsukui S, Adachi M and Ishida T 2005 Physica C **426-431** 104
[6] Ako M, Machida M, Koyama T, Ishida T and Kato M 2005 Physica C **426-431** 122
[7] Hirayama M, Machida M, Koyama T, Ishida T and Kato M 2005 Physica C **426-431** 127
[8] Koyama T, Machida M, Kato M and Ishida T 2005 Physica C **426-431** 1561
[9] Machida M, Koyama T, Kato M and Ishida T 2005 Physica C **426-431** 1566
[10] Koyama T, Machida M, Kato M and Ishida T 2006 AIP Conf. Proc. **850** 779
[11] Kato M, Hirayama M, Machida M, Koyama T and Ishida T 2006 AIP Conf. Proc. **850** 781
[12] Ishida T, Fujii M, Abe T, Miki S, Kawamata S, Satoh K, Yotsuya T, Kato M, Machida M,

Koyama T, Terashima T, Tsukui S and Adachi M 2006 Physica C **437-438** 104
[13] Kato M, Hirayama M, Suematsu H, Machida M, Koyama T and Ishida T 2006 J. Korean Phys. Soc. **48** 1074
[14] Hirayama M, Kato M, Machida M, Koyama T and Ishida T 2006 Physica C **445-448** 264
[15] Koyama T, Machida M, Kato M and Ishida T 2007 Physica C **460-462** 1305
[16] Kato M, Hirayama M, Nakajima S, Koyama T, Machida M and Ishida T 2007 J. Mag. Mag. Mat. **310** 495
[17] Nakajima S, Kato M, Koyama T, Machida M, Ishida T and Nori F 2008 Physica C **468** 769
[18] Tsuei C C and Kirtley J R 2000 Rev. Mod. Phys. **72** 969
[19] Van Harlingen D J 1995 Rev. Mod. Phys. **67** 515
[20] Hilgenkamp A, Ariando A, H.Smilde H-J, Blank D H A, Rijnders G, Rogalla H, Kirtley J R and Tsuei C C 2003 Nature **422** 50
[21] Kawamata S, Yamamoto M, Uno M, Satoh K, Yuki T, Yotsuya T and Ishida T Physica C in press.

International Conference on Theoretical Physics Dubna-Nano 2010 IOP Publishing
Journal of Physics: Conference Series **248** (2010) 012029 doi:10.1088/1742-6596/248/1/012029

Current noise and Coulomb effects in superconducting contacts

Artem V Galaktionov[1] **and Andrei D Zaikin** [2,1]

[1]I.E. Tamm Department of Theoretical Physics, P.N. Lebedev Physics Institute, 119991 Moscow, Russia

[2] Institut für Nanotechnologie, Karlsruher Institut für Technologie (KIT), 76021 Karlsruhe, Germany

E-mail: andrei.zaikin@kit.edu

Abstract. We derive an effective action for contacts between superconducting terminals with arbitrary transmission distribution of conducting channels. In the case of normal-superconducting (NS) contacts we evaluate interaction correction to Andreev conductance and demonstrate a close relation between Coulomb effects and shot noise in these systems. In the case of superconducting (SS) contacts we derive the electron-electron interaction correction to the Josephson current. At $T = 0$ both corrections are found to vanish for fully transparent NS and SS contacts indicating the absence of Coulomb effects in this limit.

1. Introduction

Low energy electron transport across the interface between normal metals and superconductors (NS) is provided by the mechanism of Andreev reflection [1]. This mechanism involves conversion of a subgap quasiparticle entering the superconductor from the normal metal into a Cooper pair together with simultaneous creation of a hole that goes back into the normal metal. Each such act of electron-hole reflection corresponds to transferring twice the electron charge $e^* = 2e$ across the NS interface and results, e.g., in non-zero conductance of the system at subgap energies [2].

Andreev reflection is also responsible for dc Josephson effect in superconducting weak links without tunnel barriers. Suffering Andreev reflections at both NS interfaces, quasiparticles with energies below the superconducting gap are effectively "trapped" inside the junction forming a discrete set of levels which can be tuned by passing the supercurrent across the system [3]. At the same time, these subgap Andreev levels themselves contribute to the supercurrent [3, 4, 5, 6] thus making the behavior of superconducting point contacts and SNS junctions in many respects different from that of tunnel barriers.

Note that the above theories remain applicable if one can neglect Coulomb effects. In small-size superconducting contacts, however, such effects can be important and should in general be taken into account. A lot is known about interplay between fluctuations and charging effects in superconducting tunnel barriers [7]. Here we examine the properties of superconducting junctions going beyond the tunneling limit. Below we will demonstrate that Coulomb blockade in such junctions weakens with increasing barrier transmissions and eventually disappears in the limit of fully open superconducting contacts. We will also argue that in superconducting systems – similarly to normal contacts [8] – there also exists a direct relation between Coulomb effects and current fluctuations.

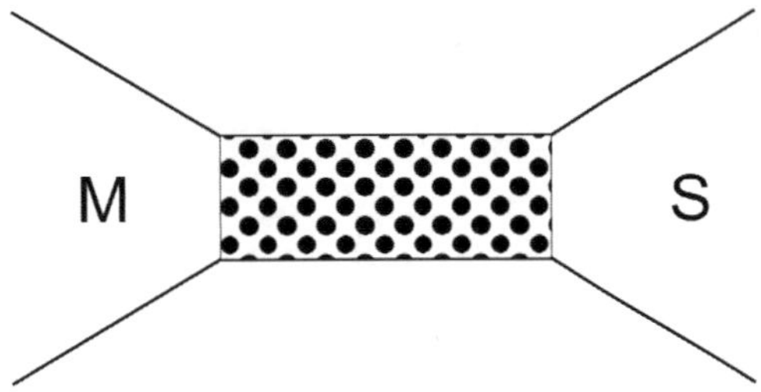

Figure 1. The system under consideration. Two big metallic reservoirs – one superconducting (S) and another one either normal or superconducting (M=N,S) – are connected by a short normal conductor.

2. The model and effective action

As it is shown in Fig. 1, we will consider big metallic reservoirs one of which is superconducting while another one could be either normal or superconducting. These two reservoirs are connected by a rather short normal bridge (conductor) with arbitrary transmission distribution T_n of its conducting modes and normal state conductance $G_N \equiv 1/R_N = (e^2/\pi)\sum_n T_n$. Both phase and energy relaxation of electrons may occur only in the reservoirs and not inside the conductor which length is assumed to be shorter than dephasing and inelastic relaxation lengths. As usually, Coulomb interaction between electrons in the conductor area is accounted for by some effective capacitance C.

In order to analyze electron transport in the presence of interactions we will make use of an approach based on the effective action formalism combined with the scattering matrix technique [8, 9, 10, 11, 12]. This approach can be conveniently generalized to superconducting systems. In fact, the structure of the effective action remains the same also in the supeconducting case, one should only replace normal propagators by 2×2 matrix Green functions which account for superconductivity, as it was done, e.g., in [7, 13, 14].

Following the standard procedure we express the kernel J of the evolution operator on the Keldysh contour in terms of a path integral over the fermionic fields which can be integrated out after the standard Hubbard-Stratonovich decoupling of the interacting term. Then the kernel J takes the form

$$J = \int \mathcal{D}\varphi_1 \mathcal{D}\varphi_2 \exp(iS[\varphi]), \tag{1}$$

where $\varphi_{1,2}$ are fluctuating phases defined on the forward and backward parts of the Keldysh contour and related to fluctuating voltages $V_{1,2}$ across the conductor as $\dot{\varphi}_{1,2}(t) = eV_{1,2}$. Here and below we set $\hbar = 1$.

The effective action consists of two terms, $S[\varphi] = S_c[\varphi] + S_t[\varphi]$, where

$$iS_c[V] = \frac{C}{2e^2}\int_0^t dt'(\dot{\varphi}_1^2 - \dot{\varphi}_2^2) \equiv \frac{C}{e^2}\int_0^t dt\dot{\varphi}^+\dot{\varphi}^- \tag{2}$$

describes charging effects and the term $S_t[V]$ accounts for electron transfer between normal and superconducting reservoirs. It reads [14]

$$S_t[\varphi] = -\frac{i}{2}\sum_n \mathrm{Tr}\ln\left[1 + \frac{T_n}{4}\left(\left\{\check{G}_M, \check{G}_S\right\} - 2\right)\right], \tag{3}$$

where $\check{G}_M$ and $\check{G}_S$ are 4×4 Green-Keldysh matrices of M- and S-electrodes which product implies time convolution and which anticommutator is denoted by curly brackets. In Eq. (2) we also introduced "classical" and "quantum" parts of the phase, respectively $\varphi_+ = (\varphi_1 + \varphi_2)/2$ and $\varphi_- = \varphi_1 - \varphi_2$.

For later purposes we also express the average current and the current-current correlator via the effective action as

$$\langle \hat{I}(t) \rangle = ie \int \mathcal{D}\varphi_\pm \frac{\delta}{\delta\varphi_-(t)} e^{iS[\varphi]}, \tag{4}$$

$$\frac{1}{2}\langle \hat{I}\hat{I} \rangle_+ = -e^2 \int \mathcal{D}\varphi_\pm \frac{\delta^2}{\delta\varphi_-(t)\delta\varphi_-(t')} e^{iS[\varphi]}, \tag{5}$$

where $\langle \hat{I}\hat{I} \rangle_+ = \langle \hat{I}(t)\hat{I}(t') + \hat{I}(t')\hat{I}(t) \rangle$.

Let us introduce the matrix $\check{X}_0[\varphi_+] = 1 - T_n/2 + (T_n/4)\left\{\check{G}_M, \check{G}_S\right\}|_{\varphi_-=0}$. As the action S_t vanishes for $\varphi_-(t) = 0$ one has $\mathrm{Tr}\ln \check{X}_0 = 0$. Making use of this property we can identically transform the action (3) to

$$S_t = -\frac{i}{2}\sum_n \mathrm{Tr}\ln\left[1 + \check{X}_0^{-1} \circ \check{X}'\right], \tag{6}$$

where $\check{X}' = 1 + (T_n/4)\left(\left\{\check{G}_M, \check{G}_S\right\} - 2\right) - \check{X}_0$. Now let us separately consider NS and SS interfaces.

2.1. NS interfaces

At temperatures and voltages well below the superconducting gap Andreev contribution to the action of NS system dominates. Hence, it suffices to consider the limit of low energies $\epsilon \ll \Delta$. In this limit we can define the Andreev transmissions [2] $\mathcal{T}_n = T_n^2/(2 - T_n)^2$ and Andreev conductance $G_A = (2e^2/\pi)\sum_n \mathcal{T}_n$. Let us assume that either dimensionless Andreev conductance $g_A = 4\sum_n \mathcal{T}_n$ is large, $g_A \gg 1$, or temperature is sufficiently high (though still smaller than Δ). In either case one can describe quantum dynamics of the phase variable φ within the quasiclassical approximation [8, 9] which amounts to expanding S_t in powers of (small) "quantum" part of the phase $\varphi_-(t)$. Employing the above equations and expanding S_t up to terms $\sim \varphi_-^2$ we arrive at the Andreev effective action [15]

$$iS_t = iS_R - S_I, \tag{7}$$

where

$$iS_R = -\frac{ig_A}{2\pi}\int_0^t dt'\, \varphi^-(t')\dot{\varphi}^+(t'), \tag{8}$$

$$S_I = -\frac{g_A}{4}\int_0^t dt' \int_0^t dt'' \frac{T^2}{\sinh^2[\pi T(t'-t'')]}\varphi^-(t')\varphi^-(t'') \times[\beta_A\cos(2\varphi^+(t') - 2\varphi^+(t'')) + 1 - \beta_A] \tag{9}$$

and

$$\beta_A = \frac{\sum_n \mathcal{T}_n(1-\mathcal{T}_n)}{\sum_n \mathcal{T}_n} \tag{10}$$

is the Andreev Fano factor defined in a complete analogy with the normal Fano factor $\beta_N = \sum_n T_n(1-T_n)/\sum_n T_n$. We observe that the action S_t is expressed in exactly the same form as that for normal conductors [8, 9] derived within the the same quasiclassical approximation for the phase variable $\varphi(t)$. In order to observe the correspondence between the action [8, 9] and that defined in Eqs. (7)-(9) one only needs to interchange normal and Andreev conductances as well as the corresponding Fano factors

$$G_N \leftrightarrow G_A, \quad \beta_N \leftrightarrow \beta_A \tag{11}$$

and to account for an extra factor 2 in front of the phase φ_+ under cos in Eq. (9). This extra factor implies doubling of the charge during Andreev reflection.

2.2. Superconducting contacts

Turning to superconducting contacts we assume that fluctuating phases $\varphi_\pm(t)$ are sufficiently small and perform regular expansion of the exact effective action in powers of these phases. Then we obtain

$$iS_t = -\frac{i}{e}\int\limits_0^t dt' I_S(\chi)\varphi_-(t') + iS_R - S_I, \tag{12}$$

where χ is the time-independent phase difference,

$$I_S(\chi) = \frac{e\Delta\sin\chi}{2}\sum_n \frac{T_n}{\sqrt{1-T_n\sin^2(\chi/2)}}\tanh\frac{\Delta\sqrt{1-T_n\sin^2(\chi/2)}}{2T}. \tag{13}$$

defines the supercurrent across the system [16, 5] and

$$S_R = \int\limits_0^t dt'\int\limits_0^t dt''\mathcal{R}(t'-t'')\varphi^-(t')\varphi^+(t''), \tag{14}$$

$$S_I = \int\limits_0^t dt'\int\limits_0^t dt''\mathcal{I}(t'-t'')\varphi^-(t')\varphi^-(t'') \tag{15}$$

with both kernels $\mathcal{R}(t)$ and $\mathcal{I}(t)$ being real functions. The complete expressions for these functions turn out to be somewhat lengthy and for this reason are not presented here. Below we only emphasize some of the properties of $\mathcal{R}(t)$ and $\mathcal{I}(t)$.

To begin with, it is straightforward to verify that in the lowest order in barrier transmissions T_n the result (12)-(15) reduces to the standard AES action [7] for tunnel barriers in the limit of small phase fluctuations. Qualitatively new features emerge in higher orders in T_n being directly related to the presence of subgap Andreev levels $\pm\epsilon_n(\chi)$ inside the contact. Consider, for instance, the kernel $\mathcal{I}(t)$. It can be split into three contributions of different physical origin

$$\mathcal{I}(t) = \mathcal{I}_1(t) + \mathcal{I}_2(t) + \mathcal{I}_3(t). \tag{16}$$

The first of these terms, $\mathcal{I}_1(t)$, represents the subgap contribution due to discrete Andreev states. The Fourier transform of this term has the form

$$\mathcal{I}_{1\omega} = \frac{\pi\Delta^4}{4}\sum_n\Bigg\{\frac{T_n^2\sin^2\chi}{2\epsilon_n^2(\chi)\cosh^2(\epsilon_n(\chi)/2T)}\delta(\omega)$$
$$+\frac{T_n^2(1-T_n)\sin^4(\chi/2)}{\epsilon_n^2(\chi)}\left[1+\tanh^2(\epsilon_n(\chi)/2T)\right]\times\left[\delta\left(\omega-2\epsilon_n(\chi)\right)+\delta\left(\omega+2\epsilon_n(\chi)\right)\right]\Bigg\}. \tag{17}$$

It is obvious that this contribution is not contained in the AES action at all. The second term $\mathcal{I}_2(t)$ can be interpreted as the "interference term" between subgap Andreev levels and quasiparticle states above the gap. In the low temperature limit $T \to 0$ the Fourier transform of this term $\mathcal{I}_{2\omega}$ differs from zero only at sufficiently high frequencies $|\omega| > \Delta + \epsilon_n(\chi)$. At higher temperatures $T > \epsilon_n(\chi)$, however, $\mathcal{I}_{2\omega}$ vanishes only for $|\omega| < \Delta - \epsilon_n(\chi)$ and remains non-zero otherwise. In the limit of small barrier transmissions this term scales as $\mathcal{I}_2 \propto T_n^{3/2}$ and, hence, is not contained in the AES action either. Finally, the third term $\mathcal{I}_3(t)$ accounts for the contribution of quasiparticles with energies above the gap. In the high frequency limit $\omega \gg \Delta$ or for $\Delta \to 0$ this term reduces to the standard result for a normal conductor $\mathcal{I}_{3\omega} \to (\omega/2e^2 R_N)\coth(\omega/2T)$.

Turning now to the function $\mathcal{R}(t)$ in Eq. (14) we note that its Fourier transform can be represented as $\mathcal{R}_\omega = \mathcal{R}'_\omega + i\mathcal{R}''_\omega$, where both $\mathcal{R}'_\omega$ and $\mathcal{R}''_\omega$ are real functions. The function $\mathcal{R}'_\omega$ is even in ω while $\mathcal{R}''_\omega$ is an odd function of ω, thus implying that the function $\mathcal{R}(t)$ is real.

The functions $\mathcal{R}(t)$ and $\mathcal{I}(t)$ are not independent. For instance, the Fourier transform $\mathcal{R}''_\omega$ is related to $\mathcal{I}_\omega$ by means of the fluctuation-dissipation relation $\mathcal{R}''_\omega = 2\mathcal{I}_\omega \tanh(\omega/2T)$. The two functions $\mathcal{R}'_\omega$ and $\mathcal{R}''_\omega$ are in turn linked to each other by the causality principle: the function $\mathcal{R}(t)$ should vanish for $t < 0$.

Finally we would like to point out that with the aid of the above Gaussian effective action one can easily evaluate the phase-phase correlation functions for our problem. Combining Eqs. (12)-(15) with (2) one finds (cf., e.g. [17])

$$\langle \varphi_+(t_1)\varphi_+(t_2)\rangle = -\int_{-\infty}^{\infty} \frac{d\omega}{2\pi} \mathrm{Im}\left(\frac{1}{C\omega^2/e^2 + \mathcal{R}_\omega}\right)\coth\frac{\omega}{2T} e^{-i\omega(t_1-t_2)}, \tag{18}$$

$$\langle \varphi_+(t_1)\varphi_-(t_2)\rangle = i\int_{-\infty}^{\infty} \frac{d\omega}{2\pi}\left(\frac{1}{C\omega^2/e^2 + \mathcal{R}_\omega}\right) e^{-i\omega(t_1-t_2)},$$

$$\langle \varphi_-(t_1)\varphi_-(t_2)\rangle = 0.$$

Now we will employ the above results in order to describe the effect of electron-electron interactions on transport properties of superconducting contacts.

3. Coulomb blockade of Andreev reflection

We start from NS systems. In this case in the absence of interactions we set $\dot{\varphi}_+ = eV$ and trivially recover the standard result $I = G_A V$. For the current fluctuations $\delta I(t)$ from Eqs. (7)-(9) and (5) analogously to [8] we obtain

$$\frac{\langle |\delta I|^2_\omega\rangle}{G_A} = (1-\beta_A)\omega\coth\frac{\omega}{2T} + \frac{\beta_A}{2}\sum_{\pm}(\omega \pm 2eV)\coth\frac{\omega \pm 2eV}{2T}. \tag{19}$$

This equation fully describes current noise in NS structures at energies well below the superconducting gap. For $eV \gg T, \omega$ Eq. (19) reduces to the result [18] while in the diffusive regime the correlator (19) matches with the semiclassical result [19].

Let us now turn on interactions. In this case one should add the charging term (2) to the action and account for phase fluctuations. Proceeding along the same lines as in [8], for $g_A \gg 1$ or $\max(T, eV) \gg E_C = e^2/2C$ we get

$$I = G_A V - 2e\beta_A T \mathrm{Im}\left[w\Psi\left(1+\frac{w}{2}\right) - iv\Psi\left(1+\frac{iv}{2}\right)\right]. \tag{20}$$

where $\Psi(x)$ is the digamma function, $w = g_A E_C/\pi^2 T + iv$ and $v = 2eV/\pi T$.

The last term in Eq. (20) is the interaction correction to the I-V curve which scales with Andreev Fano factor β_A in exactly the same way as the shot noise. Thus, we arrive at an important conclusion: *interaction correction to Andreev conductance of NS structures is proportional to the shot noise power in such structures.* This fundamental relation between interaction effects and shot noise goes along with that established earlier for normal conductors [8] extending it to superconducting systems. In both cases this relation is due to discrete nature of the charge carriers passing through the conductor.

Another important observation is that the interaction correction to Andreev conductance defined in Eq. (20) has exactly the same functional form as that for normal conductors, cf. Eq. (25) in [8]. Furthermore, in a special case of diffusive systems we have $G_N = G_A$, $\beta_N = \beta_A = 1/3$ and the only difference between the interaction corrections to the I-V curve in normal and NS systems is the charge doubling in the latter case. As a result, the Coulomb dip on the I-V curve of a diffusive NS system at any given T is exactly *2 times narrower* than that in the normal case. We believe that this narrowing effect was detected in normal wires attached to superconducting electrodes [20], cf. Fig. 3c in that paper.

4. Interaction correction to supercurrent

Let us now turn to the electron-electron interaction correction to the equilibrium Josephson current (13). Previously such correction was analyzed in the case of Josephson tunnel barriers in the presence of linear Ohmic dissipation [7]. The task at hand is to investigate the interaction correction to the supercurrent in contacts with arbitrary transmission distribution.

In order to evaluate the interaction correction it is necessary to go beyond the Gaussian effective action (12)-(15) and to evaluate the higher order contribution $\sim \varphi^3$. It is easy to observe that the interaction correction to the supercurrent is provided by the following non-Gaussian terms in the effective action:

$$\delta(iS_t) = \int\int\int dt_1dt_2dt_3Y(t_1,t_2,t_3)\varphi_-(t_1)\varphi_+(t_2)\varphi_+(t_3) + \int\int\int dt_1dt_2dt_3Z(t_1,t_2,t_3)\varphi_+(t_1)\varphi_-(t_2)\varphi_-(t_3). \tag{21}$$

The function $Y(t_1, t_2, t_2)$ can be written as

$$Y(t_1,t_2,t_3) = \int\int \frac{d\omega_1}{2\pi}\frac{d\omega_2}{2\pi}Y(\omega_1,\omega_2)e^{-i\omega_1(t_1-t_2)}e^{-i\omega_2(t_1-t_3)}, \tag{22}$$

where $Y(\omega_1, \omega_2) = Y(\omega_2, \omega_1)$. The function $Z(t_1, t_2, t_3)$ can be expressed in a similar way.

Adding the non-Gaussian terms (21) to the action and employing Eq. (4) we arrive at the following expression for the interaction correction

$$\delta I_S(\chi) = ie\int\limits_{-\infty}^{\infty}\frac{d\omega}{2\pi}Y(\omega,-\omega)\langle\varphi_+\varphi_+\rangle_\omega + 2ie\int\limits_{-\infty}^{\infty}\frac{d\omega}{2\pi}Z(0,-\omega)\langle\varphi_+\varphi_-\rangle_\omega, \tag{23}$$

where the phase-phase correlators are defined in Eq. (18).

Let us consider the first term in the right-hand side of Eq. (23). It is easy to see that in the limit of low temperatures only frequencies $|\omega| > \Delta + \epsilon_n(\chi)$ contribute to the integral in Eq. (18) for $\langle\varphi_+\varphi_+\rangle$ while the contribution from the frequency interval $|\omega| < \Delta + \epsilon_n(\chi)$ vanishes. Furthermore, the leading contribution from the first term in Eq. (23) is picked up logarithmically from the interval $2\Delta \ll |\omega| \ll 1/R_NC$ where

$$\langle\varphi_+\varphi_+\rangle_\omega \simeq \frac{e^2R_N}{|\omega|} \tag{24}$$

and the function $Y(\omega, -\omega)$ tends to a frequency independent value.

After a straightforward but tedious calculation in the interesting frequency range $\omega \gg \Delta$ from Eq. (3) one finds

$$Y(\omega,-\omega)=\frac{i\Delta\sin\chi}{4}\sum_n\frac{T_n(1-T_n)(2-T_n\sin^2(\chi/2))}{(1-T_n\sin^2(\chi/2))^{3/2}}F(\epsilon_n(\chi)). \tag{25}$$

This high-frequency term involves the factor $1-T_n$, i.e. it vanishes for fully open conducting channels. Combining Eqs. (24), (25) with (23), we arrive at the expression for the supercurrent

$$I(\chi)=I_S(\chi)+\delta I_S(\chi). \tag{26}$$

In the limit of low temperatures the interaction correction reads

$$\delta I_S(\chi)=-\frac{e\Delta}{2g_N}\ln\left(\frac{1}{2\Delta R_N C}\right)\sin\chi\sum_n\frac{T_n(1-T_n)}{(1-T_n\sin^2(\chi/2))^{3/2}}\left(2-T_n\sin^2\frac{\chi}{2}\right), \tag{27}$$

where $g_N=2\pi/(e^2R_N)$ is the dimensionless normal state conductance of the contact. This result is justified as long as the Coulomb correction $\delta I_S(\chi)$ remains much smaller than the non-interacting term $I_S(\chi)$ (13). Typically this condition requires the dimensionless conductance to be large $g_N \gg \ln(1/2\Delta R_N C)$.

Note that Eq. (27) was derived only from the first term in Eq. (23). The second term in this equation involving the function $Z(0,-\omega)$ and the correlator $\langle\varphi_+\varphi_-\rangle$ can be treated analogously. It turns out to be smaller than that of the first term by the logarithmic factor $\sim \ln(1/2\Delta R_N C)$.

Let us emphasize again an important property of the result (27): The interaction correction contains the factor $1-T_n$ and, hence, vanishes for fully open barriers. In other words, *no Coulomb blockade of the Josephson current is expected in fully transparent superconducting contacts.*

The expression for the interaction correction (27) can further be specified in the case of diffusive contacts. In the absence of interactions the Josephson current in such contacts follows from (13) and takes the form corresponding to the zero-temperature limit of a well known Kulik-Omelyanchuk formula for a short diffusive wire

$$I_S(\chi)=\frac{\pi\Delta}{2eR_N}\cos\frac{\chi}{2}\ln\frac{1+\sin\frac{\chi}{2}}{1-\sin\frac{\chi}{2}}. \tag{28}$$

Including interactions and averaging (27) with the bimodal transmission distribution

$$P(T_n)\propto\frac{1}{T_n\sqrt{1-T_n}}. \tag{29}$$

one finds

$$\delta I_S(\chi)=-\frac{e}{8}\Delta\ln\left(\frac{1}{2\Delta R_N C}\right)\cot(\chi/2)\left[\left(\sin\frac{\chi}{2}+\sin^{-1}\frac{\chi}{2}\right)\ln\frac{1+\sin(\chi/2)}{1-\sin(\chi/2)}-2\right]. \tag{30}$$

Note that the result (27) can formally be reproduced if one substitutes $T_n\to T_n+\delta T_n$ into Eq. (13), where

$$\delta T_n=-\frac{2}{g_N}\ln\left(\frac{1}{2\Delta R_N C}\right)T_n(1-T_n), \tag{31}$$

and then expands the result to the first order in δT_n. Interestingly, the same transmission renormalization (31) follows from the renormalization group (RG) equations [10, 11, 12]

$$\frac{dT_n}{dL}=-\frac{T_n(1-T_n)}{\sum_k T_k},\quad L=\ln\left(\frac{1}{\epsilon R_N C}\right) \tag{32}$$

derived for *normal* conductors. In order to arrive at Eq. (31) one should just start the RG flow at $\epsilon = 1/R_N C$ and stop it at $\epsilon = 2\Delta$. Thus, the result (27) can be interpreted in a very simple manner: Coulomb interaction provides high frequency renormalization $T_n + \delta T_n$ (31) of the barrier transmissions which should be substituted into the classical expression for the supercurrent (13). It should be stressed, however, that the last step would by no means appear obvious without our rigorous derivation since the Coulomb correction to the Josephson current originates from the term $\sim \varphi_- \varphi_+^2$ in the effective action which is, of course, totally absent in the normal case.

5. Summary

In this paper we derived a general expression for the effective action of superconducting contacts with arbitrary transmissions of conducting channels. In the case of NS systems we described the interplay between Coulomb blockade and Andreev reflection and demonstrated a direct relation between shot noise and interaction effects in these structures. The fundamental physical reason behind this relation lies in discrete nature of the charge carriers – electrons and Cooper pairs – passing through NS interfaces. Our results allow to explain recent experimental findings [20].

Superconducting contacts with arbitrary channel transmissions show qualitatively new features as compared to the case of Josephson tunnel barriers [7]. The main physical reason for such differences is the presence of subgap Andreev bound states inside the system. Our results for the interaction correction might explain a rapid change between superconducting and insulating behavior recently observed [21] in comparatively short metallic wires with resistances close to the quantum resistance unit ~ 6.5 KΩ in-between two bulk superconductors. Previously it was already argued [22] that such a superconductor-to-insulator crossover can be due to Coulomb effects. Our present results provide further quantitative arguments in favor of this conclusion.

Acknowledgment

This work was supported in part by RFBR grant 09-02-00886.

References

[1] Andreev A F 1964 *Sov. Phys. JETP* **19** 1228
[2] Blonder G E, Tinkham M and Klapwijk T M 1964 *Phys. Rev. B* **25** 4515
[3] Kulik I O 1970 *Sov. Phys. JETP* **30** 944
[4] Ishii C 1970 *Progr. Theor. Phys.* **44** 1525
[5] Beenakker C W J 1991 *Phys. Rev. Lett.* **67** 3836
[6] Galaktionov A V and Zaikin A D 2002 *Phys. Rev. B* **65** 184507
[7] Schön G and Zaikin A D *Phys. Rep.* 1990 **198** 237
[8] Golubev D S and Zaikin A D 2001 *Phys. Rev. Lett.* **86** 4887
[9] Galaktionov A V, Golubev D S and Zaikin A D 2003 *Phys. Rev. B* **68** 085317; *ibid.* **68** 235333
[10] Kindermann M and Nazarov Yu V 2003 *Phys. Rev. Lett.* **91** 136802
[11] Golubev D S and Zaikin A D 2004 *Phys. Rev. B* **69** 075318
[12] Bagrets D A and Nazarov Yu V 2005 *Phys. Rev. Lett.* **94** 056801
[13] Zaikin A D 1994 *Physica B* **203** 255
[14] Snyman I and Nazarov Yu V 2008 *Phys. Rev. B* **77** 165118
[15] Galaktionov A V and Zaikin A D 2009 *Phys. Rev. B* **80** 174527
[16] Kulik I O and Omel'yanchuk A N 1978 *Sov. J. Low Temp. Phys.* **4** 142
[17] Golubev D S and Zaikin A D 1999 *Phys. Rev* B **59** 9195
[18] De Jong M J M and Beenakker C W J 1994 *Phys. Rev. B* **49** 16070
[19] Nagaev K E and Büttiker M 2001 *Phys. Rev. B* **63** 081301
[20] Bollinger A T, Rogachev A and Bezryadin A 2006 *Europhys. Lett.* **76** 505
[21] Bollinger A T, Dinsmore III R C, Rogachev A and Bezryadin A 2008 *Phys. Rev. Lett.* **101** 227003
[22] Arutyunov K Yu, Golubev D S and Zaikin A D 2008 *Phys. Rep.* **464** 1

Journal of Physics: Conference Series **248** (2010) 012030 doi:10.1088/1742-6596/248/1/012030

Intermediate vortex states in nanoscale anti-dots and mesoscopic superconductors

K Hirata, A Thakur, S Ooi and T Mochiku

National Institute for Materials Science, Superconducting Materials Center, 1-2-1 Sengen, Tsukuba, 305-0047, Japan

E-mail;HIRATA.Kazuto@nims.go.jp

Abstract. We have fabricated anti-dots array in Nb and NbN, and high-T_c superconductor $Bi_2Sr_2CaCu_2O_y$ (Bi-2212), and have measured the flow-resistance of vortices perpendicular to the array and magnetic field. We could produce a variety of vortex-matching effect; the usual and the fractional matching. In low-T_c superconductors, the usual matching effect shows "dips" in the flow resistance at the matching field. However, it shows "humps" at higher matching magnetic fields to be related to the formation of giant vortices and/or interstitial vortices. In Bi-2212, the matching effect is closely related to the first order transition of vortex lattice melting in the pristine samples in the presence of the anti-dot arrays. We also have measured the vortex state in mesoscopic type-I superconductor of In. This includes the observation of topological hysteresis with a signature of occurrence in different critical fields during vortex entry and exit. We will show the existence of a metastable configuration of vortices depending on the recipes to access them, and demonstrate the manifestation of superheating and supercooling of superconducting and normal states, respectively, across the superconducting transition.

1. Introduction

Recent progress in nanosize fabrication techniques enables us to study superconductivity in nanosize superconductors. Photo-, electron beam-lithography, and Focused Ion Beam (FIB) milling are the compromising methods for fabricating nanosize superconductors from bulk samples. Small size (submicron- to nanosize) superconductors can be obtained also by electrochemical deposition. To make the size of superconductors narrower and narrower, with decreasing dimensionality, there are anticipated a competition between the superconducting coherence length and the sample size. Theoretical proposals have been made about half a century ago [1-3]. They proposed that, reducing the dimensionality, superconducting phase slip may happen by thermal fluctuations, which was confirmed by the experiment with submicron size samples [4]. Recently, Bezryadin *et al* [5] have obtained experimental results of quantum suppression in superconductivity, and Lau *et al* [6] have measured nanosize samples with changing the cross-sectional area and have obtained the results of thermally-activated phase slip and quantum phase slip. which were made by depositing MoGe compound on the carbon nanotube. To make the superconductors narrower and narrower, it may suppress the superconductivity or sometimes creates new phenomena (macroscopic quantum tunneling, for example). Changing the size of superconductors from nanosize to mesoscopic size, several interesting phenomena have been observed. Non-integer magnetic flux carrying has been observed in Al thin film disks [7]. Formation of giant vortex state has been observed [8]. Recently, Berdiyorov *et al* [9] have simulated the vortex configuration in type-I superconductors. The ground state of the vortex configuration depends on the sample geometry and magnetic field. Intermediate state of flux structure has been well produced in magnetization and free energy versus magnetic field. Here we present the experimental results on In nanowires encapsulated in insulating ZnS nanotubes [10, 11], which show the supercooling and superheating effects according to the magnetic field cycles.

On the other hand, fabrication of nanosize holes (anti-dots) or of putting nanosize magnetic dots is also possible by the nanosize fabrication, which can be used to study a matching effect of vortices in a magnetic field [12-16] and electronic properties in superconductors with anti-dots array [17-19]. In most of the experiments on the vortex physics in superconducting films/crystals with an anti-dots array, the vortices were considered as classical point defects with the pinning. However, the vortices are in fact extended objects which interact with the geometry of the underlying anti-dots lattice in many-body problem. This opens up the possibility for geometry and/or pinning strength (of the anti-dot lattice) induced novel vortex phenomena such as vortex merging, symmetry induced creation of vortex-antivortex configurations, coupling, rearrangement, etc, bringing into forefront the importance of interstitial vortices among the anti-dots also. Here we present the preliminary results of our investigations on the interplay between the vortices and the geometry of the underlying nano-engineered anti-dot lattice in Nb and NbN thin films [20]. The anti-dots are fabricated in the triangular or hexagonal shape with gradually increasing area of anti-dots such that one essentially goes over from a triangular anti-dots array to a honeycomb wire network.

The effect of the anti-dots array into high-T_c superconductors (HTSCs) is also one of the interesting subjects. In high-T_c superconductors, the first-order transition of vortex lattice melting (FOT-VLM) has been found [21]. This has been observed in relatively clean samples, which have intrinsically pinning centers caused by defects (imperfections of crystals; impurities, crystalline defects, etc). Then, VLM occurs at relatively higher temperatures below T_c, because, at lower temperatures, the pinning strength becomes stronger, and the VLM is mediated from the first order transition to the second order or sometimes disappears. Introducing the anti-dots array into HTSCs, it is interesting to investigate how the FOT-VLM will be changed, as the competition might be occurred between the vortex lattice and anti-dots array. We present here the experimental results on $Bi_2Sr_2CaCu_2O_y$ (Bi-2212) with a triangular array of anti-dots, and will show broad and steep steps at the superconducting transition depending on the magnetic field, in addition to the fractional matching effect [22].

2. Experimental

In nanowires were synthesized with ZnS nanotubes carried out in a high temperature vertical induction furnace. The details are described in Ref. 10&11. The average wire diameter was 120nm, and the length 5-20μm, while the average external diameter of the shell ZnS nanotube was about 200nm, which is shown in figure 1 Superconducting transition temperature T_c was 3.41K, which is close to the bulk T_c. Bundle of the In nanowires in almost aligning to the magnetic field was measured to put into Quantum Design SQUID MPMS XL, and the magnetization was measured.

The anti-dot array samples of Nb and NbN films were prepared on a Si substrate [23, 24]]. Initially, 100nm thick films were deposited by sputtering [20]. A probe pattern was made via photolithography followed by ion beam etching for a typical four-probe measurement with region between the voltage probes typically being 30μm x 30μm wide. The etching was performed at a typical etching rate of 0.5 nm/sec in Argon plasma [20]. Under these etching conditions, there was essentially no damage to the Nb and NbN films. In these 30μm x 30μm wide available regions, anti-dot arrays networks were patterned using the FIB milling technique utilizing an aperture of 50μm and a typical ion-beam current of 13pA. The transport measurements were carried out with the conventional four-probe technique using a home-made insert which goes into the Quantum Design SQUID MPMS XL. The data acquisition was made with the external device control (EDC) option to operate the Keithley 6430 sub-femtoamp remote sourcemeter and the Keithley 2182A nanovoltmeter.

Journal of Physics: Conference Series **248** (2010) 012030 doi:10.1088/1742-6596/248/1/012030

An additional temperature sensor was placed very close to the sample in the insert to be read with a Cryocon 62 AC resistance bridge.

In HTSC Bi-2212, high-quality single-crystals of Bi-2212 were grown by the travelling-solvent floating zone technique [25]. Thin single-crystal films with submicron thickness were prepared through repeating cleave by adhesive tapes, which is the characteristic of Bi-2212 to easily cleaved in the *ab*-plane. After thinning the crystals sufficiently on the MgO substrate, thin Au layer (20nm) was deposited on the surface as electrodes. The films were patterned for four-terminals electric resistance measurements using a photolithography and an Ar-ion milling processes. Equilateral triangular array of anti-dots was fabricated by FIB milling using Micrion JFIB-2100 to the bridge part of films between the voltage electrodes. The superconducting transition temperature of the sample determined from zero resistance with a current of 10μA was 89.0K. In the measurements, the magnetic field was always applied perpendicular to the *ab*-plane.

3. Results and Discussion

3.1 Intermediate state of In nanowires

Figure 1 shows a TEM image of In nanowire encapsulated in ZnS nanotube. Inset of figure 1 shows the selected-area electron diffraction pattern at the point indicated by arrow, which shows that the nanotube used in the measurements is a single crystalline structure. For the magnetization measurements, the bundle of the nanotubes was used. It was confirmed that the In nanowires belong to the type-I nature, showing nearly the same magnitude of the critical magnetic field with the *M-H* loops of the bulk In. Difference between nanowire and bulk In lies in that of critical fields with field sweep-up and sweep-down, and attributes to the phenomenon of supercooling and superheating of intermediate vortex state mentioned later.

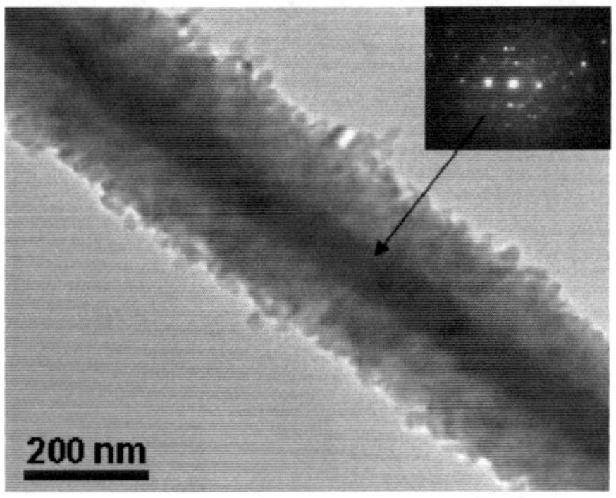

Figure 1. TEM image of the In nanowire encapsulated in ZnS nanotube. Inset shows the selected area electron diffraction pattern at the point indicated by arrow.

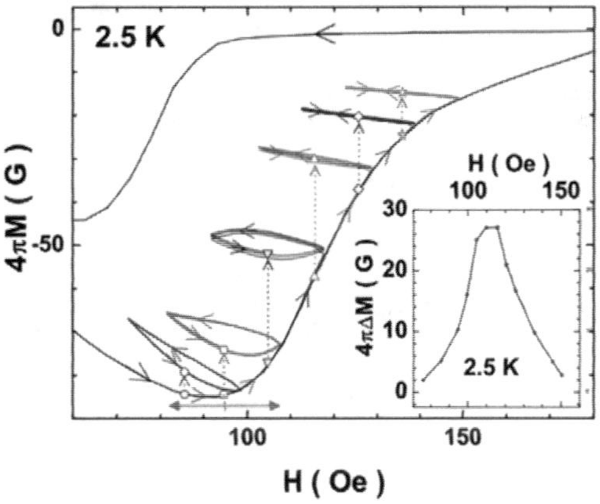

Figure 2. A part of the magnetization loop (solid black lines) in the field sweeping at 2.5K. The arrows show the field sweep directions. Minor loops also indicated in the colour loops. The dotted arrows show the initial to the final states (open symbols) of the loops. The inset shows changes in magnetization values in the field-shaking process (see the text).

Figure 2 shows the result of magnetization measurements in *M-H* loops, which partially indicates a *M-H* loop and minor loops with the field-shaking process. The field-shaking process was made during the magnetization measurements with sweeping the field up and down by an amount of ΔH, which are shown in the coloured loops. The inset shows a plot in changes in magnetization values with the field-shaking process, in which a peak can be seen as a maximum at intermediate field of about 110Oe.

In the temperature-shaking process shown in figure 3, it also shows an intermediate state of vortices. The colour scale contour in figure 3 is plotted for the evolution of magnetization values from the field cooling state for various temperature-shakings, in which the colour scale from red to blue indicates the superconductivity becomes stronger. The final state in figure 3 (b) corresponds to the starting state in the panel (a). In this case, as shown in figures 3 (a) for a negative temperature-shaking and (b) for a positive, a low temperature superconducting phase is superheated and a high field normal state is suppressed, respectively. In these hysteresis observed, there are several factors to contribute; pinning centers, geometrical barriers, topologically, superheating and supercooling across the boundary between the superconducting phase and the normal state, etc. Among of them, topological hysteresis may understand the results by assuming the presence of a surface barrier during flux entry and exit. These intermediate vortex state may be related to the intermediate vortex configurations published recently [9].

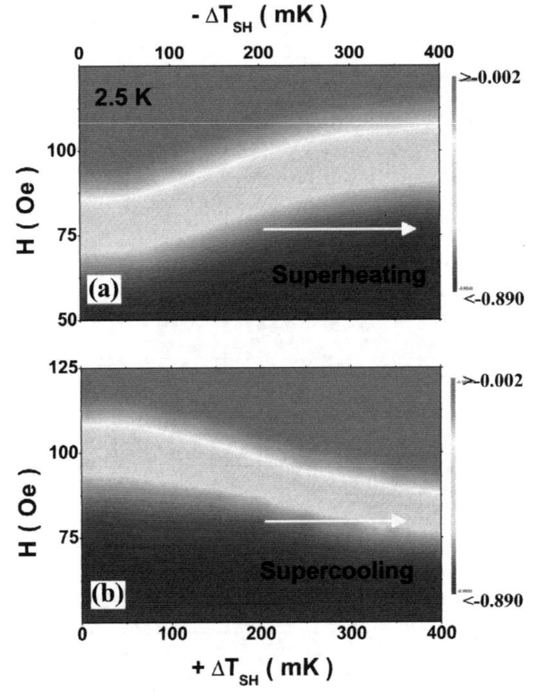

Figure 3. Colour scale contour plot of the evolution of magnetization from the field cooled state following the field-cooling temperature-sweeping process with negative temperature-shaking (a) and positive (b) at 2.5K.

3.2 Vortex state with anti-dots array in Nb and NbN

As the vortex flow-resistance measurements in Nb film with regular array of anti-dots have been well made and studied [13, 26], here we present on the experimental results on NbN film, which has a relatively higher T_c in metallic superconductors and has no reports to measure on the vortex matching effect to the anti-dots array in detail. Figures 4 (a) and (b) show the typical results of the flow-resistance versus normalized magnetic field f ($=H/H_1$, H_1 is the matching field to the array of anti-dots) measured with a current I_{dc}=100μA on the samples with changing a temperature and with a temperature T=0.932T_c with changing a current I_{dc}=1-150μA, in which the scanning ion beam images are shown in the insets. In figure 4 (a), dip structures are clearly seen at integer number of the matching field f up to 8th. This can be observed in the sample of the triangular anti-dots array [20]. The maximum number of vortices captured by an anti-dot with a diameter d is given by the saturation number (n_s) such that, $n_s=d/4\xi(t)$, where $\xi(t)$ is the coherence length at reduced temperature t ($=T/T_c$) given by $\xi(t)= \xi_0/(1-t)^{1/2}$, with being the zero temperature coherence length about 5nm. In the square anti-dots arrays, n_s lies in the range 3.05-2.90 for t in the range of 0.925-0.923, which is far from the matching periods. So, the multi-vortex states do not seem to exist in the observations. Instead of that, it may be understood in the supermatching flux-line lattices model formed by a reorganization of vortices that enter the interstitial sites at the fields larger than $n_s H_1$ [15, 16]. To confirm the model, we fabricated a hexagonal lattice of anti-dots missing the center anti-dot as shown in figure 4 (b). In this case, we can observe the integer matching effect up to the saturation number 3. However, above 3rd matching, the dip structure changes to enhance the flow-resistance, resulting to form the hump structure. This is understood that, up to the saturation number of vortices, they form multi-vortex state pinned at the anti-dots. However, exceeding to the number, the next vortices come to the center of hexagonal anti-dots, become interstitial vortices and mobile, and these weekly pinned vortices enhance the flow-resistance. The role of the interstitial vortices must be studied carefully from analytical thermodynamic calculations taking the interaction among the vortices and anti-dots into account.

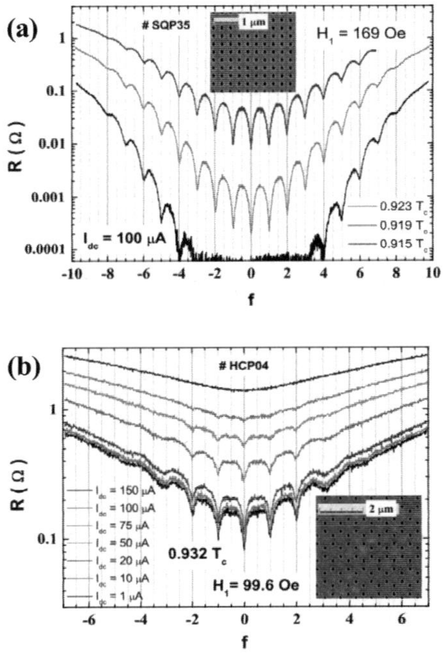

Figure 4. Flow-resistance versus a normalized magnetic field f (=H/H_1, H_1 is the matching field to the array of anti-dots) obtained with a current I_{dc}=100μA. The inset shows the Scanning Ion Beam image of the square anti-dots lattice H_1=169Oe with a pitch of 350nm and a diameter of 180nm for (a), and of the hexagonal anti-dots lattice H_1=99.6Oe with a pitch of 400nm and a diameter of 170nm for (b).

3.3 Vortex state with anti-dots array in Bi-2212

To introducing anti-dots array into Bi-2212 is a challenging issue, as there have been few report except into $YBa_2Cu_3O_{7-y}$. Scanning-ion beam image of the anti-dots in Bi-2212 between the voltage electrodes is shown in figure 5, fabricated by FIB. The diameter of the anti-dots is 300nm, the pitch is 1.0μm, and the matching field H_0 is about 23.9Oe. Figure 6 shows the results of the flow-resistance measurements. At the matching field H_0, the flow-resistance shows a dip structure as seen in low temperature superconducting film (figures 4 (a) and (b)). Furthermore, fractional matching effect can be clearly observed. Experimentally, we obtained the fractional matching at 8/9, 6/7, 3/4, 2/3, 1/2, 1/3. 1/4, 1/7, 1/9H_0. The fractional matchings of 1/3. 1/4, 1/7, 1/9H_0 can be explained well with commensurate matching of the Abrikosov vortex-lattice to anti-dots arrays. The 1/3H_0 matching is most remarkable, which corresponds to match to the second nearest neighbor anti-dots. However, 1/2H_0 matching is hard to explain at present, which may be related to the interstitial vortices as mentioned in previous subsection.

Another point of view to study vortex states in the anti-dots array in Bi-2212 is that, in Bi-2212 pristine sample without anti-dots, as the FOT of VLM has been observed (which has also been confirmed in this experiment before the fabrication), it is very interesting to see what it will be

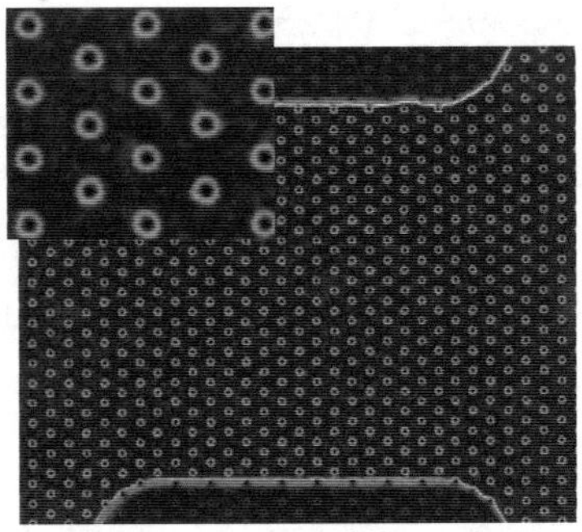

Figure 7. Scanning-ion beam image of the anti-dots in Bi-2212 between the voltage electrodes, fabricated by FIB. The diameter of the anti-dots is 300nm, and the pitch is 1.0μm.

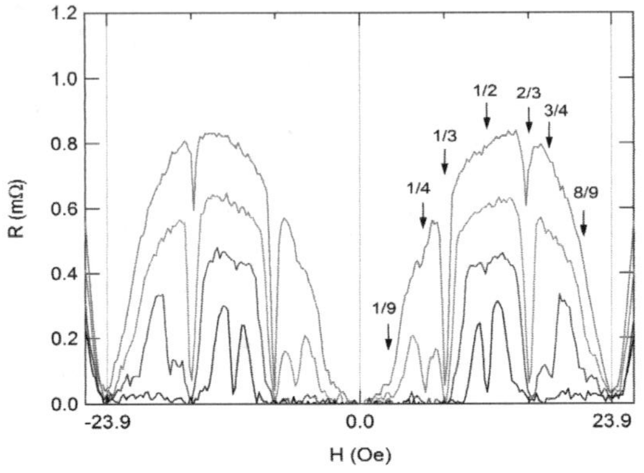

Figure 8. Flow-resistance versus magnetic field measured on the sample shown in figure1. Fractional matching can be clearly seen symmetrically against magnetic field direction.

happened in the presence of the anti-dots. Figure 9 shows resistance versus temperature at the fixed magnetic field from 0.5 to 50Oe with an interval of 0.5Oe, measured on the same sample of figure7. At lowest magnetic field, the superconducting transition is relatively sharp, however, at the fractional matching field, it becomes much sharper and, at the matching fields, it is reduced. To see more clearly, we plot it as three-dimensional gradient map ($d\log R/dT$) in figure 10. At the fractional matching fields, the steep step can be clearly observed. This may be considered as the commensurability between Abrikosov vortex lattice and the anti-dots arrays. At the fractional matching, it corresponds to incommensurate case and the FOT still remains as solid-liquid melting transition. At the matching fields, which becomes commensurate, it may cause the floating solid to show a broad transition.

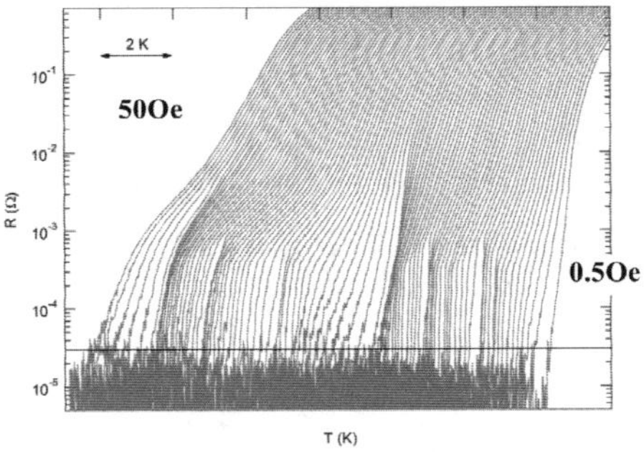

Figure 9. TEM image of the In nano-wire encapsulated in ZnS nanotube. Inset shows the selected area electron diffraction pattern at the point indicated by arrow.

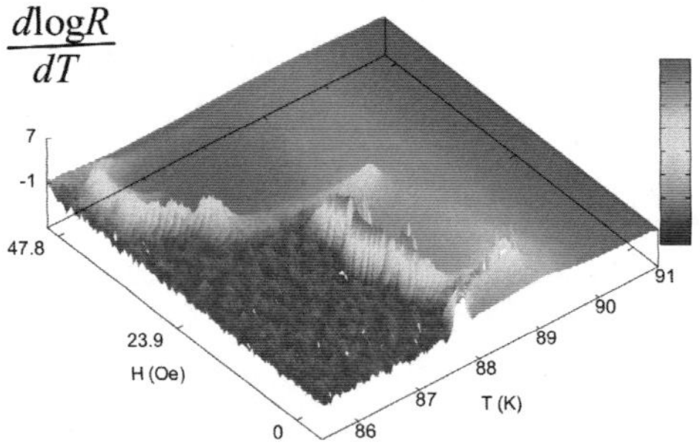

Figure 10. Three-dimensional map of $d\log R/dT$ obtained from R-T measurements shown in figure 9. At the fractional matching field, the slope becomes much sharper.

4. Summary

We have fabricated anti-dots structure into low- and high T_c superconductors and have measured the flow-resistance of vortices. We could produce a variety of vortex-matching effect; the usual and the fractional matching. In low-T_c superconductors, the usual matching effect shows "dips" in the flow resistance at the matching field. However, multi-vortex state explains hardly at higher matching magnetic field. We have confirmed that interstitial vortices play a role to accommodate more vortices, as it has been shown as "humps" at higher matching magnetic fields. In Bi-2212, the matching effect is closely related to the first order transition of vortex lattice melting in the pristine samples in the presence of the anti-dots arrays. We also have measured the vortex state in mesoscopic type-I superconductor of In. This includes the observation of topological hysteresis with a signature of occurrence in different critical fields during vortex entry and exit. We showed the existence of a metastable configuration of vortices depending on how to access them, and demonstrate the manifestation of superheating and supercooling of superconducting and normal states, respectively, across the superconducting transition.

References

[1] Hohenberg P C 1967 Existence of long-range orger in one and two dimensions *Phys. Rev. Lett.* **158** 383
[2] Langer JS and Ambegaokar MR 1967 Intrinsic resistive transition in narrow superconducting channels *Phys. Rev.* **164** 498
[3] McCumber D E and Halperin B I 1970 Time scale of resistive fluctuations in thin superconducting wires *Phys. Rev.* B **1** 1054
[4] Newbower R S, Beasley M R and Tinkham M 1972 Fluctuation effects on the superconducting transition of Tin whisker crystals *Phys. Rev.* B **5** 864
[5] Bezryadin A, Lau C N and Tinkham M 2000 Quantum supression of superconductivity in ultrathin nanowires *Nature* **404** 971
[6] Lau C N, Markovic N, Bockrath M, Bezryadin A and Tinkham M 2001 Quantum phase slips in superconducting nanowires *Phys. Rev. Lett.* **87** 217003
[7] Geim A K, Dubson S V, Grigorieva L V, Novoselov K S, Peeters F M and Schweigert V A 2000 Non-quantized penetration of magnetic flux in the vortex state of superconductors *Nature* **407** 55
[8] Kanda A, Baelus B J, Peeters F M, Kadowaki K and Ootsuka Y 2004 Experimental evidence of giant vortex states in a mesoscopic superconducting disk *Phys. Rev. Lett.* **93** 257002
[9] Berdiyorov G R, Hernandez A D and Peeters F M 2009 Confinement effects on intermediate-state flux patterns in mesoscopic type-I superconductors *Phys. Rev. Lett.* **103** 267002
[10] Sheet G, Gautam U K, Thakur A D, Hirata K, Bando Y and Nakayama T 2009 Clean superconducting In nanowires encapsulated within insulating ZnS nanotubes *Appl. Phys. Lett.* **94** 053108
[11] Thakur A D, Gautam U K, Sheet G, Nakayama T, Bando Y, Goldberg D, Ooi S and Hirata K 2010 History and metastability effects in the intermediate state of mesoscopic type I superconducting Indium *cond-mat.super-con* arXiv:1001.3903v1
[12] Fiory A T, Hebard A F and Somekh S 1978 Critical currents associated with the interaction of commensurate flux-line sublattices in a perforated Al film *Appl. Phys. Lett.* **32** 73
[13] Harada K, Kamimura O, Kasai H, Matsuda T, Tonomura A and MOshchalkov V V 1996 Direct observation of vortex dynamics in superconducting films with regular arrays of defects *Science* **274** 1167
[14] Martin J I, Veliez M, Nogues J and Schuller I K 1997 Flux pinning in a superconductor by an array of sibmicroneter magnetic dots *Phys. Rev. Lett.* **79** 1929
[15] Martin J I, Veliez M, Hoffmann A, Schuller I K and Vincent J L 1999 Artificially induced reconfiguration of vortex lattice by arrays of magnetic dots *Phys. Rev. Lett.* **83** 1022
[16] Metlushko V, Welp U, Crabtree G W, Zhang Z, Brueck R J, Watkins B, Delong L E, Chung B

IIlc K and Hesketh P J 1999 Nonlinear flux-line dynamics in vanadium films with square lattices of submicron holes *Phys. Rev.* B **59** 603

[17] Baturina T 2007 Nano-size antidots TiN thin film *Phys. Rev. Lett.* **98** 257003

[18] Mironov A Y, Baturina T, Vinokur V M, Postolova S V, Kropotin P N, Baklanov M R, Nasimov D A and Latyshev A V 2010 Diorder and vortex matching effects in nanoperforated ultlathin TiN film *Physica* C to be published

[19] Baturina T, Mironov A Y, Vinokur Chtchelkatchev N M, Glatz A, Nasimov D A and Latyshev A V 2009 Resonant Andreev reflection in two-dimensional array of SNS junctions *Physica* C to be published

[20] Thakur A D, Ooi S, Chockalingam S P, Jesudasan J, Raychaudhuri P and Hirata K 2009 Vortex matching effect in engineered thin films of NbN *Appl. Phys. Lett.* **94** 262501

[21] Zeldov E, Majer D, Konczykowski M, Geshkenbein V B, Vinokur V M and Strikman H 1995 Thermodynamic observation of first-order vortex-lattice melting transition in $Bi_2Sr_2CaCu_2O_8$ *Nature* **375** 373

[22] Ooi S, Mochiku T and Hirata 2009 Fractional matching effect in single-crystal films of $Bi_2Sr_2CaCu_2O_{8+y}$ with antidot lattice *Physica* C **469** 1113

[23] Chokalingam S P, Chand M, Jesudasan J, Tripathi V and Raychaudhuri P 2008 Superconducting properties and Hall effect of epitaxial NbN thin films *Phys. Rev.* B **77** 214503

[24] Chokalingam S P, Chand M, Kamlapure A, Jesudasan J, Mishra A, Tripathi V and Raychaudhuri P 2009 Tunneling studies in a homogeneously disordered s-wave superconductor: NbN *Phys. Rev.* B **79** 094509

[25] Mochiku T, Hirata K and Kadowaki K 1997 Crystalline improvement of $Bi_2Sr_2CaCu_2O_{8+\delta}$ single crystal by TSFZ method *Physica* C **282-287** 475

[26] Thakur A D, Ooi S and Hirata K 2009 Triangular antidot array to honeycomb wire network – Role of onterstitial vortices *Physica* C **469** 1071

International Conference on Theoretical Physics Dubna-Nano 2010 IOP Publishing
Journal of Physics: Conference Series **248** (2010) 012031 doi:10.1088/1742-6596/248/1/012031

Anomalous non-equilibrium electron transport in one-dimensional quantum nano wire at half-filling: time dependent density renormalization group study

M. Okumura[1,4,*], **H. Onishi**[2], **S. Yamada**[3,4], **and M. Machida**[3,4]

[1] Computational Condensed Matter Physics Laboratory, RIKEN, Wako, Saitama 351–0198, Japan
[2] Advanced Science Research Center, Japan Atomic Energy Agency, Tokai, Ibaraki 319–1195, Japan
[3] CCSE, Japan Atomic Energy Agency, 6–9–3 Higashi-Ueno, Taito-ku, Tokyo 110–0015, Japan
[4] CREST(JST), 4–1–8 Honcho, Kawaguchi, Saitama 332–0012, Japan

E-mail: * `okumura@riken.jp`

Abstract. We study non-equilibrium properties of one-dimensional Hubbard model by the density-matrix renormalization-group method. First, we demonstrate stability of "doublon", which characterized by double occupation on a site due to the integrability of the model. Next, we present a kind of anomalous transport caused by the doublons created under strong non-equilibrium conditions in an optical lattice system regarded as an ideal testbed to investigate fundamental properties of the Hubbard model. Finally, we give a result on development of the pair correlation function in a strong non-equilibrium condition. This can be understood as a development of coherence among many excited doublons.

1. Introduction

Recently, photoinduced insulator-metal transition has attracted a lot of attention since the experimental observation of the surprisingly ultrafast non-equilibrium phase-transition [1, 2, 3]. Several theoretical studies has been made to understand the mechanism of the ultrafast dynamics during the phase transition. However, all properties of the phase transition have been not fully resolved [4]. On the other hand, optical lattice system is a new fascinating avenue toward "quantum simulator" [5, 6]. This system gives us an almost pure Hubbard model because of its perfectly clean feature. Thus, one can easily observe dynamical properties of the Hubbard model in various non-equilibrium conditions. Namely, one can systematically examine unsolved non-equilibrium properties of the Hubbard model.

The theoretical description of non-equilibrium phenomena is one of the most challenging problems in many-body physics. In weak non-equilibrium cases, we have a very powerful tool to evaluate physical quantities, i.e., the linear response theory [7]. But, the theory is limited to "weak" non-equilibrium (near equilibrium) system. Thus, "strong" non-equilibrium system is still a frontier in many-body Physics. Recently, non-equilibrium phenomena in one-dimensional (1D) system attract much interest of theorists, because some kinds of 1D systems show curious behaviors, i.e., no relaxation to thermal equilibrium if the system's Hamiltonian is almost integrable. This was demonstrated by not only numerical simulations [8] but also

Journal of Physics: Conference Series **248** (2010) 012031 doi:10.1088/1742-6596/248/1/012031

surprisingly cold atom experiments [9]. The experiments also indicate that the cold atom system is an ideal testbed for 1D non-equilibrium phenomena. For example, the doublon tunneling was indeed observed in both the bosonic optical well [10] and the center of mass motion of attractive interacting fermions in an optical lattice [11], while the doublon decay was found out in 3D fermionic optical lattice [12] as our expectation.

In this paper, we present non-equilibrium dynamics of 1D Hubbard model obtained by using time-dependent density-matrix renormalization group (TDDMRG) method [13, 14, 15] which enables us to access "strong" non-equilibrium condition in addition to strong correlation. This method is an extension of the density-matrix renormalization method, [16, 17, 15] which is a very powerful tool for investigation of static and dynamical properties of 1D Hubbard model.

2. Model and Method

Let us start with introducing the model employed throughout this paper. The main part of the model is the 1D Hubbard model, which is the most fundamental model describing strongly correlated fermions. The Hamiltonian is given

$$H_{\mathrm{H}} = -t \sum_{\langle i,j\rangle,\sigma} c^{\dagger}_{i\sigma} c_{j\sigma} + U \sum_{i} n_{i\uparrow} n_{i\downarrow} \,, \tag{1}$$

where $n_{i\sigma} = c^{\dagger}_{i\sigma} c_{i\sigma}$, $\sigma = \uparrow, \downarrow$, and $\langle i,j\rangle$ represents summation with nearest neighbor sites. t is the hopping parameter and U is the on-site interaction strength.

In the rest of the paper, we add time-dependent perturbations to this Hamiltonian to observe non-equilibrium behaviors. In order to study the time evolution of the 1D Hubbard model, we use TDDMRG method [16, 17, 15].

3. Stability of Doublon

First, we examine the doublon creation and its stability. We study the following two conditions, one of which is the half-filling 1D Hubbard chain lying inside a flat 1D space with open boundary condition described by H_1 and another of which is 1D Hubbard chain confined by a harmonic well potential H_2.

$$H_1 = H_{\mathrm{H}} + H_{\delta t(\tau)} \,, \tag{2}$$

$$H_2 = H_{\mathrm{H}} + H_V + H_{\delta t(\tau)} \,, \tag{3}$$

$$H_{\delta t(\tau)} = \delta t \sin(\omega\tau) \sum_{\langle i,j\rangle,\sigma} c^{\dagger}_{i\sigma} c_{j\sigma} \,, \tag{4}$$

$$H_V = \sum_{i,\sigma} V(x_i - x_{\mathrm{c}})^2 n_{i\sigma} \,, \tag{5}$$

where δt is the modulation strength of the hopping parameter, ω is the modulation angular frequency, τ is the normalized time by $1/t$, $x_i = ai$ is the position of the i-th site, where a is the lattice constant, and x_{c} is the position of the trap center. In both cases, we use the following parameters, the system size $U/t = 8$, $\omega/t = 8$, and $\delta t/t = 0.1$. The total system size $L = 40$ for the flat potential case and $L = 60$ for the trapped potential case. In the latter case, we choose the trap strength $Va^2/t = 0.01$. The initial states in these Hamiltonians H_1 and H_2 are shown in Figs. 1 and 2, respectively. In both cases, the perturbation $H_{\delta t(\tau)}$ is switched on at $\tau = 0$ and off at $\tau = 20$, and the number of the doublons $N_{\mathrm{d}} = \sum_i \langle n_{i\uparrow} n_{i\downarrow}\rangle$ is measured.

First, let us study the model H_1. We find that the doublons are created by the perturbation $H_{\delta t}$ until it is switched off (Fig. 3, upper blue line). After switching off ($\tau > 20$), the number of the doublons does not change at all (see Fig. 4, upper blue line). This reflects the absence of

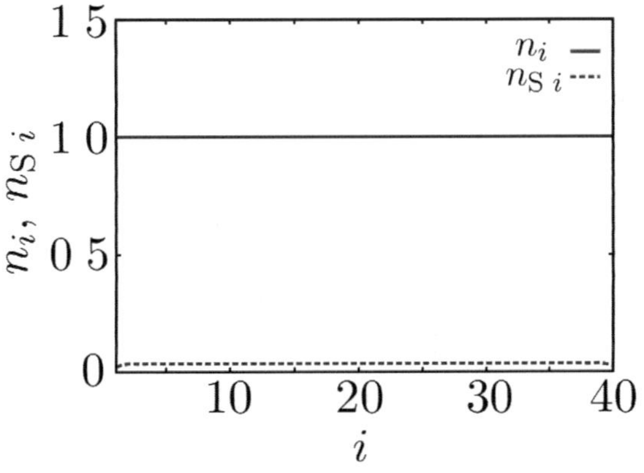

Figure 1. An initial state in the H_1 model. The horizontal and vertical axes represent the site number and both the density of the fermions and the spin polarization.

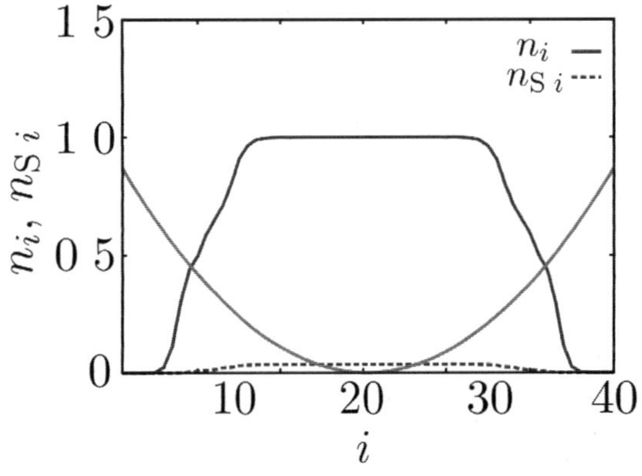

Figure 2. An initial state in the H_2 model. The horizontal and vertical axes represent the site number and both the density of the fermions and the spin polarization.

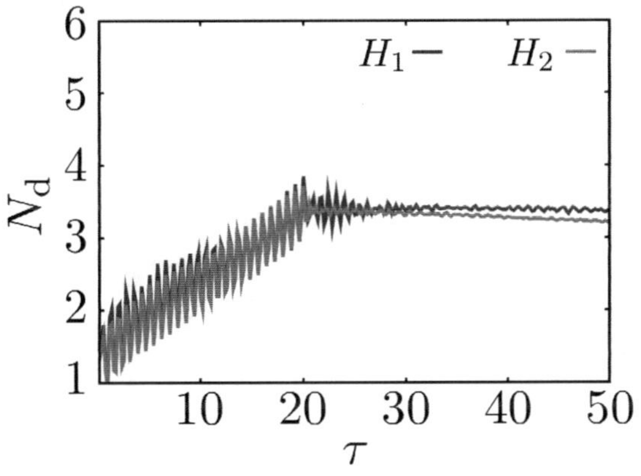

Figure 3. The time evolution of the number of doublons in the H_1 (the upper blue line) and H_2 (the lower red line) models. The horizontal and vertical axes represent the normalized time and the number of doublons, respectively.

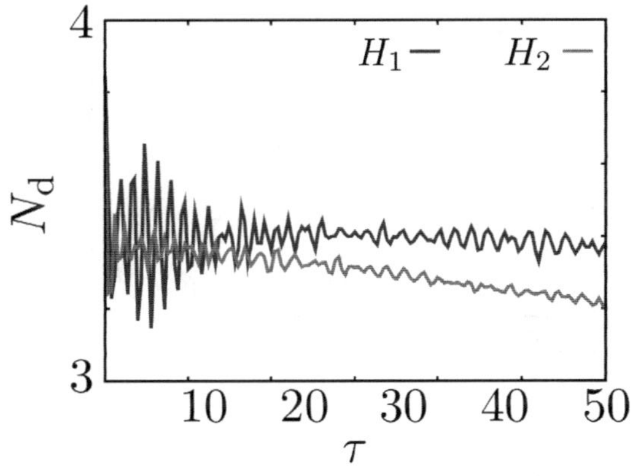

Figure 4. The enlarged figure of the time evolution of the number of doublons in H_1 (the upper blue line) and H_2 (the lower red line) models. The horizontal and vertical axes represent the normalized time and the number of doublons, respectively.

thermal equilibrium due to the integrability. Then, the doublon is a stable excited particle in the 1D Hubbard model if it is once created.

On the other hand, in the model H_2 (Fig. 3, the lower red line), the number of doublons N_d slightly decreases. This is because that the integrability is partly broken by the confinement harmonic potential H_V in H_2. However, the decay speed is not so fast (see Fig. 4, lower red line). Then, the doublon can be regarded as a quasi-stable particle. These strong doublon stability is clearly contrast to the experimental results in 3D optical lattice [12]. This difference is reasonable because the spatial dimension plays a crucial role in the non-equilibrium process as mentioned above.

In this section, we have shown that the doublon is a (quasi-)stable particle in 1D Hubbard chain if it is once created. We have demonstrated that the created doublons play an important role in transport in the following sections.

4. Anomalous Center of Mass Motion in Optical Lattice System

In this section, we examine the center of mass (CoM) motion of fermionic atoms loaded on 1D optical lattice system in order to study transport properties of 1D Hubbard model. The reason why we pay attention to such a specific CoM motion is that the confinement potential is an inevitable setup in the optical lattice systems. However, we would like to note that the optical lattice is perfectly pure without any impurities. Such a system is quite excellent for studies of the transport (non-equilibrium) physics intrinsic to the Hubbard model.

The CoM motion is caused by a sudden shift of the confining potential (see Fig. 5). Then, the target Hamiltonian is given by

$$H_3 = H_{\rm H} + H_{V(\tau)}\,, \tag{6}$$

$$H_{V(\tau)} = \sum_{i,\sigma} V\left(x_i - x_{\rm c} + d\theta(\tau)\right)^2 n_{i\sigma}\,, \tag{7}$$

where d is a shift distance of the trap center (Fig. 5), and $\theta(\tau)$ is the step function. In this study, we choose the system size $L = 40$, the number of particles $N_\uparrow = N_\downarrow = 12$, the trap strength $Va^2/t = 2.63 \times 10^{-3}$, the trap center $x_{\rm c}/a = 20.5$, and the shift distance $d/a = 3$. We prepare four initial states with $U/t = 1$, 2, 3, and 4 respectively (Fig. 6), where one finds that the phases are metallic for all U's. We define CoM as $X = \sum_i ian_i/N$ and we also define $\mathcal{X} = X - x_{\rm c} + d$ for convenience.

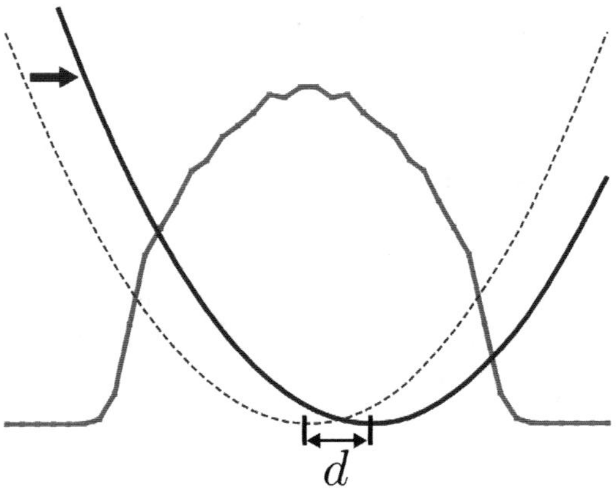

Figure 5. A cartoon picture of the way to drive the center of mass motion in optical lattice system.

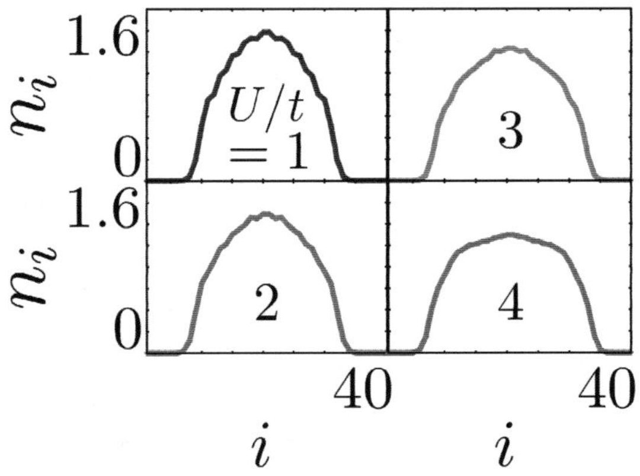

Figure 6. Initial states with $U/t = 1$, 2, 3, and 4 in the H_3 model. The horizontal and vertical axes represent the site number and the density of the fermions, respectively.

We display the time evolution of $\mathcal{X}$ (Fig. 7) and the number of doublons $N_{\rm d}$ (Fig. 8). One finds damped oscillating behaviors when $U/t = 1$ and 2 (Fig. 7). In addition, one also finds the damped oscillation of $N_{\rm d}$ (Fig. 8). From these doublon behaviors, we can interpret that the CoM motion is mainly governed by single fermion tunneling, because the number of doublons vary during the single fermion tunneling (Fig. 9). On the other hand, one finds the over-damped motion when $U/t = 3$ and 4 (Fig. 7). In these cases, $N_{\rm d}$ does not change oscillatory (Fig. 8). This implies that the over-damped motion is caused by the doublon tunneling, because this process does not effectively change $N_{\rm d}$ (Fig. 10). From these consideration, we expect that the doublon causes the over-damped motion. In fact, such a motion was also observed in attractive interaction cases, where the doublons are obviously stable due to the interaction. This scenario can be easily confirmed by a comparison between experiments and simulations in the attractive cases.

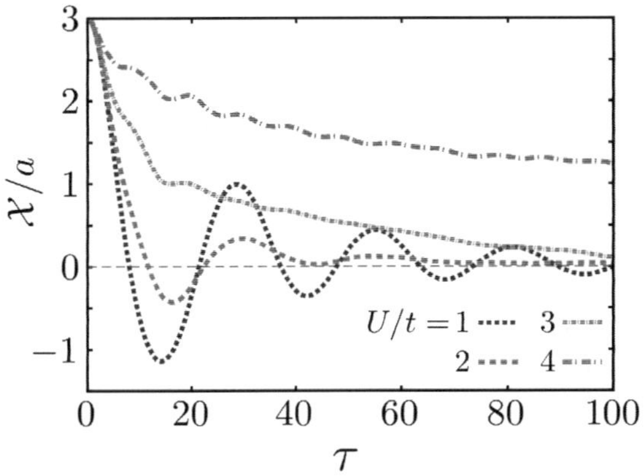

Figure 7. The time evolution of the CoM for the H_2 model with $U/t = 1$, 2, 3, and 4. The horizontal and vertical axes represent the normalized time and $\mathcal{X}$ defined in the text, respectively.

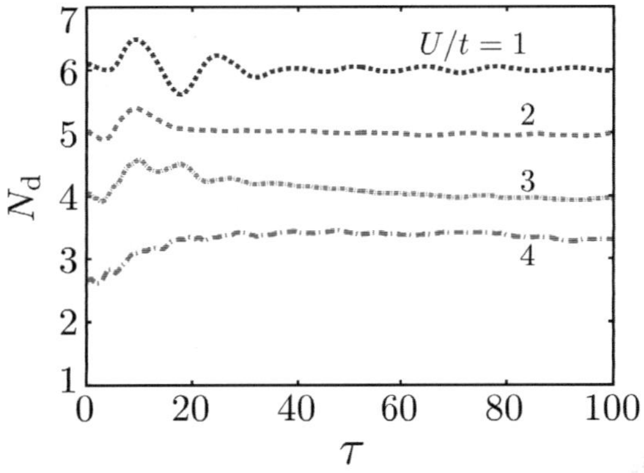

Figure 8. The time evolution of the number of the doublons for the H_2 model with $U/t =$ 1, 2, 3, and 4. The horizontal and vertical axes represent the normalized time and the number of the doublons defined in the text, respectively.

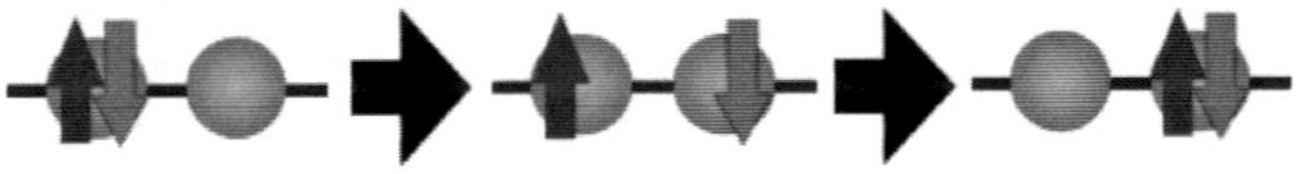

Figure 9. A cartoon picture of the single tunneling.

Figure 10. A cartoon picture of the doublon tunneling.

5. Time Evolution of Pair Correlation

In the previous section, the doublon, i.e., a bosonic excitation plays a crucial role in non-equilibrium physics in 1D fermionic Hubbard model. However, we have not examined "coherence" among many doublons yet. Then, we investigate the coherence of the doublons in this section. In order to study the time evolution of the coherence of doublons created by some strong disturbance, we examine the following Hamiltonian with AC field.

$$H_4 = H_{\mathrm{H}} + H_{W(\tau)}\,, \tag{8}$$

$$H_{W(\tau)} = W \sum_{i,\sigma} i\theta(\tau)\sin(\omega\tau) n_{i\sigma}\,, \tag{9}$$

where W is the amplitude of the AC field and ω is the angular frequency of the AC field (Fig. 11). In this study, we choose the following parameters, the system size $L = 10$, the number of particles $N_\uparrow = N_\downarrow = 5$, the interaction strength $U/t = 8$, the angular frequency $\omega/t^{-1} = 8$,

Journal of Physics: Conference Series **248** (2010) 012031 doi:10.1088/1742-6596/248/1/012031

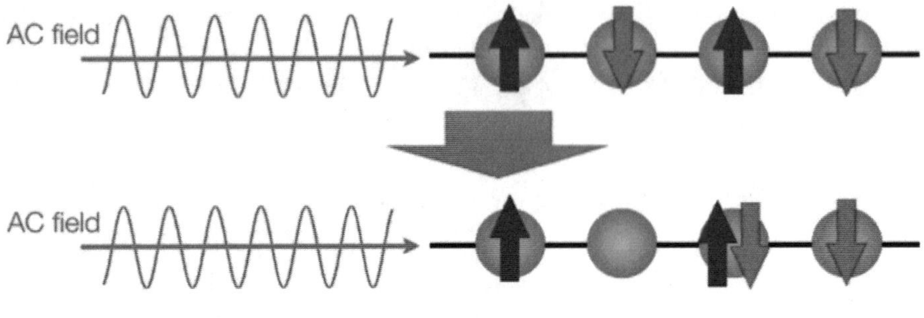

Figure 11. A cartoon picture of carriers (doublons and holons) created by the irradiation of AC field.

and the AC field strength $W/t = 1$. In such a system, we evaluate the time evolution of the magnetic (spin) and superconductivity (pair) correlation functions defined by

$$\Delta(r) = \frac{\sum_{|i-j|=r} \langle \Delta_i^\dagger \Delta_j \rangle}{\sum_{|i-j|=r}}, \tag{10}$$

$$S(r) = \frac{\sum_{|i-j|=r} \langle \boldsymbol{S}_i \cdot \boldsymbol{S}_j \rangle}{\sum_{|i-j|=r}}, \tag{11}$$

where

$$\Delta_i = c_{i\uparrow} c_{i\downarrow}, \tag{12}$$

$$\boldsymbol{S}_i = \begin{pmatrix} c_{i\uparrow}^\dagger & c_{i\downarrow}^\dagger \end{pmatrix} \boldsymbol{\sigma} \begin{pmatrix} c_{i\uparrow} \\ c_{i\downarrow} \end{pmatrix}. \tag{13}$$

Here, $\boldsymbol{\sigma}$ is the vector representation of the Pauli matrix. The following is the simulated situation. Firstly, we prepare the ground state of half-filled 1D Hubbard model H_{H} with L=10, U/t=8, which is the Mott insulating state. Secondly, we add the AC field from $\tau = 0$ to create carriers, i.e., doublons as well as holons (Fig. 11). In this paper, we choose just a resonant AC field frequency ($\omega/t = U/t = 8$). The doublons and holons are not created, if we take off-resonant frequency (not shown in this paper). During the carrier creation process, the number of the doublons increases as shown in Fig. 12. This implies that the system is finally excited far from the ground state by the AC field. Then, it is expected that the correlation in the excited state is considerably different from the ground state. Indeed, one finds that the pair correlation drastically develops (Fig. 13) in the far non-equilibrium-state while the spin-spin correlation does not vary significantly during the addition of the AC field (Fig. 12). This result indicates that the created doublons gains a coherence in non-equilibrium excited states. This non-trivial result can be obtained by the TDDMRG method allowing us to trace the time evolution of strongly correlated systems.

6. Summary

In this paper, we clarified some fundamental properties of 1D Hubbard model in non-equilibrium conditions. First, we demonstrated that the doublon is a (quasi-)stable particle when it is created once by pumping externally. Second, we exhibited that the doublon plays a crucial role in CoM motion of strongly correlated fermions in trapped optical lattices. Indeed, the doublon tunneling (co-tunneling) causes the overdamped motion of CoM. Finally, we presented that the created doublons gains coherence in non-equilibrium situations. Some of these results are detectable in the future optical lattice experiments. We believe that non-equilibrium condition in strongly correlated matters is a quite promising frontier of advanced physics.

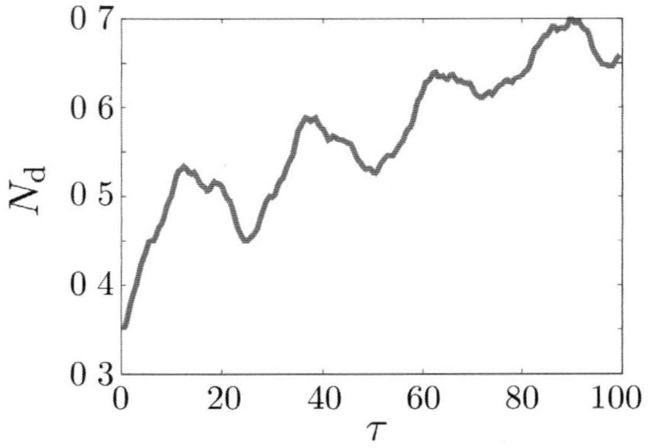

Figure 12. The time evolution of the number of the doublon by the irradiation of AC field. The horizontal and vertical axes show the normalized time and the number of doulons, respectively.

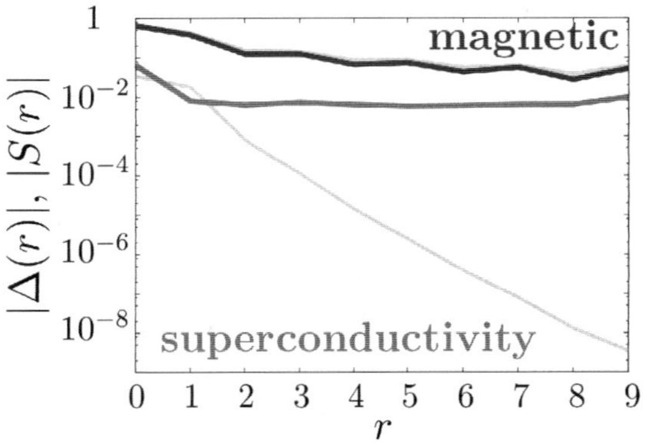

Figure 13. The magnetic (spin) (the upper blue line) and the superconductivity (pair) (the lower red line) correlation functions at $\tau = 100$. The horizontal and vertical axes show the distance r and the absolute value of the correlation functions, respectively. The value of these correlation functions at $\tau = 0$ are shown as green lines for comparison.

Acknowledgments
The authors wish to acknowledge CCSE stuff members for their supports on computer simulations in JAEA. M. M. was partially supported by Grant-in-Aid for Scientific Research on Priority Area "Physics of new quantum phase in superclean materials"(No. 20029019) from the Ministry of Education, Culture, Sports, Science, and Technology of Japan. S. Y. was also partially supported by Grant-in-Aid for Scientific Research (C) from Japan Society for the Promotion of Science (No. 20500044).

References

[1] Iwai S, Ono M, Maeda A, Matsuzaki H, Kishida H, Okamoto H and Tokura Y 2003, *Phys. Rev. Lett.* **91**, 057401
[2] Cavalleri A, Doors Th, Chong H H W, Kieffer J C and Schoenlein R W 2004 *Phys. Rev.* B **70** 161102(R)
[3] Okamoto H, Matsuzaki H, Wakabayashi T, Takahashi Y and Hasegawa T 2007 *Phys. Rev. Lett.* **98** 037401
[4] See for recent developments in theoretical study, e.g., Tsuji N, Oka T, and Aoki H, 2009 *Phys. Rev. Lett.* **103** 047403
[5] Bloch I, Dalibard J and Zwerger W 2008 *Rev. Mod. Phys.* **80** 885
[6] Lewenstein M, Sanpera A, Ahufinger V, Damski B, Sen(De) A and Sen U 2007 *Adv. Phys.* **56** 243
[7] Kubo R 1957 *J. Phys. Soc. Jpn.* **12** 570
[8] For recent development, see e.g., Rigol M (2009) *Phys. Rev. Lett.* **103** 100403
[9] Kinoshita T, Wenger T, and Weiss D S 2006 *Nature* **440** 900
[10] Fölling S, Trotzky S, Cheinet P, Feld M, Saers R, Widera A, Müller T, and Bloch I (2007) *Nature* **448** 1029
[11] Strohmaier N, Takasu Y, Günter K, Jördens R, Köhl M, Moritz H, Esslinger T (2007) *Phys. Rev. Lett.* **99** 220601
[12] Strohmaier N, Greif D, Jördens R, Tarruell L, Moritz H, Esslinger T, Sensarma R, Pekker D, Altman E, and Demler E (2010) *Phys. Rev. Lett.* **104** 080401
[13] Daley A J, Kollath C, Schollwöck U and Vidal G (2004) *J. Stat. Mech.* P04005
[14] White S R and Feiguin A E (2004) *Phys. Rev. Lett.* **93** 076401
[15] Schollwöck U (2005) *Rev. Mod. Phys.* **77** 25
[16] White S R (1992) *Phys. Rev. Lett.* **69** 2863
[17] White S R (1993) *Phys. Rev.* B **48** 10345

Journal of Physics: Conference Series **248** (2010) 012032 doi:10.1088/1742-6596/248/1/012032

$Mo_6S_3I_6$ molecular wires: from one-dimensional electron fluids to a self-organised critical self-assembled network

C Gadermaier, J Strle, M Uplaznik, D Vengust, B Bercic and D Mihailovic

Complex Matter Department, Jozef Stefan Institute, Jamova 39, 1000 Ljubljana, Slovenia

christoph.gadermaier@ijs.si

Abstract. We have studied the charge transport through flexible $Mo_6S_3I_6$ molecular wires of different diameters and found that it is dominated by two main phenomena. Tunnelling from Fermi liquid electrodes into the Tomonaga-Luttinger liquid electron system of the one-dimensional wires determines the behaviour of about half the measured samples, while for the other half environmental Coulomb blockade arising from deformations or imperfections is the main mechanism. The unique S-bonding chemistry of $Mo_6S_3I_6$ wire ends and gold nanoparticles enables self-assembly of large molecular-scale inorganic networks. These show an intrinsic tendency to self-organize into a scale-invariant critical state as found also in biological neural networks.

1. Introduction

Progress in molecular electronics faces two major challenges: finding the adequate molecular building blocks and assembling them into functional structures. While big advances have been made in molecular transistors and switches, molecular connectors are still elusive. In this paper we report on the electronic transport properties of $Mo_6S_3I_6$ molecular wires (MoSI MWs) [1,2] and show how they self-assemble into scale-invariant self-ordered critical networks upon co-dispersion with gold nanoparticles (GNPs).

2. Methods

MoSI MWs are prepared as described in [3], yielding stable dispersions containing MW bundles with diameters ranging from 1 to a few 100 nm. For transport measurements nickel electrodes were prepared by electron beam lithography. The MW bundles were deposited via dielectrophoresis - by placing a drop of solution over the electrodes and applying a 50 Hz AC electric field to the electrodes. The entire circuits were then annealed in vacuum at 700 °C for 1 h. AFM images of typical samples after annealing are shown in the insets of figures 1a and 2a.

For the network formation a dispersion in water was diluted and centrifuged to contain only thin bundles of 1-3 nm diameter. A GNP dispersion of 5 nm nominal diameter (British Biocell International) was added and the mixture was left to react and form a network for 3 days at 4 °C. Thus prepared samples are drop-cast onto muscovite mica and rapidly heat dried with a lamp. Mica was chosen since water has a low contact angle on the mica surface and during drying MWs and MW networks are deposited rather homogeneously, while on other surfaces they tend to form large aggregates, which compromises the network analysis. Samples were also prepared on a holey carbon substrate for examination of the MoSI-GNP connections by HRTEM.

3. Charge transport

We measured the I/V characteristics of about 30 samples for different temperatures from 20 K to room temperature and find that about half the samples consistently display a behaviour as depicted in figure 1 as an example of this group, while the others behave as shown in figure 2. For both groups we examine whether their behaviour can be described by one of the nonlinear transport processes known from the literature on low-dimensional conductors: tunnelling from Fermi liquid contacts into a Tomonaga-Luttinger liquid (TLL) [4,5], environmental Coulomb blockade (ECB) [6], and variable range hopping (VRH) [7].

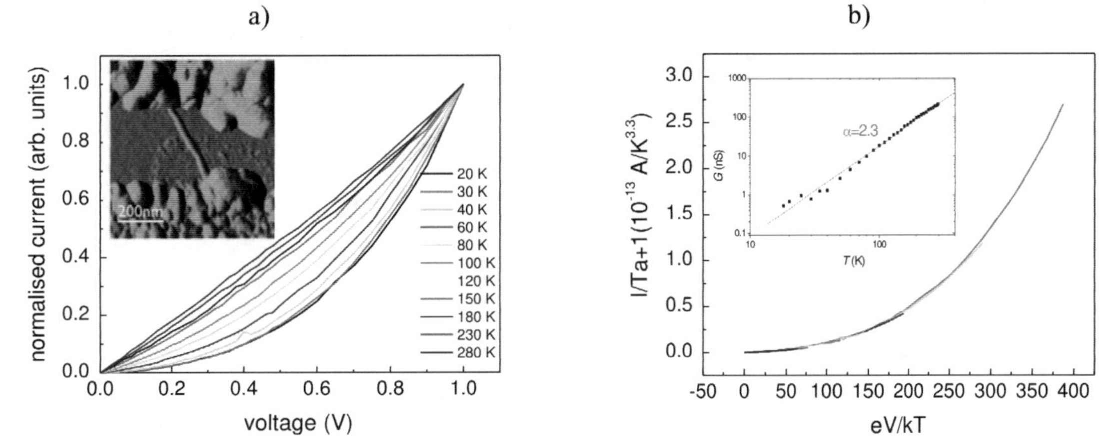

Figure 1. a) Normalized I-V characteristics at different temperatures. The current was arbitrarily normalized at 1 V for all curves. At 295 K, I_0=230 nA. The inset shows an AFM topographic image of the annealed circuit. b) A plot of $I/T^{\alpha+1}$ vs eV/kT, α=2.3. All curves collapse onto one (same colour code as a). The inset shows the zero-bias conductance (slope of $I-V$ curve at low bias, V=±20 mV) as a function of temperature and a fit to T^{α}.

TLL behaviour has been invoked in describing the transport in circuits containing carbon nanotubes [8], conjugated polymers at high currents [9], and $Li_2Mo_6S_6$ nanowires [10], the latter being close relatives of MoSI MWs. The characteristic temperature and field dependence of tunnelling into a TLL (see [4] for a detailed analytic description) makes the set of curves shown in figure 1a collapse onto one single curve when $I/T^{\alpha+1}$ (with the exponent α obtained from the power-law temperature dependence of the zero-field differential conductivity, see inset) is plotted against eV/kT, as shown in figure 1b. Given the excellent agreement with TLL and the poor agreement with both ECB and VRH we conclude that in the first group of samples the current limiting step, which defines the T-dependent I/V behaviour, is tunnelling from metallic contacts into the one-dimensional electron liquid of the MWs. For increasing bundle diameters we find a linear increase of the conductance rather than the quadratic one we expect if all MWs in the bundle conducted. We therefore conclude that only the outermost MWs contribute to charge transport and there is no transversal conduction (e.g. inter-wire hopping) that would make the inner bundles available for transport.

ECB is observed if an element in a circuit acts as a quantum dot, which accepts electrons only at the expense of some electrostatic energy. This leads to a characteristic kink in the I/V curve (see figure 2a), which translates a crossover between two regimes for the differential conductance around the kink (right panel). The second group of samples shows a behaviour that is dominated by ECB (while neither TLL nor VRH can describe the data). From the position of the kink we can estimate the capacity of the quantum dot, which for the sample shown in figure 2 is $C=e/2V \sim 5\times10^{-19}$ F. This is about 10 % of the capacity of the whole bundle between the two contacts [11], indicating that only a segment of the bundle forms the quantum dot. This segmentation can either arise from strong bending, or from stacking faults during the growth, as seen in the HRTEM image in figure 2b.

a)

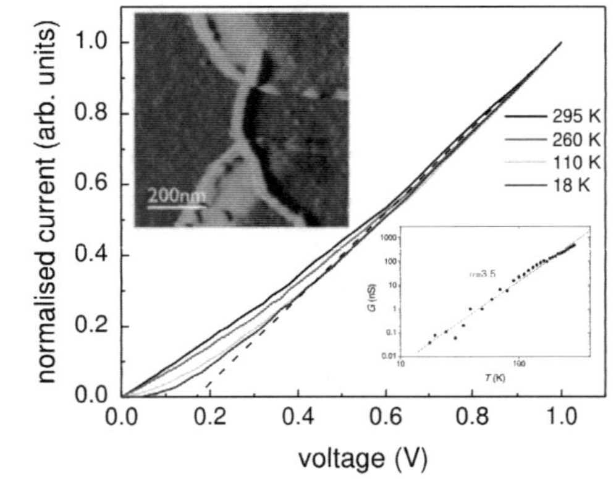

b)

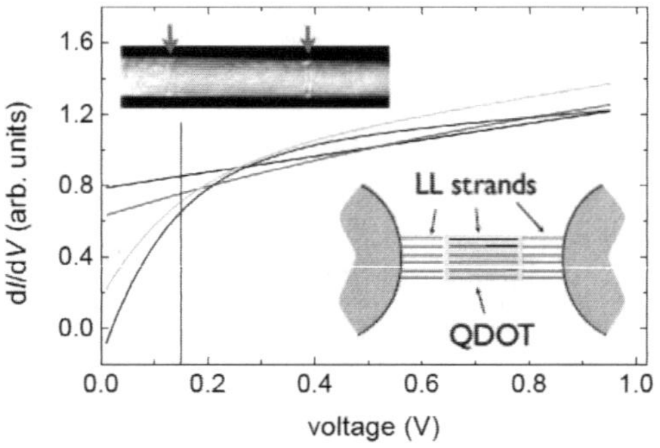

Figure 2. a) Normalized I-V at different temperatures. The current was arbitrarily normalized at 1 V for all curves. The dashed line is the tangent to the low temperature curve. The inset shows an AFM topographic image of the annealed circuit and slope of the $I-V$ curve at low bias, $V=\pm20$ mV as a function of temperature and a fit to T^{α}, $\alpha=3.5$. At 295 K, $I_0=620$ nA.

b) The derivatives dI/dV obtained from the normalized $I-V$ data. The derivative at low V increases as T increases from 18 to 300 K. The insert shows a high-resolution transmission electron microscope image of a MW bundle with a diameter of 124 nm. The defects in the structure are clearly visible as streaks across the wire which may cause the formation of a quantum dot. The vertical line marks the I=0 intercepts from the tangent in a).

We conclude that the MoSI MWs in a bundle are channels of TLL conductors. Only a fraction of the channels (probably the MWs on the surface) contribute to the conduction mechanism. In one group of bundles tunnelling from the Fermi liquid electrodes into the TLL is the rate-limiting step and therefore dominates the electronic transport properties. In the other group there are breaks in the conducting channels, which modify the TLL behaviour through ECB.

4. Network formation

The self-assembly of MW-GNP networks proceeds in a solution that contains approximately 1 thin NW bundle and 1 GNP per 10-100 μm^3 of water. The GNPs are more mobile and first attach to the molecular wire ends or defects, after which they act as anchoring spots for other wires and a three-dimensional (3D) network is formed. After deposition on the substrate, the water evaporates and the 3D network formed of flexible MWs collapses into a 2D shape that can be imaged with an AFM (see figure 3a). Many MWs are bent between vertices, with the convexity preferentially pointing towards the lower right corner. This indicates that bending is caused by a directed flow as the water droplet shrinks while the GNPs anchor on the surface. The anchoring also prevents orientation of the MWs along the three mica axes, as is observed for isolated bundles without GNPs. Once the solvent is removed, the MWs themselves also stick strongly to the substrate.

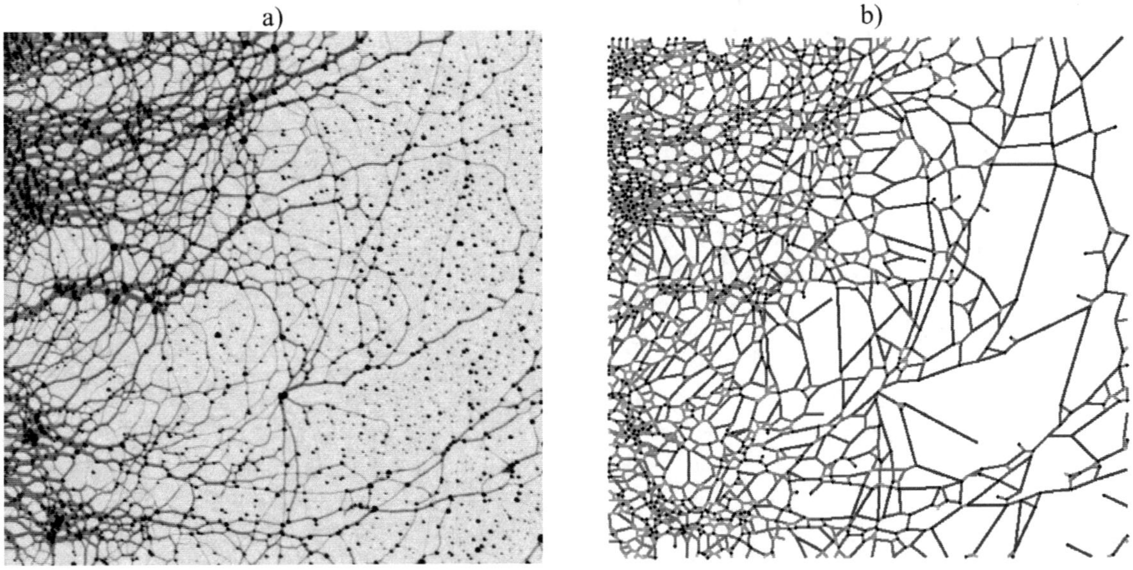

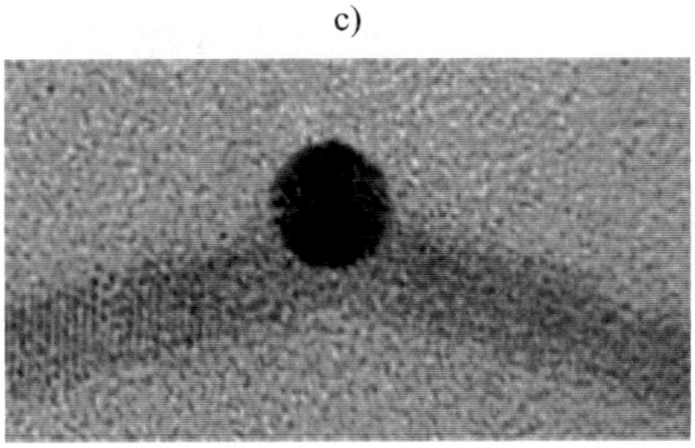

Figure 3. a) a) The raw AFM topographical image of a large self-assembled network on mica. The GNPs have a height of 5 nm and are connected with nanowires whose height ranges from 0.9 to 2.5 nm. The small dots believed to be individual MoSI clusters have a height of 1.4 nm and are not of interest here. b) A computer generated network based on the data shown in panel a is used for further analysis of the network properties. Line colours signify length of each segment and vary from light green (short) to blue (long). Vertexes with particles are marked with black dots. c) High-resolution transmission electron microscopy images of a GNP connected to a break in the MoSI bundle.

For statistical analysis the topographical AFM image was transformed into a mathematical graph, recording the vertex coordinates and connections between them (figure 3b). The graph consists of more than 2000 vertices and approximately 4000 edges connecting them across an area of 10 × 10 μm^2. Slightly less than half of these vertices contain a GNP, the others arise simply from branched bundles and from crossing of MWs as the 3D network formed in solution collapses into 2D.

From the analysis of the mathematical graph we extract the distribution of edge lengths (see figure 4b). While individual bundles show a log-normal distribution function (see figure 4a), the edge lengths in the graph follow a power law $P(x) \sim x^{\alpha}$ with $\alpha = -2.3 \pm 0.1$ for most of the measured length range, from 100 to around 3000 nm. The cut-off below 100 nm is a result of the finite lateral resolution of the AFM tip. Although we cannot measure the original 3D network properties in solution, we expect it to exhibit a similar power-law length distribution, while the exponent α may be different from what we obtain in 2D.

a)

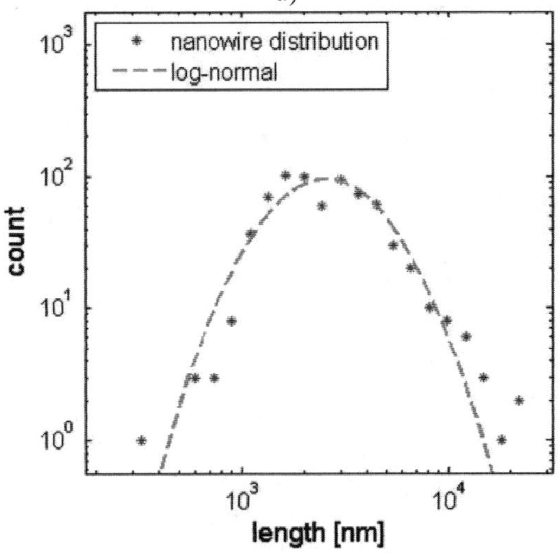

b)

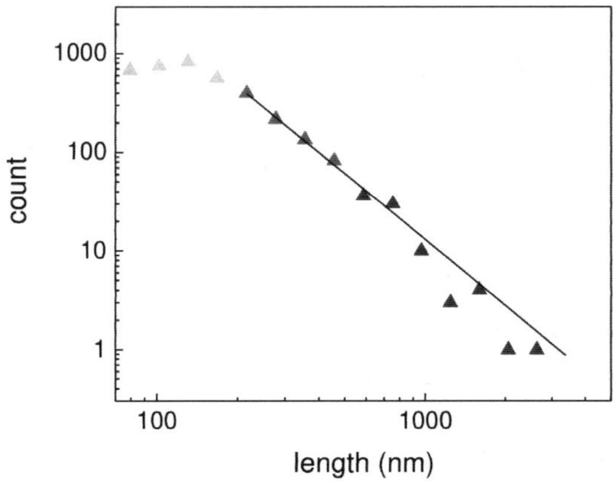

Figure 4. a) The length distribution of individual nanowires in an isopropanol solution can be described with log-normal distribution.

b) The length distribution of MW segments in the analyzed network with fitted power-law tail: P(x) $\sim x^{\alpha}$, $\alpha = -2.3 \pm 0.1$. The colour of the dots in the graph corresponds to the colour coding of the graph edges in figure 3b.

Since the AFM lateral resolution is limited to >20 nm, edge lengths significantly below 100 nm are not accurate. However, by closely examining smaller sections of the network we observed that some of the thicker bundles in the network are actually composed of numerous thinner wires forming an intricate web of their own [12]. Although the AFM resolution precludes mapping of these features, they suggest that shorter edges are much more frequent than they appear in our statistics and the observed power-law dependence may continue well below 100 nm.

The power-law behaviour of the edge lengths indicates that the network is scale-invariant over a large range. Such self-organized critical behaviour occurs in neural networks [13] and leads to the functionality of brains [14]. Neurons operate via electrical signals (action potentials) propagating through the network, where the vertices (the somas) act as nonlinear elements that process the signals, while the axons transmit the signals. A superficial analogy suggests that the GNPs might act as nonlinear elements (or be replaced other nanoparticles or molecules which do), while the MWs can transmit the electrical signals between them. However, the MoSI-GNP circuits are much smaller than biological neural networks, and charge transport is much faster since it is governed by Coulomb blockade at the nodes rather than the action of chemical neurotransmitters on neuron receptor sites. The faster charge transport and the shorter signal paths make the MoSI-GNP (or MoSI combined with other active nodes) networks potentially much faster than biological networks of equivalent complexity.

5. Conclusions

The present experiments show that non-linear low-dimensional transport dominates the behaviour of MoSI MWs and bundles formed from them. The TLL model with its distinctive data collapse of the I-V characteristics at different temperatures holds well for a significant proportion of the thin wire circuits. Another subset of thin wires shows clear signatures of the kinked I-V curves predicted by

ECB. Discontinuities in the wires either as a result of bending, and/or structural stacking faults within the wire are believed to cause the formation of quantum dots, which lead to the occurrence of ECB behaviour. MoSI MWs can be dispersed to thin enough bundles to allow covalent S bonding of their ends to individual molecules [15], so for the construction of molecular scale circuits, where thin and relatively short wires are of interest, they may potentially revolutionize molecular electronics. Our analysis of the first self-assembled inorganic conducting network shows that the system is in a self-ordered critical state. Such systems are amenable to sophisticated data processing, not least in the human brain. The MoSI-GNP networks are much smaller than typical neural networks, which bears the scope for low energy consumption and fast electronic processing speeds. The unique chemical connectivity enables to use a wide variety of nonlinear vertex elements with different functionalities.

Acknowledgements

C.G. acknowledges the Marie-Curie Reintegration Grant ERG-230975.

References

[1] Remskar M, Mrzel A, Skraba Z, Jesih A, Ceh M, Demsar J, Stadelmann P, Levy F and Mihailovic D 2001 *Science* **292** 479
[2] Mihailovic D 2009 *Prog. Mat. Sci.* **54** 309
[3] Nicolosi V, Vrbanic D, Mrzel A, McCauley J, O'Flaherty S, McGuinness C, Compagnini G, Mihailovic D, Blau W J and Coleman J N 2005 *J. Chem. Phys.* B **109** 7124
[4] Voit J 1995 *Rep. Prog. Phys.* **58** 977
[5] Deshpande V V, Bockrath , Glazman L I and Yacoby A 2010 *Nature* **464** 209
[6] Devoret M H and Grabert H 1992 *Single Charge Tunneling (NATO Advanced Study Institute, Series B: Physics* vol 294*)* ed H Grabert and M H Devoret (New York: Plenum)
[7] Fogler M M, Teber S and Shklovskii B I 2004 *Phys. Rev.* B **69** 035413
[8] Bockrath M, Cobden D H, Lu J, Rinzler A G, Smalley R E, Balents L, and McEuen P L 1999 *Nature* **397** 598
[9] Yuen J D, Menon R, Coates N E, Namdas E B, Cho S, Hannahs S T, Moses D and Heeger A J, 2009 *Nature Mater.* **8** 572
[10] Venkataraman L, Hong Y S and Kim P 2006 *Phys. Rev. Lett.* **96** 076601
[11] Uplaznik M, Bercic B, Remskar M and Mihailovic D 2009 *Phys. Rev.* B **80** 085402
[12] Strle J, Vengust D and Mihailovic D 2009 *Nano Lett.* **9** 1091
[13] Da Silva L, Papa A R R and De Souza A M C 1998 *Phys. Lett.* A **242** 343
[14] Eguiluz V M 2005 *Phys. Rev. Lett.* **94**, 018102
[15] Ploscaru M I, Jenko Kokalj S, Uplaznik M, Vengust D, Turk D, Mrzel A and Mihailovic D 2007 *Nano Lett.* **7** 1445

Tunneling and transport dynamics of trapped Bose-Einstein condensates

V O Nesterenko[1], A N Novikov [1], E Suraud [2] and J Kvasil [3]

[1] Bogoliubov Laboratory of Theoretical Physics, Joint Institute for Nuclear Research, Dubna, Moscow region, 141980, Russia

E-mail: nester@theor.jinr.ru

[2] Laboratoire de Physique Thoretique, Université Paul Sabatier, 118 Route de Narbonne, 31062 cedex, Toulouse, France

[3] Institute of Particle and Nuclear Physics, Charles University, CZ-18000, Praha, Czech Republic

Abstract. We present particular examples of tunneling and transport dynamics of Bose-Einstein condensate (BEC) in a two-well trap: Josephson oscillations (JO), macroscopic quantum self-trapping (MQST) and complete adiabatic transport. Being quite general, these dynamical regimes have close analogies with Josephson junctions in superconductors and transport problems in nanosystems. The calculations are performed within the time-dependent mean-field prescription within (transport) and beyond (JO and MQST) two-mode approximation. For the transport a universal adiabatic population transfer scheme is proposed, allowing a robust and complete transport even under strong nonlinear effects.

1. Introduction

Nowadays the trapped Bose-Einstein condensate (BEC) is widely recognized as a source of new fascinating physics and remarkable cross-over with other areas, see e.g. monographs [1, 2] and recent reviews [3, 4, 5, 6]. In this broad field, the dynamics of weakly bound BECs in two-well systems is now of a keen interest. Its fundamental features and numerous links to other subjects (Josephson junctions [3, 7, 8, 9, 10, 11], optical lattices [3, 4, 6], transport problems [12, 13, 14, 15, 16, 17], topological states [18, 19] etc.) are indeed of a general interest.

In this paper we present some particular examples of tunneling and transport dynamics of BEC in a two-well trap, which might be interesting not only for BEC community but also for experts from various branches of nanophysics. First, Josephson oscillations (JO) of BEC are considered and compared with those in superconductor Josephson junctions (SJJ) [9]. Then the macroscopic quantum self-trapping (MQST) is exhibited [7, 8]. This effect takes place in Bose Josephson junction (BJJ) but not in the SJJ [9]. Finally, the adiabatic transport of BEC between the trap wells is considered. Usually such transport heavily suffers from the nonlinearity caused by the interaction between BEC atoms [20, 21]. To overcome this trouble, we propose some universal adiabatic transport protocols [16, 17] which work even at strong nonlinearity. The protocols are developed on the basis of Landau-Zener [22] and Rosen-Zener [23] schemes.

The calculations were performed with the Gross-Pitaevski equation [24]. It was directly solved in 3D coordinate grid space for JO and MQST and treated in the two-mode approximation [7, 8, 9] for more complicated transport problem.

2. Formalism

The calculations were performed in the mean-field approximation by solving the time-dependent Gross-Pitaevski equation [24]

$$i\hbar\dot{\Psi}(\vec{r},t) = [-\frac{\hbar^2}{2m}\nabla^2 + V(\vec{r},t) + g_0|\Psi(\vec{r},t)|^2]\Psi(\vec{r},t) \qquad (1)$$

where $\Psi(\vec{r},t)$ is the classical order parameter of the system, $V(\vec{r},t)$ is the double-well trap potential, $g_0 = 4\pi\hbar^2 a_s/m$ is the parameter of interaction between BEC atoms, a_s is the s-scattering length and m is the atomic mass.

For JO and MQST, the static confinement $V(\vec{r})$ is used and Eq. (1) is directly solved by implementation of Hartree-Fock (HF) static and time-dependent algorithms for 3-dimensional systems [25]. Only the lowest level occupied by BEC is involved into HF procedure. Note, that in most of the previous JO and MQST studies [7, 8, 9, 10] performed within the two-mode approximation (TMA), the energy of the ground state E_1 and corresponding barrier penetrability K were approximated by constants independent on the number of BEC atoms N and interaction g_0. In the present approach, the energy and penetrability are calculated self-consistently and depend on N and g_0. Note that the difference between TMA and HF results at a strong interaction can be essential [26].

For more complicated transport problem with the time-dependent confinement in a double-well trap, we however are enforced to use the TMA [7, 8, 9]:

$$\Psi(\vec{r},t) = \sqrt{N}(\psi_1(t)\Phi_1(\vec{r}) + \psi_2(t)\Phi_2(\vec{r})) \qquad (2)$$

where $\Phi_k(\vec{r})$ is the static ground state solution of (1) for the isolated k-th well ($k = 1, 2$) and

$$\psi_k(t) = \sqrt{N_k(t)}e^{i\phi_k(t)} \qquad (3)$$

is the expansion amplitude expressed via the normalized population $N_k(t)$ and phase $\phi_k(t)$. The total number of atoms N is fixed: $\int d\vec{r}|\Psi(\vec{r},t)|^2/N = N_1(t) + N_2(t) = 1$.

Substituting (2)-(3) to (1) and integrating out the spatial distributions $\Phi_k(\vec{r})$ we get the system of equations for the populations $N_k(t)$ and phases $\phi_k(t)$ [8]:

$$\dot{N}_k = -2\Omega(t)\sqrt{N_j N_k}\sin(\phi_j - \phi_k)\,, \qquad (4)$$

$$\dot{\phi}_k = -[E_k(t) + UN_k] + \Omega(t)\sqrt{\frac{N_j}{N_k}}\cos(\phi_j - \phi_k) \qquad (5)$$

with $j \neq k$. Here

$$\Omega(t) = -\frac{1}{\hbar}\int d\vec{r}\,[\frac{\hbar^2}{2m}\nabla\Phi_1^* \cdot \nabla\Phi_2 + \Phi_2^* V(t)\Phi_1] \qquad (6)$$

is the coupling between BEC fractions (= barrier penetrability),

$$E_k(t) = \frac{1}{\hbar}\int d\vec{r}\,[\frac{\hbar^2}{2m}|\nabla\Phi_k^*|^2 + \Phi_k^* V(t)\Phi_k] \qquad (7)$$

is the depth of the k-th well and

$$U_k = \frac{g_0}{2\hbar}\int d\vec{r}\,|\Phi_k|^4 \qquad (8)$$

labels the interaction between BEC atoms in the same well (here $U_1 = U_2 = U$). The values $\Omega(t)$, $E_k(t)$ and U have the dimension of frequency.

Considering N_k and $-\phi_k$ as conjugate variables and using the linear canonical transformation [21], $z = N_1 - N_2$ and $\theta = \frac{1}{2}(\phi_2 - \phi_1)$, one may extract the integral of motion N and the total phase, thus reducing the problem to finding the population imbalance z and phase difference θ:

$$\dot{z} = -\bar{\Omega}(t)\sqrt{1-z^2}\sin 2\theta \quad (9)$$

$$\dot{\theta} = \frac{1}{2}[\Delta(t) + \Lambda z + \bar{\Omega}(t)\frac{z}{\sqrt{1-z^2}}\cos 2\theta] \quad (10)$$

where

$$\bar{\Omega}(t) = \frac{\Omega(t)}{K}, \quad \bar{E}_k(t) = \frac{E_k(t)}{2K}, \quad \Delta(t) = \bar{E}_1(t) - \bar{E}_2(t), \quad \Lambda = \frac{UN}{2K} \quad (11)$$

are the dimensionless coupling, well depths, detuning and interaction, scaled by the coupling amplitude K. The time in (9)-(10) is also scaled by $2K$ and so is dimensionless. The key parameter Λ representing the interaction-coupling ratio is decisive in the tunneling dynamics. In JO and MQST problems, we use the static symmetric trap with $\bar{\Omega}(t)$=1 and $\Delta(t)$=0 while for transport we deal with time-dependent $\bar{\Omega}(t)$ and $\Delta(t)$.

Eqs. (9)-(10) can be also derived as canonical equations for the conjugate variables z and θ and the classical Hamiltonian

$$\mathrm{H_{cl}} = \frac{1}{2}[\Delta(t)z + \frac{\Lambda}{2}z^2 - \bar{\Omega}(t)\sqrt{1-z^2}\cos 2\theta] \ . \quad (12)$$

3. JO and MQST

Josephson Oscillations and Macroscopic Quantum Self-Trapping represent the main regimes of BEC tunneling dynamics in a double-well trap. A weak coupling of BEC fractions through the barrier is assumed. Here we consider the static trap $V(x)$ of symmetric configuration (Δ=0). The static barrier penetrability is $\Omega = K$, i.e. $\bar{\Omega}$=1. JO are explored for the ideal BEC (no interaction, $U = 0$) while MQST is discussed for the repulsive interaction between BEC atoms ($U > 0$). The calculations are performed for N=100 BEC atoms. The initial state is computed with $\sim$ 15000 static iterations.

Figure 1a) demonstrates the double-well confinement $V(x)$, the energy spectrum E_n and the initial density $\rho(x)$ of the ideal BEC. Since the isolated wells are described by equal oscillator potentials with the frequency ω_0 and oscillator length a_0, it is convenient to give the values in Fig. 1 in natural units of length a_0, frequency ω_0 and time ω_0^{-1}. The h.o. levels below the barrier are known to be split into two sublevels [27]. Magnitude of the splitting ΔE is determined by the barrier penetrability K at the level energy, and related to the JO frequency. In our case of weak coupling, the splitting of the lowest h.o. level, $\Delta E = E_2 - E_1 \sim 0.01\omega_0$, is very small and not visible in the figure. Actually the sublevels E_n, n=1-5, look in Fig. 1a as 2 levels.

In Fig. 1, the time evolution starts from the non-stationary state ($N_1 = 1, N_2 = 0, z = N_1 - N_2 = 1$) and the tunneling gives the non-stop Josephson or Rabi oscillations, see Fig. 1b. The non-stop regime arises for the energy conservation reason. If the system is isolated and BEC energy (12) is conserved, then the non-stationary state with initial $z(0) = 1, \theta(0) = 0$ is unable to relax into the stationary state ($N_1 = N_2 = 0.5, z = 0$) with another energy, hence endurant oscillations. For ideal BEC in the symmetric trap, the JO frequency is $\omega_{BJJ} = 2K$ [8], i.e. is determined by the barrier penetrability at $E_1 \approx E_2$. Fig. 1b demonstrates JO for two well separations, $d = 5$ and $6a_0$. For $d = 6a_0$, the barrier at $E_1 \approx E_2$ is wider and so ω_{BJJ} is smaller.

It is instructive to compare the dynamics in Bose and superconductor Josephson junctions, i.e. BJJ vs SJJ [3, 28]. There are obviously some similarities: in both cases we deal with tunneling between left and right states and get the alternating current (ac). At the same time, there are significant differences. First, in the above example, BJJ gives ac for Δ=0, i.e. without

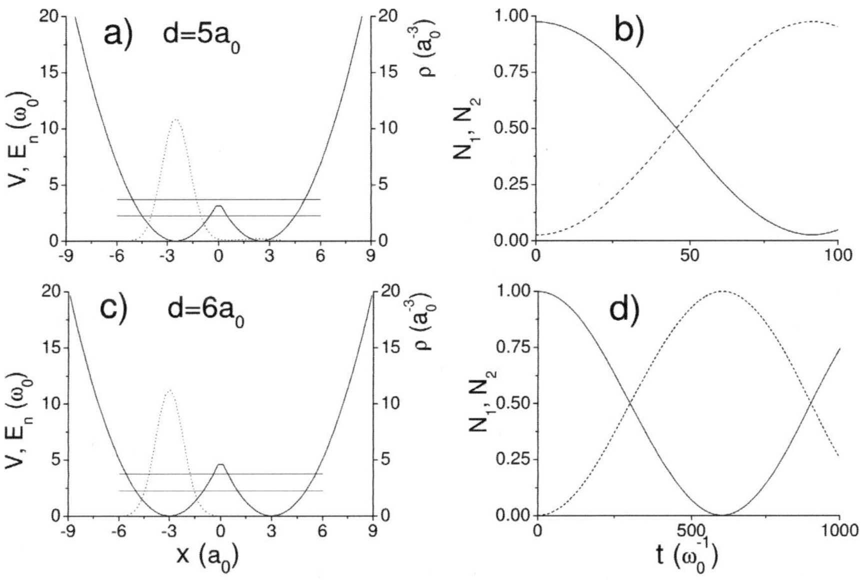

Figure 1. Mean-field characteristics (left) and Josephson oscillations (right) of the ideal BEC in a double-well trap with well separations $d = 5a_0$ (upper) and $d = 6a_0$ (bottom). The left plots exhibit confinement field $V(x)$ (solid curve), lowest 5 energy levels E_n (solid horizontal lines) and initial BEC density $\rho(x)$ (dash curve). The right plots give populations $N_1(t)$ (solid curve) and $N_2(t)$ (dotted curve) of the fist and second wells, respectively.

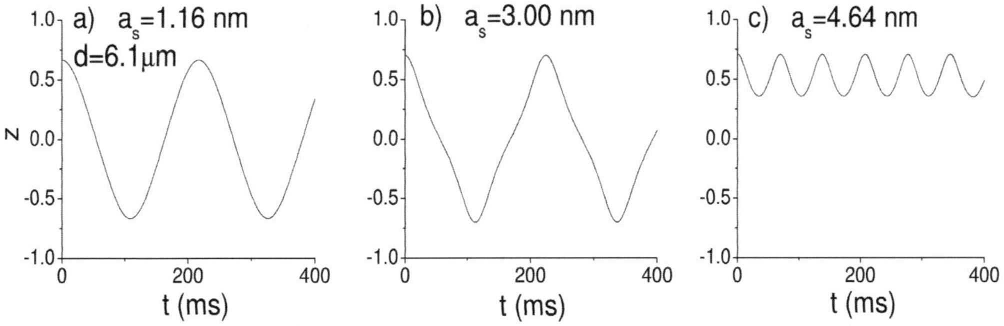

Figure 2. The population imbalance $z(t)$ in quasiharmonic (a), anharmonic (b), and MQST (c) tunneling regimes for scattering lengths a_s as indicated. The well separation is $d = 5a_0 = 6.1\mu m$ for all the plots.

the potential difference. Instead, SJJ has ac only at nonzero applied voltage $eV_s = E_1 - E_2$ and gives the direct current (dc) for V_s=0 , see e.g. [30, 31]. Second, the BJJ ac frequency $\omega_{BJJ} = 2K$ is determined by the barrier penetrability, i.e. by the system configuration. In contrast, the SJJ ac frequency $\omega_{SJJ} = 2eV_s/\hbar$ is determined by the applied voltage and does not depend on the system configuration.

Actually, the difference between BJJ and SJJ oscillations can be put away by considering Eqs. (9)-(10) in the limit $K \ll \Delta$ instead of $K \gg \Delta$ used above [8]. Then the BJJ frequency is reduced to $\omega_{BJJ} = E_1 - E_2$ in accordance to the SJJ case. Note that the limit $K \ll \Delta$ is typical for SJJ while BJJ can deal with both $K \ll \Delta$ and $K \gg \Delta$ options. The more detailed comparison of BJJ and SJJ ac can be found in the recent study [32].

Another physical situation arises in BJJ if the interaction between BEC atoms comes to play. This is demonstrated in Fig. 2 for different values of the repulsive interaction characterized by the scattering length a_s. Here, to give an idea of the typical scales, the values a_s, d and time t are given in standard units. Fig. 2 shows that increasing the interaction leads to anharmonicity of oscillations (plot b)) and finally to the self-trapping regime (MQST) with the time average $< z > \approx 0.5$ (plot c)). In the later, BEC oscillates not around the stationary state with $< z >=0$ like in plots a)-b) but around the non-stationary initial population with nonzero $< z >$. Thus the main BEC part is permanently trapped in the left well. The JO and MQST for the repulsive BEC were observed in experiment [11]. MQST takes place at sufficiently strong interaction or, at fixed $\Lambda > 0$, at a large initial population imbalance $|z(0)|$ [7, 8, 9]. Though the self-trapping or self-locking is a general phenomenon existing in many physical systems [29], it occurs in the BJJ but not in SJJ. In BJJ the effect depends on the initial phase difference, e.g. is quite different for $\theta(0)=0$ (present case) and π [9].

4. Adiabatic transport

The transport of BEC in a two-well trap assumes that BEC, being initially in one of the wells, is then completely transferred in a controllable way to the target well and kept there. In principle such a transport can be produced by many methods, see discussion in [16]. However, most of these methods do not work even at modest nonlinearity. In this connection, we have recently proposed a novel adiabatic transport protocol [16] based on the generalization of Landau-Zener [22] and Rosen-Zener [23] schemes. The protocol is robust to a strong nonlinearity and can be applied to both repulsive and attractive BEC.

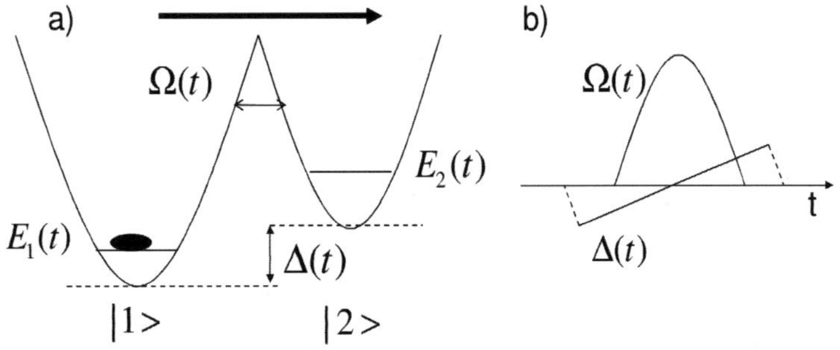

Figure 3. a) The double-well trap with time dependent ground state energies $E_1(t)$ and $E_2(t)$, detuning $\Delta(t)$, and coupling (barrier penetrability) $\Omega(t)$. The BEC is initially placed at the first (left) well as indicated by the bold dot. Then it is transferred to the 2nd (right) well as indicated by the bold arrow. b) Time evolution of the coupling $\Omega(t)$ and detuning $\Omega(t)$.

The protocol scheme is illustrated in Fig. 3. As seen from the plot b), the transport is driven by simultaneous varying the well depths and coupling (barrier penetrability). At the initial time, the trap is asymmetric ($E_1 < E_2$) and the barrier is actually impenetrable (Ω=0). Then the well depths evolve to get the opposite asymmetry $E_1 > E_2$. Simultaneously the barrier narrows and becomes penetrable, providing the coupling between BEC fractions in the neighboring wells. Just that time the BEC transfer occurs. Finally, the well evolution stops, the barrier becomes again impenetrable (Ω=0) and the transport is over. In the present study, the evolution of wells and coupling are linear and Gaussian, respectively,

$$\Delta(t) = \bar{E}_1(t) - \bar{E}_2(t) = \alpha t, \quad \bar{\Omega}(t) = \exp\{-\frac{(\bar{t}-t)^2}{2\Gamma^2}\}, \tag{13}$$

where α is the detuning rate and $\bar{t}$ and Γ are the centroid and width parameters. Obviously, this protocol is a generalization of the Landau-Zener (linear change of $\Delta(t)$ and constant $\bar{\Omega}$)

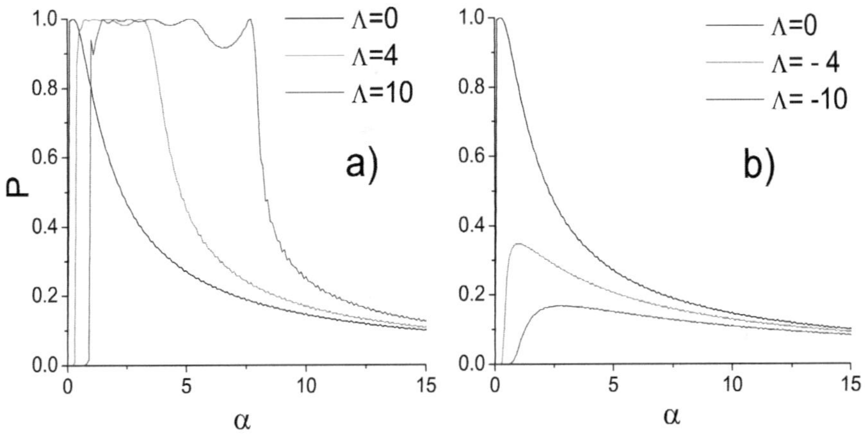

Figure 4. Dependence of the final adiabatic population P of the 2nd well on the detuning rate α and nonlinearity Λ for the repulsive (a) and attractive (b) BEC.

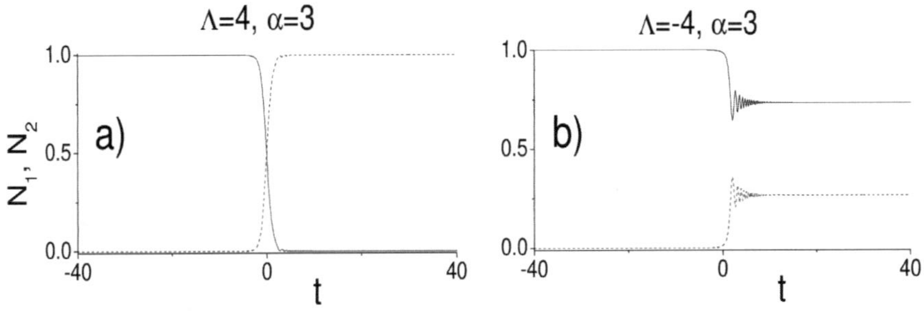

Figure 5. Time evolution of populations N_1 (solid curve) and N_2 (dash curve) of the first and second wells for the repulsive (a) and attractive (b) BEC at the detuning $\alpha = 3$.

and Rosen-Zener (constant Δ and pulse $\bar{\Omega}(t)$) schemes [22, 23]. In the experimental setup, the drives (13) can be routinely produced by imposing an additional modulation field and changing the separation between the wells.

Results of the calculations in the two-mode approximation are given in Figs. 4 and 5 [16]. Fig. 4 shows the final transport result, the population $P = N_2(t \to \infty)$ of the 2nd well, at various detuning rate α and nonlinearity Λ. It is seen that for ideal BEC (Λ=0) the complete adiabatic transport is possible only at very slow detuning rate, $\alpha \to 0$. The repulsion interaction ($\Lambda > 0$) creates a plateau in $P(\alpha)$ dependence where the transport is robust and complete. The stronger the BEC interaction, the wider the plateau. Hence in the present protocol, the interaction and related nonlinearity even favor the transport by enlarging the active α interval. At larger rates, say $\alpha > 7$ for $\Lambda = 10$, the process becomes too fast and the adiabatic transfer vanishes. Note a narrow $P = 0$ window at small rates α in the interacting BEC. A sharp rise of the transfer at the right window boundary can be used for switching the transport by a small tuning of α.

The robustness of the transport of the repulsive BEC is justified in Fig. 5a for the particular case of $\Lambda = 4$ and $\alpha = 3$. It is seen that indeed the population $N_1 = 1, N_2 = 0$ at early time is transformed to the population $N_1 = 0, N_2 = 1$ at late time.

An opposite situation takes place for the attractive BEC with $\Lambda < 0$. Figs. 4b and 5b show that the complete transport fails, and the stronger the nonlinearity, the more the failure. There is no any plateau, the final population P of the interacting BEC is small, and main part of BEC still remains in the initial well.

However, due to the symmetry of the problem (9)-(10)

$$\Lambda \to -\Lambda, \quad \alpha \to -\alpha, \quad \theta \to -\theta + \frac{\pi}{2} \tag{14}$$

or

$$\alpha \to -\alpha, \quad z \to -z, \quad \theta \to -\theta\,, \tag{15}$$

the transport protocol can be recast and provide again the robust and complete transfer [16]. Actually, (14) relates the transport protocols for repulsive and attractive BEC in one direction while (15) connects the transport in opposite directions (with corresponding interchange of the initial conditions). Specifically the left-right transport $|1> \to |2>$ with the initial populations $N_1 = 1, N_2 = 0$ should use $\alpha > 0$ for $\Lambda > 0$ and $\alpha < 0$ for $\Lambda < 0$ where $\alpha < 0$ means that the initially occupied left well should be higher than the right one. Instead the right-left transport $|1> \leftarrow |2>$ with the initial populations $N_1 = 0, N_2 = 1$ should use $\alpha < 0$ for $\Lambda > 0$ and $\alpha > 0$ for $\Lambda < 0$.

The above discussion demonstrates a universal character of our transport protocol. It can be also used for other scenarios, e.g. for the transfer of BEC between its two components in a single-well trap. In this case BEC contains atoms in two different hyperfine levels, thus forming two components. The time-dependent laser pulse $\Omega(t)$ couples the components and initiate the population transfer. The evolution of the detuning is produced by the additional Stark pulse. Such scheme is already used in the Stark Chirped Rapid Adiabatic Passage (SCRAP) method [33] for the population transfer in atoms and simple molecules.

5. Conclusions

The examples of tunneling dynamics in a double well system (Josephson oscillations, macroscopic quantum self-trapping and adiabatic transport) were illustrated and discussed. Being very general, these kinds of dynamics can be met in various systems including those of nano size. In particular, they take place in two-level systems with the laser coupling where they are called as Rabi oscillations, self-locking and adiabatic population transfer. Comparison of these dynamical regimes in different physical situations can be extremely useful for better understanding the basic physics behind. In this connection, the trapped Bose-Einstein condensate which allows an effective control on the process parameters and thus accesses various physical scenarios can serve as a unique laboratory of fascinating physics with numerous crossover with other topics.

Acknowledgments

The work was supported by the grants 08-0200118 (RFBR, Russia), 684 (Université Paul Sabatier, Toulouse, France, 2008), 202/06/0363 and MSM 0021620859 (Ministry of education of the Czech Republic), and Votruba - Blokhintsev (Czech Republic - BLTP JINR). We are grateful to A Yu Cherny and V I Yukalov for help and useful discussions.

References

[1] Pitaevskii L and Stringari S 2003 *Bose-Einstein Condensation* (Oxford: Oxford University Press).
[2] Petrick C J and Smith H 2002 *Bose-Einstein Condensation in Dilute Gases*, (Cambridge: Cambridge University Press).
[3] Pitaevskii L 2006 *Usp. Fiz. Nauk* **176** 345
[4] Morsch O and Oberthaler M 2006 *Rev. Mod. Phys.* **78** 235
[5] Bloch I, Dalibard J, and Zwerger W 2008 *Rev. Mod. Phys.* **80** 885
[6] Yukalov V I 2009 *Laser Phys.* **19** 1
[7] Milburn G J, Corney J, Write E M, and Walls D F 1997 *Phys. Rev. A* **55** 4318
[8] Smerzi A, Fantoni S, Giovanazzi S, and Shenoy S R 1997 *Phys. Rev. Lett.* **79** 4950
[9] Raghavan S, Smerzi A, Fantoni S, and Shenoy S R 1999 *Phys. Rev. A* **59** 620
[10] Holthaus M and Stenholm S 2001 *Eur. Phys. J.* **20** 451

[11] Albiez M, Gati R, Fölling J, Hunsmann S, Cristiani M and Obrthaler M K 2005 *Phys. Rev. Lett.* **95** 010402
[12] Liu J, Fu L, Ou B-Y, Chen S-G, Wu B and Niu Q 2002 *Phys. Rev. A* **66** 023404
[13] Weiss C and Jinasundera Th 2005, *Phys. Rev. A* **72** 053626; Weiss C 2006, *Laser Phys. Lett.* **3** 212
[14] Zhang Q, Hanggi P and Gong J B 2008 *Phys. Rev. A* **77** 053607
[15] Nistazakis H E, Rapti Z, Frantzeskakis D J, Kevrekidis P G, Sodano P and Trombettoni A 2008 Phys. Rev. A **78** 023635
[16] Nesterenko V O, Novikov A N, Cherny A Yu, Souza Cruz F F and Suraud E 2009 *J. Phys. B: At. Mol. Opt. Phys.* **42** 235503
[17] Nesterenko V O, Novikov A N and Suraud E 2010 *Laser Phys.* **20** 1149
[18] Williams J, Walser R, Cooper J, Cornell E A and Holland M 2000 *Phys. Rev. A* **61** 033612
[19] Yukalov V I, Marzlin K-P and Yukalova E P 2004 *Phys. Rev. A* **69** 023620
[20] Graefe E M, Korsch H J and Witthaut D 2006 *Phys. Rev. A* **73** 013617
[21] Nesterenko V O, Novikov A N, de Souza Cruz F F and Lapolli E L 2009 *Laser Phys* **19** 616
[22] Landau L D 1932 *Phys. Z. U. S. S. R.* **2** 46 Zener C 1932 Proc. *R. Soc. London, Ser. A* **137** 696
[23] Rosen N and Zener C 1932 *Phys. Rev.* **40** 502
[24] Pitaevskii L P 1961 *Sov. Phys. JETF* **13** 451; Gross E P 1961 *Nuovo Cim.* **20** 454
[25] Calvayrac F, Reinhard P-G, Suraud E and Ullrich C A 2000 Phys. Rep. **337** 493
[26] Ananikian D and Bergeman T 2006 *Phys. Rev. A* **73** 013604
[27] Landau L D and E.M. Lifshitz E M *Quantum mechanics: non-relativistic theory*, 3rd ed. (Pergamon Press)
[28] Legett A J 2001 *Rev. Mod. Phys.* **73** 307
[29] Eilbeck J C, Lomdahl P S and Scott A C 1985 *Physica D* **16** 318
[30] Barone A and Paterno G 1982 *Physics and Applications of the Josephson Effect* (Wiley, New York)
[31] Zharkov G F *Sov. Phys. Usp.* **9** 198
[32] Radzihovsky L and Gurarie V 2010 *Phys. Rev. A* **81** 063309
[33] Yatsenko L P, Shore B W, Halfmann T, Bergmann K and Vardi A 1999 *Phys. Rev. A* **60** R4237

Journal of Physics: Conference Series **248** (2010) 012034 doi:10.1088/1742-6596/248/1/012034

Persistent current noise

Andrew G. Semenov[1] **and Andrei D. Zaikin**[2,1]

[1] I.E.Tamm Department of Theoretical Physics, P.N.Lebedev Physics Institute, 119991 Moscow, Russia

[2] Institut für Nanotechnologie, Karlsruher Institut für Technologie (KIT), 76021 Karlsruhe, Germany

E-mail: semenov@lpi.ru

Abstract. We demonstrate that persistent current in meso- and nanorings may fluctuate down to zero temperature provided the current operator does not commute with the total Hamiltonian of the system. For a model of a quantum particle on a ring we explicitly evaluate PC noise power which has a form of sharp peaks which become broadened for multi-channel rings or in the presence of dissipation. PC noise can be tuned by an external magnetic flux which is a fundamental manifestation of quantum coherence in the system.

Persistent currents (PC) in normal meso- and nanorings pierced by external magnetic flux is one of fundamental consequences of quantum coherence of electrons. While the average value of PC was intensively investigated in the literature during last decades, not much is known about equilibrium fluctuations od PC. At non-zero T it is quite natural to expect non-vanishing thermal fluctuations of PC. But at $T \to 0$ the system approaches its (non-degenerate) ground state and, hence, at the first sight no PC fluctuations would be possible. Below we will demonstrate that this naive conclusion is in general not quite correct: No PC fluctuations are expected in the zero temperature limit only provided the current operator commutes with the total Hamiltonian of the system, otherwise fluctuations of persistent current occur even in the ground state exactly at $T = 0$.

Let us consider a simple model of a quantum particle with mass M on a 1d ring of radius R pierced by magnetic flux Φ_x. The particle position on the ring is parameterized by the angle θ, which is the quantum mechanical variable of interest in our problem. The Hamiltonian for this system reads

$$\hat{H} = \frac{(\hat{\phi} + \phi_x)^2}{2MR^2} + U(\theta) + \hat{H}_{int}(\theta, X), \tag{1}$$

where $\hat{\phi} = -i\frac{\partial}{\partial\theta}$ is the (dimensionless) flux operator, $U(\theta)$ defines the potential profile for our particle, $\phi_x = \Phi/\Phi_0$ and Φ_0 is the flux quantum. The collective variable X represents all other degrees of freedom interacting with the particle. Note that within our model the interaction term involves only the coordinate (not momentum) operator. Such models were previously studied in a number of papers [1, 2, 3, 4, 5, 6] in the context of various dissipative environments and, in addition, could be of interest for the problem of PC in superconducting nanorings in the presence of quantum phase slips [7].

Let us define the current operator in the Schrödinger representation

$$\hat{I} = \frac{e}{2\pi}\dot{\hat{\theta}} = \frac{ie}{2\pi}[\hat{H}, \hat{\theta}] = \frac{e(\hat{\phi} + \phi_x)}{2\pi MR^2}. \tag{2}$$

In the Heisenberg picture the current operator is given by standard formula $\hat{I}(t) = e^{it\hat{H}}\hat{I}e^{-it\hat{H}}$. Let us also introduce the equilibrium current-current correlation function $\langle \hat{I}(t)\hat{I}(0) + \hat{I}(0)\hat{I}(t)\rangle$ and define PC noise power

$$S(t) = \langle \hat{I}(t)\hat{I}(0) + \hat{I}(0)\hat{I}(t)\rangle - \langle \hat{I}\rangle^2 = \int \frac{d\omega}{2\pi} S_\omega e^{-i\omega t}. \tag{3}$$

Employing the full set of eigenstates $\hat{H}|m\rangle = \varepsilon_m(\phi_x)|m\rangle$ after a straightforward calculation we obtain

$$S(t) = \frac{1}{\mathcal{Z}}\sum_{m,n} e^{it(\varepsilon_m - \varepsilon_n)}|\langle m|\hat{I}|n\rangle|^2(e^{-\beta\varepsilon_m} + e^{-\beta\varepsilon_n}) - 2\langle\hat{I}\rangle^2 \tag{4}$$

and

$$S_\omega = 4\pi\mathcal{P}\delta(\omega) + \frac{2\pi}{\mathcal{Z}}\sum_{m\neq n}|\langle m|\hat{I}|n\rangle|^2(e^{-\beta\varepsilon_m} + e^{-\beta\varepsilon_n})\delta(\omega + \varepsilon_m - \varepsilon_n), \tag{5}$$

where $\mathcal{Z}$ is the grand partition function for our system, $\mathcal{P}$ is some constant.

We observe that PC noise power has the form of peaks at frequencies equal to the distance between the energy levels with non-zero matrix elements of the current operator plus an additional peak at zero frequency. In the zero temperature limit $T \to 0$ the amplitude of this peak tends to zero and Eq. (5) reduces to

$$S_\omega = 2\pi\sum_{m>0}|\langle m|\hat{I}|0\rangle|^2\left(\delta(\omega + \varepsilon_0 - \varepsilon_m) + \delta(\omega + \varepsilon_m - \varepsilon_0)\right) \tag{6}$$

This result demonstrates that PC fluctuations indeed persist down to $T = 0$ in which case peaks of PC noise power S_ω occur at frequencies corresponding to transitions between energy levels for which the matrix elements of the current operator differ from zero. In other words, PC correlator differs from zero dwn to zero temperature provided the current operator does not commute with the total Hamiltonian of the system. A specific feature of PC noise is the dependence of S_ω on the external magnetic flux ϕ_x. This dependence occurs due to the presence of quantum coherence in the system and disappears if this coherence gets destroyed. Hence, such sensitivity of PC noise spectrum to the flux can be used as a measure of quantum coherence in our system.

It is important to emphasize that, while the above results hold for a single particle on a ring, in other situations PC noise spectrum might look differently. For instance, broadening of such peaks inevitably occurs in ensembles of rings or individual rings with many conducting channels. In this case the total PC noise produced by the system is given by the sum of a large number of very close peaks effectively resulting in a much smoother and broader noise spectrum. Likewise, in the presence of dissipation due to interaction of the particle with other (quantum) degrees of freedom the energy levels acquire a finite width, the peaks get broadened and the noise power should differ from zero also in a wider range of frequencies. Two examples will be considered below.

Our first example is a particle on the ring in periodical potential. Let us employ the imaginary time technique and define the generating functional

$$\mathcal{Z}[\zeta] = \int_0^{2\pi} d\theta_0 \sum_{m=-\infty}^{\infty} e^{2\pi i m\phi_x} \int_{\theta_0}^{\theta_0+2\pi m} \mathcal{D}\theta\, e^{-\int_0^\beta d\tau\left(\frac{MR^2\dot{\theta}^2}{2} + U(\theta) - i\zeta\dot{\theta}\right)}. \tag{7}$$

This functional enables one to evaluate all imaginary-time current cumulants by taking the derivatives with respect to the source field ζ and, hence, to establish "full-counting statistics" of PC in our problem. Real-time correlators might be obtained by analytic continuation.

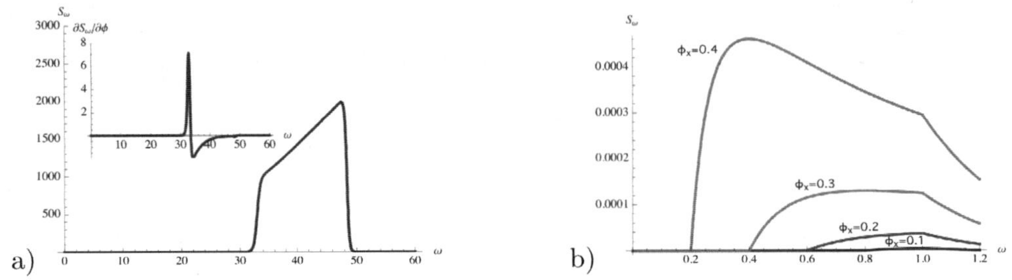

Figure 1. (a)Zero temperature PC noise spectrum S_ω (arbitrary units) and its derivative with respect to the flux $\partial S_\omega/\partial\phi$ (arbitrary units) as functions of ω (measured in units of $1/2MR^2$) for an ensemble of rings (or for a ring containing many independent channels) with U_0 uniformly distributed within the interval from $30/MR^2$ to $65/MR^2$.(b) Zero temperature PC noise spectrum S_ω (arbitrary units) was a function of ω (measured in units of $1/(2MR^2)$) for a dissipative system (11) for different fluxes ϕ_x.

For the sake of definiteness below we will set $U(\theta) = U_0(1 - \cos(\kappa\theta))$, i.e. we will assume that the particle on a 1D ring is moving in a periodic potential with the distance $2\pi/\kappa$ between the adjacent minima. The potential barriers between these minima will be assumed high, $U_0 \gg \kappa^2/(MR^2)$, in which case the particle moves around the ring due to hopping from one minimum to another. Semiclassically, these hops are described by multi-instanton trajectories $\Theta(\tau) = \sum_j \nu_j\tilde{\theta}(\tau - \tau_j)$ with $\nu_j = \pm 1$, which dominate the path integral (7). Here $\tilde{\theta}(\tau) = 4\arctan(e^{\Omega\tau})/\kappa$ is well known kink solution, describing the particle tunneling with the amplitude $\Delta/2 = 4(\Omega U_0/\pi)^{1/2}e^{-\frac{8U_0}{\Omega}}$, where $\Omega = \kappa\sqrt{U_0/(MR^2)}$. Substituting the trajectories $\Theta(\tau)$ into Eq. (7) and performing Gaussian integration we get

$$\mathcal{Z}[\zeta] = \mathcal{Z}_0[\zeta]\sum_{k=1}^{\kappa} e^{\Delta\int_0^\beta d\tau\cos\left(\frac{2\pi(\phi_x-k)}{\kappa}+\int_0^\beta \zeta(\tau_1)\dot{\tilde{\theta}}(\tau-\tau_1)d\tau_1\right)}. \tag{8}$$

Here $\mathcal{Z}_0[\zeta]$ is the generation functional for a harmonic oscillator. Employing the above results at $T \ll \Omega$ we obtain

$$S_\omega = 4\pi\mathcal{P}\delta(\omega) + \frac{e^2\Omega^3}{4\pi\kappa^2U_0\mathcal{Z}}\sum_{k=1}^{\kappa} e^{\beta\Delta\cos\left(\frac{2\pi(\phi_x-k)}{\kappa}\right)}\left(\delta(\omega-\Omega-\epsilon_k)+\delta(\omega+\Omega+\epsilon_k)\right), \tag{9}$$

where we defined $\epsilon_k = (32U_0\Delta/\Omega)\cos\left(2\pi(\phi_x-k)/\kappa\right)$ and

$$\mathcal{P} = \frac{e^2\Delta^2}{\kappa^2}\frac{\sum\limits_{k=1}^{\kappa}\sin^2\left(\frac{2\pi(\phi_x-k)}{\kappa}\right)e^{\beta\Delta\cos\left(\frac{2\pi(\phi_x-k)}{\kappa}\right)}}{\sum\limits_{k=1}^{\kappa}e^{\beta\Delta\cos\left(\frac{2\pi(\phi_x-k)}{\kappa}\right)}} - \frac{e^2\Delta^2}{\kappa^2}\left(\frac{\sum\limits_{k=1}^{\kappa}\sin\left(\frac{2\pi(\phi_x-k)}{\kappa}\right)e^{\beta\Delta\cos\left(\frac{2\pi(\phi_x-k)}{\kappa}\right)}}{\sum\limits_{k=1}^{\kappa}e^{\beta\Delta\cos\left(\frac{2\pi(\phi_x-k)}{\kappa}\right)}}\right)^2 \tag{10}$$

As we already discussed, in the case of many channel rings or an ensemble of rings with varying parameter U_0 PC noise power consists of many close peaks, so that the noise spectrum gets smoother, as it is shown in Fig. 1a.

Our second example is a particle interacting with the dissipative bath formed by electrons in a disordered metal. In this case the action takes the form [3, 4]

$$S[\theta] = \frac{MR^2}{2}\int_0^\beta d\tau\left(\frac{\partial\theta}{\partial\tau}\right)^2 + \alpha\int_0^\beta d\tau\int_0^\beta d\tau'\frac{\pi^2T^2K(\theta(\tau)-\theta(\tau'))}{\sin^2(\pi T(\tau-\tau'))} \tag{11}$$

Here $K(z) = 1 - \frac{1}{\sqrt{4r^2\sin^2(z/2)+1}}$, $\alpha = 3/(8k_F^2 l^2)$ and $r = R/l$, l is the electron mean free path. Below we will employ the perturbation in α technique developed by Golubev and one of the present authors [8]. Following this approach let us express the partition function in the following form:

$$\mathcal{Z} = \sum_{n=-\infty}^{\infty} Z[\beta, \phi_x - n] \tag{12}$$

where the Laplace transform of Z is defined as

$$Z_p[q] = \int_0^\infty d\tau e^{-p\tau} Z[\tau, q] = \frac{1}{p + q^2/(2MR^2) - \Sigma_p[q]} \tag{13}$$

Here $\Sigma_p[q]$ is the self-energy defined as a sum of all irreducible diagrams. Here we proceed perturbatively taking into account only the simplest one loop diagram. We also restrict our analysis to zero temperature limit. In that case the self-energy reads

$$\Sigma_p[q] = \frac{\alpha}{2}\sum_{n=1}^{r}\alpha_n \sum_{s=\pm1}(p + (q+sn)^2/(2MR^2))\ln(p + (q+sn)^2/(2MR^2)) - \\ -\alpha\sum_{n=1}^{r}\alpha_n(p + (q^2+n^2)/(2MR^2))\ln(p + (q^2+n^2)/(2MR^2)) \tag{14}$$

where $\alpha_n = \frac{2}{\pi r}\ln(r/n)$. Making use of this expression it is straightforward to derive the current-current correlator. Neglecting vertex corrections we obtain

$$\langle\dot\theta(0)\dot\theta(\tau)\rangle = -\frac{1}{(MR^2)^2}\frac{\sum\limits_{n=-\infty}^{\infty}(n-\phi_x)^2 Z[\tau, n-\phi_x]Z[\beta-\tau, n-\phi_x]}{\sum\limits_{n=-\infty}^{\infty} Z[\beta, n-\phi_x]} \tag{15}$$

After analytic continuation to real time we arrive at the following expression for zero-temperature PC noise power:

$$S_\omega = -\frac{e^2\phi_x^2}{2\pi^2(MR^2)^2}\left(\rho_{\phi_x^2/(2MR^2)+\omega}[-\phi_x] + \rho_{\phi_x^2/(2MR^2)-\omega}[-\phi_x]\right) \tag{16}$$

where $\rho_p[q] = \Im Z_{-p+i0}[q]$. This result is depicted in Fig. 1b.

In summary, we investigated equilibrium fluctuations of persistent current in nanorings and demonstrated that these fluctuations do not vanish even at $T = 0$ provided the current operator does not commute with the total Hamiltonian of the problem. A specific feature of PC noise is its quantum coherent nature implying that the noise spectrum can be tuned by an external magnetic flux inside the ring. We believe that the key features captured by our analysis will survive also in other models and can be verified in future experiments.

References

[1] Golubev D S and Zaikin A D 1998 *Physica B* **255** 164
[2] Cedraschi P, Ponomarenko V V and Büttiker M 2000 *Phys. Rev. Lett.* **84** 346
[3] Guinea F 2002 *Phys. Rev. B* **65** 205317
[4] Golubev D S, Herrero C P and Zaikin A D 2003 *Europhys. Lett.* **63** 426
[5] Horovitz B and Le Doussal P 2006 *Phys. Rev. B* **74** 073104
[6] Semenov A G and Zaikin A D 2009 *Phys. Rev. B* **80** 155312
[7] Arutyunov K Yu, Golubev D S and Zaikin A D 2008 *Phys. Rep.* **464** 1
[8] Golubev D S and Zaikin A D 1994 *Phys. Rev. B* **50**8736

Field-emission current from quantum system

G.V.Shpatakovskaya

Keldysh Institute of Applied Mathematics Russian Academy of Sciences, Miusskaya sq.4, 125047 Moscow, Russia

E-mail: shpagalya@yandex.ru

Abstract. A universal semiclassical approach is developed to calculate an emission current from a quantum system (atom, ion, metal cluster, metal surface, graphene ribbon, etc.) in a stationary electric field. The same expression is in use for a spherical emitter (atom, negative ion, metal cluster) and a plane metal surface. It is shown the results agree with those of other methods.

1. Introduction The interest in the field-emission from a quantum system has quickened because of a new nano-dimension objects (metal cluster, grapfene, nanotubes and so on) appearance, the geometry of the systems being various. The semiclassical approach is used to calculate the emission current from the spherical atom and ion [1]. In the paper we attempt to construct an universal field-emission current expression [2] for a various emitter geometry and to verify it by solving the well-known problems and comparing the results with those of other methods.

It's assumed that the emitter electron with an energy E_{em} is tunneling through a wide barrier

$$l \gg \hbar/\sqrt{2mE_{em}} \tag{1}$$

2. Semiclassical approach Semiclassical expression for the tunneling current is derived using the semiclassical 3D wave function [3]

$$\Psi(\mathbf{r}) = \Psi_0(\mathbf{r}_0)\left|\frac{p_0(\mathbf{r}_0)}{p(\mathbf{r})}\right|^{1/2} \exp\left(\frac{i}{\hbar}S(\mathbf{r}_0,\mathbf{r}) - \frac{1}{2}\int_{\mathbf{r}_0}^{\mathbf{r}} dl \,\mathrm{div}\mathbf{e}\right) \tag{2}$$

Here the points $\mathbf{r}_0$ and $\mathbf{r}$ belong to the same trajectory of the particle motion with the boundary conditions $\mathbf{r} = \mathbf{r}_0, \mathbf{p} = \mathbf{p}_0(\mathbf{r}_0)$, dl is an arc element along the trajectory, $\mathbf{p}(\mathbf{r}) = \nabla S(\mathbf{r})$ is a momentum, $S(\mathbf{r}_0, \mathbf{r}) = \int_{\mathbf{r}_0}^{\mathbf{r}} \mathbf{dl}\,\mathbf{p}$ is a classical action, $\mathbf{e} = \mathbf{p}/p$ is an unit momentum vector, $p(\mathbf{r}) = \sqrt{2m(E-U(\mathbf{r}))}$ is an absolute value of the momentum vector, E and $U(\mathbf{r})$ are the electron total and potential energies. Generally there is a sum of a few waves in the right part of Eq.(2). In the case Ψ_0 is the partial wave function, which phase gradient is equal to the initial momentum vector $\mathbf{p}_0(\mathbf{r}_0)$.

The corresponding current

$$\mathbf{j}(\mathbf{r}) = \frac{1}{m}\mathbf{e}(\mathbf{r})|\mathbf{p}_0(\mathbf{r}_0)||\Psi_0(\mathbf{r}_0)|^2 \exp\left(\int_{\mathbf{r}_0}^{\mathbf{r}} dl \,\mathrm{div}\mathbf{e}\right) \tag{3}$$

satisfies a continuity equation $\mathrm{div}\mathbf{j} = 0$ in the classical permitted region.

The tunneling probability in a unit time is equal to $w = \int_\Sigma \mathbf{ds}\,\mathbf{j}(\mathbf{r})$. Here Σ is the surface that embraces the barrier region. Every point $\mathbf{r}$ of the surface Σ has the initial point $\mathbf{r}_0$ along the same trajectory. Whole of these points forms a surface Σ_0 which we set inside the barrier region far away from the inner barrier surface to use the WKB approximation of an electron wave function. The surface Σ_0 is also assumed to be far from the outer barrier surface to minimize the outer field effect. To satisfy these demands one needs a wide barrier that agrees with the condition Eq.(1).

In this case the action in Eq.(2) is equal to $S(\mathbf{r}_0, \mathbf{r}) = i\,\mathrm{Im}\,S(\mathbf{r}_0, \mathbf{r}_{ex}) + S(\mathbf{r}_{ex}, \mathbf{r})$ with $\mathbf{r}_{ex}$ being the barrier exit point. Here the imaginary unit is connected with an absolute value of the momentum vector. The current vector is then given by

$$\mathbf{j}(\mathbf{r}) = \frac{1}{m}\mathbf{e}(\mathbf{r})|\mathbf{p}_0(\mathbf{r}_0)||\Psi_0(\mathbf{r}_0)|^2 \exp\left(\int_{\mathbf{r}_0}^{\mathbf{r}} dl\,\mathrm{div}\,\mathbf{e} - \frac{2}{\hbar}\,\mathrm{Im}\,S(\mathbf{r}_0, \mathbf{r}_{ex})\right). \qquad (4)$$

The current Eq.(4) outside the barrier satisfies a continuity equation too. It is evident from the equations $d\mathbf{r}_0/dl = d\mathbf{r}_{ex}/dl = 0$ for the same trajectory.

One can prove that the Eq.(4) describes a permanent quantity when the point $\mathbf{r}$ being into the barrier region and the magnitude $\mathrm{Im}\,S(\mathbf{r}_0, \mathbf{r}_{ex})$ relating to the points $\mathbf{r}_0, \mathbf{r}_{ex}$ as before. It is because of the same trajectory connects the points $\mathbf{r}_0, \mathbf{r}, \mathbf{r}_{ex}$ with each other regardless of their order. The Eq.(4) in this case doesn't of course coincide with a true current.

Using the conservation of the quantity Eq.(4) one can replace the surface Σ to superpose the surface Σ_0, that is to unite the points $\mathbf{r}_0, \mathbf{r}$. It means an absence of the term with $\mathrm{div}\,\mathbf{e}$ in the current expression Eq.(4).

The total charge current from the emitter at the temperature T is equal to

$$I = 2e\sum_{em} w_{em}\, f(E_{em}) = 2e\sum_{em} f(E_{em}) \int_{\Sigma_0} \mathbf{ds}\,\mathbf{j}(\mathbf{r}_0) \exp\left(-\frac{2}{\hbar}\,\mathrm{Im}\,S(\mathbf{r}_0, \mathbf{r}_{ex})\right). \qquad (5)$$

with the Fermi distribution function $f(E) = (1 + e^{(E-\mu)/T)})^{-1}$, where μ is a chemical potential. Here $\mathbf{j}(\mathbf{r}_0)$ looks like the field-free current $\mathbf{j}(\mathbf{r}_0) = \mathbf{e}(\mathbf{r}_0)|\mathbf{p}_0(\mathbf{r}_0)||\Psi_0(\mathbf{r}_0)|^2/m_e$

4. Tunneling probability from a spherical system Let us calculate the tunneling probability from the electron bound E, l, m - state in a radial symmetric system in an outer electric field $\mathbf{F}(z) \parallel z$ axis ($\mathbf{F}(z) = F(z)\mathbf{k}$) within the semiclassical theory Eq.(5). We have no need to know a specific view of the inner field. One assumes only that the outer field is small in comparison with the inner field inside and nearby the emitter.

It is evident that only a little surface environment of z axis gives a main contribution to the integral in the Eq.(5), that is a region of small angles $\theta \ll 1$. One can prove that in this case the WKB approximation of the angular part of the wave function $\Psi_0(\mathbf{r}) = R_{El}(r)Y_{lm}(\theta, \varphi)$ is right for the small values l, m too and write

$$Y_{lm}(\theta, \varphi) = A_{lm}\theta^{|m|}e^{im\varphi}, \ \ |A_{lm}| = \frac{1}{2^{|m|}|m|!}\sqrt{\frac{2l+1}{4\pi}\frac{(l+|m|)!}{(l-|m|)!}}$$

Furthermore the small values θ mean approximately $\mathbf{j}(\mathbf{r}_0) \parallel \mathbf{r}_0$.

Let us divide action in the Eq.(5) into two parts

$$S(\mathbf{r}_0, \mathbf{r}_{exit}) = \int_{\mathbf{r}_0}^{\mathbf{r}} \mathbf{dl}\,\mathbf{p} = \int_{z_0}^{z_{exit}} dz\, p_z + \int_{\mathbf{r}_{0\perp}}^{\mathbf{r}_{exit\perp}} \mathbf{dr}_\perp\,\mathbf{p}_\perp, \qquad (6)$$

where

$$p_z = \sqrt{p^2(r) - p_\perp^2}, \ \ \mathbf{dr}_\perp = \frac{1}{m_e}\mathbf{p}_\perp dt, \ p_\perp = p\sin\theta.$$

Because of assumed smallness of the inner field in the barrier region $\mathbf{p}_\perp \simeq \mathbf{p}_{0\perp}$. So the second term in the Eq.(6) is equal to the value $p_{0\perp}^2\,\tau_E(F)$, where

$$\tau_E(F) = \int_{z_0}^{z_{ex}} \frac{dz}{|p_z(\theta = 0)|} \tag{7}$$

is the time of the motion between the points z_0 and z_{ex} that is equal to the time of the motion between the points $\mathbf{r}_{0\perp}$ and $\mathbf{r}_{ex\perp}$.

After accounting for a small value $p_{0\perp} = p_0\,\theta$ in the Eq.(6) we have the following expressions for the action

$$S(\mathbf{r}_0, \mathbf{r}_{ex}) = i\sigma_E(F) + i\frac{|p_0|^2\,\tau_E(F)}{2}\,\theta^2,$$

$$\sigma_E(F) = \int_{z_0}^{z_{ex}} dz\,|p_z(\theta = 0)|. \tag{8}$$

Here the quantities $\sigma_E(F)$, $\tau_E(F)$ depend on the particle energy E and outer field F.

After integration over θ the tunneling probability is given by

$$w(E;l,m) = \pi\, r_0^2\, R_{El}^2(r_0)\,|m|!\,|A_{lm}|^2\,\frac{p_{0E}}{m_e} \times \frac{\exp\left(-2\sigma_E(F)/\hbar\right)}{\left(p_{0E}^2\,\tau_E(F)/\hbar\right)^{|m|+1}} \tag{9}$$

If the sum of fields in the point r_0 are small one can write

$$|p_{0E}(r_0)| \approx p_E = \sqrt{2m_e|E|}.$$

Let us use the common expression Eq.(9) to calculate the ionization probability of the atom or negative ion by the uniform electric field F. An asymptotic form of the radial function is then given by

$$R_{El}(r_0) = C_{El}\,p_E^{3/2}(p_E\, r_0/\hbar)^{\lambda-1}\exp(-p_E\, r_0/\hbar).$$

Here C_{El} is a numeral coefficient, $\lambda = m_e\alpha/(\hbar p_E)$, $U(r \to \infty) = -\alpha/r$. For the short-range potential (negative ion) $\alpha = 0$. The condition Eq.(1) holds true under $\hbar m_e F \ll p_E^3$. As a result of the integration in Eqs.(7),(8) and substitution into Eq.(9) we easily get a well-known expression [1] (in the atomic units)

$$w(E;l,m) = |C_{El}|^2\frac{p_E^2}{2}\frac{2l+1}{|m|!2^{|m|}}\frac{(l+|m|)!}{(l-|m|)!} \times \left(\frac{2p_E^3}{F}\right)^{2\lambda-|m|-1} \exp\left(\frac{2p_E^3}{3F}\right) \tag{10}$$

5. Field-emission current from a metal planar surface The Bardeen transfer Hamiltonian (BTH) method [4] and jellium model have been used in the paper [5] to calculate the field-emission current from a metal surface at the temperature $T = 0$. Let us derive the corresponding expression within the semiclassical approach Eq.(5) and compare the results.

Assuming that the jellium surface normal and the electron emission direction are along the z axis ($\mathbf{F}(z) \parallel z$ axis), the wave function $\Psi_0(\mathbf{r})$ can be written as

$$\Psi_0(\mathbf{r}) = \psi_n(z)\,e^{i\mathbf{k}_\perp\mathbf{r}_\perp},$$

where $\mathbf{r}_\perp$, $\mathbf{k}_\perp$ are the vectors normal to z direction. The eigenvalues E may be written as

$$E_n(k_\perp) = \varepsilon_n + \frac{\hbar^2}{2m_e}k_\perp^2.$$

Here ε_n are a "normal" energy spectrum. In this case a surface Σ_0 in Eq.(5) is the plane $z = z_0$.

Let us calculate the current through the cross-sectional area σ of this plane

$$I = 2e\,\sigma \sum_n \int \frac{d\mathbf{k}_\perp}{4\pi^2}\theta(E_F - E_n(\mathbf{k}_\perp))\; j_n(z_0)\, \exp\left(-\frac{2}{\hbar}\,\mathrm{Im}\, S_n[z_0, z_{ex}]\right) \tag{11}$$

The integration over $\mathbf{k}_\perp$ can be written as $(E_F - \varepsilon_n)/(2\pi\hbar^2)$.

Using the semiclassical asymptotic form for $\psi_n(z)$ in the barrier region we have a following expression for the current at point z_0

$$j_n(z_0) = \frac{|C_n|^2}{4m_e}\,\exp\left(-\frac{2}{\hbar}\,\mathrm{Im}\, S_n[z_n, z_0]\right),$$

where z_n is the inner turning point.

The expression for the field-emission current density $J = I/\sigma$ is then given by the same equation as in the paper [5]

$$J = \frac{e}{4\pi\hbar^2}\sum_n (E_F - \varepsilon_n)|C_n|^2 \times \exp\left(-\frac{2}{\hbar}\int_{z_n}^{z_{ex}} dz'\,|p_n(z')|\right), \quad |p_n(z)| = \sqrt{2m|\varepsilon_n - U(z)|}. \tag{12}$$

6. About effective potential in the barrier region One needs to know an effective potential $U(\mathbf{r})$ in the barrier region to calculate an action $S(\mathbf{r}_0, \mathbf{r}_{ex})$ in Eq.(5). If an emitter is considered to be a conducting system (that is right for a metal surface in the section 5, for a metal cluster when using Eq.(9) and so on) it is necessary to account for an image potential in addition to the sum of inner and outer potentials. Let us cite here some image potentials for a various emitter geometry [6].

The classical image potentials at a distance z of the conductive sphere of radii R and metal surface are correspondingly given by $U_{sph}(z) = -eR/2(z^2 - R^2)$, $U_{surf}(z) = -e/4z$. The interaction energy of the charge e with the conductive half plane ($z = 0,\ x < 0$), that may be a model of a wide graphene ribbon, is equal to

$$U_{h-pl}(x, z) = -\frac{e^2}{4\pi}\left[\frac{1}{\sqrt{x^2+z^2}} + \frac{1}{z}\arcsin\frac{z}{\sqrt{x^2+z^2}}\right]. \tag{13}$$

Here x is the distance of the ribbon edge and z is the perpendicular distance from the ribbon plane.

7. Conclusion We have developed the common semiclassical approach to calculate the field-emission current from quantum systems with a various geometry provided that a barrier is wide and the outer field doesn't distort the eigenstates of the emitter. The method requires the knowledge of the free-field emitter electron energy spectrum and WKB wave functions on some surface in the barrier region and effective potential in the region between this surface and barrier exit surface.

We have solved two known problems to verify the method.

Acknowledgments
This work was supported in part by Union State Programme SKIF-GRID (project 209P420) and the Russian Foundation for Basic Research (project 08-01-00291)

References
[1] Smirnov B M and Chibisov M I 1965 *JETP* **49** 841
[2] Shpatakovskaya G V 2010 *Physics of Extreme States of Matter-2010* ed V E Fortov et al (Chernogolovka: IPCP RAS) pp 24-27
[3] Van Horn H 1965 *Thesis* (Cornell University) CRSR 211
[4] Bardeen J 1961 *Phys. Rev. Lett.* **6** 57
[5] Ramprasad R, Fonseca L R C and von Allmen Paul 2000 *Phys. Rev.* B **62** 5216
[6] Landau L D and Lifshits E M 1959 *Electrodynamics of Continious Media* (Moscow: FIZMATLIT)

Journal of Physics: Conference Series **248** (2010) 012036 doi:10.1088/1742-6596/248/1/012036

Electrodynamics and intrinsic Josephson effects in multi-gap superconductors

Tomio Koyama[1,3], Yukihiro Ota[2,3] and Masahiko Machida[2,3]

[1]Institute for Materials Research, Tohoku University, Sendai 980-8577, Japan
[2]CCSE, Japan Atomic Energy Agency, Higashi-Ueno, Tokyo 110-0015, Japan
[3]CREST-JST, Kawaguchi 332-0012, Japan

E-mail: tkoyama@imr.tohoku.ac.jp

Abstract.
We develop a theory for the Josephson effects in 2-gap intrinsic Josephson junction stacks (IJJ's). The coupled dynamical equations for the phase differences are derived from the low-energy effective Lagrangian. The equations can describe the longitudinal Josephson plasma and the Josephson-Leggett (JL) mode propagating in the direction perpendicular to the junctions. Numerical results for the $I-V$ characteristics are presented. The $I-V$ characteristics shows multiple-branch structure similar to that in Bi-2212 IJJ's. When the Josephson frequency is approached to the JL mode frequency in non-uniform voltage branches, the JL mode is resonantly excited. At the resonant voltage a step-like structure appears in the $I-V$ curves in low-voltage branches.

1. Introduction

Since the discovery of high-temperature superconductivity in layered iron-pnictide compounds with several disconnected Fermi surfaces [1], multi-gap superconductors have attracted a great interest. Recently, new types of layered iron-pnictide superconductors with extremely thick blocking layers, $Sr_2ScFePO_3$ and $(Fe_2As_2)(Ca_{n+1}(Sx,Ti)_nO_y)$ with $n = 3-5$, have been synthesized by Ogino et al. [2,3]. In these new iron-based superconductors the distance between iron layeres ranges from $\sim 15\mathring{A}$ to $\sim 25\mathring{A}$, which is much wider than that in Bi-2212 high-T_c cuprate forming the intrinsic Josephson junction stacks (IJJ's). Reflecting the large inter-layer width, a large value, $\gamma \simeq 250$, has been predicted for the anisotropy parameter γ in $Sr_2ScFePO_3$ [4] on the basis of the first-principles calculations for the electronic states, which is comparable to that in $Bi_2Sr_2CaCu_2O_8$ IJJ's. From this fact one may expect that the iron-based superconductors, $Sr_2ScFePO_3$ and $(Fe_2As_2)(Ca_{n+1}(Sx,Ti)_nO_y)$, form the IJJ's composed of a stack of multi-gap superconducting layers. In this paper we develop a theory describing the dynamics of the superconducting phase differences in multi-gap IJJ's.

In previous papers we presented a theory for the Josephson effect in S-I-S tunneling junctions made of 2-gap superconductors, in which the effective Lagrangian is derived from a microscopic model Hamiltonian for the 2-gap system [5-7]. In this paper the theory is extended to the case of multi-gap IJJ's. It is known that in 2-gap superconductors in which the inter-band pairing interaction is not so strong ($g_1g_2 - g_{12}^2 > 0$ with g_i and g_{12} being the intra- and inter-band BCS coupling constants) a collective excitation mode called the Leggett mode appears in the low-energy region [8]. The Leggett mode is the out-of-phase oscillation mode of the phases of the superconducting gaps. This excitation mode remains in the low-energy region in the charged system too [9], though the in-phase mode is absorbed into the plasma by the Anderson-Higgs mechanism. This result comes from the fact that the Leggett mode is neutral in the uniform limit, which is understood from the longitudinal dielectric function in the low-energy and long-wavelength region,

$$\epsilon_L(\omega, q) = 1 - \frac{\omega_{\rm pl}^2(\omega^2 - \omega_{\rm LG}^2) - Cq^2}{(\omega^2 - Aq^2)(\omega^2 - \omega_{\rm LG}^2 - Bq^2)}, \tag{1}$$

where $\omega_{\rm pl}$ and $\omega_{\rm LG}$ are, respectively, the plasma and the Leggett mode frequencies, i.e., $\omega_{\rm pl}^2 = 8\pi e^2(N_1(0) + N_2(0))$ and $\omega_{\rm LG}^2 = 4g_{12}\Delta_1\Delta_2(N_1(0) + N_2(0))/(g_1g_2 - g_{12}^2)N_1(0)N_2(0)$ with Δ_n and $N_n(0)$ $(n = 1, 2)$ being the gap amplitude and the density of states at the Fermi level in the nth band. Note that the dielectric function is reduced to the well-known form,

$$\epsilon_L(\omega, 0) = 1 - \frac{\omega_{\rm pl}^2}{\omega^2}, \tag{2}$$

in the $q \to 0$ limit, which indicates that the uniform Leggett mode does not couple with an electric field. Thus, one needs a spatially inhomogeneous electric field to excite the Leggett mode in 2-gap superconductors. The Leggett mode was observed in the Ramann scattering experiment for the 2-gap system with weak anisotropy, MgB_2 [10]. In this paper we investigate the effect of the Leggett mode on the AC Josephson effects in the 2-gap IJJ's.

In 2-gap IJJ's one can define two kinds of gauge-invariant phase differences between consecutive superconducting layers, since there are two channels of the Cooper-pair tunneling in the 2-band system. The Leggett mode propagating in the direction perpendicular to the layers ($\perp$ c-axis) in the IJJ's, which we call the Josephson-Leggett mode hereafter, is related to the relative motion of the gauge-invariant phase differences. In this paper we clarify the dynamics of the phase differences and calculate the $I-V$ characteristics in the 2-gap IJJ's. It is shown that the Josephson-Leggett mode can be resonantly excited by the Josephson oscillations in non-uniform branches in which spatially inhomogeneous electric field distribution is realized along the c-direction.

2. Formulation

Consider the IJJ 's in which 2-gap superconducting layers with an atomic-scale layer thickness, s, and insulating layers with a thickness, d, are alternately stacked along the c-axis. Let $\varphi_\ell^{(a)}$ be the phase of the superconducting order parameter at ℓth superconducting layer in the a-band $(a = 1, 2)$. The effective Lagrangian in this 2-gap system is obtained by extending the previous theory to the multi-junction system with small in-plane area as

$$L = \sum_\ell \Big\{ \frac{s}{8\pi\mu_1^2}\Big(A_\ell^0 + \frac{\hbar}{e^*}\partial_t\varphi_\ell^{(1)}\Big)^2 + \frac{s}{8\pi\mu_2^2}\Big(A_\ell^0 + \frac{\hbar}{e^*}\partial_t\varphi_\ell^{(2)}\Big)^2 + \frac{d\epsilon}{8\pi}E_{\ell+1,\ell}^2$$
$$+\frac{\hbar}{e^*}j_c^{(1)}\cos\Big(\varphi_{\ell+1}^{(1)} - \varphi_\ell^{(1)} - \frac{e^*d}{\hbar c}A_{\ell+1,\ell}^z\Big) + \frac{\hbar}{e^*}j_c^{(2)}\cos\Big(\varphi_{\ell+1}^{(2)} - \varphi_\ell^{(2)} - \frac{e^*d}{\hbar c}A_{\ell+1,\ell}^z\Big)$$
$$+\frac{\hbar}{e^*}J_{\rm in}\cos\Big(\varphi_\ell^{(1)} - \varphi_\ell^{(2)}\Big)\Big\}, \tag{3}$$

where A_ℓ^0 is the scalar potential on the ℓth superconducting layer, $A_{\ell+1,\ell}^z$ is the z-component ($\parallel$ c-axis) of the vector potential in the insulating layer with the dielectric constant ϵ sandwiched between ℓth and $(\ell+1)$th superconducting layers and $E_{\ell+1,\ell}$ is the electric field given as

$$E_{\ell+1,\ell} = -\frac{1}{c}\partial_t A_{\ell+1,\ell}^z - \frac{A_{\ell+1}^0 - A_\ell^0}{d}. \tag{4}$$

The terms in the first line in Eq.(3) comes from the Coulomb energy of condensed electrons in the two energy-bands 1 and 2 having the charge screening lengthes, respectively, μ_1 and μ_2. Since the thickness s is comparable to μ_1 and μ_2 in IJJ's ($s/\mu_a \sim 1$), these terms cannot be neglected and they cause the capacitive coupling between junctions in the present system. The terms in the second line represent the Josephson coupling energy, i.e., $j_c^{(1)}$ and $j_c^{(2)}$ denote the Josephson critical current densities in the two bands, which are generally different from one another, $j_c^{(1)} \neq j_c^{(2)}$. In 2-gap superconductors the inter-band pair-tunneling is generally allowed to take place. This interaction is incorporated into the Lagrangian (3) by the last term, where $J_{\rm in}$ denotes its coupling constant with the dimension of current per unit area. The coupling constant $J_{\rm in}$ can take both signs, reflecting the gap symmetry. When the gaps have same (opposite) signs, we find $J_{\rm in} > 0$ ($J_{\rm in} < 0$).

Let us now define the mean and relative phases at ℓth superconducting layer as

$$\Phi_\ell = (\varphi_\ell^{(1)} + \varphi_\ell^{(2)})/2, \qquad \Psi_\ell = \varphi_\ell^{(1)} - \varphi_\ell^{(2)}, \tag{5}$$

and introduce the mean phase difference $\theta_{\ell+1,\ell}(t)$ and the relative phase difference, $\psi_{\ell+1,\ell}(t)$, as follows,

$$\theta_{\ell+1,\ell} = \Phi_{\ell+1} - \Phi_\ell - \frac{e^* d}{\hbar c} A^z_{\ell+1,\ell}, \qquad \psi_{\ell+1,\ell} = \Psi_{\ell+1} - \Psi_\ell. \tag{6}$$

These phase differences obey the coupled dynamical equations derived from the Euler-Lagrange equations for the Lagrangian (3) as

$$\frac{\mathrm{d}^2\theta_{\ell+1,\ell}}{\mathrm{d}\tau^2} = -[1 - \frac{\alpha_1+\alpha_2}{4}\Delta^{(2)}]W^{(+)}_{\ell+1,\ell} - \frac{|\alpha_1-\alpha_2|}{4}\Big(\Delta^{(2)}W^{(-)}_{\ell+1,\ell} - 2\tilde{J}_{\rm in}(\sin\Psi_{\ell+1} - \sin\Psi_\ell)\Big), \tag{7}$$

$$\frac{\mathrm{d}^2\psi_{\ell+1,\ell}}{\mathrm{d}\tau^2} = \frac{|\alpha_1-\alpha_2|}{2}\Delta^{(2)}W^{(+)}_{\ell+1.\ell} + \frac{\alpha_1+\alpha_2}{2}\Big(\Delta^{(2)}W^{(-)}_{\ell+1,\ell} - 2\tilde{J}_{\rm in}(\sin\Psi_{\ell+1} - \sin\Psi_\ell)\Big), \tag{8}$$

with

$$W^{(\pm)}_{\ell+1,\ell} = \sin(\theta_{\ell+1,\ell} + \frac{1}{2}\psi_{\ell+1,\ell}) \pm (\frac{j_c^{(2)}}{j_c^{(1)}})\sin(\theta_{\ell+1,\ell} - \frac{1}{2}\psi_{\ell+1,\ell}), \tag{9}$$

and $\Delta^{(2)}$ being the 2nd-rank difference operator, i.e., $\Delta^{(2)} f_\ell = f_{\ell+1} - 2f_\ell + f_{\ell-1}$. In Eqs.(7) and (8) the coupling constants, α_1, α_2 and $\tilde{J}_{\rm in}$, are given as $\alpha_1 = \epsilon\mu_1/sd$, $\alpha_2 = \epsilon\mu_2/sd$ and $\tilde{J}_{\rm in} = J_{\rm in}/j_c^{(1)}$ and τ is the dimensionless time normalized as $\tau = \omega_{\rm pl}^{(1)} t$ with $\omega_{\rm pl}^{(1)} = \sqrt{4\pi e^* d j_c^{(1)}/\epsilon\hbar}$, $\omega_{\rm pl}^{(1)}$ being the Josephson plasma frequency in the band 1. Note that Eqs.(7) and (8) have two uniform(ℓ-independnt) oscillating solutions with small amplitude, i.e., (1) $\theta_{\ell+1,\ell}(\tau) = \theta(\tau)$, $\psi_{\ell+1,\ell} = 0$ and (2) $\theta_{\ell+1,\ell} = 0$, $\psi_{\ell+1,\ell}(\tau) = \psi(\tau)$, which satisfy, respectively, the equations

$$\frac{\mathrm{d}^2\theta}{\mathrm{d}\tau^2} = -\frac{j_c^{(1)} + j_c^{(2)}}{j_c^{(1)}}\theta, \qquad \psi_{\ell+1,\ell} = 0, \tag{10}$$

and

$$\theta_{\ell+1,\ell} = 0, \qquad \frac{\mathrm{d}^2\psi}{\mathrm{d}\tau^2} = -(\alpha_1+\alpha_2)|\tilde{J}_{\rm in}|\psi. \tag{11}$$

The oscillation modes satisfying Eqs. (10) and (11) correspond, respectively, to the longitudinal Josephson plasma and the Josephson-Leggett mode. Their frequencies are, respectively, given as

$$\omega_{\rm pl}/\omega_{\rm pl}^{(1)} = \sqrt{1 + (j_c^{(2)}/j_c^{(1)})}, \qquad \omega_{\rm JL}/\omega_{\rm pl}^{(1)} = \sqrt{(\alpha_1+\alpha_2)(|J_{\rm in}|/j_c^{(1)})}. \tag{12}$$

Hence, one understands that the Lagrangian (3) correctly describes the collective modes propagating along the c-direction in the 2-gap systems.

3. $I-V$ characteristics

Let us investigate the $I-V$ characteristics of the 2-gap IJJ's. We consider here the N-junction system composed of a stack of $(N+1)$ equivalent 2-gap superconducting layers. In this case we have N mean phase differences, $\theta_{10},\cdots,\theta_{N,N-1}$, and N relative phase differences, $\psi_{10},\cdots,\psi_{N,N-1}$. These $2N$ variables obey $2N$-coupled equations given in Eqs.(7) and (8) in the case without damping. Note that Eqs.(7) and (8) include $(N+1)$ relative phases Ψ_ℓ with $\ell=0,\cdots,N$ in the N junction system, which can be expressed in terms of $\psi_{\ell,\ell-1}$ by using the inverse relation to Eq.(6),

$$\Psi_\ell = \sum_{m=1}^{\ell} \psi_{m,m-1} + \Psi_0, \tag{13}$$

for $\ell=1,\cdots,N$. In the present system made of finite number of junctions Ψ_0 cannot be determined dynamically. Then, in this paper we fix the value of Ψ_0 as the boundary condition and choose its value as $\Psi_0=0$ for $J_{\rm in}>0$ and $\Psi_0=\pi$ for $J_{\rm in}<0$. It is easy to see that Eqs.(7) and (8) give same dynamics in these two cases, that is, the $I-V$ characteristics does not depend on the gap symmetry (s++ or s+-) in the present system. It is noted that the combination $(\Psi_0=0, J_{\rm in}<0)$ is also possible. In this case the bottom $(\ell=0)$ superconducting layer is understood to be a single-gap one, that is, the system becomes a hetero junction composed of 2-gap IJJ's with s+- gap symmetry and a single-gap superconductor. Analysis of the AC Josephson effect in a conventional type hetero junction made of 2-gap and single-gap superconductors is given in Ref.[11]. In this paper we confine ourselves to the case of homogeneous 2-gap IJJ's.

To construct a set of dynamical equations to calculate the $I-V$ characteristics in the 2-gap IJJ's we extend Eqs.(7) and (8) by adding the damping terms, $\beta_1 {\rm d}\theta_{\ell,\ell-1}/{\rm d}\tau$ and $\beta_2 {\rm d}\psi_{\ell,\ell-1}/{\rm d}\tau$, and a bias current $I_{\rm ex}$ as follows.

$$\frac{{\rm d}^2\theta_{\ell,\ell-1}}{{\rm d}\tau^2} = -[1-\frac{\alpha^{(+)}}{4}\Delta^{(2)}]W^{(+)}_{\ell,\ell-1} - \frac{\alpha^{(-)}}{4}\Big(\Delta^{(2)}W^{(-)}_{\ell,\ell-1} - 2\tilde{J}_{\rm in}(\sin\Psi_\ell - \sin\Psi_{\ell-1})\Big) -\beta_1\frac{{\rm d}\theta_{\ell,\ell-1}}{{\rm d}\tau} + \tilde{I}_{\rm ex}, \tag{14}$$

$$\frac{{\rm d}^2\psi_{\ell,\ell-1}}{{\rm d}\tau^2} = \frac{\alpha^{(-)}}{2}\Delta^{(2)}W^{(+)}_{\ell,\ell-1} + \frac{\alpha^{(+)}}{2}\Big(\Delta^{(2)}W^{(-)}_{\ell,\ell-1} - 2\tilde{J}_{\rm in}(\sin\Psi_\ell - \sin\Psi_{\ell-1})\Big) - \beta_2\frac{{\rm d}\psi_{\ell,\ell-1}}{{\rm d}\tau}, \tag{15}$$

with $\alpha^{(\pm)}=|\alpha_1\pm\alpha_2|$ and $\tilde{I}_{\rm ex}=I_{\rm ex}/j_c^{(1)}$. We suppose that the damping term $\beta_1{\rm d}\theta_{\ell,\ell-1}/{\rm d}\tau$ in Eq.(14) comes from the dissipation due to a quasi-particle tunneling current between superconducting layers, while $\beta_2{\rm d}\psi_{\ell,\ell-1}/{\rm d}\tau$ in Eq.(15) is generated by the decay of a Cooper-pair into a single-particle pair in the inter-band tunneling process. It is noted that Eqs.(14) and (15) are a simple generalization of the dynamical equations in single-gap IJJ's given in Ref.[12].

Let us now present numerical results for the $I-V$ characteristics. In this paper we solve numerically Eqs.(14) and (15) for a stack of 20 junctions, i.e., $N=20$. The parameter values used in our calculations are summarized as

$$j_c^{(2)}/j_c^{(1)}=0.5, \quad \tilde{J}_{\rm in}=16.0, \quad \alpha_1=0.6, \quad \alpha_2=0.2, \quad \beta_1=0.1, \quad \beta_2=0.05.$$

In this case, noting Eq.(12), we find the Josephson plasma and the Josephson-Leggett mode frequencies as

$$\omega_{\rm pl}/\omega_{\rm pl}^{(1)}=1.22, \quad \omega_{\rm JL}/\omega_{\rm pl}^{(1)}=3.58.$$

Figure 1 shows the $I-V$ characteristics of the 2-gap IJJ's obtained by solving numerically Eqs.(14) and (15) for $N=20$. It is seen that the $I-V$ characteristics takes a multiple-branch structure, which is similar to that in a single-gap IJJ's such as Bi-2212. In this figure we plot only the highest and the two lowest voltage branches. Since in the present 2-gap system the Josephson critical current density is given by $j_c^{(1)} + j_c^{(2)}$ in the zero voltage state, the switching to the voltage state takes place at $I_{\rm ex} = 1.5$ for the present choice of the parameter values. In the highest voltage branch all the junctions are in the voltage state and show synchronized motion, i.e., $\theta_{10}(\tau) = \theta_{21}(\tau) = \cdots = \theta_{20,19}(\tau)$ and $\psi_{\ell,\ell-1}(\tau) = 0$. Then, one understands that the Josephson-Leggett mode cannot be excited in this uniform branch.

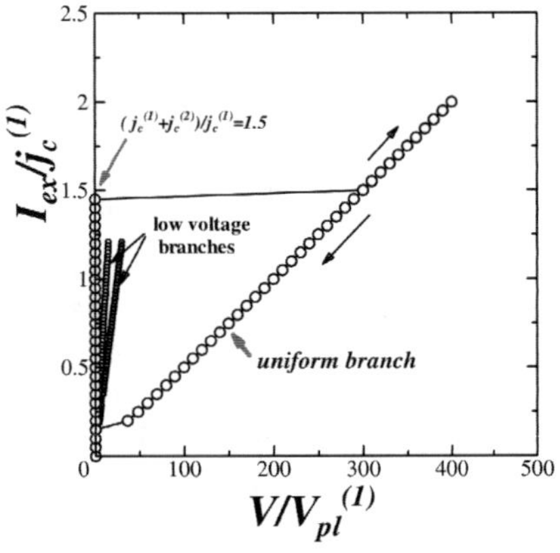

Figure 1. $I-V$ characteristics of 2-gap intrinsic Josephson junction stacks. The number of junctions in the stack is 20. The voltage V is normalized in terms of the plasma frequency of the band 1, $V_{\rm pl}^{(1)} = \hbar\omega_{\rm pl}^{(1)}/e^*$. The $I-V$ curves show hysteresis as indicated by arrows. In this figure only two lower voltage branches are shown. The total number of voltage branches is nearly equal to the number of junctions in the IJJ's, which is similar to the single-gap IJJ's such as Bi-2212 high T_c cuprates.

Figure 2 is the enlarged view of the $I-V$ curve of the lowest voltage branch. In this branch only one junction is in the voltage state. The phase differences in the other non-voltage junctions show oscillating motion of small amplitude. The voltage corresponding to the Josephson-Leggett mode frequency, $V_{\rm JL}/V_{\rm pl}^{(1)} = \omega_{\rm JL}/\omega_{\rm pl}^{(1)} = 3.58$, is indicated by the red arrow. It is clearly seen that a step-like structure, which is formed by the generation of an extra DC Josephson current, and the hysteretic behavior appear around this voltage. The analysis for our numerical results indicates that the motion of the relative phase difference at the voltage junction site is strongly enhanced at voltages near $V_{\rm JL}$, i.e., those on the $I-V$ curve from B to C. From this fact one understands that the Josephson-Leggett mode is resonantly excited by the Josephson oscillations when the Josephson frequency is close to the Josephson-Leggett mode frequency. The resonance between the Josephson-Leggett mode and the Josephson oscillations induces a step-like structure which is similar to the Shapiro steps in conventional Josephson junctions under microwave irradiation. Hence, one may call the step structure originating from the resonance with the Josephson-Leggett mode the *internal Shapiro step*. As discussed in Sec. I, the Josephson-Leggett mode can be excited by a *spatially inhomogeneous* AC electric field. In the 2-gap IJJ's spatially inhomogeneous distribution of an AC electric field is easily realized along the c-axis in the low voltage branches by applying a DC bias current. This is the reason why the Josephson-Leggett mode is resonantly excited only in the low voltage branches but is not in the highest voltage one. From this observation one may understand that the 2-gap IJJ's are the most suitable system to detect the Leggett mode in multi-gap superconductors. The detailed analysis of the $I-V$ characteristics in 2-gap IJJ's will be published in a separate paper.

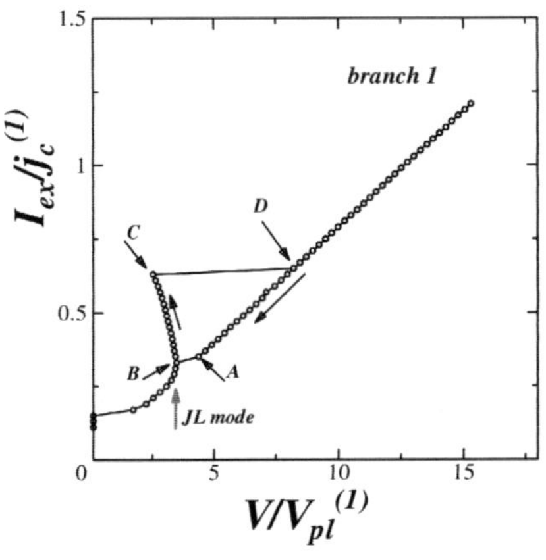

h

Figure 2. $I - V$ characteristics in the lowest voltage branch. When the bias current is increased from a value below that at point B a voltage jump appears at point C. In the current decreasing process on the $I - V$ curve including pont D the voltage jumps from A to B. Below the current value at point B no hysteresis appears in the $I - V$ curve.

In summary we have formulated a theory for the dynamics of the superconducting phase differences in 2-gap IJJ's. The coupled dynamical equations for the mean-phase differences and the relative phase differences were derived from the low energy effective Lagrangian. The equations describe correctly the two collective modes propagating in the c-direction in the 2-gap system, i.e., the longitudinal Josephson plasma and the Josephson-Leggett mode. It was shown that the Josephson oscillations can excite resonantly the Josephson-Leggett mode in spatially inhomogeneous voltage state.

Acknowledgments
The authors thank H. Matsumoto for discussions. This work was partially supported by Grant-in-Aid for Scientific Research (C) (No. 22540358) from Japan Society for the Promotion of Science.

References
[1] Kamihara Y, Watanabe T, Hirano M and Hosono H 2008 *J. Am.Chem. Soc.* **130** 3296
[2] Ogino H, Matsumura Y, Katsura Y, Ushiyama K., Horii S, Kishio K and Shimoyama J 2009 *Supercond. Sci.Technol.*. **22** 075008
[3] Ogino H, Sato S, Kishio K, Shimoyama J, Tohei T and Ikuhara Y 2010 *ArXiv* 10006.2355
[4] Nakamura H, Machida M, Koyama T and Hamada N 2009 *J. Phys. Soc. Jpn.* **78** 123712
[5] Ota Y, Machida M, Koyama T and Matsumoto H 2009 *Phys. Rev. Lett.* **102** 237003
[6] Ota Y, Machida M and Koyama T 2009 *J. Phys. Soc. Jpn.* **78** 103701
[7] Ota Y, Machida M and Koyama T 2010 *Phys. Rev.* B **81** 014502
[8] Leggett A J 1966 *Prog. Theor. Phys.* **36** 901
[9] Sharapov S G, Gusynin V P and Beck H 2002 *Eur. Phys. J* B **30** 45
[10] Blumberg G, Mialitsin A, Dennis B S, Klein M V, Zhigadrlo and Karpinski J 2007 *Phys. Reb. Lett.* **99** 227002
[11] Ota Y, Machida M and Koyama T 2010 *ArXiv*:1006.2184
[12] Koyama T and Tachiki M 1996 *Phys. Rev.* B **54** 16183

International Conference on Theoretical Physics Dubna-Nano 2010 IOP Publishing
Journal of Physics: Conference Series **248** (2010) 012037 doi:10.1088/1742-6596/248/1/012037

Theory and Simulation of Electromagnetic Wave Emission from Intrinsic Josephson Junctions

M Machida[1,3,4], Y Ota[1,3], N. Sasa[1], T Koyama[2,3] and H Matsumoto[2,3]

[1] CCSE, Japan Atomic Energy Agency, 6-9-3 Higashi-Ueno Taito-ku, Tokyo, 110-0015, Japan
[2] Institute for Materials Research, Tohoku University, 2-1-1 Katahira Aoba-ku, Sendai, 980-8577, Japan
[3] CREST(JST), 4-1-8 Honcho Kawaguchi, Saitama, 332-0012, Japan
[4] JST-TRIP, 5 Sambancho Chiyoda-ku, Tokyo, 102-0075, Japan

E-mail: machida.masahiko@jaea.go.jp

Abstract. Concerning the spontaneous THz electromagnetic wave emission from intrinsic Josephson junctions, we revisit the theory for intrinsic Josephson effects from standpoints of the couplings between stacked junctions and consequent superconducting phase dynamics. After reviewing the previous simulation works based on the theory, we present recent multi-physics simulation studies suggested by authors to systematically reproduce the emission experiments. Some typical simulation results are exhibited and the future perspective of the simulation is also given.

1. Introduction

The discovery of THz emission from layered High-T_c cuprate $Bi_2Sr_2CaCu_2O_{8+\delta}$ [1] has significantly renewed our research style on intrinsic Josephson effects, because the experimental stage generating the large emission was rather different from the previous idea [2, 3, 4, 5, 6, 7, 8]. The employed sample size was much beyond our expectation size fully guaranteeing sample uniformity. Indeed, the size belongs to a long junction class, but the whole sample surprisingly behaves just as a cavity. The other surprise was the zero magnetic-field condition for the strong spontaneous emission. Since the transverse excitation is required to generate the electromagnetic wave, a key factor for large spontaneous emission has been believed to be the application of the magnetic field. Consequently, the most of the previous works concentrated on the Josephson vortex flow [2, 3, 4, 5, 6, 7, 8]. However, the large emission was observed with no magnetic field [1], i.e., the transverse component is spontaneously excited without any external induction.

Since the seminal observation of the strong emission [1, 9, 10, 11], the central interest in the community studying intrinsic Josephson effects has been focused on mechanism of the strong emission and technique to practically enhance the emission power. The mechanism has been intensively pursued by successive experimental efforts. Now, it is well-known that when AC Josephson oscillation resonates with the cavity mode the strong spontaneous emission occurs [1]. Recently, a non-equilibrium heated range of I-V characteristics [12] has been also newly added to the strong emission stage, and the internal excitation pattern was found to be not fixed but rather controllable by dynamics of hot spot induced close to the electrode [10]. On the other hand, the technical advancement to enhance the emission power has been experimentally and theoretically examined. Consequently, not a situation attached on a large crystal substrate but a simply suspended one is suggested as one of the most promising one [13].

The first aim of the present paper is to revisit the theoretical framework to approach the emission mechanism in intrinsic Josephson junctions. Next, we review theoretical important concepts, i.e., the couplings between neighboring stacked junctions and the other dynamical issues. [14]. The basic framework has been still kept and used after the observation of the strong emission. The essence of the emission is believed to be inside the framework. Finally, we briefly touch the phase dynamics resonating with the cavity excitation numerically examined within the framework [13]. We suggest the so-called "multi-physics simulation" to directly simulate the electromagnetic wave emission and exhibit its typical simulation results.

2. Theoretical Framework on Intrinsic Josephson Junction

In this section, we begin with a microscopic Euclidean action, which is created by Bardeen-Cooper-Schrieffer (BCS) microscopic Hamiltonian for layered High-T_c superconductors [14]. By integrating out the fermion degree of freedom on the action [15], we have an effective action, which describes the superconducting phase dynamics in intrinsic Josephson junctions as shown in Fig. 1. We also have the equations of motion of both the superconducting phase and the charge density from the effective action. In the equations, an approximation, in which the charge screening length goes to zero, reduces to the coupled sine-Gordon equation derived by Sakai *et al.* [16] and Bulaevskii *et al.* [17], and another assumption, in which the in-plane variation is negligible, leads to the equation of motion suggested by Machida *et al.* [14]. Finally, the equation arrives at the simplest Koyama-Tachiki model [18] if the diffusion current via the capacitive coupling is dropped. The assumption is fully valid in under-damped junctions. These equations have been numerically and analytically examined by several authors to study the intrinsic Josephson effects repeatedly [4, 5, 6, 19, 20, 21, 22, 23, 24, 25, 26]. We briefly touch the typical results.

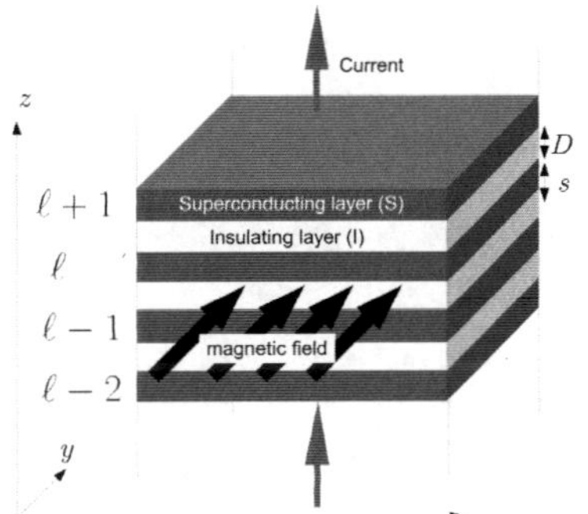

Figure 1. A schematic picture for an intrinsic Josephson junction.

2.1. Modeling and Effective Action

Let us start with a microscopic description for the stacked Josephson junction schematically depicted in Fig. 1. We confine ourselves a case, where the current is biased parallel to the layer stacking direction and the magnetic field is applied only parallel to the junction plane as shown in Fig. 1, i.e., the magnetic field piercing the junction plane creating pancake vortices are not taken into account.

Employing the BCS model for each single superconducting tunneling-junction and extending the model into stacked junction systems, the system's grand partition function is given by the following imaginary-time functional integral over the Grassmann fields, $\bar{\psi}_{\sigma\ell}$, $\psi_{\sigma\ell}$, for electrons, the gauge field, φ_ℓ, $\mathbf{A}_\ell$, $A^z_{\ell+1,\ell}$, and the auxiliary field corresponding to the superconducting order parameter, Δ_ℓ, where ℓ is the layer index and the z-component of the vector potential is defined on the link as $A^z_{\ell+1,\ell} = \int_\ell^{\ell+1} dz A^z$,

$$Z = \prod_\ell \int \mathcal{D}\bar{\psi}_{\sigma\ell}\mathcal{D}\psi_{\sigma\ell}\mathcal{D}\mathbf{A}_\ell\mathcal{D}A^z_{\ell+1,\ell}\mathcal{D}\varphi_\ell\mathcal{D}\Delta_\ell \mathrm{e}^{-\frac{1}{\hbar}S[\bar{\psi}_{\sigma\ell},\psi_{\sigma\ell},\mathbf{A}_\ell,A^z_{\ell+1,\ell},\varphi_\ell,\Delta_\ell]}. \tag{1}$$

In Eq. (1), S is the Euclidean action defined as $S = S_{\text{matter}} + S_{\text{field}}$, where

$$\begin{aligned} S_{\text{matter}} = \sum_\ell \int_0^{\beta\hbar} \mathrm{d}\tau \int \mathrm{d}\mathbf{r} \Big\{ & \frac{|\Delta_\ell|^2}{g} + \bar{\psi}_{\sigma\ell}(\mathbf{r},\tau)(\hbar\partial_\tau + \mathrm{i}e\varphi_\ell)\psi_{\sigma\ell}(\mathbf{r},\tau) \\ & + \bar{\Delta}_\ell\psi_{\downarrow\ell}\psi_{\uparrow\ell} + \Delta_\ell\bar{\psi}_{\uparrow\ell}\bar{\psi}_{\uparrow\ell} - \bar{\psi}_{\sigma\ell}(\mathbf{r},\tau)\left[\frac{\hbar^2}{2m}\left(\nabla - \mathrm{i}\frac{e}{\hbar c}\mathbf{A}_\ell\right)^2 + \mu_\ell\right]\psi_{\sigma\ell}(\mathbf{r},\tau) \\ & + (T_{\ell+1,\ell}\mathrm{e}^{\frac{e}{\hbar c}A^z_{\ell+1,\ell}}\bar{\psi}_{\sigma\ell+1}\psi_{\sigma\ell} + \text{c.c}) \Big\}, \end{aligned} \tag{2}$$

and

$$S_{\text{field}} = \sum_\ell \int_0^{\beta\hbar} \mathrm{d}\tau \int \mathrm{d}\mathbf{r} \frac{D}{8\pi}\left[\epsilon_c(E^z_{\ell+1,\ell})^2 + (B^x_{\ell+1,\ell})^2 + (B^y_{\ell+1,\ell})^2\right], \tag{3}$$

where, $T_{\ell+1,\ell}$ is the tunneling matrix element between ℓth and $(\ell+1)$th superconducting layers, $E^z_{\ell+1,\ell}$ is the electric field $(\equiv (1/cD)\partial_\tau A^z_{\ell+1,\ell} - (\varphi_{\ell+1} - \varphi_\ell)/D)$ along the z-axis, ϵ_c is the dielectric constant of the insulating barriers, and $B^x_{\ell+1,\ell}$ $(\equiv (1/D)\partial_y A^z_{\ell+1,\ell} - (A^y_{\ell+1} - A^y_\ell)/D)$ and $B^y_{\ell+1,\ell}$ $(\equiv (A^x_{\ell+1} - A^x_\ell)/D - (1/D)\partial_x A^z_{\ell+1,\ell})$ are the components of the magnetic field parallel to the junctions.

Following the standard technique to integrating out the fermion degree of freedom [15], we have the following effective action for the gauge invariant superconducting phase difference with electromagnetic fields,

$$\begin{aligned} S_{\text{eff}} = \sum_\ell \int_0^{\beta\hbar} \mathrm{d}\tau \int \mathrm{d}\mathbf{r}_\parallel \Big\{ & \frac{D}{8\pi}\left[\epsilon_c(E^z_{\ell+1,\ell})^2 + (B^x_{\ell+1,\ell})^2 + (B^y_{\ell+1,\ell})^2\right] + msn_s {v_{s}}_\ell^{\parallel 2} \\ & + \frac{s}{8\pi\mu_f^2}\left(\frac{\phi_0}{2\pi c}\frac{\partial\theta_\ell}{\partial\tau} + \varphi_\ell\right)^2 + J\cos P_{\ell+1,\ell}(\tau,\mathbf{r}_\parallel)\Big\} \\ + \sum_\ell \int_0^{\beta\hbar} \mathrm{d}\tau \int_0^{\beta\hbar} \mathrm{d}\tau' \int \mathrm{d}\mathbf{r}_\parallel & \left[-\alpha(\tau-\tau')\cos\frac{P_{\ell+1,\ell}(\tau,\mathbf{r}_\parallel) - P_{\ell+1,\ell}(\tau',\mathbf{r}_\parallel)}{2}\right], \end{aligned} \tag{4}$$

where $\mu_f \equiv \sqrt{\lambda^2_{\text{TF}}/4\pi\epsilon}$ is a constant related to the Thomas-Fermi screening length λ_{TF}, s is the thickness of the superconducting layers, n_s is the superfluid density, and $P_{\ell+1,\ell}$ and ${v_s}^\parallel_\ell$ are, respectively, the gauge-invariant phase difference and the in-plane superfluid velocity defined as $P_{\ell+1,\ell} \equiv \theta_{\ell+1} - \theta_\ell - (2\pi/\phi_0)A^z_{\ell+1,\ell}$ and ${v_s}^\parallel_\ell \equiv (\hbar/2m)[\nabla_\parallel\theta_\ell - (2\pi/\phi_0)\mathbf{A}^\parallel_\ell]$. In obtaining Eq. (4), we utilize a local approximation for the integral kernels coming from the intra-layer current, charge density, and Cooper pair tunneling current, i.e. msn_s, $2/8\pi\mu_f^2$, and J are given as local coefficients [27, 28]. These approximations are valid for the low energy phenomena, with which the tunneling process is relevant. The integral kernels $\alpha(\tau-\tau')$ in Eq. (4) describes the frequency-dependent quasi-particle tunneling, whose explicit formula is given in Ref. [15] for example. It should be noted that $\alpha(\tau)$ contributes to the phase dynamics as Ohmic dissipation discussed in the Caldera-Leggett type model [29].

2.2. Dynamical Equations and Couplings between Junctions

From the action, Eq. (4), one can get the following coupled equations for the gauge-invariant phase difference and the charge density,

$$\frac{\epsilon_c\phi_0}{2\pi c^2 D}\frac{\partial^2 P_{\ell+1,\ell}}{\partial t^2}+\frac{4\pi}{cD}(\epsilon_c\mu_f^2-\lambda_{ab}^2)(\frac{\partial\rho_{\ell+1}}{\partial t}-\frac{\partial\rho_\ell}{\partial t})-\frac{\phi_0}{2\pi D}\frac{\partial^2 P_{\ell+1,\ell}}{\partial x^2}$$
$$+\frac{4\pi}{c}(\sin P_{\ell+1,\ell}+\beta\frac{dP_{\ell+1,\ell}}{dt})-\frac{4\pi}{c}\frac{\lambda_{ab}^2}{sD}(\sin P_{\ell+2,\ell+1}+\sin P_{\ell,\ell-1}-2\sin P_{\ell+1,\ell})$$
$$-\frac{4\pi}{c}\frac{\lambda_{ab}^2}{sD}\beta(\frac{dP_{\ell+2,\ell+1}}{dt}+\frac{dP_{\ell,\ell-1}}{dt}-2\frac{dP_{\ell+1,\ell}}{dt})=0, \tag{5}$$

$$\frac{\phi_0\epsilon_c}{2\pi cD}(\frac{\partial P_{\ell+1,\ell}}{\partial t}-\frac{\partial P_{\ell,\ell-1}}{\partial t})+\frac{4\pi\epsilon_c\mu_f^2}{D}(\rho_{\ell+1}+\rho_{\ell-1}-2\rho_\ell)=4\pi s\rho_\ell, \tag{6}$$

In deriving Eqs. (5) and (6), one uses the modified Josephson relations,

$$\frac{\phi_0}{2\pi c}\frac{\partial P_{\ell+1,\ell}}{\partial t}=-4\pi\mu_f^2(\rho_{\ell+1}-\rho_\ell)+DE^z_{\ell+1,\ell}, \tag{7}$$

$$\frac{\phi_0}{2\pi}\frac{\partial P_{\ell+1,\ell}}{\partial x}=\frac{4\pi\lambda_{ab}^2}{c}(j^x_{\ell+1}-j^x_\ell)+DB_{\ell+1,\ell}. \tag{8}$$

It is noted that Eq. (7) and (8) contain two parameters, μ_f and λ_{ab}, which differ extremely in their length-scales, i.e., $\mu_f \ll \lambda_{ab}$. The terms with the coefficient λ_{ab} originate from the inductive coupling between junctions [16, 17], while the capacitive coupling between junctions, which arises from the incomplete charge-screening along the stacking direction, yields the terms with the coefficient μ_f [18, 30]. The difference between these two couplings can be understood by linearizing the sinusoidal function and taking a continuum limit on the discretized difference terms. The approximated equations corresponds to the extended versions of the London equations for anisotropic superconductors, which clarify characteristic lengths of both the couplings. Since $\epsilon_c\mu_f^2 \ll \lambda_{ab}^2$, one finds that the inductive coupling is overwhelmingly dominant in the presence of the parallel magnetic field. However, only from this fact, one cannot conclude that the capacitive coupling between junctions can be always neglected in intrinsic Josephson junctions systems, since the inductive coupling vanishes in the absence of Josephson vortices. This is just the case, in which the electromagnetic wave emission is experimentally observed.

3. Numerical Simulations and Superconducting-Phase Dynamics

In the previous section, we derived the general equations describing both the inductive and capacitive couplings. In this section, we revisit the intrinsic Josephson effects based on each coupling by separating out the other couplings. First, we begin with the simplest situation, in which the spatial variation of the superconducting phase along the junction plane is dropped and only the superconducting phase dynamics relevant to the capacitive coupling is taken into account. We also touch recent theoretical advancement on the superconducting phase dynamics. Next, we turn to the case of the so-called stacked long-junctions, in which the inductive coupling is a dominant coupling and the vortex dynamics play a key role.

3.1. Koyama-Tachiki Model and Multiple Branches in I-V Characteristics

Under the assumption of uniformity of the superconducting phase along the junction plane, the terms originated from the inductive coupling diminish in Eq. (5) and (6), and a simple equation appears as follows.

$$\frac{\mathrm{d}^2}{\mathrm{d}t'^2}P_{\ell+1,\ell}+\gamma\frac{\mathrm{d}}{\mathrm{d}t'}P_{\ell+1,\ell}+\sin P_{\ell+1,\ell} = \frac{I}{j_c}+\frac{\epsilon_c\mu^2}{sD}\Big[\sin P_{\ell+2,\ell+1}-2\sin P_{\ell+1,\ell}+\sin P_{\ell,\ell-1}\Big]$$
$$+\frac{\epsilon_c\mu^2}{sD}\gamma\Big[\frac{\mathrm{d}P_{l+2,l+1}}{\mathrm{d}t'}-2\frac{\mathrm{d}P_{l+1,l}}{\mathrm{d}t'}+\frac{\mathrm{d}P_{l,l-1}}{\mathrm{d}t'}\Big], \tag{9}$$

where $t' = \omega_p t$ and ω_p is the Josephson-plasma frequency, $\omega_p = \sqrt{8\pi^2 cDj_c/\epsilon_c\phi_0}$. This equation was firstly derived by Machida *et al.* [14], through the microscopic treatment as briefly described above. We immediately notice that if one drops the third term in the right-hand side, then Eq. (9) coincides with Koyama-Tachiki model [18] derived by their phenomenological treatment. The model has been extensively investigated initially by Machida *et al.* [30], and recently by Shukrinov *et al.* [25, 26]. The main physics of the model is the existence of the localized rotating mode, which has been intensively studied in nonlinear physics community motivated by a counter-intuitive concept, in which fully stable localized excitations emerge even on non-integrable discrete lattices like solitonic excitations. The experimental confirmation of such localized excitations has been long an important issue for the nonlinear physics community [32, 33]. Machida *et al.* [31] argued that intrinsic Josephson junction is an excellent reality of the localized excitation.

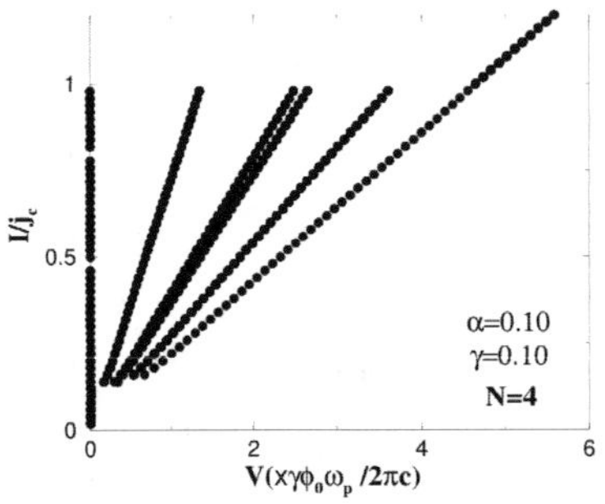

Figure 2. A typical numerical simulation result (*I*-*V* characteristics) of the Koyama-Tachiki model. $\alpha(\equiv \epsilon_c\mu_f^2/sD) = 0.10$, and $\gamma = 0.10$ in Eq. (9). The number of the stacked junctions, $N = 4$.

Here, let us display a typical example of the simulation studies on Koyama-Tachiki model. Fig. 2 is typical *I*-*V* characteristics obtained by ramping up/down the current. The *I*-*V* characteristics clearly show the branching behaviors, which are always measured in well-fabricated small samples. The numerical simulation reveals that the branching arises from the varietical excitation nature of the localized rotating mode. On the other hand, the instability of the AC Josephson oscillation due to the resonance with the longitudinal plasma mode has been recently examined through numerical studies done by Shukrinov *et al.* [25]. The instability is called "breakpoint", which is a unique feature characteristic to capacitively coupled intrinsic Josephson junctions. This can be assigned as a cavity resonance of AC Josephson oscillation along the longitudinal stacked direction. The breakpoint phenomenon is dual to the electromagnetic wave emission, in which AC Josephson oscillation resonates with a cavity mode excited along the in-plane direction. Shukrinov *et al.* [26] also revealed that the longitudinal resonance brings about charge-density oscillation dynamics as a consequence of the resonant instability. The experimental confirmation has been challenged by some experimental groups. However, it still remains unexplored fully.

3.2. Coupled sine-Gordon Equation and Resonant Vortex Dynamics

By eliminating the constraint of the uniformity along the in-plane direction and taking into account of only the inductive coupling in Eqs. (7) and (8), we have the coupled sine-Gordon equation derived by Sakai *et al.* [16] and, Bulaevskii *et al.* [17] as follows.

$$\begin{aligned}\frac{\partial^2 P_{\ell+1,\ell}}{\partial x'^2} &= -\frac{\lambda_{ab}^2}{sD}\left(\frac{\partial^2 P_{\ell+2,\ell+1}}{\partial t'^2} + \frac{\partial^2 P_{\ell,\ell-1}}{\partial t'^2} - 2\frac{\partial^2 P_{\ell+1,\ell}}{\partial t'^2}\right) \\ &- \frac{{\lambda_{ab}}^2}{sD}(\sin P_{\ell+1,\ell} + \sin P_{\ell,\ell-1} - 2\sin P_{\ell+1,\ell})\end{aligned}$$

$$- \frac{{\lambda_{ab}}^2}{sD}\gamma(\frac{\partial P_{\ell+2,\ell+1}}{\partial t'} + \frac{\partial P_{\ell,\ell-1}}{\partial t'} - 2\frac{\partial P_{\ell+1,\ell}}{\partial t'}) + \sin P_{\ell+1,\ell} + \gamma\frac{\partial P_{\ell+1,\ell}}{\partial t'} + \frac{\partial^2 P_{\ell+1,\ell}}{\partial t'^2}, \quad (10)$$

where $x' = x/\lambda_c$ with $\lambda_c = \sqrt{c\phi_0/8\pi^2 D j_c}$. This equation was initially examined by Sakai *et al.*, [16] in their analytical way, while a pioneering simulation on the Josephson vortex dynamics was made by Kleiner [20]. The main issues of the equation are the existence of the multiple plasma excitation modes propagating mainly along the in-plane direction, called the transverse plasma, and the resonant vortex dynamics with the multiple plasma modes. The latter issue has been numerically investigated by several groups repeatedly. To our knowledge, the most surprising physics is the emergence of the in-phase locked flux flow as shown in Fig. 3 due to the resonance between the in-phase like propagating mode and flowing vortex lattice. This flow state was expected to emerge in a high-bias voltage range because the resonating mode frequency is the highest among all the transverse modes, and the predicted stability was numerically confirmed by Ustinov *et al.*[21] and Machida *et al.*[14]. Afterwards, several experiments have been made on the equivalent situations, and indirect evidence has been actually obtained [7, 8]. However, since the application of the magnetic field, whose magnitude is an order of Tesla, complicates the experimental setups, the advancement speed is slow compared to the recent zero-field radiation.

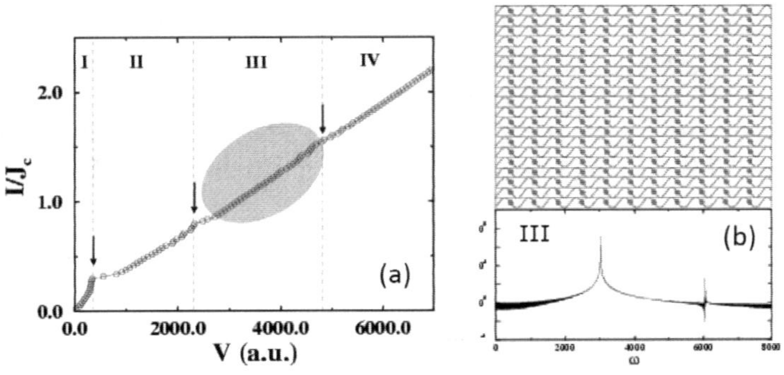

Figure 3. A typical numerical result of the coupled sine-Gordon equation, Eq. (10). In the high voltage region III, which is pointed by the pink color, the in-phase locked flux flow state emerges.

4. Multi-physics Modeling for Electromagnetic Wave Emission

In this section, we briefly present our recent theoretical and numerical works to understand the mechanism of the strong spontaneous electromagnetic-wave emission in the zero field application [13]. Our strategy of the simulation work is as follows. Since the emission is sensitively dependent on the experimental setups, a full consideration on the experimental situations is of importance to understand the mechanism and investigate device design including further power enhancement. Thus, we take account of the external vacuum, attached electrodes, superconducting or normal substrate as well as internal capacitively and inductively coupled junctions. Initially, we made two-dimensional frameworks, so-called xy- and xz-models, which assume uniformity in z-(c-axis) and y-(an in-plane) axis directions, respectively, to avoid an enormous computational degree of freedoms in three-dimensional direct cases. At the present, we also start to perform three-dimensional full simulations by parallelizing the simulation code and employing a multi-scale technique to save the computational degree of freedom through consideration of the scale difference between the internal junction and external vacuum regions.

4.1. Two-dimensional xy-model Simulation

Here, we concentrate on only xy-model due to the space limitation. For xz-model, see Ref. [13]. Figure 4 shows typical simulation results for the excited magnetic field inside and outside the junction. The central regions dominated by red arrows are intrinsic Josephson junctions, and the surrounded ones are the external vacuum. The emitted electromagnetic wave passes through the edges of the external vacuum and does not reflect at all. For no reflection, we put a specific region called "perfect matched layer" at the edges. See Refs. [13, 34] for the details of the perfect matched layer.

The Maxwell equations inside the junction region for the xy-model are given as [13]

$$\frac{1}{c}\frac{\partial \epsilon E_z}{\partial t} = \frac{\partial B_y}{\partial x} - \frac{\partial B_x}{\partial y} - \frac{4\pi}{c} j_c \sin\phi - \sigma E_z, \tag{11}$$

$$\frac{1}{c}\frac{\partial B_x}{\partial t} = -\frac{\partial E_z}{\partial y}, \quad \frac{1}{c}\frac{\partial B_y}{\partial t} = \frac{\partial E_z}{\partial x}, \tag{12}$$

where σ is the quasi-particle conductivity, j_c is the critical Josephson current density, and ϕ is the gauge-invariant phase difference. The external c-axis current is introduced by the Ampere's law. The external vacuum region outside the junction obeys the Maxwell equations, which coincides with $\sigma = j_c = 0$, and $\epsilon = \epsilon_0$ in Eq.(11) and (12). Typical simulation results are displayed in Fig. 4, where the snapshots of the spatial distribution of the inductive field are displayed. Through this kind of simulation, we can obtain the angle dependence of the power of the emitted electromagnetic wave with I-V characteristics. The spatial pattern depends on the sample shape and size and the internal excitation pattern varying with the voltage. For more details of the results and their analysis, see Ref. [13].

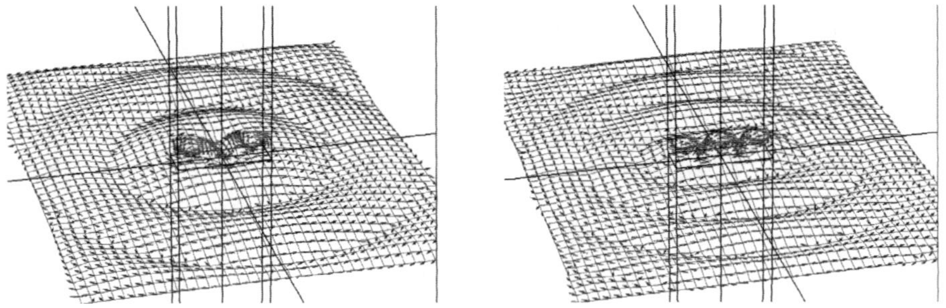

Figure 4. Spatial distributions of emitted magnetic fields depending on the applied voltage. The junction regions are drawn by red arrows.

5. Summary

We reviewed the theoretical framework for the intrinsic Josephson junction systems. We started with the microscopic model and derived the general equations of the superconducting phase and the charge density. The obtained equations were reduced to the Koyama-Tachiki model only for the capacitive coupling and the coupled sine-Gordon equation describing only the most dominant inductive coupling, and typical intrinsic Josephson effects were presented. Finally, we introduced our recent numerical works on the strong electromagnetic wave emission. The used framework is basically the same as the previous works, but the device reality including the external vacuum, attached electrodes, and substrates is taken into account. With further advancement of supercomputers, the employed simulation style, i.e., multi-physics simulation will have an more important role in the future development of intrinsic Josephson junctions as a THz light source.

Acknowledgments

The authors wish to acknowledge valuable discussions with K. Kadowaki, R. Kleiner, H-B. Wang, A. E. Koshelev, and Y. Shukrinov. MM was partially supported by Grant-in-Aid for Scientific Research on Priority Area "Physics of new quantum phases in superclean materials" (No. 20029019) from the Ministry of Education, Culture, Sports, Science and Technology of Japan (MEXT). TK was partially supported by Grant-in-Aid for Scientific Research (C) (No. 22540358) from Japan Science and Technology Agency.

References

[1] Ozyuzer L, Koshelev A E, Kurter C, Gopalsami N, Li Q, Tachiki M, Kadowaki K, Yamamoto T, Minami H, Yamaguchi H, Tachiki T, Gray K E, Kwok W K and Welp U 2007 *Science* **318** 1291
[2] Koyama T and Tachiki M 1995 *Solid State Commun.* **96** 367
[3] Hechtfischer G, Kleiner R, Ustinov A V and Müller P 1997 *Phys. Rev. Lett.* **79** 1365
[4] Machida M, Koyama T, Tanaka A and Tachiki M 2000 *Physica* C **330** 85
[5] Tachiki M, Iizuka M, Minami K, Tejima S and Nakamura H 2005 *Phys. Rev.* B **71** 134515
[6] Savel'ev S, Yampol'skii V, Rakhmanov A and Nori F 2005 *Phys. Rev.* B **72** 144515
[7] Kadowaki K, Kakeya I, Yamamoto T, Yamazaki T, Kohri M and Kubo Y 2006 *Physica* C **437-438** 111
[8] Bae M H, Lee H J and Choi J H 2007 *Phys. Rev. Lett.* **98** 027002
[9] Minami H, Kakeya I, Yamaguchi H, Yamamoto T and Kadowaki K 2009 *Appl. Phys. Lett.* **95** 232511
[10] Wang H B, Gue'non S, Gross B, Yuan J, Jiang Z G, Zhong Y Y, Grünzweig M, Iishi A, Wu P H, Hatano T, Koelle D and Kleiner R 2010 *Phys. Rev. Lett.* **105** 057002
[11] Tsujimoto M, Yamaki K, Deguchi K, Yamamoto T, Kashiwagi T, Minami H, Tachiki M, Kadowaki K and Klemm R A 2010 *Phys. Rev. Lett.* **105** 037005
[12] Lee K, Wang W, Iguchi I, Tachiki M, Hirata K and Mochiku T 2000 *Phys. Rev.* B **61** 3616
[13] Koyama T, Matsumoto H, Machida M and Kadowaki K 2009 *Phys. Rev.* B **79** 104522
[14] Machida M, Koyama T, Tanaka A and Tachiki M 2000 *Physica* C **331** 85
[15] Šimánek E 1994 *Inhomogeneous Superconductors: Granular and Quantum Effects* (New York: Oxford University Press).
[16] Sakai S, Bodin P and Pedersen N F 1993 *J. Appl. Phys.* **73** 2411
[17] Bulaevskii L N, Zamora M, Baeriswyl D, Beck H and Clem J R 1994 *Phys. Rev.* B **50** 12831
[18] Koyama T and Tachiki M 1996 *Phys. Rev.* **B** 16183
[19] Kleiner R, Gaber T and Hechtfischer G 2000 *Phsy. Rev.* B **62** 4086
[20] Kleiner R 1994 *Phys. Rev.* B **50** 6919
[21] Ustinov A V and Sakai S 1998 *Appl. Phys. Lett.* **73** 686
[22] Bulaevskii L N and Koshelev A E 2007 *Phys. Rev. Lett.* **99** 057002
[23] Lin S and Hu X 2008 *Phys. Rev. Lett.* **100** 247006
[24] Koshelev A E 2008 *Phys. Rev.* B **78** 174509
[25] Shukrinov Yu M and Mahfouzi F 2007 *Phys. Rev. Lett.* **98** 157001
[26] Shukrinov Yu M, Hamdipour M and Kolahchi M R 2009 *Phys. Rev.* B **80** 014512
[27] Ambegaokar V, Eckern U and Schön G 1982 *Phys. Rev. Lett.* **48** 1745
[28] Eckern U, Schön G and Ambegaokar V 1984 *Phys. Rev.* B **30** 6419
[29] Caldeira A O and Leggett A J 1983 *Ann. Phys. (N.Y.)* **149** 374
[30] Machida M, Koyama T and Tachiki M 1999 *Phys. Rev. Lett.* **83** 4618
[31] Machida M and Koyama T 2004 *Phys. Rev.* B **70** 024523
[32] Zheng Z, Hu B and Hu G 1984 *Phys. Rev.* E **57** 1139
[33] Takeno S and Peyard M 1996 *Physica* D **92** 140
[34] Berenger J P 1994 *J. Comput. Phys.* **114** 184

Journal of Physics: Conference Series **248** (2010) 012038 doi:10.1088/1742-6596/248/1/012038

Experimental and theoretical investigation on high-Tc superconducting intrinsic Josephson junctions

Alexander Grib[1,2], **Yury Shukrinov**[3], **Frank Schmidl**[1], **Paul Seidel**[1]

[1] Institut für Festkörperphysik, Friedrich-Schiller-Universität Jena, D-07743 Jena, Germany
[2] Physics Department, V. N. Karazin Kharkiv National University, 61077 Kharkiv, Ukraine
[3] Joint Institute for Nuclear Research, BLTP, Dubna, 141980, Russia

E-mail: "Paul Seidel" <Paul.Seidel@uni-jena.de>

Abstract. Within the last years many groups have realized and investigated different types of intrinsic Josephson junction (IJJ) arrays out of high-temperature superconducting single crystals or thin films. We tried to improve the synchronization between the junctions by external shunts. Mesa structures as well as microbridges on vicinal cut substrates showed multi-branch behaviour in their IV characteristics and random switching between branches. Theoretical modelling was done investigating phase dynamics and stability numerically as well as analytically. Branch structure in current voltage characteristics of IJJ is studied in the framework of different models, particularly, in capacitevely coupled Josephson junctions (CCJJ) model and CCJJ model with diffusion current. Results of modelling of return current in IV characteristics for stacks with different number of IJJ are presented. We discussed the possible mechanisms of synchronization and the ranges of stability. Conclusions with respect to application of such arrays such as radiation sources were given.

1. Introduction

Since the discovery of the intrinsic Josephson effects by Kleiner et al. [1] there are a lot of works concerning the physics of the intrinsic junctions, e.g. [2-13]. Thus we will restrict ourself here to facts which are relevant for application. The intrinsic Josephson effects in the cuprates where the superconducting CuO_2 - planes are coupled via non-superconducting barrier layers within the unit cell of the materials can be used to produce series junction arrays. The number of junctions depends on the thickness of the high-Tc material between the electrodes. Optimization of technology and a complex control of the preparation process allows to reduce the number of junctions to a few or even a single one. There are different technologies using mesa geometry of vertical stacks, stacks grown at steps in the substrate, focussed ion beam cutted single crystal or film structures and planar "horizontal stacks" grown on vicinal cutted substrates. Additional doping or external shunting of the intrinsic arrays can be used to adjust the junction parameters or to enhance the synchronization of the junctions within the array. In this way the weak coupling is realized in a periodic stack of Josephson junctions with atomic dimensions along the crystallographic c-axis. These intrinsic Josephson junctions can be described by the Josephson or more general weak link physics but the small dimensions as well as the series connection of

Journal of Physics: Conference Series **248** (2010) 012038 doi:10.1088/1742-6596/248/1/012038

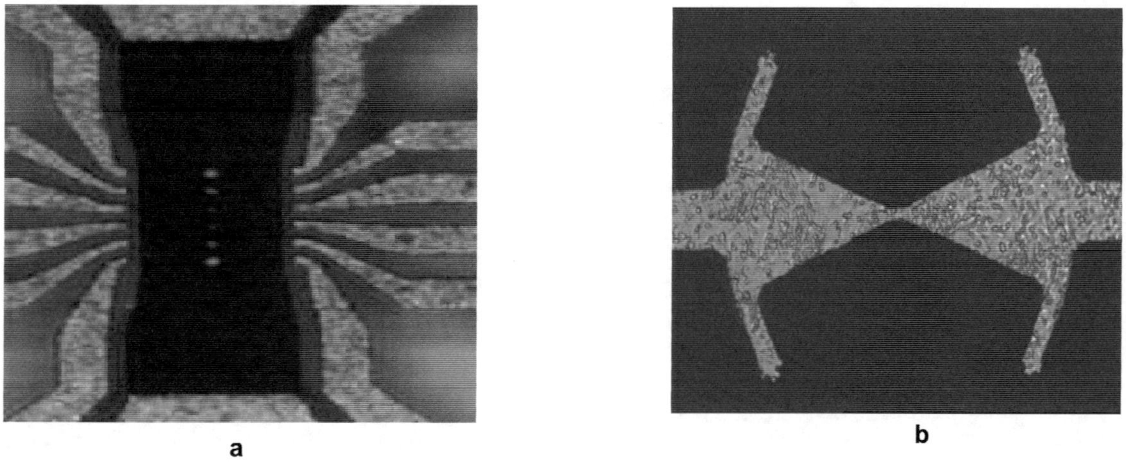

Figure 1. Two types of thin film intrinsic Josephson junctions. (a) 6 Tl-2212 mesas with areas from 2x2 μm^2 to 5x5 μm^2 in the center of the substrate before they were covered by a gold contact which is cut by ion etching to get separate contacts for 4-point measurement; (b) the microbridge on vicinal cut substrate with a 2x2 μm^2 *planar* stack imbedded in an antenna structure (Tl-2212 on 20 degrees vicinal cut $LaAlO_3$)

many junctions gives rise to some special behaviour and parameters. Modeling of the junctions has to take into account these special properties. Some examples will be given here.

2. Preparation of junctions

While the first investigations on the intrinsic junctions used single crystals or etched parts out of them, different thin film technologies have been developed meanwhile, too. Thin film structures used a mesa-type geometry where the high-Tc film is patterned with lateral dimensions in the μm-range, e.g. [14, 15] and references therein. The problem of metal electrodes to the mesas was solved even for a four-point-measurement which requires two separated contacts on top of each mesa like in Fig. 1a, see Seidel et al. [16] . A very interesting technology of preparation of quite homogeneous and well-defined junction arrays was proposed by Kim et al. [17] for single crystal whiskers and later adapted to single crystal pieces mounted on substrates [6]. Focused ion beams were used to etch the small well-defined stack of junctions out of the material from two sides in a way that the remaining high-Tc materials act as electrodes to the stack. By this "double-sided fabrication method" or "flip-chip technique" the quality and homogeneity of the intrinsic junction arrays was dramatically improved. On the other side technologies were developed to realize a very small junction number down to the single junction limit, see e.g. [4], [18], and [19]. Koval et al. [20] showed that T_C, I_C and R_n of intrinsic Bi-2212 Josephson junctions can be tuned in a large range by current injection. This carrier injection effect is reversible and persistent. An alternative way to realize a thin film intrinsic stacked junction array uses substrates with a surface not parallel to the CuO_2-planes [21]. On such vicinal cutted substrates the film growths with CuO_2-planes tilted with respect to the surface. Patterning of a microbridge results in a nearly horizontal stack of junctions and the length of the microbridge corresponds to the number of junctions in series, see Fig. 1b. Intrinsic Josephson behavior of the microbridges was observed for misorientation angles equal or larger than 15 degrees [22]. This kind of arrays offers new possibilities for synchronization [23].

3. Experiments and results

The superconducting CuO_2 -planes in IJJ are separated by coupling layers of some tenth of a nanometer. This leads to many differences compared to artificial prepared planar barrier junctions with quite compact superconducting electrodes and much thicker single barrier layers. The other main difference is that intrinsic junctions are naturally series arrays instead of single junctions. Thus the observed IV-characteristic is a sum of single junctions characteristics leading to many branches up to high voltages, see Fig. 2a. For very high voltages the heat dissipation leads to non-equilibrium effects and negative differential resistance parts in the IV. While the number of junctions in the intrinsic arrays is quite easy to control by thickness of the stack of superconducting unit cells their homogeneity is still a problem. Thus there is a quite large spread in single junction parameters. If the spread can be reduced the internal synchronization of the junctions improves the dynamic of these arrays. Additional shunting or resonance environment can further improve the synchronization leading to a collective many-junction behavior [24-26]. This is of relevance e.g. for radiation sources realized by intrinsic arrays. The lateral dimensions of intrinsic Josephson junctions play a crucial role, too. Perpendicular to the atomic arrays there is flux-flow of Josephson vortices corresponding to these dimensions. This results in plasma waves and additional dynamic effects. On one side such effects can be applied for radiation sources [10, 11] on the other side they lead to a complex behavior and additional noise contributions, see e.g. [8] , [10], [27], [28] and references therein.

4. Synchronization

One of the most important applications of the intrinsic Josephson effect is the construction of sub-microwave sources of coherent radiation. The common synchronizing ac current in all junctions is the condition of strong synchronization. Radiation of synchronizing junctions itself can produce this ac current if there is a feedback in the system. The existence of the current resonant mode in the system provides both the strong ac current produced by radiation of junctions and a strong feedback. Thus, an application of a superconducting resonator is a solution to obtain synchronization of Josephson junctions. It was shown in [24] that the synchronizing ac current is largest in serial arrays of junctions. Intrinsic Josephson junctions satisfy the requested conditions and small spread of critical currents and normal resistances can be in principle achieved. In the present paper we present results of numerical and analytical investigations [23, 25, 28] and experiments [23] for stacks of junctions in a resonator. In our 'lumped' analysis the Josephson junction is always the parallel connection of the source of Josephson current I_c, the resistance R_n and the capacitance C (this is a so-called extended resistively shunted model of a junction). The resonator is always modeled as a parallel connection of some capacitance C_{res} and an inductance L_{res}, so the resonance frequency is $\frac{1}{\sqrt{L_{res}C_{res}}}$. At first, we present results for synchronization of stacks of 'classical' junctions, i.e. junctions which are separated by large distances of superconductor which have the constant order parameter, and then we discuss the specific features of synchronization of intrinsic junctions. The system of differential equations for phase dynamic of N junctions with different loads can be represented schematically as follows:

$$\left(1-(-1)^k\,\delta\right)\left(\beta_C\ddot{\varphi}_k\left(\tau\right)+\dot{\varphi}_k+sin\left(\varphi_k\right)\right)=i_b-\tilde{i_k}$$
$$\sum\dot{\varphi}_k=D\left(\tilde{i_k}\right), \qquad (1)$$

where k is the number of the junction, δ is spread of critical currents, $\tilde{i_k}$ is the ac current flowing in the junctions due to a feedback, φ_k, $\dot{\varphi}_k$ and $\ddot{\varphi}_k$ are the phase difference across k-th junction and its first and second derivative with respect to the dimensionless time $\tau=\frac{2\pi R_n I_c t}{\Phi_0}$ with Φ_0 as the quantum of magnetic flux, i_b is the dc bias current (all small letters here and further denote

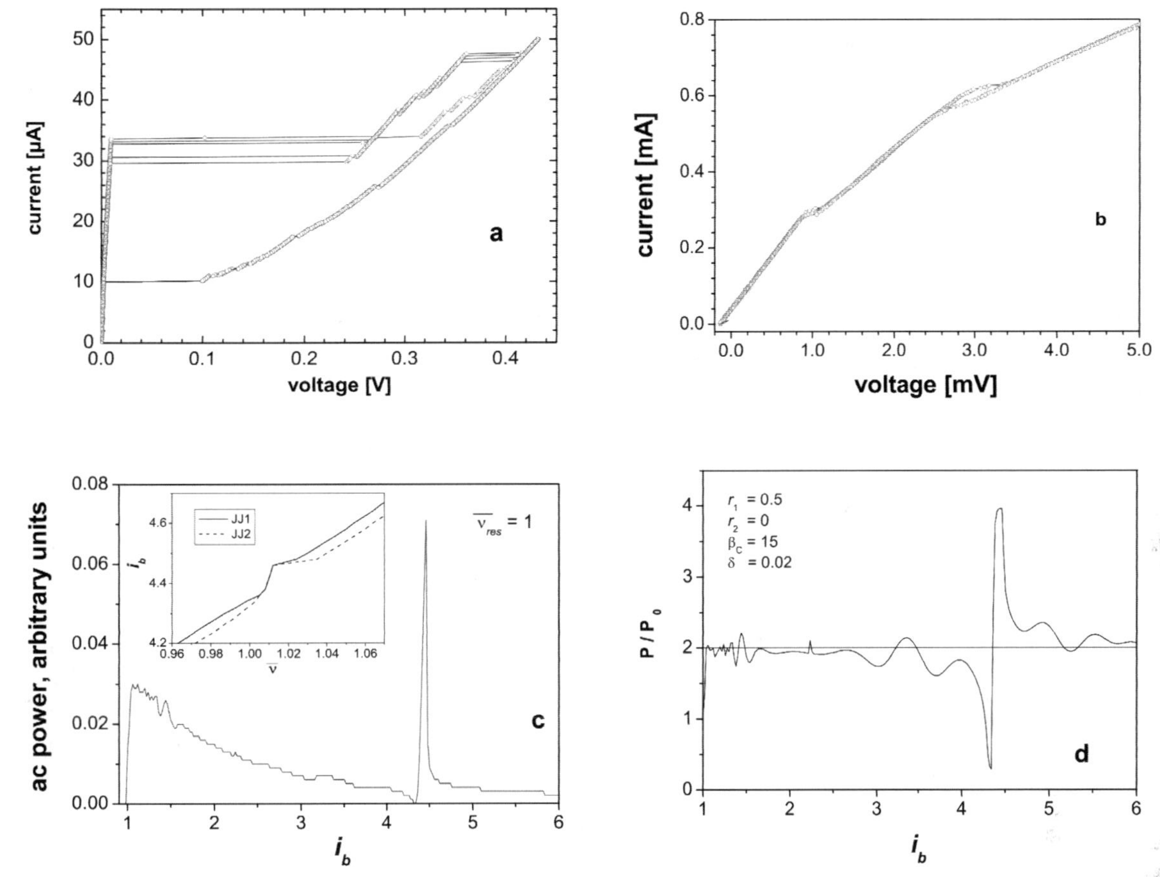

Figure 2. IV-characteristics measured with the use of the current-biased scheme for the microbridge in the resonator without the shunt (a) and with the gold shunted cover (b); the calculated dependence of emitted power plotted as a function of the bias current for the shunted chain of two junctions in the resonator (c) and the emitted power normalized with respect to the power of the single junction (d). In the inset in (c) the IV-characteristics of both junctions are shown.

normalized units, see Ref.[24], impedances are normalized with respect to R_n and voltages across junctions measured in units of I_cR_n are $\dot{\varphi}_k$), β_C is the McCumber parameter and $D\left(\tilde{i_k}\right)$ is the differential equation which determines the ac current in the load which is different for various types of the load. With the use of Eqs. (1) we modeled phase dynamic of a stack of Josephson junctions connected on ends with several types of loads, namely the resistive matched shunt, the sequential LC-resonant contour and a resonant transmission line with open ends. We found that the LC-contour and a resonant transmission line provide the maximal tolerant spread of critical currents (about 15 %), whereas for the matched load this value is only about 7 % [23, 25]. The parameters of transmission line appeared to be hard to regulate and, therefore, the most optimal choice for the synchronization system is the superconducting resonator. We note that some reasonable resistive losses in the resonator decrease the maximal tolerant spread to 10-12 %.

The optimal parameters provide strong synchronization for the stack of Josephson junctions with the resonator up to 15% is $\beta_C = 0.8 - 1.3$, $\beta_{Cres} = 0.4 - 0.6$ and $\beta_{Lres} = 3 - 5$ (here $\beta_{Cres} = \frac{2\pi I_c R_n^2 C_{res}}{\Phi_0}$ and $\beta_{Lres} = \frac{2\pi I_c L_{res}}{\Phi_0}$ are the reduced capacitance and the reduced inductance of the contour, correspondingly). These values are far from those of intrinsic junctions ($\beta_C = 10 - 30$). Therefore, we tried to decrease the McCumber parameter by the external shunting. We covered the chain of intrinsic junctions by the resistive shunt to decrease the McCumber parameter. For the experiments we used $Tl_2Ba_2CaCu_2O_8$ films grown epitaxially on 20 degrees misaligned $LaAlO_3$ in a two-step process [23]. The system was implemented into a resonant structure with a resonant frequency of about 500 GHz, see Fig. 1b. Microbridges were covered by a 10 nm shunting gold layer. The thickness of the shunt in first measurements is about 30 nm that provided a resistance of the shunt per junction of about 1 Ohm. Each of the microbridges was consisted of about 400 junctions. In Fig. 2a we show IV- characteristics for microbridges which gold layer is etched away (i.e. the microbridge was not shunted). This IV-characteristic shows the multibranch behaviour. The effect of the multibranch behaviour is almost absent in the IV-characteristic of the shunted microbridge (Fig. 2b). We modeled the emitted power of the shunted microbridge in the resonator for two junctions with the small spread of critical currents and found that the line of the coherent generation can be very narrow (Fig. 2c). The normalized ac power P/P_0 of radiation showed in-phase synchronization of junctions (Fig. 2d, values of P/P_0 are equal to 0, 2 and 4 if there is anti-phase synchronization, no synchronization and in-phase synchronization, correspondingly). However, there are optimal parameters for the in-phase synchronization. We calculated the optimal parameters of the system for the shunting microbridges. The maximal tolerant spread of critical currents in this case is equal to about 26 % [24]. It is possible for junctions with the value of the McCumber parameter of about 15 when the resistances of the shunt and the interface between the shunt and the junctions are equal to 0.5 and 0.01 parts of the resistance of the junction and the resonant frequency is about 1/3 of the characteristic frequency of the junction. In our experiment, the resonant frequency and the resistance of the interface is close to this optimal frequency, though the resistance of the shunt was too low. Despite of some disparity of parameters, the measured IV-characteristic of the microbridge (Fig.2b) shows the absence of branches and the break which is characteristic for the behaviour of IV-characteristics in the resonator. The absence of branches shows that the junctions can be synchronized, though the check of this fact requires the straightforward measurements of the emitted radiation. In our calculations, we found that the effect of the multibranch behaviour impedes synchronization and one should avoid to apply the resonant frequency in the region of the multibranch behaviour. The narrow-band sychronization of two junctions is calculated in Figs. 2c, d for the region of the IV-characteristic above the hysteretic region. There is a way to use the successive set of capacitances of junctions as a capacitance of the resonator. In this case junctions with the spread of critical currents of about 10% are strongly synchronized in the hysteretic region [28]. We would like to note that the frequency spectrum of radiation in the hysteretic region contains many harmonics, and the same principle as used in Ref.[28] gives the best results in the non-hysteretic region, i.e. above the characteristic frequencies of the junctions. In this case the line of the in-phase generation has only one main harmonic.

5. Return current

The temperature dependence of the return current of single junction is determined by the temperature dependence of the critical current and the McCumber parameter β_C. For the coupled system of Josephson junctions it is additionally determined by the number of junctions in the stack, by the coupling between junctions and boundary conditions. The temperature dependence of β_C in its turn depends on the temperature dependence of the critical current $I_C(T)$, junction resistance $R_J(T)$ and capacitance $C_J(T)$. In present paper we assume that the capacitance of the junctions is the temperature independent geometrical capacitance

$C_J = \varepsilon_r\varepsilon_0 S/D$, where S is the area of the junctions, ε_r is dielectric constant, ε_0 is electric constant, and D is the thickness of the insulating layer. In the capacitively coupled Josephson junctions model with diffusion current (CCJJ+DC model),[29, 30] the stack with N intrinsic Josephson junctions is described by a system of dynamical equations for the gauge-invariant phase differences $\varphi_l(t) = \theta_{l+1}(t) - \theta_l(t) - \frac{2e}{\hbar}\int_l^{l+1} dz A_z(z,t)$ between superconducting layers (S-layers). Here θ_l is the phase of the order parameter in S-layer l, A_z is the vector potential in the barrier. To investigate the temperature dependence of different characteristics of IJJ, and to compare results at different temperatures, it's suitable to normalize all values to the parameters at $T = 0$, because I_c, ω_p, R_J, and β depend on temperature. In this case the system of equations has a form

$$\frac{d}{dt}V_l = I - J_c \sin\varphi_l - \tilde{\beta}\frac{d\varphi_l}{dt}$$
$$\frac{d}{dt}\varphi_l = V_l - \alpha(V_{l+1} + V_{l-1} - 2V_l), \qquad (2)$$

where time is $\tau * \omega_{p0}$ ($\omega_{p0}^2 = 2eI_{c0}/\hbar C$), current is normalized to I_{c0}, the voltage - to the value $V_0 = \hbar\omega_{p0}/2e$, and $\tilde{\beta} = 1/(\omega_{p0}R_jC) = \beta\sqrt{J_c}$. I.e., the behavior of system is determined by parameters α, J_c and $\tilde{\beta}$. Compared to the single Josephson junction, the system of the coupled Josephson junctions has a multiple branch structure and the definition of the return current is more general now: the system can return to this state from any branch and the value of the return current depends on from which one the system has returned to the zero voltage state. In our simulation we calculate the return current from the first branch. We fix the temperature and determine the corresponding value of the dissipation parameter β. Then, at these values of β, we solve numerically [31, 32] the system of dynamical equations (2). It allow us to get the CVC of the first branch and find the value of the return current. Then we repeat the procedure for another value of temperature.

We compare the results of numerical simulation of the temperature dependence of the return current with the dependence of the Zappe model [33, 34] calculated by formula

$$I_r(T) = I_c(T)\frac{-(\pi-2) + \sqrt{(\pi-2)^2 + 8\beta_c}}{2\beta_c} \qquad (3)$$

with the same fitted dissipation parameter β. Zappe model with Heine resistivity [34] can not be used to explain strong temperature dependence of the return current in $Bi_2Sr_2CaCu_2O_x$ intrinsic Josephson junctions because it does not include the coupling between junctions in the stack. The influence of the coupling between junctions on the return current can be investigated directly by numerical methods.

Here we present two results on simulation of the temperature dependence of the return current. One is based on the fitting of the temperature dependence tunnel resistivity of the Josephson junction. For critical current is assumed the Ambegaokar-Baratoff dependence [35]. Another one used fitting of the experimental results on of critical current and McCumber parameter. In the simple parallel resistance model [36] a single junction resistivity $\rho_J(T)$ at subgap voltage region is given by $\rho_J^{-1}(T) = \rho_{sg}^{-1} + \rho_C^{-1}(T)$, where ρ_{sg} is the temperature independent tunnel resistivity of the junction, and $\rho_C(T) = a\exp(b/T) + cT + d$ is the empirical Heine formula of the c-axis resistivity [34] with a, b, c, d as fitting parameters. At low temperature c-axis resistance freezes out, and the junction resistance is dominated by the temperature independent term ρ_{sg}.

Estimating the tunnel resistivity by $\rho_{sg} = \Delta(0)S/eDI_c(0)$, the energy gap Δ from the expression $2\Delta(0)/kT_c = 6$ and using the equation for $\rho_C(T)$, we can find the temperature

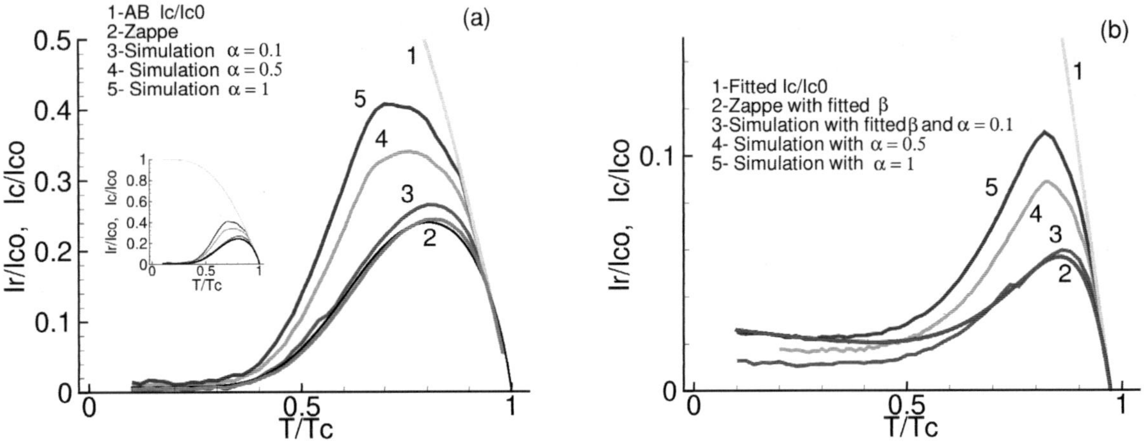

Figure 3. (Color online) (a) - the simulated temperature dependence of the return current I_r at different values of the coupling parameter α together with the Ambegaokar-Baratoff dependence of the critical current I_c. (b) - the simulated temperature dependence of the return current I_r at different values of the coupling parameter α with fitted critical current I_c.

dependence of the junction resistance $R_J = \frac{\rho_{sg}\rho_c}{(\rho_{sg}+\rho_c)}\frac{D}{S}$ and the temperature dependence of the dissipation parameter by formula

$$\beta^2 = 1/\beta_c = \hbar/2eC_J R_N^2 I_C. \tag{4}$$

In our simulations we choose $S = 2.32 * 10^{-10} m^2$ for the area, $T_c = 90K$ for the critical temperature, and $j_c(0) = 9 * 10^6 A/m^2$ at $T = 0$ for the density of critical current. In Ref.[37] the fitting parameters were chosen as $a = 6 * 10^{-4}\Omega m$, $b = 273K$, $c = 24 * 10^{-6}\Omega m/K$, $d = 1.23 * 10^{-2}\Omega m$ as parameter values. Results of simulation are presented in Fig. 3 a. They demonstrate a plateau at small temperatures, maximum at intermediate temperatures and decrease at temperature close to the critical one. An increase in α increases the value of the maximum and shifts it to the smaller temperatures. At small temperature the Zappe curve lies above the $\alpha = 1$ curve, then they coincides practically in some intervals. At high temperature coupling results in high values of return current.

In the second method we used the fitting of the temperature dependence of the dissipation parameter found from the experimentally determined values of C_J, I_c and R_N by formula (4) resulting in $\beta = a + b * \exp(T/c * T_c)$ with $a = 0.01861, b = 5.10414 * 10^{-5}, c = 0.09989$. Result of fitting of the experimental temperature dependence of the critical current by the same formula give us for the same sample $I_c(T) = 1.41667 - 0.39391 * \exp(T/0.79944)$. In Fig. 3 b we present the results of the simulation of the I_r temperature dependence using this fitting formula at three values of coupling parameter: $\alpha = 0.5, 1, 2$. The main features are the same as in previous case, but the Zappe curve shows qualitatively different behavior at low temperatures.

6. Summary and conclusions

Intrinsic Josephson junctions open new ways to realize compact synchronized radiation sources for the THz region. There are many experimental problems which can be solved using results of simulation and modeling. We have reported some results to improve synchronization and the power of radiation by means of adapted shunting. We have also discussed the influence of the return current. Further calculations are necessary for better description of this system to optimize the parameters for applications.

Acknowledgments
This research was partially supported by Heisenberg-Landau Program and the German Research Foundation (DFG). Yu.M.Shukrinov thanks S.Flach and MPIPCS, Dresden, where part of this paper was written.

References
[1] Kleiner R, Steinmeyer F, Kunkel G, Müller P 1992 *Phys. Rev. Lett.* **68** 2394
[2] Kleiner R, Müller, P 1994 *Phys. Rev.* B **49** 1327
[3] Kim S -J, Latyshev Y I, Yamashita T 1999 *Appl. Phys. Lett.* **74** 1156
[4] Yurgens A 2000 *Supercond. Sci. Technol.* **13** R85
[5] Wang H B, Wu P H, Yamashita T 2001 *Phys. Rev. Lett.* **87** 107 002
[6] Wang H B, Wu P H, Yamashita T 2001 *Appl. Phys. Lett.* **78** 4010
[7] Wang H B, Hatano T, Yamashita T, Wu P H, Müller P 2005 *Appl. Phys. Lett.* **86** 023504
[8] Wang HB, Guénon S, Yuan J, Jishi A, Arisawa S, Hatano T, Yamashita T, Koelle D, Kleiner R 2009 *Phys. Rev. Lett.* **102** 017006
[9] Tachiki M, Iizuka M, Minami K, Tejima S, and Nakamura H 2005 *Phys. Rev.* B **71** 34515
[10] Tachiki M, Fukuya S, and Koyama T 2009 *Phys. Rev. Lett.* **102** 127002
[11] Gray K E, Ozyuzer L, Koshelev A E, Kurter C, Kadowaki K, Yamamoto T, Minami H, Yamaguchi M, Tachiki M, Kwok W-K, Welp U 2009 *IEEE Trans. Appl. Supercond.* **19** 886
[12] Saval'ev S, Yampol'skii V A, Rakhmanov A L, Nori F 2010 *Rep. Prog. Phys.* **73** 026501
[13] Hu X, Lin S-Z 2010 *Supercond. Sci. Technol.* **23** 053001
[14] Schmidl F, Pfuch A, Schneidewind H, Heinz E, Dörrer L, Matthes A, Seidel P, Hübner U, Veith M, Steinbeiss E 1995 *Supercond. Sci. Technol.* **8** 740
[15] Haruta M, Kume E and Sakai S 2009 *Supercond. Sci. Technol.* **22** 125004
[16] Seidel P, Schmidl F, Pfuch A, Schneidewind H, Heinz E 1996 *Supercond. Sci. Technol.* **9** A9
[17] Kim S-J, Latyshev YI, Yamashita T 1999 *Appl. Phys. Lett.* **74** 1156
[18] You L X, Torstensson M, Yurgens A, Winkler D, Lin C T, and Liang B 2006 *Appl. Phys. Lett.* **88** 222501
[19] Yurgens A, Torstensson M, You L X, Bauch T, Winkler D, Kakeya I, Kadowaki K 2008 *Physica C* **468** 674
[20] Koval Y, Jin X, Bergmann C, Simsek Y, Ozyuzer L, Müller P, Wang H, Behr G, Büchner B 2010 *Appl. Phys. Lett.* **96** 082507
[21] Chana O S, Kuzhakhmetov A R, Warburton P A, Hyland D M C, Dew-Hughes D, Grovenor C R M 2000 *Appl. Phys. Lett.* **76**, 3603
[22] Mans M, Schneidewind H, Buenfeld M, Schmidl F, Seidel P 2006 *Phys. Rev.* B **74** 214514
[23] Grib A N, Mans M, Scherbel J, Buenfeld M, Schmidl F, Seidel P 2006 *Supercond. Sci. Technol.* **19** S 200
[24] Grib A N, Seidel P, Scherbel J 2002 *Phys. Rev.* B. **65** 94508
[25] Seidel P, Grib A N, Shukrinov Yu M, Scherbel J, Hübner U, Schmidl F 2001 *Physica* C **362** 102
[26] Wang H B, Aruga Y, Chen J, Nakajima K, Yamashita T, Wu P H 2000 *Appl. Phys. Lett.* **77** 1017
[27] Tachiki M, Iizuka M, Minami K, Tejima S, and Nakamura H 2005 *Phys. Rev.* B **71** 134515
[28] Grib A, Seidel P 2009 *Physica Status Solidi RRL* **3** 302
[29] Machida M, Koyama T, Tanaka A and Tachiki M 2000 *Physica* C **85** 330
[30] Shukrinov Yu M, Mahfouzi F, Seidel P 2006 *Physica* C **62** 449
[31] Shukrinov Yu M, Mahfouzi F 2007 *Phys.Rev.Lett.* **98** 157001
[32] Shukrinov Yu M, Mahfouzi F, Pedersen N F 2007 *Phys. Rev.* B **75** 104508
[33] Zappe HH 1973 *J. Appl. Phys.* **44** 1371
[34] Heine G, Lang W, Wang X L, and Dou S X 1999 *Phys. Rev.* B **59** 11179
[35] Ambegaokar V and Baratoff A 1963 *Phys. Rev. Lett.* **10** 486
[36] Yurgens A, Winkler D, Zavaritsky N V, and Claeson T 1997 *Phys. Rev. Lett.* **79** 5122
[37] Okanoue K and Hamasaki K 2005 *Appl. Phys. Lett.* **87** 252506

Journal of Physics: Conference Series **248** (2010) 012039 doi:10.1088/1742-6596/248/1/012039

A tunneling Hamiltonian theory of 0-π transition in d-wave superconductor/ferromagnetic-insulator heterostructures

Shiro Kawabata[1,2], Satoshi Kashiwaya[3], Yukio Tanaka[4], Alexander A. Golubov[5], Yasuhiro Asano[6]

[1]Nanosystems Research Institute (NRI), National Institute of Advanced Industrial Science and Technology (AIST), Tsukuba, Ibaraki, 305-8568, Japan
[2]CREST, Japan Science and Technology Corporation (JST), Kawaguchi, Saitama 332-0012, Japan
[3]Nanoelectronics Research Institute (NRI), National Institute of Advanced Industrial Science and Technology (AIST), Tsukuba, Ibaraki, 305-8568, Japan
[4]Department of Applied Physics, Nagoya University, Nagoya, 464-8603, Japan
[5]Faculty of Science and Technology and MESA+ Institute of Nanotechnology, University of Twente, 7500 AE, Enschede, The Netherlands
[6]Department of Applied Physics and Center for Topological Science & Technology, Hokkaido University, Sapporo, 060-8628, Japan

E-mail: s-kawabata@aist.go.jp

Abstract. We report a theory of the Josephson transport through ferromagnetic insulators (FIs). Recently the appearance of the atomic scale 0-π transition in such junctions has been predicted based on a numerical simulation. In order to understand the physical mechanism of this anomalous phenomena, we have analytically calculated the Josephson current using the tunneling Hamiltonian method and found that the spin dependent π-phase shift in the FI barrier gives the atomic scale 0-π transition. We also show an experimental setup for observing the atomic-scale 0-π transition based on a d-wave junction with a LBCO barrier.

1. Introduction

The developing field of superconducting spintronics subsumes many fascinating physical phenomena with potential applications that may complement conventional spintronics devices [1, 2]. Moreover there is an increasing interest in the novel properties of junctions between superconductors and magnetic materials [3, 4]. One of the peculiar phenomena is the Josephson π-junction formation in superconductor/ferromagnetic-metal/superconductor (S/FM/S) heterostructures [5, 6]. In the ground-state phase difference between two coupled superconductors is π instead of 0 as in the ordinary 0-junctions. In terms of the Josephson relationship $I_J = I_C \sin \phi$, where ϕ is the phase difference between the two superconductor layers, a transition from the 0 to π states implies a change in sign of I_C from positive to negative. This phenomenon is related with the damping oscillatory behavior of the Cooper pair wave function in a FM. Experimentally the existence of the π-junction in S/FM/S systems has been confirmed by Ryanzanov et al. [7] and Kontos et al. [8] Until recently, however, investigations on the π junction have been mainly focused on the S/FM/S systems.

We have predicted a possibility of the π-junction formation in Josephson junctions through *ferromagnetic insulators* (FIs) by numerically solving the Bogolibov-de Genne equation [9, 10] and the Nambu Green's function [11, 12, 13, 14]. The formation of the π junction using such an insulating barrier is very promising for future qubit application [15, 16, 17, 18] because of it's low decoherence nature [19, 20]. More interestingly, we found that the ground state of such junction alternates between 0- and π-states when thickness of FI is increasing by a single atomic layer. In this paper in order to understand the physical mechanism of this atomic scale 0-π transition, we analytically calculate the Josephson current based on the tunneling Hamiltonian method and show that the spin-dependent π-phase shift of the tunneling matrix element of the FI layer is the origin of this anomalous transition.

2. Electronic properties of a ferromagnetic insulator

In this section, we briefly describe the electronic structure of a representative of FI materials, i.e., La_2BaCuO_5 (LBCO) [21, 22, 23]. The typical DOS of LBCO for each spin direction is shown schematically in Fig. 1(a). The exchange splitting $V_{\rm ex}$ is estimated to be 0.34 eV by a first-principle band calculation using the spin-polarized local density approximation [24]. Since the exchange splitting is large and the bands are originally half-filled, the system becomes FI. Based on the band structure [Fig. 1(a)], the spin-dependent transmission coefficient T_σ for the FI barrier can be calculated as [9]

$$T_\sigma = \alpha_{L_F}\left(\frac{\rho_\sigma t}{g}\right)^{L_F}, \tag{1}$$

by using the transfer matrix method, where $\rho_{\uparrow(\downarrow)} = -(+)1$ and α_{L_F} is a spin-independent complex constant, t is the hopping integral in the FI, g is the gap between the up- and down-

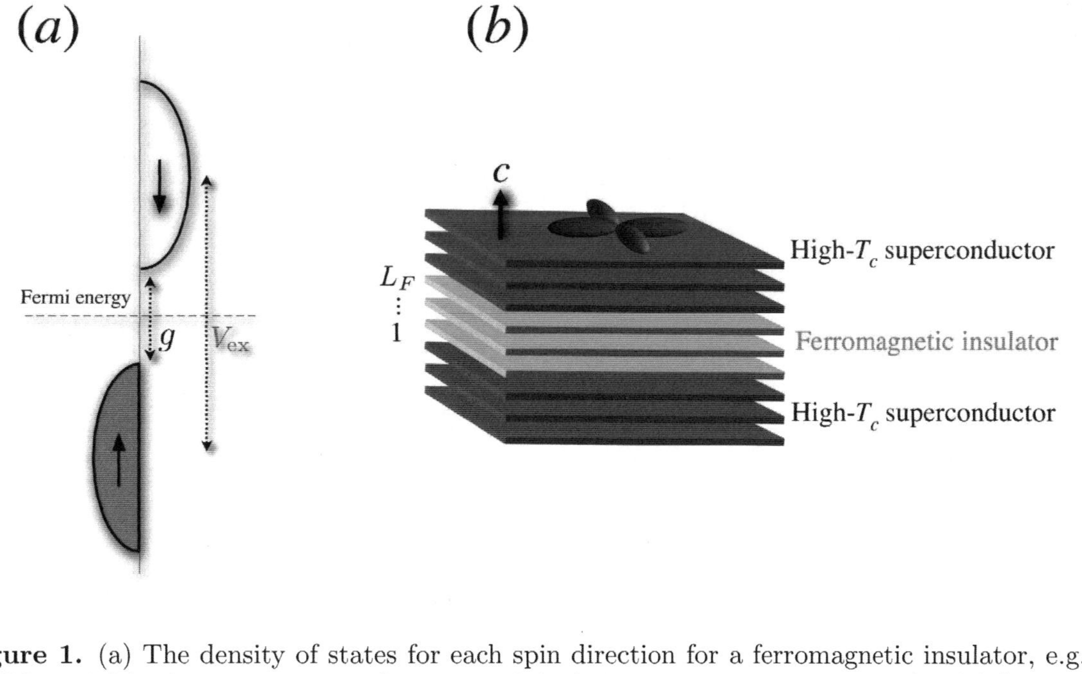

Figure 1. (a) The density of states for each spin direction for a ferromagnetic insulator, e.g., LBCO, and (b) schematic picture of c-axis stack high-T_c d-wave superconductor/LBCO/high-T_c d-wave superconductor Josephson junction.

spin band, and L_F is the layer number of the FI. We immediately find

$$\frac{T_\uparrow}{T_\downarrow} = (-1)^{L_F}, \tag{2}$$

so the relative phase of T_σ between spin up and down is π (0) for the odd (even) number of L_F. Next section, we will calculate the Josephson current through such a FI analytically.

It is important to note that the usage of high-T_c cuprate superconductors, e.g., $YBa_2Cu_3O_{7-\delta}$ and $La_{2-x}Sr_xCuO_4$ is desirable from the perspectives of the high-temperature device-operation. Moreover, recent development of the pulsed laser deposition technique enables layer-by-layer epitaxial-growth of such oxide superlattices [25]. Thus in this paper we mainly focused on the high-T_C d-wave junction with a FI barrier [Fig. 1(b)]. We note that the qualitatively same result can be obtained for the case of s-wave junctions.

3. Theory

In this section, we develop a analytical calculation method for the Josephson current of S/FI/S junctions based on the tunneling Hamiltonian approach. Let us consider a three-dimensional S/FI/S junction as shown in Fig. 1(b). The Hamiltonian of the system is given by

$$\mathcal{H} = \mathcal{H}_1 + \mathcal{H}_2 + \mathcal{H}_T + \mathcal{H}_Q, \tag{3}$$

where $\mathcal{H}_1$ and $\mathcal{H}_2$ are Hamiltonians describing the d-wave superconductors:

$$\begin{aligned} \mathcal{H}_1 &= \sum_\sigma \int d\boldsymbol{r}\, \psi_{1\sigma}^\dagger(\boldsymbol{r}) \left(-\frac{\hbar^2 \nabla^2}{2m} - \mu \right) \psi_{1\sigma}(\boldsymbol{r}) \\ &- \frac{1}{2} \sum_{\sigma,\sigma'} \int d\boldsymbol{r} \int d\boldsymbol{r}' \psi_{1\sigma}^\dagger(\boldsymbol{r})\, \psi_{1\sigma'}^\dagger(\boldsymbol{r}')\, g_1(\boldsymbol{r}-\boldsymbol{r}')\, \psi_{1\sigma'}(\boldsymbol{r}')\, \psi_{1\sigma}(\boldsymbol{r}), \end{aligned} \tag{4}$$

where μ is the chemical potential and ψ ($\psi^\dagger$) is the fermion field operator and $g(\boldsymbol{r}-\boldsymbol{r}')$ is the attractive interaction. The tunneling Hamiltonian with a spin-dependent tunneling matrix element t_σ of the FI barrier is given by

$$\mathcal{H}_T = \sum_\sigma \int d\boldsymbol{r} \int d\boldsymbol{r}' \left[t_\sigma(\boldsymbol{r},\boldsymbol{r}')\, \psi_{1\sigma}^\dagger(\boldsymbol{r})\, \psi_{2\sigma}(\boldsymbol{r}') + \text{h.c.} \right], \tag{5}$$

and

$$\mathcal{H}_Q = \frac{(Q_1 - Q_2)^2}{8C} \tag{6}$$

is the charging Hamiltonian, where C is the capacitance of the junction and $Q_{1(2)}$ is the operator for the charge on the superconductor 1 (2), which can be written as

$$Q_1 = e \sum_\sigma \int d\boldsymbol{r} \psi_{1\sigma}^\dagger(\boldsymbol{r})\, \psi_{1\sigma}(\boldsymbol{r}). \tag{7}$$

By using the functional integral method [19, 26, 27, 28, 29, 30] and taking into account the spin dependence of t_σ explicitly, the ground partition function for the system can be written as follows

$$\mathcal{Z} = \int \mathcal{D}\bar{\psi}_1 \mathcal{D}\psi_1 \mathcal{D}\bar{\psi}_2 \mathcal{D}\psi_2 \exp\left[-\frac{1}{\hbar} \int_0^{\hbar\beta} d\tau \mathcal{L}(\tau) \right], \tag{8}$$

where $\beta = 1/k_B T$, $\psi(\bar{\psi})$ is the Grassmann field which corresponds to the fermionic field operator $[\psi(\psi^\dagger)]$ in Eq. (2), and the Lagrangian $\mathcal{L}$ is given by

$$\mathcal{L}(\tau) = \sum_{\sigma} \sum_{i=1,2} \int d\boldsymbol{r} \bar{\psi}_{i\sigma}(\boldsymbol{r},\tau)\, \partial_\tau \psi_{i\sigma}(\boldsymbol{r},\tau) + \mathcal{H}(\tau). \tag{9}$$

In order to remove the ψ^4 term in the Hamiltonian $\mathcal{H}(\tau)$, we will use the Hubbard-Stratonovich transformation,

$$\begin{aligned}
&\exp\left[-\frac{1}{\hbar}\int_0^{\hbar\beta} d\tau \int d\boldsymbol{r} \int d\boldsymbol{r}' \bar{\psi}_\uparrow(\boldsymbol{r}',\tau)\, \bar{\psi}_\downarrow(\boldsymbol{r},\tau)\, g(\boldsymbol{r}-\boldsymbol{r}')\, \psi_\downarrow(\boldsymbol{r},\tau)\, \psi_\uparrow(\boldsymbol{r}',\tau)\right] \\
&\quad = \int \mathcal{D}\bar{\Delta}(\boldsymbol{r},\boldsymbol{r}';\tau) \int \mathcal{D}\Delta(\boldsymbol{r},\boldsymbol{r}';\tau) \exp\left[\frac{1}{\hbar}\int_0^{\hbar\beta} d\tau \int d\boldsymbol{r} \int d\boldsymbol{r}' \right. \\
&\qquad \times \left\{ -\frac{|\Delta(\boldsymbol{r},\boldsymbol{r}';\tau)|^2}{g(\boldsymbol{r}-\boldsymbol{r}')} + \bar{\Delta}(\boldsymbol{r},\boldsymbol{r}';\tau)\psi_\downarrow(\boldsymbol{r},\tau)\, \psi_\uparrow(\boldsymbol{r}',\tau) \right. \\
&\qquad\qquad \left.\left. + \bar{\psi}_\uparrow(\boldsymbol{r},\tau)\, \bar{\psi}_\downarrow(\boldsymbol{r}',\tau)\, \Delta(\boldsymbol{r},\boldsymbol{r}';\tau) \right\}\right].
\end{aligned} \tag{10}$$

This introduces a complex pair potential field

$$\Delta(\boldsymbol{r},\boldsymbol{r}';\tau) = |\Delta(\boldsymbol{r},\boldsymbol{r}';\tau)| \exp\left[i\phi(\boldsymbol{r},\boldsymbol{r}';\tau)\right]. \tag{11}$$

The resulting action is only quadratic in the Grassmann field, so that the functional integral over this number can readily be performed explicitly. The functional integral over the modulus of the pair potential field is taken by the saddle-point method. Then the partition function is reduced to a single functional integral over the phase difference $\phi = \phi_1 - \phi_2$. To second order in the tunneling matrix element, one finds

$$\mathcal{Z} = \int \mathcal{D}\phi(\tau) \exp\left[-\frac{\mathcal{S}_{\text{eff}}[\phi]}{\hbar}\right], \tag{12}$$

where the effective action is given by

$$\mathcal{S}_{\text{eff}}[\phi] \;=\; \int_0^{\hbar\beta} d\tau \left[\frac{C}{2}\left(\frac{\hbar}{2e}\frac{\partial\phi(\tau)}{\partial\tau}\right)^2 - \frac{\hbar}{2e} I_C \cos\phi(\tau)\right]. \tag{13}$$

In the calculation we have ignored the irrelevant quasiparticle tunneling term for simplicity. Here

$$I_C = -\frac{2e}{\hbar}\int_0^{\hbar\beta} d\tau \beta(\tau) \tag{14}$$

is the Josephson critical current, and then the Josephson current is given by

$$I_J(\phi) = I_C \sin\phi, \tag{15}$$

where the negative (positive) I_C corresponds to the π (0) junction. The Josephson kernel $\beta(\tau)$ is given in terms of the off-diagonal Nambu Green's function for two superconductor $\mathcal{F}_i$ $(i = 1, 2)$, *i.e.*,

$$\beta(\tau) = -\frac{2}{\hbar}\sum_{\boldsymbol{k},\boldsymbol{k}'} t_\downarrow^*(\boldsymbol{k},\boldsymbol{k}')t_\uparrow(\boldsymbol{k},\boldsymbol{k}')\mathcal{F}_1(\boldsymbol{k},\tau)\,\mathcal{F}_2^\dagger(\boldsymbol{k}',-\tau). \tag{16}$$

The Nambu Green functions is given by

$$\mathcal{F}_i(\boldsymbol{k},\omega_n) = \frac{\hbar\Delta_i(\boldsymbol{k})}{(\hbar\omega_n)^2 + \xi_{\boldsymbol{k}}^2 + \Delta_i(\boldsymbol{k})^2}, \tag{17}$$

where $\xi_{\boldsymbol{k}} = \hbar^2\boldsymbol{k}^2/2m - \mu$ is the single particle energy relative to the Fermi energy and $\hbar\omega_n = (2n+1)\pi/\beta$ is the fermionic Matsubara frequency (n is an integer). In the case of the cuprate high-T_c superconductors (the $d_{x^2-y^2}$ symmetry) [31], the order parameter is given by

$$\Delta_i(\boldsymbol{k}) = \Delta_0 \cos 2\theta. \tag{18}$$

Below we will calculate the Josephson critical current I_C analytically and discuss the possibility of the atomic scale 0-π transition.

4. Josephson critical current

The Josephson critical current I_C for a c-axis d-wave junction through the ferromagnetic insulator [Fig. 1(b)] can be expressed as

$$I_C = \frac{2e^2}{\hbar\beta}\sum_{\boldsymbol{k},\boldsymbol{k}'}\sum_{\omega_n} t_\downarrow^*(\boldsymbol{k},\boldsymbol{k}')t_\uparrow(\boldsymbol{k},\boldsymbol{k}')\mathcal{F}(\boldsymbol{k},\omega_n)\mathcal{F}(\boldsymbol{k}',\omega_n) \tag{19}$$

by assuming $\Delta_1 = \Delta_2 = \Delta_0\cos 2\theta$ and thus $\mathcal{F}_1 = \mathcal{F}_2 = \mathcal{F}$. We also assume that the tunneling matrix element $t_\sigma(\boldsymbol{k},\boldsymbol{k}')$ is given in terms of the coherent tunneling in which the momentum $\boldsymbol{k}_\parallel$ parallel to the layer is conserved [26, 29] and has the same L_F dependence on the transmission coefficient T_σ as Eq. (1) [9], *i.e.*,

$$t_\downarrow^*(\boldsymbol{k},\boldsymbol{k}')t_\uparrow(\boldsymbol{k},\boldsymbol{k}') = |t_0|^2(-1)^{L_F}\delta_{\boldsymbol{k}_\parallel\boldsymbol{k}'_\parallel}, \tag{20}$$

we obtain an analytical expression of I_C for $T = 0$ K as

$$I_C = (-1)^{L_F}\frac{\Delta_0 G_N}{2\pi e}, \tag{21}$$

where

$$G_N = \frac{4\pi e^2}{\hbar}|t_0|^2 N_0^2, \tag{22}$$

is the normal conductance with N_0 being the density of states at E_F. The sign of I_C becomes negative for the odd number of L_F and positive for the even number of L_F as was numerically found in [9, 10, 11, 12, 13, 14]. Therefore, the spin dependent π-phase shift of the tunneling matrix element t_σ in the FI barrier gives rise to the atomic scale 0-π transition.

Finally would like to show an experimental set-up for observing the π-junction using LBCO. The experimental detection of the π-junction is possible by using a superconducting ring which contains two Josephson junctions as shown in Fig. 2. When both junctions are in 0- (or π-) state at the same time, the critical current of the ring reaches its maximum at zero external magnetic flux. On the other hand, the critical current reaches its minimum at zero magnetic flux when the 0 state is stable in one junction and π is stable in the other [32]. Experimentally, the half-periodic shifts in the interference patterns of the HTSC ring can be used as a strong evidence of the π-junction. Such a half flux quantum shifts have been observed in a s-wave ring made with a S/FM/S [33] and a S/quantum dot/S junction. [34]

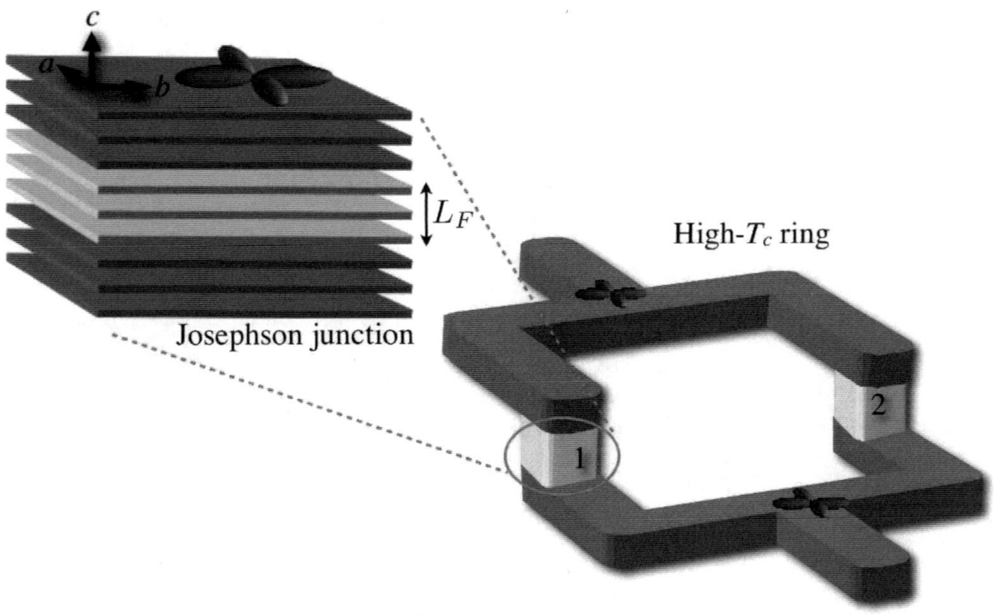

Figure 2. Schematic picture of a superconductor ring with high-T_c superconductor/LBCO /high-T_c superconductor Josephson junctions which can be used in experimental observations of the π-junction.

Note that in the case of c-axis stack high-T_c Josephson junction [35], no zero-energy Andreev bound-states [31] which give a strong Ohmic dissipation [27, 28, 36] are formed. Moreover, the influence of nodal-quasiparticles due to the d-wave order-parameter symmetry on the macroscopic quantum dynamics in such junctions is found to be week both theoretically [26, 37, 38, 39] and experimentally [40, 41, 42, 43, 44]. Therefore we can safely apply such high-T_c superconductor/LBCO/high-T_c superconductor π-junctions to qubits.

5. Summary

To summarize, we have studied the Josephson effect in S/FI/S junction by use of the tunneling Hamiltonian method. We have analytically calculated the Josephson current and showed the possibility of the formation of the atomic scale 0-π transition in such systems. This observation is consistent with previous numerically results. We hop that such FI based π-junctions become an element in the architecture of quantum information devices.

Acknowledgements

This work was supported by CREST-JST, and a Grant-in-Aid for Scientific Research from the Ministry of Education, Science, Sports and Culture of Japan (Grant No. 19710085).

[1] I. Zutic, J. Fabian, and S. D. Sarma, Rev. Mod. Phys. **76**, 323 (2004).
[2] J. Fabian, A. Matos-Abiague, C. Ertlera, P Stanoa, and I. Zutic, Acta Phys. Slov. **57**, 565 (2007).
[3] A. A. Golubov, M. Y. Kupriyanov, and E. Il'ichev, Rev. Mod. Phys. **76**, 411 (2004).
[4] A. I. Buzdin, Rev. Mod. Phys. **77**, 935 (2005).
[5] L. N. Bulaevskii, V. V. Kuzii, and A. A. Sobyanin, JETP Lett. **25**, 291 (1977).
[6] A. I. Buzdin, L. N. Bulaevskii, and S. V. Panyukov, JETP Lett. **35**, 179 (1982).
[7] V. V. Ryazanov, V. A. Oboznov, A. Y. Rusanov, A. V. Veretennikov, A. A. Golubov, and J. Aarts, Phys. Rev. Lett. **86**, 2427 (2001).

[8] T. Kontos, M. Aprili, J. Lesueur, F. Genêt, B. Stephanidis, and R. Boursier, Phys. Rev. Lett. **89**, 137007 (2002).
[9] S. Kawabata, Y. Asano, Y. Tanaka, A. A. Golubov, and S. Kashiwaya, Phys. Rev. Lett. **104**, 117002 (2010).
[10] S. Kawabata, Y. Tanaka, and Y. Asano, Physica E (2010) in press.
[11] S. Kawabata, Y. Asano, Int. J. Mod. Phys. B **23** (2009) 4329.
[12] S. Kawabata, Y. Asano, Y. Tanaka, S. Kashiwaya, Physica C **469** (2009) 1621.
[13] S. Kawabata, Y. Asano, Y. Tanaka, S. Kashiwaya, Physica E **42** (2010) 1010.
[14] S. Kawabata, Y. Asano, Low Temp. Phys. (2010) in press.
[15] S. Kawabata, S. Kashiwaya, Y. Asano, Y. Tanaka, Physica C **437-438** (2006) 136.
[16] S. Kawabata, S. Kashiwaya, Y. Asano, Y. Tanaka, A. A. Golubov, Phys. Rev. B **74** (2006) 180502(R).
[17] S. Kawabata, A. A. Golubov, Physica E **40** (2007) 386.
[18] S. Kawabata, Y. Asano, Y. Tanaka, S. Kashiwaya, A. A. Golubov, Physica C **468** (2008) 701.
[19] G. Schön, A. D. Zaikin, Phys. Reports **198** (1990) 237.
[20] T. Kato, A. A. Golubov, Y. Nakamura, Phys. Rev. B **76** (2007) 172502.
[21] F. Mizuno, H. Masuda, I. Hirabayashi, S. Tanaka, M. Hasegawa, and U. Mizutani, Nature **345**, 788 (1990).
[22] H. Masuda, F. Mizuno, I. Hirabayashi, and S. Tanaka, Phys. Rev. B **43**, 7881 (1991).
[23] W. Ku, H. Rosner, W. E, Pickett, and R. T. Scalettar, Phys. Rev. Lett. **89**, 167204 (2002).
[24] V. Eyert, K. H. Höc, and P. S. Riseborough, Europhys. Lett. **31**, 385 (1995).
[25] W. Prellier, P. Lecoeur and B. Mercey, J. Phys.: Cond. Matter **13**, R915 (2001).
[26] S. Kawabata, S. Kashiwaya, Y. Asano, and Y. Tanaka, Phys. Rev. B **70**, 132505 (2004).
[27] S. Kawabata, S. Kashiwaya, Y. Asano, and Y. Tanaka, Phys. Rev. B **72**, 052506 (2005).
[28] S. Kawabata, A. A. Golubov, Ariando, C. J. M. Verwijs, H. Hilgenkamp, and J. R. Kirtley, Phys. Rev. B **76**, 064505 (2007).
[29] T. Yokoyama, S. Kawabata, T. Kato, and Y. Tanaka, Phys. Rev. B **76**, 134501 (2007).
[30] S. Kawabata, A. A. Golubov, Ariando, C. J. M. Verwijs, H. Hilgenkamp, and J. R. Kirtley, Phys. Rev. B **76**, 064505 (2007).
[31] S. Kashiwaya and Y. Tanaka, Rep. Prog. Phys. **63**, 1641 (2000).
[32] M. Sigrist and T. M. Rice, J. Phys. Soc. Jpn. **61**, 4283 (1992).
[33] W. Guichard, M. Aprili, O. Bourgeois, T. Kontos, J. Lesueur, and P. Gandit, Phys. Rev. Lett. **90**, 167001 (2003).
[34] J. A. van Dam, Y. V. Nazarov, E. P. A. M. Bakkers, S. De Franceschi, and L. P. Kouwenhoven, Nature **442**, 667 (2006).
[35] R. Kleiner, F. Steinmeyer, G. Kunkel, and P. Müller, Phys. Rev. Lett. **68**, 2394 (1992).
[36] S. Kawabata, S. Kashiwaya, Y. Asano, Y. Tanaka, T. Kato, and A. A. Golubov, Supercond. Sci. Technol. **20**, S6 (2007).
[37] Y. V. Fominov, A. A. Golubov, and M. Kupriyanov, JETP Lett. **77**, 587 (2003).
[38] M. H. S. Amin and A. Y. Smirnov, Phys. Rev. Lett. **92**, 017001 (2004).
[39] T. Umeki, T. Kato, T. Yokoyama, Y. Tanaka, S. Kawabata, and S. Kashiwaya, Physica C **463-465**, 157 (2007).
[40] K. Inomata, S. Sato, K. Nakajima, A. Tanaka, Y. Takano, H. B. Wang, M. Nagao, H. Hatano, and S. Kawabata, Phys. Rev. Lett. **95**, 107005 (2005).
[41] X. Y. Jin, J. Lisenfeld, Y. Koval, A. Lukashenko, A. V. Ustinov, P. Müller, Phys. Rev. Lett. **96**, 177003 (2006).
[42] T. Matsumoto, H. Kashiwaya, H. Shibata, S. Kashiwaya, S. Kawabata, H. Eisaki, Y. Yoshida, and Y. Tanaka, Supercond. Sci. Technol. **20**, S10 (2007).
[43] H. Kashiwaya, T. Matsumoto, H. Shibata, S. Kashiwaya, H. Eisaki, Y. Yoshida, S. Kawabata, and Y. Tanaka, J. Phys. Soc. Jpn. **77**, 104708 (2008).
[44] H. Kashiwaya, T. Matsumoto, H. Shibata, H. Eisaki, Y. Yoshida, S. Kawabata, and S. Kashiwaya, Appl. Phys. Exp. **3**, 043101 (2010).

International Conference on Theoretical Physics Dubna-Nano 2010 IOP Publishing
Journal of Physics: Conference Series **248** (2010) 012040 doi:10.1088/1742-6596/248/1/012040

Theory of Josephson effects in iron-based multi-gap superconductor junctions

Y Ota[1,3], M Machida[1,3,4] and T Koyama[2,3]

[1] CCSE, Japan Atomic Energy Agency, 6-9-3 Higashi-Ueno Taito-ku, Tokyo, 110-0015, Japan
[2] Institute for Materials Research, Tohoku University, 2-1-1 Katahira Aoba-ku, Sendai, 980-8577, Japan
[3] CREST(JST), 4-1-8 Honcho Kawaguchi, Saitama, 332-0012, Japan
[4] JST-TRIP, 5 Sambancho Chiyoda-ku, Tokyo, 102-0075, Japan

E-mail: ohta.yukihiro@jaea.go.jp, machida.masahiko@jaea.go.jp, tkoyama@imr.tohoku.ac.jp

Abstract. We revisit the theory for Josephson effects in Josephson junctions with multiple tunneling channels, which can be realized when multi-gap superconductors, e.g., iron-based superconductors or MgB_2 are employed as the electrodes. We mainly re-examine a hetero (multi-band)superconductor- insulator-(single-band)superconductor junction. Deriving an effective Lagrangian density based on the time-dependent Ginzburg-Landau model, we discuss how the relative fluctuations between multiple gaps (i.e., Leggett's excitation modes) and the pairing symmetry modify the Shapiro steps.

1. Introduction

Josephson effect is one of the most drastic phenomena in superconductivity [1, 2]. In addition to the remarkable effects applicable to devices, it is now well-known that the Josephson junction sensitively reflect fundanmental superconducting properties like Cooper pair symmetry. Hence, one can utilize Josephson junctions to explore unsettled properties of novel superconducting materials and assess their application potential.

Multi-band superconductivity has increasingly attracted interests since the discovery of an iron-based superconductor [3, 4, 5, 6, 7, 8]. Various experiments [9, 10] revealed that multiple bands contribute to the superconductivity and multiple superconducting full (s-wave) gaps open below the transition temperature although exceptional nodal cases have been also reported [11]. Several theoretical works [12, 13, 14, 15, 16, 17, 18] have proposed that a sign change occurs between the multiple s-wave gaps when a strong repulsion works between the quasiparticles on the disconnected Fermi surfaces. The order-parameter symmetry with such a sign change has been called $\pm s$-wave. Indeed, some experiments support the type of symmetry [19, 20], but the consensus of the pairing symmetry of the iron-based superconductors have not yet been attained.

Recently, various types of Josephson junctions with iron-based superconductors have been successfully fabricated and typical Josephson effects have been confirmed [21, 22, 23, 24]. Among them, a hetero Josephson junction between an iron-based and a conventional s-wave single-gap superconductors has been regarded as a possible candidate to directly detect the pairing symmetry of iron-based superconductors. The hetero junction system is theoretically described based on multiple tunneling channels, some of which are π channels and the others are 0 ones [25, 26], depending on the pairing symmetry in multi-gap superconductors. It is now an urgent task to propose a definitive and reliable method to detect the pairing symmetry by using such Josephson junctions with iron-based superconductors. Therefore, we reveal how the multi-channels alter basic Josephson effects and how the pair symmetry are detectable.

In this paper, we review the theory of Josephson junctions whose electrodes are composed of multi-gap superconductors. We mainly focus on the hetero (multi-band)superconductor-insulator-(single-band)superconductor junction. Deriving the effective Lagrangian density based on the time-dependent Ginzburg-Landau model, we investigate AC Josephson effects. In particular, we reveal how the relative fluctuations between multiple gaps (i.e., Leggett's collective excitation modes) and the pairing symmetry modify Shapiro steps.

The paper is organized as follows. In the next section, we display the present target, i.e., the hetero Josephson junction between single-gap and two-gap superconductors. Then, we begin with the general theory for Josephson junctions between single-gap and N-gap superconductors in the subsequent two sections. Section 5 is the heart of this paper. A typical Josephson effect, i.e., Shapiro step is examined on the basis of the theory reviewed in the previous sections. We stress that the result offers a clear probe to detect the pairing symmetry of the iron-based superconductors. Section 6 is devoted to summary.

2. Model

In this paper, we focus on the hetero superconductor-insulator-superconductor (SIS) Josephson junction as shown in figure 1. The left (right) electrode is a single-(two-)gap superconductor with width $s_{\rm L}$ ($s_{\rm R}$). The insulator width and the dielectric constant are d and ϵ, respectively. The bias current and the magnetic field are applied along z and y direction, respectively. We have revealed that our theory for this system is applicable to other junction systems including multi-gap superconductors, e.g., grain-boundary junctions [27] and intrinsic Josephson junction stacks [28]) with proper modification.

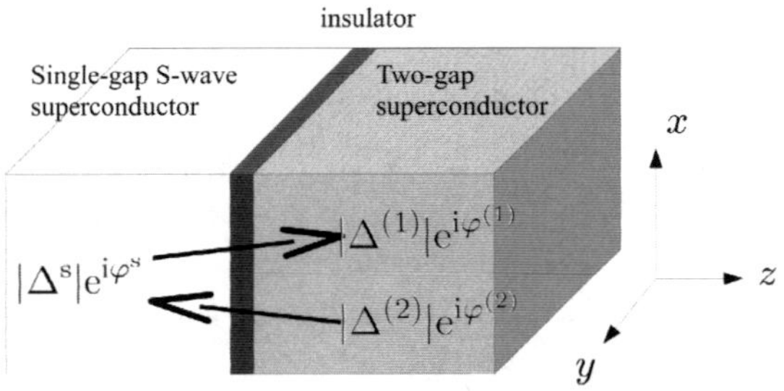

Figure 1. A schematic view for a hetero SIS Josephson junction. The left superconducting electrode is a single-gap s-wave superconductor, while the right one a two-gap superconductor.

Let us briefly summarize theoretical studies about such a two-tunneling-channel Josephson junction. A theoretical development for the present device was done by Brinkman *et al.* [29] and Agterberg *et al.* [30], as for MgB_2. The modification in the conventional Ambegaokar-Baratoff relation [31] was shown in these literatures. Moreover, proximity effects were studied in another hetero structure composed of a normal metal and a multuigap superconductor [32]. The observation of collective modes in two-gap superconductors through Josephson junctions was also proposed [33]. After the discovery of iron-pnictide superconductors, the importance of examining effects of the sign change between the superconducting gaps has grown significantly. A large amount of studies on Josephson junctions or tunneling spectroscopy have been reported in terms of several issues, e.g., the Andreev bound states [25, 34, 36, 37], the DC Josephson effect [26, 27, 35, 38, 39], the Riedel anomaly [40], the Josephson vortex [41], and the AC Josephson effect [42]. For the present junction, Ng and Nagaosa [43] studied the two tunneling channels from the viewpoint of a time-reversal symmetry breaking state.

In the subsequent section, we review the theory of the hetero SIS Josephson junction between single-gap and N-gap superconductors on the basis of the microscopic standard approach for Josephson junctions [44, 45].

3. Effective action

We begin with the microscopic description for the present junction system. In the present and the next sections, we give general formulation in which the number of the electron bands in the right electrode is an arbitrary integer N. Afterwards, the number will be reduced to two for simplicity. The system's Hamiltonian is given by $\hat{H} = \hat{H}_{\rm L} + \hat{H}_{\rm R} + \hat{H}_{\rm T}$, where

$$\hat{H}_{\rm L} = \int_{\rm L} {\rm d}^3\boldsymbol{r} \Big[\sum_{\sigma=\uparrow,\downarrow} \hat{\psi}^{{\rm s}\,\dagger}_{\sigma} \varepsilon_{\rm s}(-{\rm i}\hbar\nabla)\hat{\psi}^{\rm s}_{\sigma} - g_{\rm s}\hat{\psi}^{{\rm s}\,\dagger}_{\uparrow}\hat{\psi}^{{\rm s}\,\dagger}_{\downarrow}\hat{\psi}^{\rm s}_{\downarrow}\hat{\psi}^{\rm s}_{\uparrow} \Big], \tag{1}$$

$$\hat{H}_{\rm R} = \int_{\rm R} {\rm d}^3\boldsymbol{r} \Big[\sum_{i=1}^{N}\sum_{\sigma=\uparrow,\downarrow} \hat{\psi}^{(i)\,\dagger}_{\sigma} \varepsilon_{i}(-{\rm i}\hbar\nabla)\hat{\psi}^{(i)}_{\sigma} - \sum_{1\le i,j\le N} g_{ij}\hat{\psi}^{(i)\,\dagger}_{\uparrow}\hat{\psi}^{(i)\,\dagger}_{\downarrow}\hat{\psi}^{(j)}_{\downarrow}\hat{\psi}^{(j)}_{\uparrow} \Big], \tag{2}$$

$$\hat{H}_{\rm T} = \sum_{i=1}^{N}\int_{\rm L} {\rm d}^3\boldsymbol{r}\int_{\rm R} {\rm d}^3\boldsymbol{r}' [T^{(i)}_{\boldsymbol{r}\boldsymbol{r}'}\hat{\psi}^{{\rm s}\,\dagger}(\boldsymbol{r})\hat{\psi}^{(i)}(\boldsymbol{r}') + {\rm h.c.}]. \tag{3}$$

The Hamiltonian $\hat{H}_{\rm L}$ is for the left superconducting electrode, while $\hat{H}_{\rm R}$ for the right one. The Hamiltonian $\hat{H}_{\rm T}$ describes the tunneling between the electrons in the left and the right electrodes, in which the tunneling matrix element $T^{(i)}_{\boldsymbol{r}\boldsymbol{r}'}$ is real. We assume that $g_{\rm s}$ is a real positive constant, g_{ij} is real constants, and $\mathcal{G}(\equiv (g_{ij}))$ is a positive definite matrix. Here, $g_{ij} = g_{ji}$ $(i \neq j)$ since a positive definite matrix is a Hermitian one. We also note that the diagonal element g_{ii} (i.e., intra-band pairing) is positive while the off-diagonal one g_{ij} $(i \neq j)$ (i.e., inter-band pairing) is either positive (i.e., attractive coupling) or negative (repulsive coupling). In the present paper, the right electrode is regarded as a N-band Bardeen-Cooper-Schrieffer type superconductor. The theoretical description for $N = 2$ is given in [46, 47]. Although the momentum dependence in g_{ij} should be taken into account in iron-based superconductors, basic physical properties in the present Josephson junction, i.e., Josephson effects ascribed to the multiple tunneling channels are fully illustrated through such a simplified model.

Introducing complex auxiliary fields $\Delta^{(i)}, \Delta^{(i)*}$ and $\Delta^{\rm s}, \Delta^{{\rm s}*}$ through the Hubbard-Stratonovich transformation [44], the partition function in imaginary time formalism is given by $Z = Z_0 \int \prod_{i=1}^{N} \mathcal{D}\Delta^{(i)}\mathcal{D}\Delta^{(i)*}\mathcal{D}\Delta^{\rm s}\mathcal{D}\Delta^{{\rm s}*}\, {\rm e}^{-S_{\rm eff}/\hbar}$, where

$$S_{\rm eff} = \int_0^{\hbar\beta} {\rm d}\tau \Big(\int_{\rm R} {\rm d}^3\boldsymbol{r}\Delta^{(i)*}(\mathcal{G}^{-1})_{ij}\Delta^{(j)} + \int_{\rm L} {\rm d}^3\boldsymbol{r}\frac{|\Delta^{\rm s}|^2}{g_{\rm s}} \Big) - {\rm Tr}\ln\hat{G}_0 - {\rm Tr}\ln\hat{G}^{-1}. \tag{4}$$

β is the inverse temperature, and Z_0 is the partition function for the non-interacting part. The condition $\mathcal{G} > 0$ ensures the presence of the inverse matrix $\mathcal{G}^{-1}$ and $\det\mathcal{G} > 0$. The Green functions $\hat{G}_0$ and $\hat{G}$ are given by $\hat{G}_0^{-1}(x;x') = (-\hat{I}\hbar\partial_\tau - \hat{h}_0)\hbar\delta(x-x')$ and $\hat{G}^{-1}(x;x') = [-\hat{I}\hbar\partial_\tau - \hat{h}_0 - \hat{D}(x)]\hbar\delta(x-x') - \sum_{i=1}^{N}\hat{T}^{(i)}_{\boldsymbol{r}\boldsymbol{r}'}\hbar\delta(\tau-\tau')$, respectively. Here, we use the notation as $x = (\boldsymbol{r},\tau)$ (e.g., $\delta(x-x') = \delta(\tau-\tau')\delta^{(3)}(\boldsymbol{r}-\boldsymbol{r}')$). $\hat{h}_0$ and $\hat{D}$ are $2(N+1)\times 2(N+1)$ matrices, $\hat{h}_0 = \tau_3\xi_{\rm s} \oplus (\oplus_{i=1}^{N}\tau_3\xi_i)$ and $\hat{D} = \hat{D}_{\rm s} \oplus (\oplus_{i=1}^{N}\hat{D}_i)$, where $\xi_{\rm s} = \varepsilon_{\rm s} - \mu_{\rm L}$, $\xi_i = \varepsilon_i - \mu_{\rm R}$, $\hat{D}_{\rm s} = \tau_1\,{\rm Re}\Delta^{\rm s} - {\rm i}\tau_2\,{\rm Im}\Delta^{\rm s}$, and $\hat{D}_i = \tau_1\,{\rm Re}\Delta^{(i)} - {\rm i}\tau_2\,{\rm Im}\Delta^{(i)}$ $(i = 1,\ldots,N)$. The chemical potential on the left (right) electrode is denoted as $\mu_{\rm L}$ $(\mu_{\rm R})$, and τ_m is the mth component of the Pauli matrices $(m = 1, 2, 3)$. The matrix $\hat{I}$ is the $2(N+1)\times 2(N+1)$ unit matrix, and $\hat{T}^{(i)}_{\boldsymbol{r}\boldsymbol{r}'}$ is also $2(N+1)\times 2(N+1)$ matrix. One of the characteristic terms in the Josephson junctions with multi-gap superconductors is the inter-band Josephson coupling [48, 49, 50, 51, 52] given by the off-diagonal elements of $\mathcal{G}^{-1}$. For the case of $N=2$, this type of the coupling reads $-2(g_{12}/\det\mathcal{G})|\Delta^{(1)}||\Delta^{(2)}|\cos(\varphi^{(2)}-\varphi^{(1)})$, in which $\Delta^{(i)}=|\Delta^{(i)}|\mathrm{e}^{\mathrm{i}\varphi^{(i)}}$.

4. Dynamical equations for gauge-invariant phase differences

4.1. Effective Lagrangian density

Using the standard procedure on (4) [44, 45], we obtain the effective Lagrangian density of the superconducting phases on zx plane in the real time formalism

$$\mathcal{L}_{\text{eff}} = \frac{s_{\rm L}}{8\pi\mu_{\rm s}^2}q_{\rm s}^2+\sum_{i=1}^{N}\frac{s_{\rm R}}{8\pi\mu_i^2}q_i^2-\frac{s_{\rm L}}{8\pi\lambda_{\rm s}^2}v_{\rm s}^2-\sum_{i=1}^{N}\frac{s_{\rm R}}{8\pi\lambda_i^2}v_i^2+\sum_{i=1}^{N}\frac{\hbar j_i}{e^*}(1-\cos\theta^{(i)})$$
$$+\sum_{i<j}\frac{\hbar J_{ij}}{e^*}[1-\cos(\varphi^{(j)}-\varphi^{(i)})]+\frac{\epsilon d}{8\pi}(E_{\rm RL}^z)^2-\frac{d}{8\pi}(B_{\rm RL}^y)^2. \quad (5)$$

We note that $q_{\rm s}=(\hbar/e^*)\partial_t\varphi_{\rm s}+A_{\rm L}^0$, $v_{\rm s}=(\hbar c/e^*)\partial_x\varphi_{\rm s}-A_{\rm L}^x$, $q_i=(\hbar/e^*)\partial_t\varphi^{(i)}+A_{\rm R}^0$, $v_i=(\hbar c/e^*)\partial_x\varphi^{(i)}-A_{\rm R}^x$, and $e^*=2e$. The gauge invariant phase difference reads $\theta^{(i)}=\varphi^{(i)}-\varphi^{\rm s}-(e^*d/\hbar c)A_{\rm RL}^z$. The phase $\varphi^{\rm s}$ is defined as $\Delta^{\rm s}=|\Delta^{\rm s}|e^{i\varphi^{\rm s}}$. j_i is the Josephson critical current between ith and single-band Cooper pairs. The charge screening length and the penetration depth on the left (right) electrode are $\mu_{\rm s}$ (μ_i) and $\lambda_{\rm s}$ (λ_i), respectively. The last term in the gauge-invariant phase difference is the z component of the spatial averaged vector potential in the insulator, defined as $A_{\rm RL}^z=d^{-1}\int_{-d/2}^{d/2}A^z(z)\,\mathrm{d}z$. The electric and the magnetic fields in the insulator are defined as $E_{\rm RL}^z=-c^{-1}\partial_tA_{\rm RL}^z-d^{-1}(A_{\rm R}^0-A_{\rm L}^0)$ and $B_{\rm RL}^y=d^{-1}(A_{\rm R}^x-A_{\rm L}^x)-\partial_xA_{\rm RL}^z$, respectively. The magnitude of the coupling constant J_{ij} is proportional to $s_{\rm R}|(\mathcal{G}^{-1})_{ij}/\det\mathcal{G}|$. If $s_{\rm R}$ is much larger than d as the conventional Josephson junctions, then it allows us to regard $|J_{ij}|>j_i$. We assume that J_{ij} is a real number in (5). The form of the inter-band Josephson coupling depends on the number of the bands in the right electrode. For example, $N=2$, one can notice that J_{ij} must be real [48, 49]. We remark that the sign of J_{ij} is a matter of convention in bulk two-gap superconductors but not ture in the present hetero Josephson junction due to the presence of the single-gap superconductor in the left electrode.

4.2. Josephson relations and Maxwell equation

From (5), we find the Euler-Lagrange equation with respect to $A_{\rm RL}^z$ which gives the Maxwell equation

$$\frac{e^*d}{\hbar c}\partial_xB_{\rm RL}^y=\sum_{i=1}^{N}\frac{1}{\lambda_{{\rm J}i}^2}\sin\theta^{(i)}+\frac{\epsilon}{c^2}\frac{e^*d}{\hbar}\partial_tE_{\rm RL}^z, \quad (6)$$

where $\lambda_{{\rm J}i}^{-2}=4\pi e^*dj_i/\hbar c^2$. The first term on the right hand side of (6) is the summation of the Josephson current terms. Using the Euler-Lagrange equations with respect to A_l^0 and A_l^x ($l={\rm R},{\rm L}$), we find that the Josephson relations are given by

$$\sum_{i=1}^{N}\frac{\bar{\alpha}}{\alpha_i}\partial_t\theta^{(i)}=C\frac{e^*d}{\hbar}E_{\rm RL}^z,\quad \sum_{i=1}^{N}\frac{\bar{\eta}}{\eta_i}\partial_x\theta^{(i)}=L\frac{e^*d}{\hbar c}B_{\rm RL}^y, \quad (7)$$

where the dimensionless parameters in each electrode are defined as $\alpha_{\rm s}=\epsilon\mu_{\rm s}^2/s_{\rm L}d$, $\alpha_i=\epsilon\mu_i^2/s_{\rm R}d$, $\bar{\alpha}^{-1}=\sum_i\alpha_i^{-1}$, $\eta_{\rm s}=\lambda_{\rm s}^2/s_{\rm L}d$, $\eta_i=\lambda_i^2/s_{\rm R}d$, and $\bar{\eta}^{-1}=\sum_i\eta_i^{-1}$. The magnitude of the electric (magnetic) field coupling is characterized by α and α_i (η and η_i) [53]. The constants C and L are defined as $C=1+\alpha_{\rm s}+\bar{\alpha}$ and $L=1+\eta_{\rm s}+\bar{\eta}$, respectively.

4.3. In-phase motion

The present description is valid from very thin electrode junctions ($s_{\rm L} \sim \mu_{\rm s}$ and $s_{\rm R} \sim \mu_i$) to conventional thick ones ($s_{\rm L} \gg \mu_{\rm s}$ and $s_{\rm R} \gg \mu_i$). We stress that the junction electrode size considerably affects the dynamics of the gauge-invariant phase differences in the Josephson junction with multiple tunneling channels. Let us focus on the latter case. We find that the contributions from the charge compressibility (i.e., the first and the second terms in (5)) are negligible [54]. This case corresponds to the limit $\alpha_{\rm s}, \alpha_i \to 0$. Then, the temporal behaviors of $\varphi_{\rm s}$ and φ_i are rigidly pinned according to $q_{\rm s} = 0$ and $q_i = 0$. It means that

$$\partial_t \theta^{(1)} = \partial_t \theta^{(2)} = \cdots = \partial_t \theta^{(N)} \left(= \frac{e^* d}{\hbar} E^z_{\rm RL} \right). \tag{8}$$

Hence, in-phase motion with respect to $\{\theta^{(i)}\}$ is only allowed, i.e., $\theta^{(i)}(t) = \gamma(t) + \theta_{0,i}$, where $\theta_{0,i}$ is a constant phase. In the absence of the magnetic field, the Maxwell equation leads to $0 = \sum_{i=1}^N \lambda_{{\rm J}i}^{-2} \sin(\gamma + \theta_{0,i}) + (\epsilon/c^2)(e^* d/\hbar)\partial_t^2 \gamma$. Thus, we find that the system is described by a single phase γ. When the pairing symmetry of the right superconducting electrode is s-wave or $\pm s$-wave, the constant phases can have 0 (for s-wave) or π (for $\pm s$-wave). Therefore, the contribution from the multiple tunneling channels can be rewritten as $\sum_i \lambda_{{\rm J}i}^{-2} \cos(\theta_{0,i} - \theta_{0,1}) \sin\gamma$. We can find that a cancellation between the multiple Josephson currents occurs for the $\pm s$-wave symmetry while those are simply superimposed for the s-wave without the sign changes. According to the above arguments, we can obtain a bound of the Josephson critical current in the present Josephson junction (i.e., the Ambegaokar-Baratoff relation) and the resultant expression gives an effective criterion for the $\pm s$-wave symmetry [39]. The out-of-phase motion between $\theta^{(i)}$ and $\theta^{(j)}$, whose basic mode is called Leggett's mode [48], is excited as long as the contributions from the charge compressibility is not irrelevant to the dynamics of the present junction. We will examine this point in the subsequent section, in which we focus on the vicinity of the interface in the junction whose region is characterized by about μ_i from the interface.

4.4. Phase equations

Now, let us derive the phase equations in the hetero SIS Josephson junction. First of all, using (6)-(7), we obtain

$$\sum_{i=1}^N \frac{C\bar{\eta}}{\eta_i} \partial_x^2 \theta^{(i)} = \sum_{i=1}^N \frac{CL}{\lambda_{{\rm J}i}^2} \sin\theta^{(i)} + \sum_{i=1}^N \frac{L\bar{\alpha}}{\alpha_i} \partial_t^2 \theta^{(i)}. \tag{9}$$

Next, from the Euler-Lagrange equations about $\varphi^{(i)}$, we have

$$\frac{\epsilon}{c^2} \frac{e^*}{\hbar} \partial_t \left(\frac{q_i}{\alpha_i} - \frac{q_j}{\alpha_j} \right) = \frac{e^*}{\hbar c} \partial_x \left(\frac{v_i}{\eta_i} - \frac{v_j}{\eta_j} \right) + \frac{1}{\lambda_{{\rm J}i}^2} \sin\theta^{(i)} - \frac{1}{\lambda_{{\rm J}j}^2} \sin\theta^{(j)} - (F_i^{\rm IBJ} - F_j^{\rm IBJ}), \tag{10}$$

where

$$F_i^{\rm IBJ} = \sum_{i<k} \frac{\kappa_{ik}}{\lambda_{ik}^2} \sin(\theta^{(i)} - \theta^{(k)}) - \sum_{k<i} \frac{\kappa_{ki}}{\lambda_{ki}^2} \sin(\theta^{(k)} - \theta^{(i)}) \tag{11}$$

and

$$\frac{1}{\lambda_{ij}^2} = \frac{4\pi e^* d |J_{ij}|}{\hbar c^2}, \quad \kappa_{ij} = {\rm sgn}(J_{ij}). \tag{12}$$

Equation (10) describes the dynamics of electric and magnetic imbalance between the ith and the jth bands in the right superconducting electrode. The use of the kinematic relations $q_{\rm s}/\alpha_{\rm s} = -dE^z_{\rm RL}$ and $v_{\rm s}/\eta_{\rm s} = -dB^y_{\rm RL}$ given by the Euler-Lagrange equations with respect to

$A^0_{\rm L}$ and $A^x_{\rm L}$ make (10) a more convenient formula,

$$\begin{aligned}&\frac{\epsilon}{c^2}\partial_t^2(\alpha_i^{-1}\theta^{(i)}-\alpha_j^{-1}\theta^{(j)})-(\alpha_i^{-1}-\alpha_j^{-1})(1+\alpha_{\rm s})\frac{\epsilon}{c^2}\frac{e^*d}{\hbar}\partial_t E^z_{\rm RL}\\ =\;&\partial_x^2(\eta_i^{-1}\theta^{(i)}-\eta_i^{-1}\theta^{(j)})-(\eta_1^{-1}-\eta_2^{-1})(1+\eta_{\rm s})\frac{e^*d}{\hbar c}\partial_x B^y_{\rm RL},\\ &+\frac{1}{\lambda_{{\rm J}i}^2}\sin\theta^{(i)}-\frac{1}{\lambda_{{\rm J}j}^2}\sin\theta^{(j)}-(F_i^{\rm IBJ}-F_j^{\rm IBJ}).\end{aligned}\tag{13}$$

The second terms in the both left and right hand sides represent the in-phase component of the dynamics. They are relevant to the system's dynamics as far as $\alpha_i\neq\alpha_j$ and $\eta_i\neq\eta_j$. Using the Josephson relation (7), we find that (13) is described only by the gauge-invariant phase difference $\theta^{(i)}$. Summarizing the above arguments, the dynamics of the hetero SIS Josephson junction is given by simultaneously solving (9) and (13).

5. Josephson effects in the presence of multiple tunneling channels

Here, let us examine basic Josephson effects in the hetero SIS Josephson junction between single-gap and two-gap superconductors. In this case, we have only an inter-band Josephson coupling constant, J_{12}. We denote it as $J_{\rm in}$ in what follows. Associated with this notation, we write λ_{12} and κ_{12} as, respectively, $\lambda_{\rm in}$ and $\kappa_{\rm in}$. We focus on a typical AC Josephson effect, Shapiro steps in the hetero SIS Josephson junction betwee single- and two-gap superconductors [41].

To obtain the Shapiro step, we irradiate a microwave generating the AC volage $V_{\rm mv}\cos\Omega_{\rm mw}t$ under no external magnetic field. Thus, $\theta^{(i)}$ does not depend on the spatial variables. To understand effects ascribed to the relative phase motion, we explicitly take the finite charge compressibility inside the superconducting electrodes, in which the compressible region is confined around the interface whose thickness is characterized by the charge screening length. Namely, the parameters $\alpha_{\rm s}$ and α_i become, respectively, $\alpha_{\rm s}\simeq\epsilon\mu_{\rm s}/d$ and $\alpha_i\simeq\epsilon\mu_i/d$. The temporal evolution of $\theta^{(i)}$ reads

$$\theta^{(i)}(t)=\theta_0^{(i)}+\Lambda f(t)+(-1)^{i+1}\frac{\alpha_i}{\alpha_1+\alpha_2}\widetilde{\psi}(t),\quad f(t)=\Omega_0 t+u\sin\Omega_{\rm mw}t\tag{14}$$

where $\Omega_0=e^*V_0/\hbar$, and $u=e^*V_{\rm mw}/\hbar\Omega_{\rm mw}$. The phases $\theta_0^{(1)}$ and $\theta_0^{(2)}$ are constants satisfying the relation, $\psi_0=\theta_0^{(1)}-\theta_0^{(2)}=0\ (\pi)$ for $J_{\rm in}>0\ (<0)$. The quantity $\widetilde{\psi}(t)$ describing the relative phase fluctuation, $\widetilde{\psi}(t)=\psi(t)-\psi_0$. In this paper, $\widetilde{\psi}$ is assumed to be small. Under this assumption, we find that a linerlized equation for $\widetilde{\psi}$ from (13) [26], which is similar to the one for a forced harmonic oscillator,

$$\partial_t^2\widetilde{\psi}=-\omega_{\rm JL}^2\widetilde{\psi}+F_{\rm C}(t)+F_{\rm J}(t),\tag{15}$$

where

$$F_{\rm C}(t)=-\xi\frac{\alpha_1+\alpha_2}{2}\Omega_{\rm mw}^2 u\sin\Omega_{\rm mw}t,\quad F_{\rm J}(t)=-\nu\frac{\alpha_1+\alpha_2}{2}(\omega_1^2+\omega_2^2)\sin[\theta_0^{(1)}+\Lambda f(t)],\tag{16}$$

with $\nu=(\omega_1^2-\kappa_{\rm in}\omega_2^2)/(\omega_1^2+\omega_2^2)$, where the Josephson plasma frequency for the ith band is defined as $\omega_i=c/\sqrt{\epsilon}\lambda_{{\rm J},i}$. Similarly, we introduce a frequncy associated with the inter-band Josephson coupling as $\omega_{\rm in}=c/\sqrt{\epsilon}\lambda_{\rm in}$. The angular frequency $\omega_{\rm JL}$ is given as $\omega_{\rm JL}=\sqrt{\alpha_1+\alpha_2}\omega_{\rm in}$, which is equal to the mass of the Josephson-Leggett mode [26]. The external force terms $F_{\rm C}$ and $F_{\rm J}$ are originated from, respectively, the charge imbalance and the Cooper pair tunneling. When $F_{\rm C}$ is predominant, we can obtain an interesting approximate solution for (15) [42], $\widetilde{\psi}(t)\approx A\sin\Omega_{\rm mw}t$, in which where $A = -\xi[(\alpha_1+\alpha_2)/2][(u\Omega_{\rm mw}^2/\omega_{\rm JL}^2)/(1-\Omega_{\rm mw}^2/\omega_{\rm JL}^2)]$ with $\Omega_k = \Lambda\Omega_0 + k\Omega_{\rm mw}$ and $J_k(\Lambda u)$ being the Bessel function of the kth order. It should be noted that $\widetilde{\psi}$ has poles at $\Omega_{\rm mw}^2 = \omega_{\rm JL}^2$. Namely $\widetilde{\psi}$ is resonantly enhanced at these frequencies. The resonant condition can be satisfied by tuning the frequency of the applied microwave. In a region far from the resonance we have $\widetilde{\psi} \simeq 0$. By the use of the Fourier-Bessel expansion, we obtain a condition that DC component emerges in the superconducting current, i.e., $\Lambda\Omega_0 + n\omega_{\rm JL} = 0$, in which n is an integer. Hence, the voltages at which the Shapiro steps appear can be determined by the relation $V_0 = -n\Lambda^{-1}(\hbar\Omega_{\rm mw}/e^*)$. The step height is obtained as $I_{DC} = |j_1 J_p(x_1) + j_2\cos\psi_0 J_p(x_2)|$, where $x_1 = \Lambda u + (\alpha_1 A)/(\alpha_1+\alpha_2)$ and $x_2 = \Lambda u - (\alpha_2 A)/(\alpha_1+\alpha_2)$. Since A is large near the resonance frequency, one may have an approximate relation, $x_1 \simeq -x_2$. In this case, employing the relation $J_n(-x) = (-1)^n J_n(x)$, we find that [42]

$$I_{DC} = |j_1 + (-1)^p j_2 \cos\psi_0|\,|J_p(x_1)|\,. \tag{17}$$

The coefficient $|j_1+(-1)^p j_2\cos\psi_0|$ takes alternating values with respect to p, that is, $j_1+(-1)^p j_2$ in the s-wave case ($\psi_0 = 0$), whereas $j_1+(-1)^{p+1}j_2$ in the $\pm s$-wave case ($\psi_0 = \pi$), which leads to the fact that a larger DC current appears at even (odd) p in the s-wave ($\pm s$-wave) case. This remarkable feature of the Shapiro steps caused by the resonance with Leggett's mode is quite clear-cut for the determination of the pairing symmetry in multi-gap superconductors.

6. Summary

We reviewed the theoretical description for the hetero SIS Josephson junction between single-gap and N-gap superconductors. The phase equations have the out-of-phase motion as well as the conventional in-phase motion. Then, we studied one of the basic Josephson effects, the Shapiro steps. We revealed that when an external microwave applied to the hetero SIS Josephson junction resonantly excites Leggett's mode, the voltage dependence of the Shapiro steps is sensitive to the inter-band sign change, i.e., the pairing symmetry of the multi-based superconductor. We suggest that by tuning the microwave frequency close to the Leggett's mode one has not only novel types of Shapiro steps but also a clear probe to the gap symmetry of a multi-gap superconductor.

Acknowledgments

The authors wish to acknowledge valuable discussions with H. Matsumoto, S. Shamoto, N. Hayashi, M. Okumura, H. Nakamura, N. Nakai, R. Igarashi, Y. Nagai, and A. Yamamoto. TK was partially supported by Grant-in-Aid for Scientific Research (C) (No. 22540358) from Japan Science and Technology Agency.

References

[1] Hinken J H 1991 *Superconductor Electronics: Fundamentals and Microwave Applications* (Berlin: Springer) chapter 5
[2] Schmidt V V 1997 *The Physics of Superconductors: Introduction to Fundamentals and Applications* (Berlin: Springer) chapter 6
[3] Kamihara Y, Watanabe T, Hirano M and Hosono H 2008 *J. Am. Chem. Soc.* **130** 3296
[4] Rotter M, Tegel M and Johrendt D 2008 *Phys. Rev. Lett.* **101** 107006
[5] Sasmal K, Lv B, Lorenz B, Guloy A M, Chen F, Xue Y-Y and Chu C-W 2008 *Phys. Rev. Lett* **101** 107007
[6] Tapp J H, Tang Z, Lv B, Sasmal K, Lorenz B, Chu P C W and Guloy A M 2008 *Phys. Rev.* B **78** 060505(R)
[7] Wang X C, Liu Q, Lv Y, Gao W, Yang L X, Yu R C, Li F Y and Jin C 2008 *Solid. State. Commun.* **148** 538
[8] Hsu F-C, Luo J-Y, Yeh K-W, Chen T-K, Huang T-W, Wu P M, Lee Y-C, Huang Y-L, Chu Y-Y, Yan D-C and Wu M-K 2008 *Proc. Natl. Acad. Sci. U.S.A.* **105** 14262
[9] Ding H, Richard P, Nakayama K, Sugawara K, Arakane T, Sekiba Y, Takayama A, Souma S, Sato T, Takahashi T, Wang Z, Dai X, Fang Z, Chen G F, Luo J L and Wang N L 2008 *EPL* **83** 47001

[10] Evtushinsky D V *et al.* 2009 Momentum-resolved superconducting gap in the bulk of $Ba_{1-x}K_xFe_2As_2$ from combined ARPES and μSR measurements *New J. Phys.* **11** 055069
[11] Hicks C W, Lippman T M, Huber M E, Analytis J G, Chu J-H, Erickson A S, Fisher T R, Moler K A 2009 *Phys. Rev. Lett.* **103** 127003
[12] Mazin I I, Singh D J, Johannes M D and Du M H 2008 *Phys. Rev. Lett.* **101** 057003
[13] Kuroki K, Onari S, Arita R, Usui H, Tanaka Y, Kontani H and Aoki H 2008 *Phys. Rev. Lett.* **101** 087004
[14] Graser S, Maier T A, Hirschfeld P J and Scalapino D J 2009 *New J. Phys.* **11** 025016
[15] Wang F, Zhai H, Ran Y, Vishwanath A and Lee D-H 2009 *Phys. Rev. Lett.* **102** 047005
[16] Yu S L, Kang J and Li J-X 2009 *Phys. Rev.* B **79** 064517
[17] Cvetkovic V and Tesanovic Z 2009 *EPL* **85** 37002
[18] Kuroki K, Usui H, Onari S, Arita R and Aoki H 2009 *Phys. Rev. B* **79** 224511
[19] Chen C-T, Tsuei C C, Ketchen M B, Ren Z-A, and Zhao Z X 2010 *Nature Phys.* **6** 260
[20] Hanaguri T, Niitaka S, Kuroki K and Takagi H 2010 *Science* **328** 474
[21] Zhang X, Oh Y S, Liu Y, Yan L, Kim K H, Greene R L and Takeuchi I 2009 *Phys. Rev. Lett.* **102** 147002
[22] Zhang X *et al.* 2009 *Appl. Phys. Lett.* **95** 062510
[23] Kashiwaya H *et al.* 2010 *Appl. Phys. Lett.* **96** 202504
[24] Katase T, Ichimaru Y, Tsukamoto A, Hiramatsu H, Kamiya T, Tanabe K and Hosono H 2010 *Appl. Phys. Lett.* **96** 142507
[25] Linder J and Sudbø A 2009 *Phys. Rev.* B **79** 020501(R)
[26] Ota Y, Machida M, Koyama T and Matsumoto H 2009 *Phys. Rev. Lett.* **102** 237003
[27] Ota Y, Machida M and Koyama T 2009 *J. Phys. Soc. Jpn.* **78** 103701
[28] Koyama T, Ota Y and Machida M 2010 I-V characteristics in multi-gap intrinsic Josephson junction stacks *Physica* C doi:10.1016/j.physc.2010.05.143 (in press)
[29] Brinkman A, Golubov A A, Rogalla H, Dolgov O V, Kortus J, Kong Y, Jepsen O and Andersen O K 2002 *Phys. Rev.* B **65** 180517(R)
[30] Agterberg D F, Demler E and Janko B 2002 *Phys. Rev.* B **66** 214507
[31] Ambegaokar V and Baratoff A 1963 *Phys. Rev. Lett.* **10** 486
[32] Brinkman A, Golubov A A and Kupriyanov M Yu 2004 *Phys. Rev.* B **69** 214407
[33] Anishchanka A, Volkov A F and Efetov K B 2007 *Phys. Rev.* B **76** 104504
[34] Onari S and Tanaka Y 2009 *Phys. Rev.* B **79** 174526
[35] Sperstad I B, Linder J and Sudbø A 2009 *Phys. Rev.* B **80** 144507
[36] Nagai Y and Hayashi N 2009 *Phys. Rev.* B **79** 224508
[37] Golubov A A, Brinkman A, Tanaka Y, Mazin I I and Dolgov O V 2009 *Phys. Rev. Lett.* **103** 077003
[38] Linder J, Sperstad I B and Sudbø A 2009 *Phys. Rev.* B **80** 020503(R)
[39] Ota Y, Nakai N, Nakamura H, Machida M, Inotani D, Ohashi Y, Koyama T and Mastumoto H 2010 *Phys. Rev.* B **81** 214511
[40] Inotani D and Ohashi Y 2009 *Phys. Rev.* B **79** 224527
[41] Ota Y, Machida M, Koyama T and Matsumoto H 2010 *Phys. Rev.* B **81** 014502
[42] Ota Y, Machida M and Koyama T 2010 Novel Shapiro Steps as a Direct Probe of $\pm s$-wave Symmetry in Multi-gap Superconducting Josephson Junctions *Preprint* arXiv:1006.2184
[43] Ng T K and Nagaosa N 2009 *EPL* **87** 17003
[44] Šimánek E 1994 *Inhomogeneous Superconductors: Granular and Quantum Effects* (New York: Oxford University Press) chapter 5.
[45] Machida M, Koyama T, Tanaka A and Tachiki M 2000 *Physica* C **331** 85
[46] Suhl H, Mattias B and Walker L 1959 *Phys. Rev. Lett.* **3** 552
[47] Moskalenko V 1959 *Fiz. Metal. Metallov.* **8** 503
[48] Leggett A J, 1966 *Prog. Theor. Phys.* **36** 901
[49] Sharapov S G, Gusynin V P and Beck H 2002 *Eur. Phys. J.* B **30** 45
[50] Tanaka Y 2002 *Phys. Rev. Lett.* **88** 017002
[51] Gurevich A and Vinokur V M 2003 *Phys. Rev. Lett.* **90** 047004
[52] Goryo J, Soma S and Matsukawa H 2007 *EPL* **80** 17002
[53] Machida M and Sakai S 2004 *Phys. Rev.* B **70** 144520
[54] Machida M, Koyama T and Tachiki M 1999 *Phys. Rev. Lett.* **83** 4618

Journal of Physics: Conference Series **248** (2010) 012041 doi:10.1088/1742-6596/248/1/012041

Sheet current model for inductances extraction and Josephson junctions devices simulation

M M Khapaev[1], **M Yu Kupriyanov**[2]

[1] Dep. of Numerical Methods, Moscow State University, VMK, 119992 Vorob'evy Gory, Moscow, Russia

[2] Laboratory for Physics of Nanostructures, Institute of Nuclear Physics, Moscow State University, 119992 Vorob'evy Gory, Moscow, Russia

E-mail: vmhap@cs.msu.su

Abstract. Sheet current model for microelectronic superconductor structures, based on Maxwell and London equations, is presented. In this model, magnetic field is three-dimensional but resulting integro-differential equations for current density potential are two-dimensional. These equations are solved using finite element method and matrix of self and mutual inductances is calculated as well as current distribution in superconductors. Then, current distribution can be used for calculation of boundary conditions for Josephson junctions equations. In particular we apply this approach for calculations of bias and control line currents for flux flow oscillator simulation problem. The program and results of calculations are presented.

1. Introduction

We consider the problem of practical 3D electromagnetic modeling of superconductive multilayer, planar, multi-connected microelectronic circuits. These circuits can be digital, SQUIDs, Flux Flow Oscillators (FFO), some modern High-T_c or microwave devices.

The shape of modern circuits can be very complex [1]. Magnetic field in such layouts has essentially 3D structure [2]. Thus numerical modeling is indispensable for practical design [3]. On practice the values of interest are inductances in form of certain matrix of inductances and current distribution. In the paper, we describe the rigorous numerical approach for extracting inductances and calculating currents from design shapes. The program we evaluated, 3D-MLSI [4], is described, and results of calculations are presented.

2. Problem formulation

2.1. Typical circuits we want to consider

The technique of this paper is applicable to multilayer, planar, multi-connected structures. We want to calculate currents in conducting layers separated by layers of dielectric. The number of conducting layers can be one or more. The circuit can have or don't have groundplane that carry return current.

The shape of conductors in each layer can be very complicated. These shapes can contain holes.

The key assumption for our technique is that thickness of conducting layers should be less or of order of London penetration depth [1]. In this case, the volume current density can be accurately approximated by sheet current density.

2.2. Simple example
On Fig. 1 simple example of typical structure is presented. It is single layer thin film shape with a hole and two current terminals. Current flow in via T_1 terminal and leave circuit via T_2 terminal. The hole trap zero or non-zero flux. The inlet and outlet terminal to terminal currents as well as currents circulating around a hole should be accounted in inductance matrix.

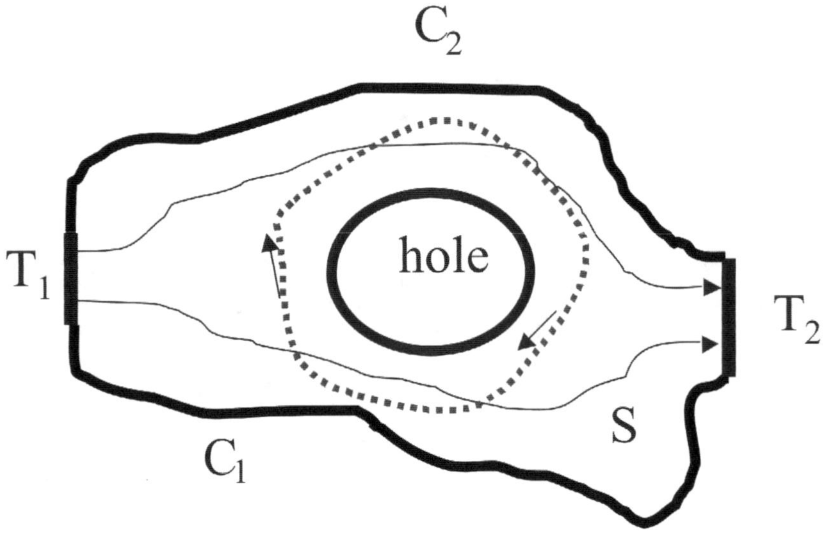

Figure 1. Circuit S with 2 terminals T_1 and T_2 and a hole. C_1 and C_2 are non-terminal boundaries. Currents can flow around a hole or from terminal to terminal

3. Integral equations

3.1. Conventional integral equations
Rigorous electromagnetic analysis should start from stationary Maxwell and London equations:

$$\lambda^2 \nabla \times \vec{j} + \vec{H} = 0, \tag{1}$$

$$\nabla \times \vec{H} = \vec{j}. \tag{2}$$

where λ is London penetration depth. Traditionally these equations are reduced to sheet current integral equations using vector potential:

$$\lambda_{\perp} \vec{J}(r) + \frac{1}{4\pi} \iint\limits_{S} \frac{\vec{J}(r')}{|r - r'|} ds = -\nabla\chi(r), \tag{3}$$

$$\nabla \cdot \vec{J}(r) = 0, \quad \Delta\chi = 0. \tag{4}$$

where where $\lambda_{\perp} = \lambda^2/t$ is London penetration depth for films, t is the thickness of film.

In (3) $\vec{J}(r)$ is unknown current, $\chi(r)$ is one more unknown function (phase). Equation (3) needs boundary conditions for function $\chi(r)$ and current $\vec{J}(r)$.

Equations similar to (3) are well known for normal conductors [5], [6] where PEEC (Partial Element Equivalent Circuit) method was evaluated. Approaches similar to PEEC for (3) for superconductors are also known [7]. For normal conductor function $\chi(r)$ has sense of voltage potential. Recently fast multipoles technique based program FASTHENRY [8] for (3) was adopted for superconductors [9].

Boundary conditions for (3) are easy for normal conductors where $\chi(r)$ is voltage. For superconductors currents excitation are more natural. Moreover, PEEC-like methods showed to be time and memory consuming in case of superconductors. It was our motivation for developing different numerical approch.

3.2. Better integral equations

Our approach is based on potential representation $\psi(r)$ of two-dimensional sheet current $\vec{J}$ which is called a *stream function*:

$$J_x(r) = \frac{\partial \psi(r)}{\partial y}, \quad J_y(r) = -\frac{\partial \psi(r)}{\partial x} \tag{5}$$

Then using stationary London and Maxwell equations (2) it is possible to obtain the following expression in terms of stream functions $\psi(r)$:

$$-\lambda_\perp \Delta \psi(r_0) + \frac{1}{4\pi} \iint\limits_S (\nabla \psi(r), \nabla_{xy} G(r, r_0))\, ds_r = -H_{z,ext}(r_0). \tag{6}$$

Here $H_{z,ext}(r_0)$ is external or exciting magnetic field. For very thin conductors $G(r, r_0)$ can be taken as

$$G(r, r_0) = \frac{1}{|r - r_0|}. \tag{7}$$

In [10] simple expressions for $G(r, r_0)$ are evaluated that take into account finite thickness of conducting objects.

Equation (6) should be completed by boundary conditions. These boundary conditions are simple first kind boundary conditions when values of stream function on the boundary are given:

$$\psi(r) = I_{h,k}, \; r \in \partial S_{h,k}, \tag{8}$$

$$\psi(r) = F(r), \; r \in \partial S_{ext}. \tag{9}$$

Here $I_{h,k}$ is the full current circulating around hole k with boundary $S_{h,k}$. On the external boundary function $F(r)$ can be easily evaluated [10].

Mathematically problem (6), (9) is very similar to boundary problem for Poisson equation [10, 16].

We prefer to solve equation (6) instead of (3) because (6) accounts important physical features of the problem and because of numerical effectivity considerations:

- Many superconductivity problems are based solely on currents and magnetic field. In these cases it is difficult to define boundary conditions for $\chi(r)$.
- Holes in S is a problem for (3). But it is an easy task for (6). Given currents circulating around holes are accounted in boundary conditions in function $I_{h,k}$ (9). Non-decaying currents circulating around holes are typical for problems in superconductivity.
- FEM for (6) has better numerical approximation then PEEC and thus can give smaller system of linear equations.

Journal of Physics: Conference Series **248** (2010) 012041 doi:10.1088/1742-6596/248/1/012041

- FEM for (6) can be adopted for solution of very large problems when matrix of equations is so large then can't be allocated in computer memory.

After function $\psi(r)$ is calculated, energy functional also can be calculated as well as inductance matrix [10], [11].

4. Numerical technique

4.1. Finite Elements Method

Equation (6) together with boundary conditions (9) are ready to solution using Finite Element Method (FEM) [15]. Finite elements method is based on bilinear form. The bilinear form $a(u,v)$ for "weak" formulation of the problem (6) is:

$$a(u,v) = \lambda_\perp \iint_S (\nabla u(r), \nabla v(r)) ds +$$

$$\frac{1}{4\pi} \iint_S ds_0 \iint_S (\nabla u(r), \nabla v(r_0)) G(r, r_0) ds. \tag{10}$$

Here the principal value integral in (6) was integrated by parts. Form $a(u,v)$ is symmetric and for thin enough conductors positively definite because for $u = v$ it gives the expression for full energy.

Then we apply conventional FEM based on triangular meshing of conductors and linear elements. There are no limitations on shape and number of conductors. More details can be found in [10, 16].

4.2. Large problems

The FEM matrix is the sum of sparse matrix resulted from Laplace operator and dense part resulted from integral operator. Dense part of the matrix can be a problem for numerical solution. If number of nodes in triangulation of S is moderate then finite element equations can be solved using Cholesky factorization. The problem arise when number of nodes is several thousands or tens thousands of nodes. It is not possible to store this matrix in dense format or to factor it. To overcome this limitation we evaluated a sparsification procedure integral equation matrix.

Consider finite element matrix elements for integral operator in (6) b_{ij}. For non-intersecting finite elements supports $S_i \bigcap S_j = \emptyset$ we have

$$b_{ij} = \iint_{S_i} ds \iint_{S_j} \frac{(\nabla u_j^h(r'), \nabla u_i^h(r))}{|r - r'|} ds' = - \iint_{S_i} ds \iint_{S_j} \frac{u_j^h(r') u_i^h(r)}{|r - r'|^3} ds'. \tag{11}$$

Here $u_j^h u_i^h$ and u_j^h are simplectic finite element shapes, $|u_j^h| \leq 1$. It is seen that b_{ij} is small if distance between S_i and S_j is large because the denominator in (11) is large and quickly grow up with the distance.

Our sparsification procedure is simple. Matrix elements below some tolerance are dropped. It makes matrix of integral equation sparse. Then, sparse Cholesky factorization can be applied. As result, large size circuits with detailed mesh can be calculated with some small accuracy trade-off.

4.3. The program

Our program, 3D-MLSI, is written on C++ with Windows GUI. Non-GUI part can be compiled for Unix. The program can be found here [4].

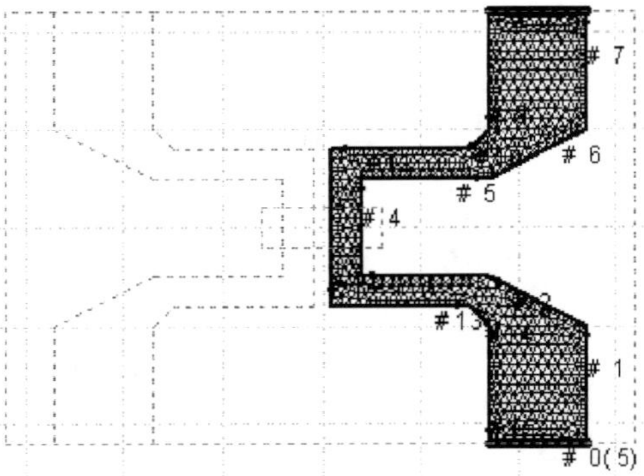

Figure 2. Transformer layout with finite element mesh on one of the conductors. Coordinate cell size 10 μm.

5. Some results of calculations

5.1. Inductances calculations for HTC transformer

As an example of a real device, a two-layer YBCO circuit [11] was calculated, see Fig. 2.

Both superconducting layers are 0.2 μm thick. The first layer is a ground plane and contains $12 \times 4\,\mu$m hole to increase the mutual inductance. The second layer contains superconducting wiring. The narrow part of wires is 3 μm wide. The thickness of the insulator between the layers is 0.19 μm. London penetration depth $\lambda = 0.18\,\mu$m. The circuit is symmetric (see Fig. 2).

We assume that the ground plane carries all return currents. It is possible to calculate full 3×3 inductance matrix: for two symmetrical current loops via top layer wires and the current around the loop in the ground plane. For the circuit analysis this matrix is unsuitable because the hole circulating currents are not excited independently on the terminal currents. It is known that for any currents flowing in the circuit the hole keeps constant trapped flux, which can be taken equal to zero, as it, in any case, does not affect the reduced inductance matrix [1]. The program calculates 2×2 inductance matrix with self inductances L and mutual inductances $-M$. For basic configuration with rectangular hole $w \times d = 12 \times 4\,\mu$m the self-inductance is calculated to be $L = 13.61$ pH, $M = 1.651$ pH (transformation ratio 0.12). For the same structure without a hole $L = 10.65$ pH, $M = 0.04$ pH (transformation ratio 0.0039).

For these calculations, a typical PC was used. The whole set of calculations took several hours in batch mode. The exact time depends on CPU but can't be large even for low performance computer.

5.2. Bias and control line currents calculations for FFO

FFO (Flux Flow Oscillator) is microwave Josephson junction device [2], [12]. Typically these devices are two-layer structures where layers of superconductor partly overlap. Part of overlapping region occupy long Josephson junction. The device is driven by two currents, bias current $I_b(t)$ that pass from one sheet to another across JJ and controlling current $I_{cl}(t)$ that flow on top or bottom sheet. Schematically FFO is presented on Fig. 4.

Simplest mathematical model of FFO is perturbed sine-Gordon equation with constant coefficients subject to Neumann boundary conditions [2]:

$$\ddot{\varphi} + \alpha\dot{\varphi} - \varphi_{xx} = \eta(x) - \sin\varphi$$
$$\frac{\partial\varphi(0,t)}{\partial x} = \frac{\partial\varphi(L,t)}{\partial x} = \Gamma. \tag{12}$$

In this equation function $\eta(x)$ and value of Γ should be defined using given full control line and bias currents as it is shown on Fig. 4.

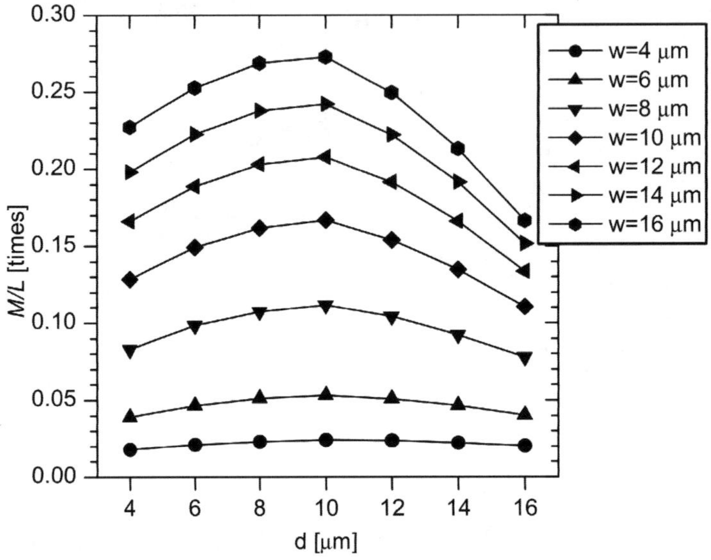

Figure 3. Transformation ratio as a function of w and d

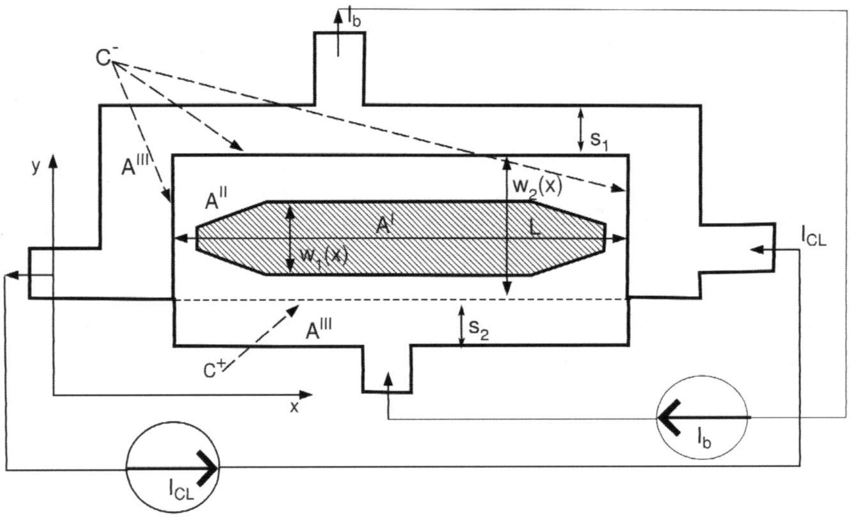

Figure 4. 2D overlap JJ, top view. A^I - junction, A^{II} - microstrip overlapping layers, A^{III} - single layer structure needed for 3D magnetic field and currents calculations

Using so called "locking principle" for superconductors, we adopted our program 3D-MLSI for calculations of right part and coefficients in (12). 3D-MLSI calculate inlet currents on the boundary of overlapping region on Fig. (4). Then this data was passed to program FFOSIM [14] where final evaluation of right part in (12) was performed. It was found that practical distribution of control line an bias currents is very different for from simple profiles used in numerical calculations of (12). For one of devices [12, 13], the graphics of control line and bias components of $\eta(x)$ is presented on Fig. 5. For results of simulation of FFO, see [14].

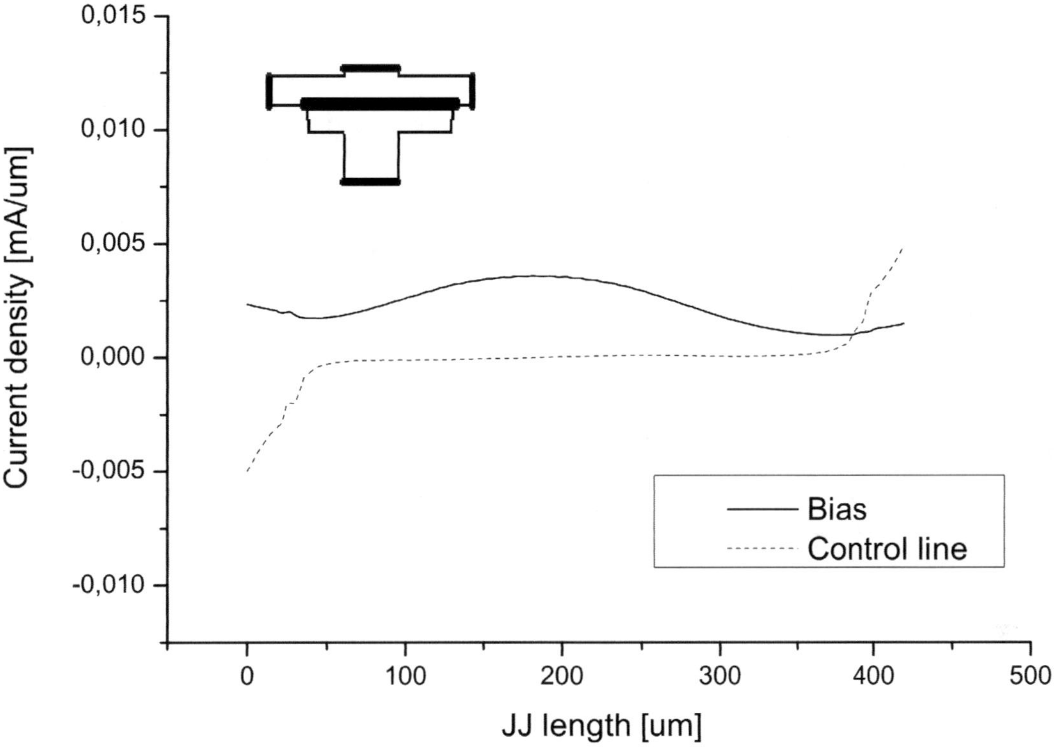

Figure 5.

6. Conclusions

We evaluated modeling technique and software 3D-MLSI for currents calculation and inductances extraction. The program can solve problems of arbitrary geometry of the circuits. The program have good user interface and visualization of current density and flow lines. Also, the program can calculate magnetic field in given set of points in space. The program can solve large problems when number of mesh points is too large to be stored in matrix in computer memory.

Our plans are to implement fluxes calculations for given contours and evaluate better triangulation of conductor shapes with external mesh generation program.

Acknowledgments

We would like to thank V. P. Koshelets and E. B. Goldobin for fruitful discussions. The work was supported by ISTC project 3174 and Russian Fund for Basic Research, grant 10-02-90014-Bel-a

References

[1] Van Duser T and Turner C W 1981 *Principles of Superconductive Devices and Circuits.* North Holland: Elsevier, 1981.
[2] Likharev K K 1986 *Dynamics of Josephson Junctions and Circuits* (New York: Gordon and Breach)
[3] Gaj K, Herr Q P, Adler V, Krasniewski A, Friedman E G and Feldman M J, 1999 *IEEE Trans. Appl. Superconductivity* **9** 18
[4] http://vm.cs.msu.su/sotr/vmhap/3dmlsi/3dmlsi.htm
[5] Brennan P A, Raver N and Ruehli A E 1979 *IBM J. Res. Develop.* **23** 661
[6] Antonini G and Ruehli A E 2003 *IEEE Trans. on Mobile Computin* **2** 288
[7] Guan B, Wengler M J, Rott P and Feldman M J 1997 *IEEE Trans. Appl. Superconductivity* **7** 2776
[8] Kamon M, Tsuk M J and White J K 1994 *IEEE Trans. Microwave Theory and Techn* **42** 1750
[9] http://www.wrcad.com/freestuff.html
[10] Khapaev M M 2001 *IEEE Trans. Microwave Theory and Techn.* **49** 217
[11] Khapaev M M, Kupriyanov M Yu, Goldobin E and Siegel M 2003 *Supercond. Sci. Technol.* **16** 24
[12] Koshelets V P and Mygind J 2001 *Studies of High Temperature Superconductors* **39** 213 NOVA Science Publishers, New York.
[13] Koshelets V P, Shitov S V, Shchukin A V, Filippenko L V, Dmitriev P N, Vaks V L, Baryshev A M, Luinge W, Golstein H and Mygind J 1999 *IEEE Trans. Appl. Superconductivity* **9** 4133
[14] Khapaev M M and Kupriyanov M Yu 2008 *J. Phys.: Conf. Ser.* **129** 012037
[15] Jin J 2002 *The Finite Element Method in Electromagnetics* (New York: Willey)
[16] Khapaev M and Kupriyanov M 2010 *Matrix Methods: Theory, Algorithms, Applications* World Scientific Publishing 510

Study of charge-phase diagrams for coupled system of Josephson junctions

M. Hamdipour[1,2] and Yu. M. Shukrinov[1]

[1] BLTP, JINR, Dubna, Moscow Region, 141980, Russia
[2] Department of Physics, Institute for Advanced Studies in Basic Sciences, P.O.Box 45195-1159, Zanjan, Iran

E-mail: hamdipur@theor.jinr.ru

Abstract. Dynamics of stacked intrinsic Josephson junctions (IJJ) in the high-Tc superconductors is theoretically investigated. We calculate the current-voltage characteristics (CVC) of IJJ and study the breakpoint region on the outermost branch of the CVC for the stacks with 9 IJJ. A method for investigation of the fine structure in CVC of IJJ based on the recording the "phase-charge" diagrams is suggested. It is demonstrated that this method reflects the main features of the breakpoint region.

1. Introduction and model

Study of intrinsic Josephson junctions in HTSc, like $Bi_2Sr_2CaCu_2O_{8+\delta}$ shows interesting physical features. In Refs.[1, 2, 3] we studied the multiple branch structure of the CVC of IJJ and showed that the branches have a breakpoint (BP) and some breakpoint region (BPR) before transition to the another branch. The BP is determined by the creation of the longitudinal plasma wave (LPW) with a definite wave number k, which depends on the coupling parameter α, dissipation parameter β, the number of junctions in the stack, and the boundary conditions.

Here we show that using the phase-charge diagrams (phase portraits) give us an additional method for investigation of the fine structure in the CVC. We use the $(CCJJ + DC)$ - model [4] to investigate the phase dynamics and CVC of coupled system of Josephson junctions. The system of equations in this model has a form

$$\frac{d^2}{dt^2}\varphi_l = (I - \sin\varphi_l - \beta\frac{d\varphi_l}{dt}) + \alpha(\sin\varphi_{l+1} + \sin\varphi_{l-1} -2\sin\varphi_l) + \alpha\beta(\frac{d\varphi_{l+1}}{dt} + \frac{d\varphi_{l-1}}{dt} - 2\frac{d\varphi_l}{dt}) \quad (1)$$

Here φ_l is the gauge-invariant phase differences $\varphi_l(t) = \theta_{l+1}(t) - \theta_l(t) - \frac{2e}{\hbar}\int_l^{l+1} dz A_z(z,t)$ between superconducting layers (S-layers), θ_l is the phase of the order parameter in S-layer l, A_z is the vector potential in the barrier. The current, voltage and time are normalized to the critical current I_c, $V_0 = \frac{\hbar\omega_p}{2e}$ and inverse of plasma frequency ω_p^{-1}, respectively. The details of simulation are available at Refs. [5, 3, 6]. We calculate the charge on S-layer as $Q_l(t) = \alpha(V_l(t) - V_{l-1}(t))$, where $V_l(t)$ is the voltage drop on insulator layer l at time moment t. We start from the current value I=1.2 and decrease it up to the transition to another branch. We record the time dependence of the phase difference between layers l and $l-1$ and the electric charge on the layer l at some selective points of the BPR.

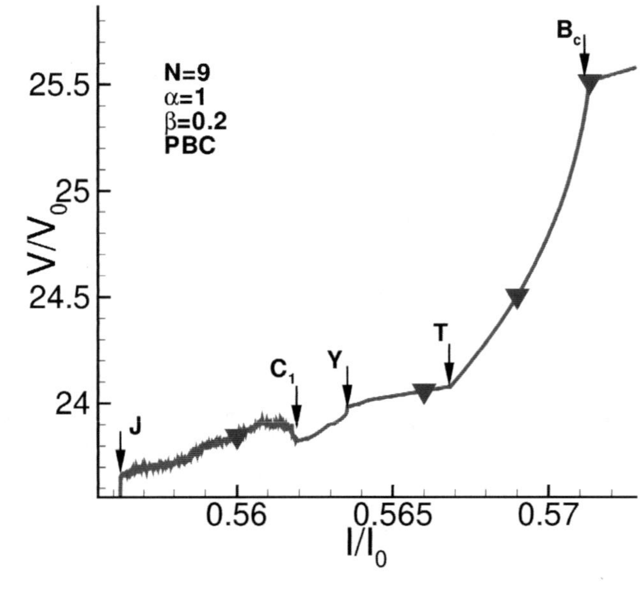

Figure 1. (Color online) The BPR part of the outermost branch for the stack with 9 IJJ at $\alpha = 1$, $\beta = 0.2$ and periodic boundary conditions. The symbol ∇ indicates the points at which the time dependence of the phase difference and the charge on the layer are investigated.

2. Phase-charge diagrams

The BPR part of the outermost branch of CVC for the stack with 9 IJJ at $\alpha = 1$ and $\beta = 0.2$ and periodic boundary conditions is presented in Fig. 1. It indicates the current values, at which we have investigated the phase-charge diagrams. The phase-charge diagram is a trajectory of junction in its phase space:(φ, Q). We suggest to use this diagrams as a tool to probe the dynamics of system. For periodic systems like simple harmonic oscillator, the phase portraits in $(\varphi,\dot{\varphi})$ space are circles with their centers at the origin of coordinate system. For the system of IJJ it is natural to investigate the phase-charge diagrams.

Fig. 2 shows such phase-charge diagrams: the variation in time of the modular of charge on the first superconducting layer Q_1 and phase difference φ_1 in the first Josephson junction of the stack. In these figures the radial axis shows the charge Q_1 and azimuthal is the phase φ_1.

The beginning part of BPR corresponds to the onset of the parametric resonance: charge on layers is growing. In this region, particularly, $I = 0.57126$ (Fig. 2a) the Josephson frequency is still high than LPW frequency, so the trajectory in phase-charge space has some small thickness. At $I = 0.569$ (see Fig. 2b) the conditions $\omega_J = 2\omega_{LPW}$ is practically ideally fulfilled, charge on layers has reached its maximum amplitude, and thickness of trajectory in phase-charge space is just thickness of line. Results of FFT analysis (not presented here) show that at this value of current we observe the peaks corresponding to the LPW only and Josephson frequency does not manifests itself. The system follows this trajectory all time. As we show below, the results of analysis of autocorrelation function, presented in Fig. 3 confirm this interpretation [6].

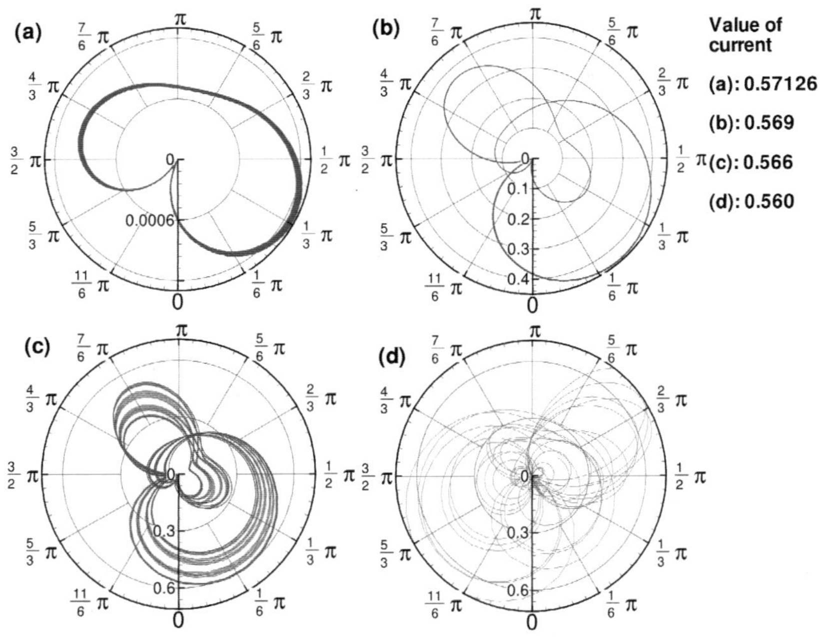

Figure 2. (Color online) The phase-charge diagram (the charge on the first superconducting layer versus the phase difference in the first junction) at different value of bias current. The radial axis is a charge, the azimuthal axis is a phase difference.

At $I = 0.566$ (Fig. 2c) the trajectory in phase-charge space is wide again. The resonance is passed: $\omega_J > 2\omega_{LPW}$. Here the amplitude of charge on the layers is merely constant but there is a beating of LPW in the stack. A new large period of system appears [7], but the system after some cycles repeats the same trajectory. Analysis of charge correlations shows that the behavior of system is regular, while Fig. 2d corresponds to a chaotic behavior of the system.

There is a correspondence between phase-charge diagrams and autocorrelation functions, presented in Fig.3. The smallest value of $\tau = \tau_0$ with $C^a(\tau = \tau_0) = C^a(\tau = 0)$ is a period of system. If the autocorrelation function has not such points, then the trajectory of system in phase-charge diagram is open. There is such $\tau = \tau_0$ in the cases presented in Figs. 3b and 3c which satisfies the mentioned condition and that's why in Figs 2b and 2c the trajectories in phase space are closed. But in Fig. 3d the autocorrelation function doesn't show such τ_0 and hence Fig. 2d show open trajectories. Behavior of the system in this case is chaotic. As we mentioned above the thickness of trajectory in Fig.3a is related to the growing charge on the superconducting layers and this is a reason of the amplitude's decreasing of autocorrelation function with time. Note that to show the very small decreasing of autocorrelation we zoomed the region in vertical axis of Fig.3a. The value τ is introduced in the definition of autocorrelation function

$$C_l^a(\tau) = \lim_{(Tf-Ti)\to\infty} \frac{1}{Tm - Ti} \int_{Ti}^{T_m} Q_l(t-\tau)Q_l(t)dt \qquad (2)$$

$C^a(\tau = 0)$ is a maximum and its value is $C_l^a(\tau = 0) =< Q_l^2(t) >$. At $I = 0.566$ (Fig. 2c) the trajectory in phase-charge space is wide again. The resonance is passed: $\omega_J > 2\omega_{LPW}$. Here the amplitude of charge on the layers is merely constant.

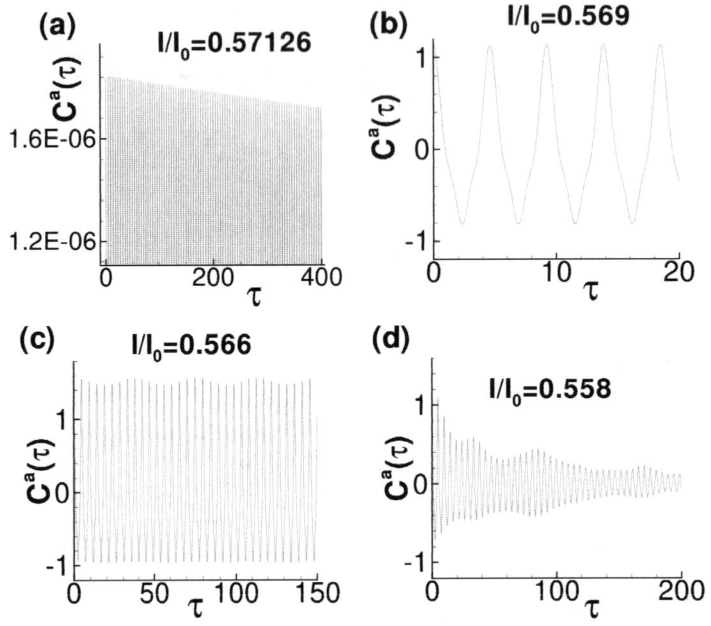

Figure 3. (Color online) The charge autocorrelation function C^a for the first layer at different values of bias current in the breakpoint region.

3. Summary

We found that the phase-charge diagram reflects the dynamical behavior of IJJ. It was shown that the results of the phase-charge diagram analysis are in agreement with the results of the autocorrelation function analysis. We demonstrated that using the phase-charge diagrams (phase portraits) give us a powerful method for investigation of the fine structure in the CVC.

Acknowledgments

We thank M.R. Kolahchi for his fruitful discussions.

References

[1] Yu. M. Shukrinov, F. Mahfouzi, Supercond. Sci.Technol., **19**, S38-S42 (2007).
[2] Yu. M. Shukrinov, F. Mahfouzi, Phys.Rev.Lett. **98**, 157001 (2007).
[3] Yu. M. Shukrinov, F. Mahfouzi, N. F. Pedersen, Phys. Rev. B **75**, 104508 (2007).
[4] Yu. M. Shukrinov, F. Mahfouzi, and P. Seidel, Physica C 449, 62 (2006).
[5] T. Koyama and M. Tachiki, Phys. Rev. B **54**, 16183 (1996).
[6] Yu. M. Shukrinov, M. Hamdipour, M. Kolahchi, Phys.Rev.B,**80**, 014512 (2009).
[7] Yu. M. Shukrinov, F. Mahfouzi, M. Suzuki, Phys. Rev. B **78**, 134521 (2008).

Journal of Physics: Conference Series **248** (2010) 012043 doi:10.1088/1742-6596/248/1/012043

Peculiarities of phase dynamics of coupled Josephson junctions in CCJJ and CCJJ+DC models

Yu. M. Shukrinov[1], **I.R. Rahmonov** [1,2] **and M. El Demery**[1,3]
[1] BLTP, JINR, Dubna, Moscow Region, 141980, Russia
[2] Physical Technical Institute, Dushanbe, Tajikistan
[3] ECTP, Modern University, Mokattum, Cairo, Egypt
E-mail: shukrinv@theor.jinr.ru

Abstract. The phase dynamics of the coupled Josephson junctions in the framework of CCJJ and CCJJ+DC models is studied. The current voltage characteristics (CVC) are numerically calculated for the stacks with different number of junctions at different model parameters. We manifest the difference of these models for the branching at I = Ic and in the hysteretic region. The essential difference is observed in the breakpoint region, where the longitudinal plasma wave is created. We discuss the main features of both models, related with the role of the diffusion current between the superconducting layers.

1. Model and method

The absence of complete screening of charge in the superconducting layer (S-layer) in HTSC leads to the generalized scalar potential Φ_l of S-layer. The Φ_l is expressed [1] through the electric scalar potential ϕ_l and the derivative of order parameter's phase $\dot{\theta}_l$ of S-layer by $\Phi_l(t) = \phi_l - V_0\dot{\theta}_l$. The Φ_l determines the charge density in S-layer $Q_l = -\frac{1}{4\pi r_D^2}\Phi_l$, where r_D is the Debye screening length. Using the Maxwell equation $div(\varepsilon\varepsilon_0 E) = Q$, where ε and ε_0 are relative dielectric and electric constants, we express the charge density Q_l in the S-layer l by the voltages V_l and V_{l+1} in the neighbor insulating layers $Q_l = Q_0\alpha(V_{l+1} - V_l)$, where $Q_0 = \varepsilon\varepsilon_0 V_0/r_D^2$.

Without screening of the charge in S-layer the Josephson equation is generalized $\dot{\varphi}_l = V_l - \alpha(V_{l+1} + V_{l-1} - 2V_l)$ and together with the expression for total current through the stack $I = \dot{V}_l + \sin\varphi_l + \beta V_l$ describe the phase dynamics of the coupled system of Josephson junctions (JJ) in the framework of CCJJ model [1]. In compare with well known RCSJ model [2], the total current in CCJJ model

$$I = \ddot{\varphi}_l + \sin\varphi_l + \beta\dot{\varphi}_l + \alpha(\sin\varphi_{l+1} - \sin\varphi_l) + \alpha(\sin\varphi_{l-1} - \sin\varphi_l) \quad (1)$$

has additional terms which depend on the coupling constant and proportional to the difference of the superconducting currents in the considered and adjusted junctions. The CVC in CCJJ model are characterized by branching at $I = I_c$ and strong branching in the hysteresis region: the number of branches in general case is equal to 2^N, where N is the number of junctions in the stack.

In CCJJ+DC model [3, 4] the diffusion current $J_D^l = \Phi_{l+1} - \Phi_l$ is added and the expression for current through the stack of junctions in this model is

$$I = \dot{V}_l + \sin\varphi_l + \beta\varphi_l \quad (2)$$

It has additional terms in compare with CCJJ model which are proportional to the product of coupling and dissipation parameters

$$I = \ddot{\varphi}_l + \sin\varphi_l + \beta\dot{\varphi}_l + \alpha(\sin\varphi_{l+1} - \sin\varphi_l) + \alpha(\sin\varphi_{l-1} - \sin\varphi_l) + \alpha\beta(\dot{\varphi}_{l+1} - \dot{\varphi}_l) + \alpha\beta(\dot{\varphi}_{l-1} - \dot{\varphi}_l) \quad (3)$$

In this paper we solve the system of dynamical equations for phase differences including the generalized Josephson relation and expression for total current at fixed value of bias current I in some time interval $(0, T_m)$ of dimensionless time $\tau = t\omega_p$ with the time step $\delta\tau$, where t is a real time. This interval is used for time averaging procedure. Then we change the bias current by step δI, and repeat the same procedure for the current $I + \delta I$ in new time interval $(T_m, 2T_m)$. The values of the phase and derivative of phase in the end of the first time interval are used as the initial conditions for second time interval and so on.

The bias current is increased from the value $I/I_c = 0.1$ till $I/I_c = 5$ with the step $\Delta I = 0.005$ and then decreased back till zeroth. In the interval $I/I_c = 0.2 \div 1.1$ the step in current is consists of $\Delta I = 0.0001$. In our simulations we add to the bias current a small noise with amplitude in the interval $\delta I = \pm 10^{-8}$. The total recorded time is calculated as $\tau + T_m(I_0 - I)/\delta I$, where I_0 is an initial value of the bias current for time dependence recording.

2. Results and discussion

2.1. Stacks with odd number of JJ

2.1.1. $\alpha = 0.1$ Let us first consider the stack with odd number of JJ with small value of coupling parameter. In Fig. 1(a) we show the one-loop CVC of the stack with 9 coupled JJ simulated in the framework of the CCJJ and CCJJ+DC models at $\alpha = 0.1$, $\beta = 0.2$ and periodic boundary conditions.

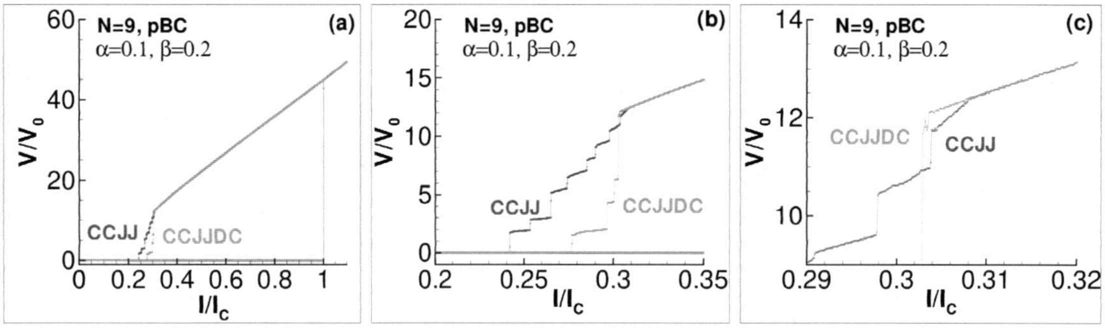

Figure 1. (Color online) (a) The one-loop CVC of the stack with 9 coupled JJ simulated in the framework of the CCJJ and CCJJ+DC models at $\alpha = 0.1$, $\beta = 0.2$ and periodic boundary conditions; (b) The enlarged branching part of CVC in hysteretic region; (c) The enlarged part of the breakpoint region on the outermost branch of CVC

As we can see, the CVC for both models are coincide with the increase in bias current. The difference appear in the branching part of CVC in hysteretic region: the enlarged part of this region is shown in Fig. 1(b). The CCJJ model demonstrates more intensive branching and larger hysteresis region in compare with CCJJ+DC model. We consider that this feature can be explained by the role of diffusion current. As it was shown in Ref.[5], in CCJJ+DC model the diffusion current has the different orientation in the neighbor junctions. It promotes for the transitions between branches and the decrease of the hysteretic region in CCJJ+DC model.

We see the different features as well on the outermost branch of CVC in the breakpoint region which enlarged parts are presented in Fig. 1(c) for both models. In the CCJJ model the breakpoint region manifest itself as a part with a slope different from the outermost branch. In CCJJ+DC we observe the typical fine structure in the BPR, but its width at $\alpha = 0.1$ is much smaller than at $\alpha = 1$ [6, 7] (see Fig. 2). The value of transition current to another branch is practically the same in both models. We note that at $\alpha = 0.1$ the difference of return currents I_r in both models is $\Delta I_r = 0.03$.

2.1.2. $\alpha = 1$ With increase in α these two models give the different results at $I = I_c$. The CCJJ model leads to the switching between branches at this value of bias current while in CCJJ+DC model we observe a jump to the outermost branch directly. In Fig. 2(a) we show the one-loop CVC of the stack with 9 coupled JJ at $\alpha = 1$ in both models.

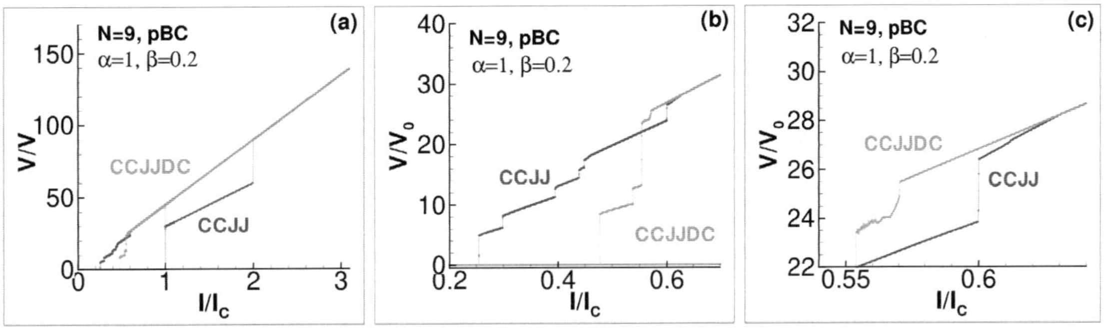

Figure 2. (Color online) The same as in Fig. 1 at $\alpha = 1$

As we can see in Fig. 2(b), there is the difference for both models in the hysteretic region as well. We note that the difference of return currents I_r in both models is increased and consists of $\Delta I_r = 0.25$ now. The value of the jump current from the outermost branch in CVC is essentially different in both models.

2.2. Stacks with even number of JJ

2.2.1. $\alpha = 0.1$ In Fig. 3(a) we show the one-loop CVC of the stack with 10 coupled JJ in CCJJ and CCJJ+DC models at $\alpha = 0.1$. We see again that at small value of coupling parameter two models give the same result at $I = I_c$.

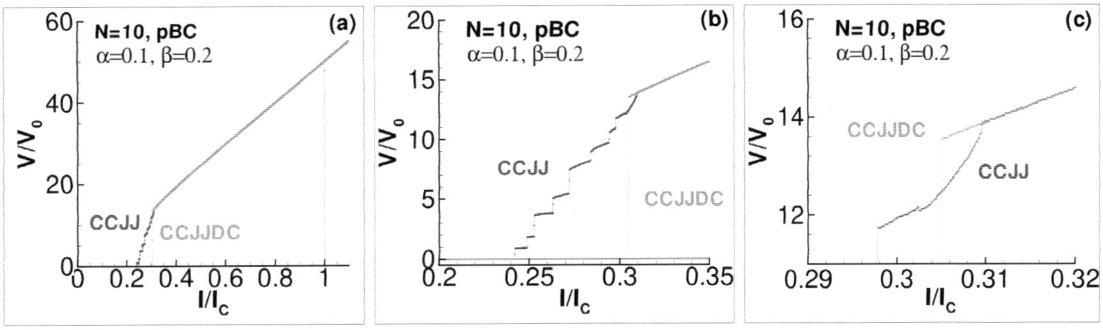

Figure 3. (Color online) The same as in Fig. 1 for the stack with 10 coupled JJ.

The hysteretic region in CCJJ model is larger than in CCJJ+DC model: the difference of the return currents for both models ΔI_r is about 0.07. There is no fine structure in the outermost branch of CVC in the stack with even number of junctions in the framework of CCJJ+DC model, but we observe some structure on the outermost branch in CCJJ model. In Fig. 3(c) we see an interesting feature: the continues transition from the outermost branch to the another one. The structure on the outermost branch of CVC in framework of CCJJ model was observed in Ref.[8] as well. This feature will be discussed in detailed somewhere else.

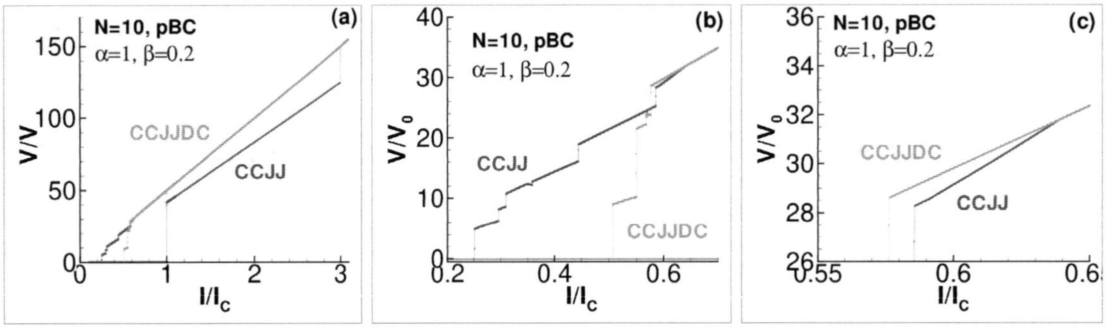

Figure 4. (Color online) The same as in Fig. 2 at $\alpha = 1$

2.2.2. $\alpha = 1$ At $\alpha = 1$ we observe the branching with the increase in bias current in CCJJ model as in the stacks with odd number of junctions. Like for stacks with odd number of junctions the hysteretic region is also increased in compare with the case $\alpha = 0.1$.

We note that continues transition from the outermost branch to another one in CCJJ model disappears at $\alpha = 1$ and such transition happens by jump in both models now. As it is in case of N=9, for N=10 the I_{jump} is as well increased in CCJJ model in compare with CCJJ+DC model.

3. Summary

We presented the CVC of the stacks with 9 and 10 coupled Josephson junctions with two values of coupling parameter $\alpha = 0.1$ and $\alpha = 1$ calculated in the framework of CCJJ and CCJJ+DC models. The CVC demonstrate the different features at critical current and in the hysteretic region. These results stress the important role of the diffusion current in formation of the CVC of intrinsic Josephson junctions in high temperature superconductors.

References

[1] T. Koyama and M. Tachiki, Phys. Rev. B, **54**, 16183 (1996)
[2] W. Buckel, R. Kleiner, Superconductivity. Fundamentals and Applications, Wiley-VCH Verlag GmbH&Co, KGaA, (2004)
[3] Yu. M. Shukrinov, F. Mahfouzi and P. Seidel, Physica C, **449**,62 (2006)
[4] M. Machida, T. Koyama, A. Tanaka and M. Tachiki, Physica C, **331**, 85 (2000)
[5] Yu. M. Shukrinov, I. R. Rahmonov. Accepted by JETP Letters.
[6] Yu. M. Shukrinov and F. Mahfouzi. Phys.Rev.Let., **98**, 157001 (2007)
[7] Yu. M. Shukrinov, F. Mahfouzi and M. Suzuki, Phys. Rev. B, **78**, 134521 (2008)
[8] M. Machida, T. Koyama, and M. Tachiki, Phys. Rev. Lett. 83, 4618 (1999).

Journal of Physics: Conference Series **248** (2010) 012044 doi:10.1088/1742-6596/248/1/012044

Influence of Josephson current second harmonic on stability of magnetic flux in long junctions

P.Kh.Atanasova[1], T.L.Boyadjiev, Yu.M.Shukrinov[1], E.V.Zemlyanaya[1], P. Seidel[2]

[1] Joint Institute for Nuclear Research, Dubna 14198, Russia.
[2] Institut für Festkörperphysik, Friedrich-Schiller-Universität Jena, D-07743 Jena, Germany

E-mail: poli@jinr.ru

Abstract. We study the long Josephson junction (LJJ) model which takes into account the second harmonic of the Fourier expansion of Josephson current. The dependence of the static magnetic flux distributions on parameters of the model are investigated numerically. Stability of the static solutions is checked by the sign of the smallest eigenvalue of the associated Sturm-Liouville problem. New solutions which do not exist in the traditional model, have been found. Investigation of the influence of second harmonic on the stability of magnetic flux distributions for main solutions is performed.

1. Motivation and model

The physical properties of the magnetic flux in Josephson junctions (JJ) are the contemporary superconducting electronics base. For the traditional JJ the current-phase dependence is a sinusoidal function. Such a model is described by the sine-Gordon equation. For a sufficiently wide class of JJ the superconducting Josephson current as a function of magnetic flux φ can be represented as a sine series [1, 2]:

$$I_S = I_c \sin\varphi + \sum_{m=2}^{\infty} I_m \sin m\varphi\,. \tag{1}$$

It was recognized recently that the higher harmonics in this expansion are important in many applications, in particular, in junctions like SNINS and SFIFS, where S is a superconductor, I is an insulator, N is a normal metal and F is a weak metallic ferromagnet [1, 3]. The interesting properties of LJJ with an arbitrarily strong amplitude of second harmonic in current phase relation were considered in [4].

Using only first two terms of the expansion (1) leads to the double sine-Gordon equation (2SG) [2].

$$-\varphi'' + a_1 \sin\varphi + a_2 \sin 2\varphi - \gamma = 0\,, x \in (-l; l)\,. \tag{2}$$

Here and below the prime means a derivative with respect to the coordinate x. The magnitude γ is the external current, l is the semilength of the junction, a_1 and a_2 are the normalized amplitudes of the first and second harmonics of the Josephson current [1, 5]. All the magnitudes are dimensionless.

The boundary conditions for (2) have the form

$$\varphi'(\pm l) = h_e, \tag{3}$$

where h_e is external magnetic field.

Numerical solution of the nonlinear boundary problem (2), (3) is solved on the basis of the continuous analog of Newton's method [7].

Stability and bifurcations of static solutions $\varphi(x,p)$, where $p = (l, a_1, a_2, h_e, \gamma)$ are analyzed on the basis of numerical solution of the corresponding Sturm-Liouville problem [6]:

$$-\psi'' + q(x)\psi = \lambda\psi, \quad \psi'(\pm l) = 0, \quad q(x) = a_1 \cos\varphi + 2a_2 \cos 2\varphi. \tag{4}$$

The minimal eigenvalue $\lambda_0(p) > 0$ corresponds to the stable solution. In case $\lambda_0(p) < 0$ solution $\varphi(x,p)$ is unstable. The case $\lambda_0(p) = 0$ indicates the bifurcation with respect to one of the parameters p. We characterize the solutions of equation (2) by number of fluxons $N(p)$ which is defined as

$$N(p) = \frac{1}{2l\pi} \int_{-l}^{l} \varphi(x)\, dx\,. \tag{5}$$

2. Deformation of the Meissner solution M_0

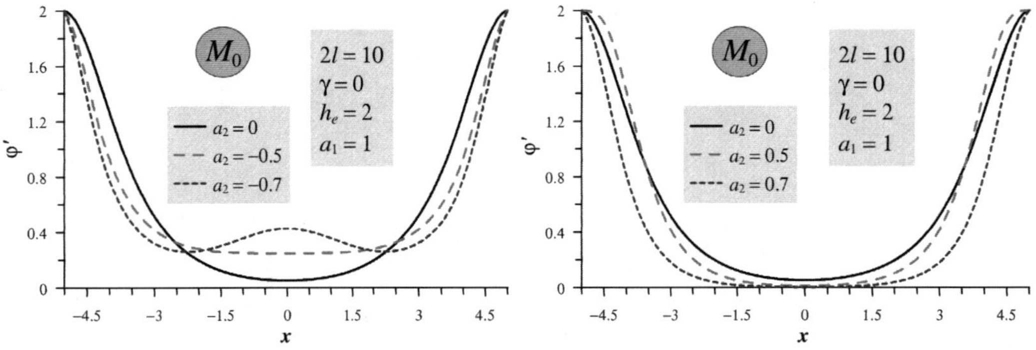

Figure 1. The internal magnetic field of the Meissner solution M_0 for LJJ with $2l = 10$, $\gamma = 0$, $h_e = 2$ and $a_1 = 1$ at different values of the parameter a_2. Left figure shows φ' at negative a_2, right one shows at positive a_2.

In the "traditional" case $a_2 = 0$ two trivial solutions $\varphi = 0$ and $\varphi = \pi$ of (2), (3) are known at $\gamma = 0$ and $h_e = 0$, which are denoted by M_0 ($N[M_0] = 0$) and M_π ($N[M_\pi] = 1$), respectively. Accounting of the second harmonic $a_2 \sin 2\varphi$ leads to the appearing of two additional solutions $\varphi = \pm\arccos(-a_1/2a_2)$ denoted as $M_{\pm ac}$ ($N[M_{\pm ac}]$ are not integer numbers and depend on the value of second harmonic). Stability properties of trivial solutions in dependence on parameter a_2 are considered in [7] and [8].

All solutions with $N[\varphi] = 0$ we denote here by M_0, even they are changed by the influence of the external magnetic field and parameter a_2.Within the LJJ model, the basic Meissner solution M_0 demonstrates the screening of the external magnetic field. Deformation of M_0 at different values of parameter a_2 under the external magnetic field $h_e = 2$ is shown in Fig. 1. At $a_2 < 0$ (left panel) the screening effect diminishes when the coefficient a_2 decreases and it amplifies at $a_2 > 0$ (right panel) with growing of a_2.

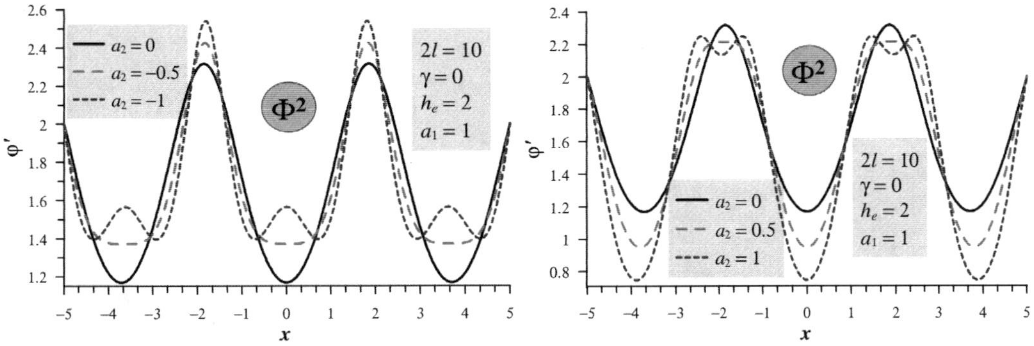

Figure 2. The internal magnetic field of the distribution Φ^2 at $2l = 10$, $\gamma = 0$, $h_e = 2$ and $a_1 = 1$ at different values of the parameter a_2. Left figure shows φ' at negative a_2, right one shows at positive a_2.

3. Deformation of the fluxon solutions Φ^n

The solutions with $N[\varphi] = 1$ which are not $\varphi = \pi$ at $h_e = 0$ and $\gamma = 0$ we denote as Φ^1. The deformation of the $\varphi'(x)$ of fluxon states under influence of the parameter $a_2 \in [-1; 1]$ was considered in [8]. We observe a qualitative change in the local minima with decrease of a_2 in the interval [-1;0] and in the local maxima with increase of a_2 in the interval [0;1]. With change of the coefficient a_2 the number of fluxons corresponding to the distribution Φ^1 is conserved [6] i.e. $\partial N/\partial a_2 = 0$ and $N[\Phi^1] = 1$. As external magnetic field h_e is growing, more complicated stable fluxon with $n = N[\varphi] = 2, 3, \ldots$ appear which are denoted by Φ^n.

The effect of the second harmonic contribution on Φ^1 was studied in [4] and [9]. It was shown in [4] that taking into account at $h_e = 0$ the second harmonics in current phase relation with $a_2 < -0.5$ leads to the appearance of the "small" fluxon state additionally to the traditional "large" one. Taking into account (5), we call the solution at $a_2 = -0.7$ in Fig. 1, left panel as M_0 (in [4] it is a "small" fluxon), because $N[M_0] = 0$. Number of fluxons of "large" fluxon is equal to $N[large] = 1$ and we denote it by Φ^1. In [9], stability properties of "large" fluxon Φ^1 at nonzero h_e have been studied. Two coexisting stable Φ^1 like fluxons were demonstrated in some region of magnetic field. Here, we investigate this effect in case of two-fluxon and three-fluxon distributions. The relation between the "small", "large" fluxons and trivial solutions under the influence of the external magnetic field and the second harmonic are the point of our further research.

In Fig. 2 the internal magnetic field of the two-fluxon distribution Φ^2 for $2l = 10$, $\gamma = 0$, $h_e = 2$ and $a_1 = 1$ at different values of the parameter a_2 is presented. We observe a strong deformation of the fluxon distributions with decrease of the parameter a_2 in the interval [-1;0]. At $a_2 = -0.5$ the curve of internal magnetic field $\varphi'(x)$ has a plateau at the points $x \approx -4$, $x \approx 0$ and $x \approx 4$ (Fig. 2, left panel). Further increase of the absolute value of a_2 leads to the transformation of these plateaus to local maxima of the internal magnetic field. Thus, accounting of the a_2 contribution qualitatively changes a shape of the fluxon distribution Φ^2. A similar deformation in the local maxima regions (points $x \approx -2$ and $x \approx 2$) is observed for $a_2 > 0$ (Fig. 2, right panel). We stress that the number of fluxons is conserved $N[\Phi^2] = 2$ for all values a_2.

4. Stability analysis of the static fluxon distribution Φ^n

The two-fluxon solution Φ^2 in LJJ with $2l = 10$ and $\gamma = 0$ is stable at $0.5 \lesssim h_e \lesssim 2.5$. Qualitatively, the behavior of the curves $\lambda_0(h_e)$ for this solution is similar to the case of one-fluxon solution Φ^1 [9]. The only difference is that in the Φ^2 case we don't have a region of

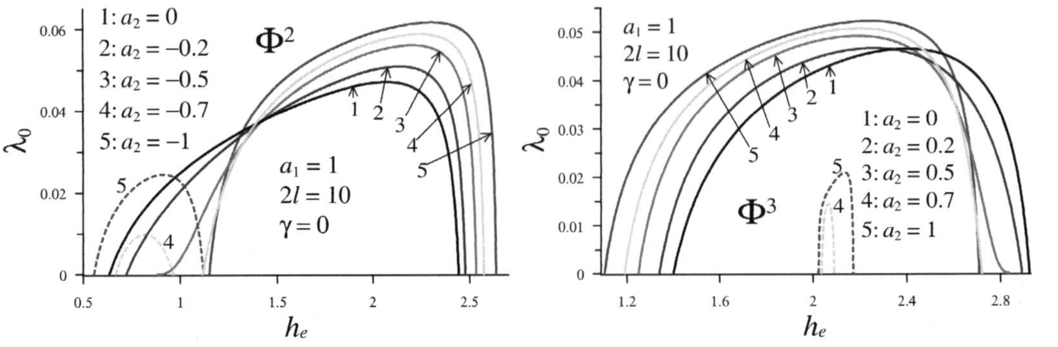

Figure 3. The dependence $\lambda_0(h_e)$ for Φ^n at $2l = 10$, $\gamma = 0$, $h_e = 2$ and $a_1 = 1$ at different values of the parameter a_2. Left figure shows $\lambda_0(h_e)$ for Φ^2 at negative a_2, right one shows $\lambda_0(h_e)$ for Φ^3 at positive a_2.

coexistence of two stable branches that was observed for negative a_2 in the Φ^1 case.

Results for $a_2 \in [-1; 0]$ are presented in Fig. 3, left panel. When a_2 decreases in $(-0.5; 0]$, the curve $\lambda_0(h_e)$ moves to the right. At $a_2 < -0.5$ the curve $\lambda_0(h_e)$ corresponding to the stable solution Φ^2 has two separate branches.

The $\lambda_0(h_e)$ curves for Φ^3 distribution for $a_2 \in [0; 1]$ are demonstrated in Fig. 3, right panel. It is seen, when a_2 is growing in the interval $[0; 1]$, the first bifurcation point moves to the left. Contrary, at $a_2 > 0.7$ the bifurcation point moves to the right. The second bifurcation point moves to the left as a_2 is growing from 0 to $a_2 \approx 0.7$. Contrary, at $a_2 > 0.7$ the second bifurcation point moves to the right. For three-fluxon state the following effect is observed. At $a_2 > 0.5$ new curves appear and we observe an existence of two different 3-fluxon states simultaneously.

For trivial solutions our calculations show the reducing of Meissner screening when the second harmonic is negative. New fluxon solutions which appear in case $a_2 \neq 0$ have a probability to be observed in the experiment. So, it would be interesting to test this.

As summary we note that our numerical investigations show that accounting of the second harmonic contributions significantly changes the shape and stability properties of trivial and fluxon static distributions in LJJ.

We thank to E. Goldobin for the stimulating discussions and important suggestions. This research was partially supported by Heisenberg-Landau Program. The work of P.Kh.A. is partially supported in the frame of the Program for collaboration of JINR-Dubna and Bulgarian scientific centers "JINR – Bulgaria". E.V.Z. was partially supported by RFFI under grant 09-01-00770-a. P.Kh.A. and E.V.Z. are thankful to I.V.Puzynin and T.P. Puzynina for support of this research.

References

[1] Golubov A A, Kypriyanov M Yu, Il'ichev E 2004 *Rev. Mod. Phys.* **76** 411–469
[2] Likharev K K 1986 *Dynamics of Josephson junctions and circuits* Gordon and Breach, Philadelphia
[3] Ryazanov V V, Oboznov V A, Rusanov A Yu et al 2001 *Phys. Rev. Lett.* **36** 2427–2430
[4] Goldobin E, Koelle D, Kleiner R and Buzdin A 2007 *Phys. Rev. B.* **76** 224523
[5] Buzdin A, Koshelev A E 2003 *Phys. Rev. B.* **67** 220504(R)
[6] Puzynin I Vet al 2007 *Physics of Particles and Nuclei* **38** No. 1 70116
[7] Atanasova P Kh, Boyadjiev T L, Shukrinov Yu M, Zemlyanaya E V 2010 *J. of Math. modeling* **22** (in print)
[8] Atanasova P Kh, Boyadjiev T L, Shukrinov Yu M, Zemlyanaya E V *e-print: cond-mat/1005.4796*
[9] Atanasova P Kh, Boyadjiev T L, Shukrinov Yu M, Zemlyanaya E V *e-print: cond-mat/1005.5691*

Journal of Physics: Conference Series **248** (2010) 012045 doi:10.1088/1742-6596/248/1/012045

Tunnel magnetoresistance of an organic molecule junction

Ashhadi Mojtaba *, **Shahtahmasebi Nasser, Vahedi Davoud, Askari Mostafa**
Department of Physics, Ferdowsi University of Mashhad, Mashhad, Iran
Nano Research Center of Ferdowsi University of Mashhad, Mashhad, Iran

E-mail: mojtaba_ash5@yahoo.com

Abstract. Coherent spin-dependent electronic transport is investigated in a molecular junction based on oligophenylene attached to two the semi-infinite ferromagnetic (FM) electrodes with finite cross sections. The work is based on the tight-binding Hamiltonian model and within the framework of a non-equilibrium Green's function (NEGF) technique. It is shown that tunnel magnetoresistance (TMR) of molecular junction can be large (over 60 %) by adjusting the related parameters, and depends on: (i) the applied voltages and (ii) the length of oligophenylele molecule.

1. Introduction

Spintronics is an exciting field of research where electron spin, in addition to charge, is exploited for electronic device application. The ability to manipulate and detect spin offers the potential of a new paradigm for electrical devices. It has been suggested that adding the spin degree of freedom to conventional charge-based electronic devices or using spin alone has the potential advantages of non-volatility, increased data processing speed, decreased electric power consumption, and increased integration densities compared with conventional semiconductor devices [1]. One of the most successful spintronic devices to date is the magnetic sensor composed of a magnetic tunnel junction exhibiting the tunneling magnetoresistance effect. Tunnel magnetoresistance effect in systems with spin-polarized transport is at the heart of spintronic [2] and has found important applications in information storage technology. A basic TMR device consists of a tunnel barrier separating two ferromagnetic (FM) metal layers which play the role of device leads. Magnetoresistance refers to a change in the electrical current when the relative magnetization of the ferromagnetic electrodes changes their alignment [3]. The TMR effect [4] originates from the quasi-particle electronic structure of the ferromagnetic electrodes which depends on their magnetization orientation. The tunneling conductance tends to be smallest when the orientations are opposite, leading to a spin valve character [4–8]. Recent experimental data have clearly demonstrated that spin-polarized injection and transport in molecular spintronic devices are not only possible, but also efficient. There have been several investigations of spin-transport through organic molecules. These include carbon nanotube spin valves [9], hot electron coherent spin transfer across molecular bridges [10], spin-injection in π-conjugated molecules [11-12] and organic tunneling junctions [13]. At the same time a few seminal theoretical studies on transport through 1, 4-benzene-dithiolate molecular spin-valves [14-15] and molecular magnets [16] have appeared. These calculations, based on density functional theory or tight-binding model, have shown that by changing the magnetic alignment of the contacts one can substantially affect the electronic current in the molecular devices. Recently charge transport of metal-molecule-metal junctions based on oligophenylenes has been investigated theoretically [17]. However, we believe that no theoretical study on spin-polarized transport through a molecular junction based on oligophenylene and its TMR effect, based on tight-binding method, has so far been reported. In this paper, we calculate the spin currents through the oligophenylene molecule, sandwiched between two the semi-infinite FM electrodes with simple cubic structure and square cross section (*xy* plane). The model of such a structure is shown schematically in figure 1. This paper is organized as follows: in Section 2, we give a model and the description of the method. The results and discussion are presented in section 3, followed by a conclusion in section 4.

Figure 1. A schematic representation of the FM/oligophenylene/FM molecular junction in two configurations Parallel (P) and Anti-Parallel (AP) magnetization.

2. Methodology

Assuming elastic scattering and neglecting spin-flip scattering and spin precession in the tunneling process because of the weak spin-orbit and hyperfine interaction expected in the π-conjugated organic layer [11], this implies that spin up electrons only tunnel to and from spin up states, and vice versa. Spin currents between two the magnetic electrodes can be written as $I=I_\uparrow+I_\downarrow$, where $I_\uparrow$ and $I_\downarrow$ are the contributions from spin-up and spin-down states. In the low-bias limit, the spin currents as a function of the applied voltage V can be calculated in the framework of the Landauer-Büttiker formula based on the NEGF method [18]

$$I_\sigma(V) = \frac{e}{h}\int_{-\infty}^{+\infty} T_\sigma(\varepsilon, V)[f(\varepsilon-\mu_L) - f(\varepsilon-\mu_R)]d\varepsilon \tag{1}$$

Here $\sigma=\uparrow,\downarrow$ is the spin index, f is the equilibrium Fermi distribution and $\mu_{L,R}=E_F\pm eV/2$ are the electro-chemical potentials of the electrodes in terms of the common Fermi energy E_F for the left (L) and right (R) electrodes. The transmission probability $T_\sigma(\varepsilon, V)$ can be expressed in terms of the Green's function of the molecule and the coupling of the system molecule with the two electrodes by the expression

$$T_\sigma(\varepsilon, V) = Tr[\Gamma_{L,\sigma} G_\sigma^r \Gamma_{R,\sigma} G_\sigma^a] \tag{2}$$

where G_σ^r and G_σ^a are, respectively, the retarded and advanced Green's function of the molecule. The $\Gamma_{L(R),\sigma}$ is the coupling terms due to the coupling of the molecule with the (L) and (R) electrodes, respectively. For a complete system, i.e., the molecule with the two electrodes, the Green's function is defined as

$$G(\varepsilon) = ((\varepsilon + i\zeta) - H)^{-1} \tag{3}$$

where ζ is a very small number which can be put as zero in the limiting approximation. The above Green's function corresponds to the inversion of an infinite matrix which consists of the finite molecule and the two semi-infinite electrodes. It can be partitioned into different sub-matrices that correspond to the individual sub-systems. The effective Green's function for the molecule can be written as

$$G_\sigma(\varepsilon, V) = \lim_{\zeta\to\infty}[(\varepsilon + i\zeta)I - H_M - \Sigma_{L,\sigma}(\varepsilon - eV/2) - \Sigma_{R,\sigma}(\varepsilon + eV/2)]^{-1} \tag{4}$$

where H_M is the Hamiltonian for the oligophenylene molecule sandwiched between the two FM electrodes. Within the non-interacting picture, the tight-binding Hamiltonian of the molecular system can be manifested as [17]

$$H_M = \sum_{i,\sigma\in M} \varepsilon_{i,\sigma} c^+_{i,\sigma} c_{i,\sigma} + \sum_{i,\sigma\in M} (t_{i,i+1} c^+_{i,\sigma} c_{i+1,\sigma} + h.c.) \tag{5}$$

where $c^\dagger_{i,\sigma}(c_{i,\sigma})$ creates (destroys) an electron with spin σ at site i of the molecular system and $\varepsilon_{i,\sigma}$ is the on-site energy and $t_{i,i+1}$ is hopping integral and is equal to t' for the nearest neighbors. In the equation (4), $\Sigma_{L,\sigma}=h^\dagger_{LM}\, g_{L,\sigma}\, h_{LM}$ and $\Sigma_{R,\sigma}=h_{RM}\, g_{R,\sigma}\, h^\dagger_{RM}$ are the self-energy terms due to the two electrodes, $g_{L,\sigma}$ and $g_{R,\sigma}$ are, respectively, the surface Green's function for the (L) and (R) electrodes, $h_{\alpha M}$ is the coupling matrice and it will be non-zero only for the adjacent points in the molecule and the electrode $\alpha(=L,R)$. The coupling terms $\Gamma_{L,\sigma}$ and $\Gamma_{R,\sigma}$ for the molecule can be calculated through the expression[18]

$$\Gamma_{\alpha,\sigma} = i[\Sigma^r_{\alpha,\sigma} - \Sigma^a_{\alpha,\sigma}] \tag{6}$$

where the advanced self-energy $\Sigma^a_{\alpha,\sigma}$ is the Hermitian conjugate of the retarded self-energy $\Sigma^r_{\alpha\,\sigma}$. Thus the coupling terms can be written as

$$\Gamma_{\alpha,\sigma} = -2\,\mathrm{Im}(\Sigma^r_{\alpha,\sigma}) \tag{7}$$

$g_{\alpha,\sigma}$ are the surface Green's functions of the uncoupled electrodes and their matrix elements are given by[19]

$$g_{\alpha,\sigma}(m,n;z) = \sum_{\mathbf{k}} \frac{\psi_{\mathbf{k}}(\mathbf{r}_m)\psi^*_{\mathbf{k}}(\mathbf{r}_n)}{z-\varepsilon_0+\sigma.\mathbf{h}_\alpha+E(\mathbf{k})} \tag{8}$$

$\sigma.\boldsymbol{h}_\alpha$ is the internal exchange energy with $\boldsymbol{h}_\alpha$ denoting the molecule field on the electrodes and σ being the conventional Pauli spin operator. m,n shown site $m^{th}(n^{th})$ in electrode α, and $r_m\equiv(x_m, y_m, z_m)$, $\boldsymbol{k}\equiv(l_x, l_y, k_z)$, $z=\varepsilon+i\zeta$,

$$\psi_{\mathbf{k}}(\mathbf{r}_m) = \frac{2\sqrt{2}}{\sqrt{(N_x+1)(N_y+1)(N_z)}} \sin\left(\frac{l_x x_m \pi}{N_x+1}\right) \sin\left(\frac{l_y y_m \pi}{N_y+1}\right) \sin(k_z z_m) \tag{9}$$

and

$$E(\mathbf{k}) = 2t\left(\cos\left(\frac{l_x\pi}{N_x+1}\right) + \cos\left(\frac{l_y\pi}{N_y+1}\right) + \cos(k_z a)\right) \tag{10}$$

Here, $l_{x,y}(=1,\ldots,N_{x,y})$ are integers, $k_z\in[-\pi/a, \pi/a]$, and N_β with $\beta=x,y,z$ is the number of lattice sites in the β direction. Note that N_x and N_y corresponds to the number of atoms at the cross-section of electrodes. ε_0 is the spin independent on-site energy in the electrodes, and will be set to $3t$ where t is the hopping strength between nearest-neighbor sites in the (L) and (R) electrodes and ε is injection energy of the transmitting electron.

By calculating the self-energies, the coupling terms $\Gamma_{L,\sigma}$ and $\Gamma_{R,\sigma}$ can be easily obtained and then the transmission probability (T_σ) and current (I_σ) will be computed from the expression as mentioned in equation (1) and (2). In the semi-infinite FM electrodes described by the single-band tight-binding model, only the central site at the cross section is connected to the molecule. In this case, tunnel magnetoresistance is defined as a relative change in the current of the system when the magnetizations of the two ferromagnetic layers switch between parallel (P) and anti-parallel (AP) configurations (figure 1), hence: $TMR\equiv(I_P\ I_{AP})/I_P$, where $I_{P,AP}$ are the total currents in the P and AP alignments of magnetizations in the FM electrodes, respectively.

3. Results and discussion

Based on the formalism described in section 2, we have investigated the coherent spin-depended transport and magnetoresistance effect of FM/oligophenylene/FM molecular junction. In our considered system, the magnetization orientation (i.e. the spins direction) in the left electrode stays fixed (in the $+y$ direction) but the other right electrode is free and may be switched back and forth by an external magnetic field (in the $+y$ or $-y$ directions) (see figure 1). Thus, when the magnetizations of two FM electrodes are parallel, spin-up and spin-down electrons see a symmetric structure, while for the anti-parallel alignment these electrons see an asymmetric structure.

We set $\varepsilon_{i,\sigma}=0$, $|\boldsymbol{h}_\alpha|=1.5\ eV$, $N_x=N_y=5$, $t=1eV$, $T=300K$, $t'=1.2eV$ [17], and consider molecule-to-electrode coupling strength equal to $0.5eV$ in the calculations.

In figure 2, we present the transmission logarithm at zero applied voltage for the P and AP alignments for the oligophenylene molecule. Study of this figures (Panels (a) and (b)) shows that each the energy peak in transmission logarithm due to matching of the energy of the incident electron with energy eigenvalue of the molecule. The curves shown an oscillatory behavior when the electrons transport through molecule it means that the molecular levels are quantized. The electrons with spin-up or spin-down can transport through the molecule only when a resonance occurs between the molecular level and the incident electronic energy. As a result, resonance peaks are produced, therefore, we can predict that this resonance peaks are a fingerprint of the electronic structure of the molecule.

The calculated current-voltage (I–V) curves for the oligophenylene molecule attached to FM electrodes are shown in figure 3 for P and AP alignment. As can be seen, the current for the P alignment reaches bigger values in comparison with the current for the AP, because of the difference between surface density of states of the FM electrodes for spin-up and spin-down electrons. This difference in the surface density of states of the FM electrodes causes the electrons with P magnetic alignment than electrons with AP magnetic alignment to have a more likely in probability transmission (figure 2).

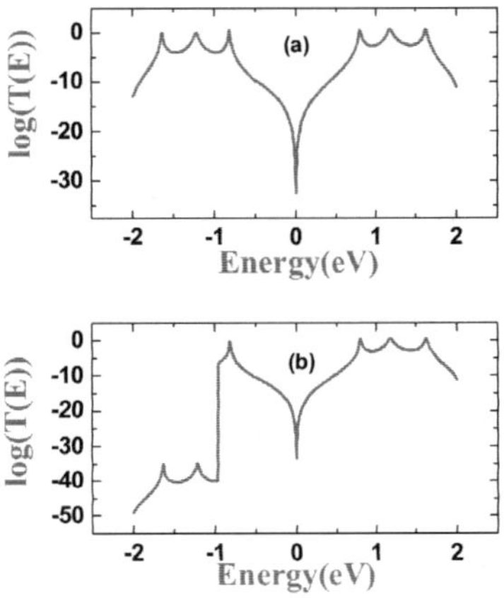

Figure 2. Panels (a) and (b) show the logarithm of the transmission coefficient at zero applied voltage versus energy for P (a) AP (b) magnetic alignment and for $N=3$,(N is the length of the oligophenylene molecule).

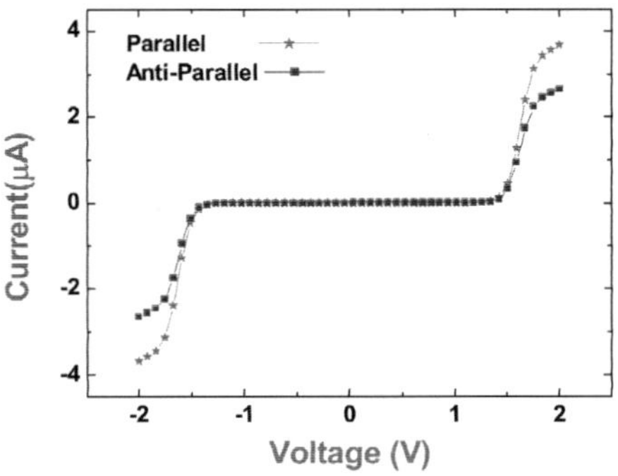

Figure 3. Spin currents for P (stars) and AP (squares) versus applied voltage and $N=3$.

The effect of an applied voltage is that of shifting the chemical potentials of the two electrodes relative to each other by eV, with e the electronic charge. Currents will flow whenever a molecular level (either the highest occupied molecular orbital or the lowest unoccupied molecular orbital) is positioned within such a bias window. The appearing of molecular levels in the bias window when the potential is increased typically leads to changes in the slope and a step-like behavior of the *I-V* characteristics.

The FM electrodes have unbalanced spin-up and spin-down electrons near its Fermi level, the tunneling channels for spin-up and spin-down electrons are explicitly separated. The electrons with magnetic moment P to the magnetization are electrons of majority-spin, and with magnetic moment AP to the magnetization are electrons of minority-spin. More specifically, in P spin configuration, the majority spin is spin-up and the minority spin is spin-down for both the electrodes, whereas in AP configuration, the majority spin is spin-up on one electrode and spin-down on the other. Consequently, the P arrangement gives much higher total current through the molecule than does the AP arrangement. This difference in the total currents is the origin of TMR effect which has been shown in figure 4.

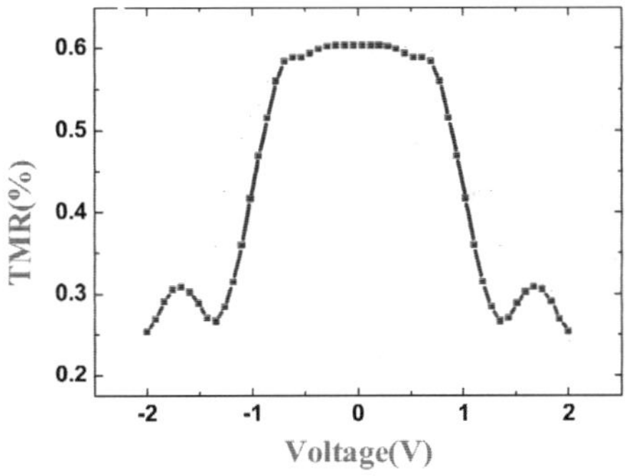

Figure 4. TMR as a function of applied voltage and for $N=3$.

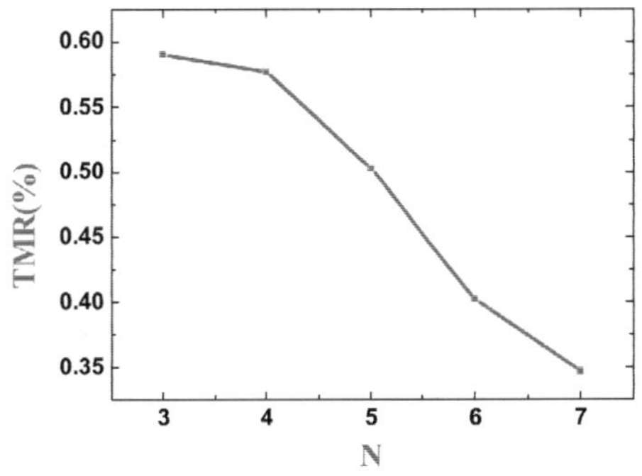

Figure 5. TMR as a function of varying lengths ($N=3... 7$) with $V=0.5eV$ attached to the ferromagnetic electrodes.

The TMR ratio has it maximum value (more than 60%) at low bias voltages. With increasing the applied voltage, we first observe that the TMR decreases and with further increase in the bias voltage, the TMR ratio increases and then, the reduction and enhancement of TMR are repeated. The reason is that with increasing bias voltage the current of tunneling through the junction increases significantly, however, the difference of current for the parallel magnetization alignment from that for alternately anti-parallel magnetization alignment increases only slightly. This leads to the decrease of TMR monotonously with increasing bias voltage. Our results are qualitatively in agreement with the experimental measurements [20, 21]. Finally, in figure 5 we show the changes of the TMR due to the increase the length of oligophenylene molecule in the junction with $V=0.5eV$. Here we can see that the TMR decrease when length of the molecule increase. The origin of the trend of decreasing TMR reflects the increasing of states in the junction and thus the current is increased. Note the Similar behavior in refs. [22-24] has been shown that TMR is decreased with increasing thickness of molecule.

4. Conclusion

Using the NEGF method and within the framework of the tight-binding Hamiltonian model, we have investigated the possibility of making an oligophenylene molecule-based magnetic tunnel junction.

We have shown that the coupling between the molecular system and magnetic electrodes in FM/oligophenylene/FM structure produces large magnetoresistance effects (greater than 60%) that can be strongly depends on: (i) the applied voltages and (ii) the length of molecule.

Throughout this study, we have ignored the effects of inelastic scattering such as: the electron-phonon interaction, electron–electron correlation, etc. These factors can affect the spin-dependent transport. Another improved method is needed for better results and we need further study for considering all these effects

References

[1] Wolf S A, Chtchelkanova A Y and Treger D M 2006 *IBM J. RES & DEV.* **50** 101
[2] Wolf S A, Awschalom D D, Buhrman R A, Daughton J M, von Molnar S, Roukes M L, Chtchelkanova A Y and Treger D M 2001 *Science* **294** 1488
[3] Zutic I, Fabian J and Das Sarma S 2004 *Rev. Mod. Phys.* **76**, 323
[4] Slonczewski J C1989 *Phys. Rev. B* **39** 6995
[5] Julliere M 1975 *Phys. Lett. A* **54** 225
[6] Bratkovsky A M 1997 *Phys. Rev. B* **56** 2344
[7] Wang B G, Wang J and Guo H 2001 *J. Phys. Soc. Jpn.* **70** 2645
[8] Mireles F and Kirczenow G 2002 *Phys. Rev. B* **66** 214415
[9] Tsukagoshi K, Alphenaar B H and Ago H 1999 *Nature* **401** 572
[10] Ouyang M and Awschalom D D 2003 *Science* **301** 1074
[11] Xiong Z H, Wu D, Valy Vardeny Z and Shi J 2004 *Nature* **427** 821
[12] Dediu V, Murgia M, Matacotta F C, Taliani C and Barbanera S 2002 *Solid State Commun.* **122** 181
[13] Petta J R, Slater S K and Ralph D C 2004 *Phys. Rev. Lett.* **93** 136601
[14] Pati R, Sanapati L, Ajayan P M and Nayak K 2003 *Phys. Rev. B* **68** 100407
[15] Emberly E G and Kirczenow G 2002 *Chem. Phys.* **281** 311
[16] Kim G H and Kim T S 2004 *Phys. Rev. Lett.* **92** 137203
[17] Viljas J K, Pauly F and Cuevas J C 2008 *Phys. Rev. B* **77** 155119
[18] Datta S 1997 *Electronic Transport in Mesocopic System* (Cambridge University Press, Cambridge)
[19] Yoshihiro A and Hidetoshi F 2005 *Phys. Rev. B* **72** 085431
[20] Moodera J S, Kinder L R, Wong T M and et al. 1995 *Phys. Rev. Lett.* **74** 3273
[21] Moodera J S, Kinder L R 1996 *J. Appl. Phys.* **79** 4724
[22] Sun J J, Sousa R C, Galvao T T P, Soares V and Freitas P P 1999 *J. Magn. Soc. Japan* **23** 55
[23] Szulczewski G, Tokuc H, Oguz K and Coey J M D 2009 *Appl. Phys. Lett.* **95** 202506
[24] Davis A H and MacLaren J M 2002 *J. Phys.: Condens. Matter* **14** 4365

Journal of Physics: Conference Series **248** (2010) 012046 doi:10.1088/1742-6596/248/1/012046

Magnetic properties of 3d transition metals adsorbed on benzene: a density functional theory study

Mahdi Sargolzaei[1,2], **Farideh Gudarzi** [1]

[1] Department of Physics, Iran University of Science and Technology, Narmak 16345, Tehran, Iran.

[2] Computational Physical Science Laboratory, Department of Nano-Science, Institute for Research in Fundamental Sciences (IPM), Tehran, Iran.

E-mail: sargolzaei@iust.ac.ir

Abstract. In this work, magnetic properties of 3d transition metal (TM) atoms (Fe, Co and Ni) add on benzene (Bz) molecule have been studied by means of density functional calculations. We estimated adsorption site, adsorption energy, spin and orbital moments of TM adatom and magnetic anisotropy energy (MAE) of these systems. We found that hollow sites of Bz molecule is the most stable geometry for adsorption of TM adatoms. Calculation of adsorption energy per atom confirms that TMBz molecules show high stability. Bz molecule reduces spin magnetic moment of TM adatom by a factor of two compare to their free states. The largest MAE value concluded for CoBz structure, that makes this molecule as a potential building block for long-term magnetic data storage.

1. Introduction

In the last decade, the interests in the interaction between transition metal atoms and benzene (C_6H_6) grow up with observation of one-dimensional multi-decker sandwich TM-benzene complexes ($TM_n(Bz)_{n+1}$) [1]. Demonstration of ferromagnetic ordering in Vanadium-benzene clusters, $V_n(Bz)_{n+1}$, was very stimulating [2]. These complexes are supposed to be applied as intriguing one-dimensional nanomagnetic building blocks in applications such as recording media or spintronic devices.

An important property of nanomagnets is preservation of magnetic moment of TM atom as it bound to the medium. An still continuous challenge in nanomagnets is to find a medium that does not deteriorate magnetic moment of TM adatoms. Another crucial parameter that makes nanomagnets so useful is magnetic anisotropy energy (MAE), which is the intrinsic tendency of magnetization to lie along a preferred spatial direction (easy axis). High MAE is vital in nanomagnets, to prevent magnetization reversal by means of thermal (superparamagnetic) or magnetic fluctuations. As suggested by Van Vleck [3] spin-orbit coupling is responsible for magnetic anisotropy. Magnetocrystalline anisotropy energy of elemental ferromagnets, e.g., Fe, Co and Ni, is quite small due to the quenching of the orbital angular momentum by the crystal field. On the other hand, in low dimension systems, the reduction of symmetry yields enhancement of orbital moment and as a result an additional anisotropy effect gives rise.

The aim of the present work, is to investigate the magnetic properties of Fe, Co and Ni atoms as they add on a Bz molecule. The paper is organized as follows: Section II contains details about the molecule structure and the numerics. Section III presents the results and discussion

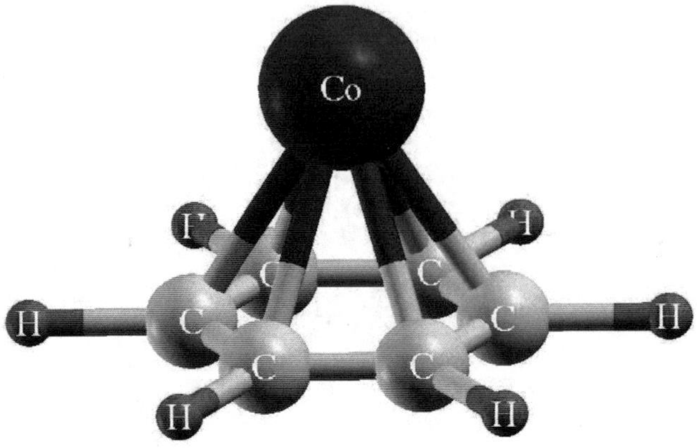

Figure 1. (Color online). TMBz molecule. the red, yellow and blue spheres indicate the H atoms, C atoms and TM atom (here Co atom) respectively.

(structural properties, calculated spin and orbital magnetic moments, and magnetic anisotropy energies). Finally, the paper is summarized in Section IV.

2. Computational Method

We employed an all-electron full relativistic full-potential local orbital scheme, FPLO-9 [4], for the DFT [5] calculations. The presented data were obtained using the generalized gradient approximation (GGA) with a parameterized exchange-correlation functional [6]. To approach an upper estimate, an orbital polarization correction (OPC), was applied to the incompletely filled d-shells of TM adatoms, that was suggested by Brooks [7] and later extended works by Eriksson [8], and Eschrig [9, 10]. The OPC term which added to the GGA exchange and correlation (XC) energy functional is:

$$E^{OPC} = -\frac{1}{2}\sum_{\sigma} B_{\sigma} L_{\sigma}^{2} ,$$

where L_{σ} is the orbital moment of the spin-σ d-subshell of TM adatom and B_{σ} are related Racah parameters obtained from the so-called Slater integrals $F^{k}_{d\sigma}$ via [11]:

$$B_{\sigma} = (9F^{2}_{d\sigma} - 5F^{4}_{d\sigma})/441 ,$$

$$F^{k}_{d\sigma} = \int dr_1 r_1^2 \int dr_2 r_2^2 \phi^2_{d\sigma}(r_1) \frac{r^k_{<}}{r^{k+1}_{>}} \phi^2_{d\sigma}(r_2) ,$$

where $r_< = \min (r_1, r_2)$, $r_> = \max (r_1, r_2)$ and ϕ are radial wave functions. We used a mesh of $30 \times 30 \times 30$ k-points in the full Brillouin zone. Geometry optimization were carried out with a scalar-relativistic but other data was obtained by means of self consistent full-relativistic calculations [12]. The valence basis set comprised 3d transition metals (3s, 3p, 3d, 4s, 4p, 4d, 5s), carbon (1s, 2s, 2p, 3s, 3p, 3d) and hydrogen (1s, 2s, 2p). All of atomic coordinates are fully relaxed and the residue forces are less than 0.001 eV/A for all of the atoms, which converges the total energy within 0.001 meV.

3. Results and discussion

In order to optimized the molecule structure, the TM adatom first located on the symmetry axis of Bz molecule (001) with an initial distance from the plane of the Bz molecule, as shown in Fig 1., then all atomic coordinates were fully relaxed. From physical point of view, one can estimate that the most stable position for the TM adatom is on axial direction perpendicular to the plane of Bz molecule. Optimized distances of the considered TM adatoms from Bz plane (d_{TM-Bz}), are given in Table.I. It can be stimulate from Table.I, that d_{TM-Bz} increases slightly with the atomic number of TM adatoms. To evaluated relative stability of TMBz molecule, adsorption energy calculations have heen carried out. The adsorption energy is defined by the formula:

$$E_{adsorption} = E_{complex} - E_{adatom} - E_{Bz} \ ,$$

where $E_{complex}$ is the total energy of optimized configuration of the TMBz molecule, E_{Bz} is the total energy of the isolated Bz molecule, and E_{adatom} is the total energy of the corresponding adsorption TM atom in its ground state. The result of calculated adsorption energies for TMBz molecule are shown in Table.I.

Table 1. The calculated distance of TM adatom from Bz plane after atomic relaxation (It is so-called height). The adsorption energy is defined in the text. Presented adsorption energies (eV) acquired from full-relativistic calculations.

TM adatom	Height ($\AA$)	Adsorption energy (eV/atom)
Fe	1.45	-1.28
Co	1.46	-1.14
Ni	1.47	-1.56

Relatively high adsorption energy concluded for TMBz structure leads to high tendency of TM atom to bound on the surface of Bz molecule to form a stable structure. An important aspect of adatom adsorption on Bz molecule is the magnetic moment of the TM adatom as it bound to the Bz molecule. When TM adatoms adsorb on the surface of Bz molecule, 3d and 4s orbitals of the TM adatom hybridize with six ligand orbitals ($2p_z$ orbitals in benzene) to compose the valence orbitals of these complexes. Table.II shows the spin moment of TM adatoms. It can be seen that Bz host reduces magnetic moment of TM adatom approximately by a factor of two, compare to their free states. Another important feature of this system is magnetic anisotropy energy (MAE), that is defined as:

$$MAE = E_{\parallel} - E\perp \ ,$$

where $E_{\parallel}$ and $E\perp$ are the calculated full relativistic total energies of the system in parallel and perpendicular magnetization of TM with respect to the Bz plane, respectively. It is worth noting that the CoBz molecule shows a huge MAE value. This value of MAE concluded for CoBz, makes it promising candidate to be used as magnetic data storage, in the modern perpendicular magnetic recording technology, which may achieve a storage density of 1 Tbits/in. In Table.III different MAE values per Co atom in different structures are presented. It has to be clarified that our estimated MAE value for CoBz molecule is very large compare to the mentioned MAE values for Co atom in other mentioned hosts (see Table.III). In order to use a nanomagnet in a magnetic data storage, the stabilization of magnetization axes versus thermal fluctuations has to be estimated. Blocking temperature as a construction parameter, at which a nanomagnet will loose its preferred magnetization axis, can be defined as a function of magnetic anisotropy energy. The upper and lower estimated blocking temperatures for CoBz molecule are found 1100 K and 220 K, respectively. For other considered structures, the calculated blocking temperatures are very below to the room temperature. This finding guarantees that the easy magnetization axis of CoBz molecule can be stabilized in high temperatures.

Table 2. The calculated spin and orbital magnetic moments of the TM adatoms. The MAE is defined in the text, and calculated by means of full relativistic (FR) and full relativistic including an extra OP correction (FR+OPC).

TM adatom	TM spin moment (μ_B)	TM orbital moment (μ_B) FR (FR+OPC)	MAE (meV) FR (FR+OPC)
Fe	2.22	0.10 (0.23)	-0.79 (-2.38)
Co	1.08	0.90 (1.13)	19.14 (94.72)
Ni	0.00	0.00 (0.00)	0.00 (0.12)

Table 3. Magnetic anisotropy energy of Co atoms in different structures: ferromagnetic hcp (hexagonal closest packing) Co bulk (a), perpendicular magnetization of alternating packed single layers ($L1_0$ structure) of Co and Pt (b), SmCo permanent magnet (c), single Co atom on the Pt (111) surface (d) and CoBz molecule, our simulation (e).

	different structures of Co	MAE (meV/Co atom)
a)	Bulk hcp Co [13]	0.06
b)	Bulk $L1_0$ CoPt [14]	1.00
c)	$SmCo_5$ [15]	1.80
d)	Co atom on Pt [15]	9.30
e)	CoBz	94.72:upper estimate (19.14:lower estimate)

4. Conclusion

By means of density functional calculations based on general gradient approximation, we studied adsorption of 3d transition metal atoms (Fe, Co,Ni) on the surface of benzene molecule. We find that hollow site of the benzene is the most stable geometry for TM adatoms. On the other hand calculation of adsorption energy shows that TMBz can be formed as a stable molecule. The calculated magnetic anisotropy energy for TMBz molecule indicates that CoBz shows a large MAE, which means it may be a secure bit to magnetic data storage.

We are thankful to Manuel Richter, and Samara Keshavarz for fruitful discussions. It made use of computing facilities of Computational Physical Science Laboratory, Department of Nano-Science, Institute for Research in Fundamental Sciences (IPM), Iran. This work has been financially supported by Research Devision at Iran University of Science and Technology.

References

[1] Kurikawa T, Takeda H, Hirano M, Judai K, Arita T, Nagao S, Nakajima A, and Kaya K, 1999 Organometallics **18**, 1430.
[2] Miyajima K, Nakajima A, Yabushita S, Knickelbein M B, and Kaya K, 2004 J. Am. Chem. Soc. **126**, 13202.
[3] Van Vleck J H, 1937 Phys. Rev. **52**, 1178.
[4] FPLO-9; improved version of the original FPLO code by K. Koepernik and H. Eschrig, Phys. Rev. B **59**, 1743 (1999); http://www.FPLO.de.

[5] Hohenberg P, and Kohn W, 1964 Phys. Rev. **136**, B864
[6] Perdew J P, Burke K, and Ernzerhof M, 1996 Phys. Rev. Lett. **77**, 3865.
[7] Brooks M S S, 1985 Physica B **130**, 6.
[8] Eriksson O, Brooks M S S and Johansson B, 1990 Phys. Rev. B **41**, R7311.
[9] Eschrig H., Sargolzaei M., Koepernik K., and Richter M., 2005 Europhys. Lett. **72**, 611.
[10] Sargolzaei M, "Orbital Polarization in Relativistic Density Functional Theory", Lambert Academic Publishing, (2010).
[11] Judd B R, "Operator Techniques in Atomic Spectroscopy", McGraw-Hill, New York (1963).
[12] Opahle I, Ph.D thesis, Technische Universität Dresden, 2001. H. Eschrig, M. Richter, and I. Opahle, in Relativistic Electronic Structure Theory-Part II: Applications, edited by P. Schwerdtfeger (Elsevier, Amsterdam, 2004), pp. 723776.
[13] Trygg J, Johansson B, Eriksson O, and Wills J M, 1955 Phys. Rev. Lett. **75**, 2871.
[14] Ravindran P, Kjekshus A, Fjellvag H, James P, Nordstrom L, Johansson B, and Eriksson O, 2001 Phys Rev B **63**, 144409.
[15] Gambardella P, Rusponi S, Veronese M, Dhesi S S, Grazioli C, Dallmeyer A, Cabria I, Zeller R, Dederichs P H, Kern K, Carbone C, Brunel H, 2003 Science **300**, 1130.

International Conference on Theoretical Physics Dubna-Nano 2010 IOP Publishing
Journal of Physics: Conference Series **248** (2010) 012047 doi:10.1088/1742-6596/248/1/012047

The application of adiabatic method for the description of impurity states in quantum nanostructures

A. A. Gusev[1], O. Chuluunbaatar[1], S. I. Vinitsky[1], E. M. Kazaryan[2], H. A. Sarkisyan[2]

[1]Joint Institute for Nuclear Research, Dubna, Russia
[2]Russian-Armenian (Slavonic) University, Yerevan, Armenia

E-mail: gooseff@jinr.ru

Abstract. In the framework of effective mass approximation the application of adiabatic method for the description of impurity states in quantum dots, wires and wells with parabolic confinement potential as well as rectangular infinitely-high potential is presented. A rate of convergence of the method and efficiency of the proposed program complex for solving a boundary value problem, realized by the finite element method, is demonstrated on examples of calculation of spectral and optical characteristics of the considered quantum nanostructures.

1. Introduction

Investigations of the Coulomb systems in quantum nanostructures are important from the point of view of application of obtained results in a construction of semiconductor devices of a new generation. It is well known that by changing a dimension of the semiconductor structures the impurity binding energy is increasing. For this reason the influence of impurity states is more essential for low dimensional semiconductors. The impurity energy spectra in quantum well (QW), wire (QWr) and dot (QD) can be manipulated by external fields too [1]. Different methods of calculations of spectral characteristics and wave functions of impurities were applied in many papers.

The eigenvalue problem with both Coulomb potential and magnetic field is analytically not solvable, because the Schrödinger equation with appropriate boundary conditions is beyond the problem of confluent hypergeometric equations. Therefore it is necessary to use different approximation methods: variational, perturbation theory method, etc. In this regard it should be mentioned, that Maksym and Chakraborty on the base of numerical calculations obtained many-particle states in parabolic QD in external magnetic field and demonstrate the generalization of Kohn theorem for quantum nanostructures [2]. Jia-Lin Zhu et al using the method of series expansion have obtained the exact forms of series in different regions of the radial equation (a) for donor states in rectangular QW [3], (b) for two-dimensional hydrogenic donor states in magnetic field [4], (c) for two electrons confined by two-dimensional and three-dimensional QDs with parabolic potentials [5]. One of the powerful methods of the analytical solvation of similar problems is the adiabatic approximation. For the first time hydrogen-like system in adiabatic approximation was considered in connection with analysis of such system behavior in extremely strong magnetic field [6]. A similar problem, but for electron with the relativistic dispersion law was solved in [7]. In [8] within the framework of the adiabatic theory the hydrogen-like impurity levels in spherical GaAs/$Ga_{1-x}Al_xAs$ QD with parabolic confinement potential under the influence of strong magnetic field are investigated. Variational and adiabatic methods were applied for study of impurity states in parabolic QWs [9, 10] and resonance effects of transmission and reflection at axial channeling of ions in parabolic QWrs or in magnetic field [11].

In this paper we continue the previous investigation in the framework of adiabatic method using elaborated symbolic-numerical algorithms [12, 13, 14] and program complex for solving the boundary value problems (BVPs), realized by the finite element method [15, 16, 17]. A rate of convergence of the method and efficiency of the proposed program complex is demonstrated on examples of calculation of spectral and optical characteristics of the considered quantum nanostructures.

2. Boundary value problems

In the effective mass approximation the Schrödinger equation for impurity electron in magnetic field in QD, QWr, QW is read as

$$\left\{\frac{1}{2\mu}\left(\hat{\vec{p}}-\frac{e}{c}\vec{A}\right)^2+U(\vec{r})-\frac{qe^2}{\kappa r_c}\right\}\Psi=E\Psi, \tag{1}$$

where $\mu = \beta m_e$ is the effective mass of electron, q is the Coulomb charge of impurity, $r_c = \sqrt{x^2+y^2+(z-z_c)^2}$ is the distance between electron and Coulomb charge of impurity center, $z_c \in [-z_0/2, z_0/2]$ is the shift of charge position q along z axis in the case of QW ($z_c = 0$ for QD and QWr considered below), κ is the static dielectric constant of semiconductor (for GaAs $\beta = 0.067$, $q = 1$, $\kappa = 13.18$), $U(\vec{r})$ is the parabolic potential with frequency $\omega = \gamma_{r_0}\hbar/(\mu r_0^2)$, $\gamma_{r_0} \sim \pi^2/3$ is an adjustable parameter (Model A):

$$U(\vec{r}) \equiv U^{\rm A}(\vec{r}) = \mu\omega^2(\zeta_1(x^2+y^2)+\zeta_3 z^2)/2, \tag{2}$$

$r_0 = \sqrt{\zeta_1(x_0^2+y_0^2)+\zeta_3 z_0^2}$ is the radius of QD ($\zeta_1 = 1$, $\zeta_3 = 1$), QWr ($\zeta_1 = 1$, $\zeta_3 = 0$) and the depth of QW ($\zeta_1 = 0$, $\zeta_3 = 1$). For comparison we consider model QD, and QW ($\zeta_3 = 1/2$) with potential $U(\vec{r})$ (Model B)

$$\begin{aligned} U(\vec{r}) &\equiv U^{\rm B}(\vec{r}) = \{0, 0 \le |\vec{r}| < r_0; +\infty, |\vec{r}| \ge r_0\}, \\ U(z) &\equiv U^{\rm B}(z) = \{0, |z| < z_0/2; +\infty, |z| \ge z_0/2\}. \end{aligned} \tag{3}$$

Choosing direction of z axis along the magnetic field $\vec{H}$ with vector potential $\vec{A} = \frac{1}{2}\vec{H}\times\vec{r}$ and using dimensionless parameter for magnetic field $\gamma = H/H_0^*$, and reduced atomic units (a.u.) (for GaAs $H_0^* = 6$ T $a_B^* = 102$ Å, $E_{au} = 2E_R^*$, $E_R^* \equiv Ry^* = 5.2$meV), we can rewrite (1) for given magnetic quantum number m in the following form

$$\begin{aligned} &\left(\frac{1}{g_{3s}(x_s)}\hat{H}_2(x_f;x_s)+\hat{H}_1(x_s)-2E\right)\Psi(x_f,x_s)=0, \\ &\hat{H}_2=-\frac{1}{g_{1f}(x_f)}\frac{\partial}{\partial x_f}g_{2f}(x_f)\frac{\partial}{\partial x_f}+\hat{V}_f(x_f)+\hat{V}_{fs}(x_f,x_s), \\ &\hat{H}_1=-\frac{1}{g_{1s}(x_s)}\frac{\partial}{\partial x_s}g_{2s}(x_s)\frac{\partial}{\partial x_s}+\hat{V}_s(x_s), \quad V_{fs}(x_f,x_s)=-\frac{2q}{\sqrt{\rho^2+(z-z_c)^2}}=-\frac{2q}{r_c}. \end{aligned} \tag{4}$$

Here $\hat{H}_2 \equiv \hat{H}_2(x_f;x_s) = g_{3s}(x_s)H_2(x_f;x_s)$ and $\hat{H}_1 \equiv \hat{H}_1(x_s) = H_1(x_s)$ are the Hamiltonians of so-called fast and slow subsystems, $\hat{V}_{fs}(x_f,x_s) = g_{3s}(x_s)V_{fs}(x_f,x_s)$ is the Coulomb potential in cylindrical or spherical coordinates (CC or SC). In the CC for QD and QWr we use the notations $x_f = \rho$, $x_s = z$, $g_{1f}(x_f) = g_{2f}(x_f) = \rho$, $g_{1s}(x_s) = g_{2s}(x_s) = g_{3s}(x_s) = 1$ (Model A):

$$\hat{V}_f(x_f) = m^2/\rho^2 + \gamma m + \gamma^2\rho^2/4 + \zeta_1\omega^2\rho^2, \quad \hat{V}_s(x_s) = \zeta_3\omega^2 z^2.$$

while for QW $x_f = z$, $x_s = \rho$, $g_{1f}(x_f) = g_{2f}(x_f) = 1$, $g_{1s}(x_s) = g_{2s}(x_s) = \rho$, $g_{3s}(x_s) = 1$,

$$\hat{V}_f(x_f) = \zeta_3\omega^2 z^2, \quad \hat{V}_s(x_s) = m^2/\rho^2 + \gamma m + \gamma^2\rho^2/4 + \zeta_1\omega^2\rho^2.$$

In the SC we use the notation $x_f = \eta_c$, $x_s = r_c$, $g_{1f}(x_f) = 1$, $g_{2f}(x_f) = (1-\eta_c^2)$, $g_{1s}(x_s) = g_{2s}(x_s) = r_c^2$, $g_{3s}(x_s) = r_c^2$ (Model A)

$$\hat{V}_f(x_f) = m^2/(1-\eta^2) + 2p\gamma m + \alpha(1-\eta_c^2) - b\eta_c + f, \quad \hat{V}_s(x_s) = \beta r_c^4,$$

where $\alpha = p^2$, $p = \gamma r_c^2/2$, $b = f = 0$, $\beta = \omega^2$ for QD ($\beta = 0$ for QWr) and $\alpha = -c^2$, $c = \omega r_c^2$, $b = -2\omega z_c r_c^3$, $f = \omega z_c r_c^2$, $\beta = 0$, $\gamma = 0$ for QW.

The required solution of (4) we find in the form of expansion over a set of solutions, $\Phi_j(x_f;x_s)$,

$$\Psi_i(x_f,x_s) = \sum_{j=1}^{j_{\max}} \Phi_j(x_f;x_s)\chi_j^{(i)}(x_s), \tag{5}$$

of the parametric eigenvalue problem of the fast subsystem in domain $\Omega_f = (x_f^{min}, x_f^{max})$

$$\left\{\hat{H}_2(x_f;x_s) - \hat{E}_i(x_s)\right\}\Phi_i(x_f;x_s) = 0, \quad \langle\Phi_i|\Phi_j\rangle_{\Omega_f} = \int_{x_f^{\min}}^{x_f^{\max}} \Phi_i(x_f;x_s)\Phi_j(x_f;x_s)g_{1f}(x_f)dx_f = \delta_{ij}. \tag{6}$$

Substituting expansion (5) in eq. (4) and taking into account (6), we come to the BVP for a set of the differential equations of the slow subsystem related to the vector-function $\boldsymbol{\chi}^{(i)}(x_s) = \{\chi_j^{(i)}(x_s)\}_{j=1}^{j_{\max}}$:

$$\mathbf{H}\boldsymbol{\chi}^{(i)}(x_s) = 2E_i\,\mathbf{I}\boldsymbol{\chi}^{(i)}(x_s), \tag{7}$$

$$\mathbf{H} = -\frac{1}{g_{1s}(x_s)}\mathbf{I}\frac{d}{dx_s}g_{2s}(x_s)\frac{d}{dx_s} + \hat{V}_s(x_s)\mathbf{I} + \mathbf{U}(x_s) + \frac{g_{2s}(x_s)}{g_{1s}(x_s)}\mathbf{Q}(x_s)\frac{d}{dx_s} + \frac{1}{g_{1s}(x_s)}\frac{dg_{2s}(x_s)\mathbf{Q}(z)}{dx_s}.$$

Here $\mathbf{I}$, $\mathbf{U}(x_s) = \mathbf{U}^T(x_s)$ and $\mathbf{Q}(x_s) = -\mathbf{Q}^T(x_s)$ are **unit, symmetric and nonsymmetric** matrixes with dimensions $j_{\max} \times j_{\max}$:

$$U_{ij}(x_s) = \frac{1}{g_{3s}(x_s)}\hat{E}_i(x_s)\delta_{ij} + \frac{g_{2s}(x_s)}{g_{1s}(x_s)}W_{ij}(x_s), \tag{8}$$

$$Q_{ij}(x_s) = -\langle\Phi_i(x_s)|\frac{\partial\Phi_j(x_s)}{\partial x_s}\rangle_{\Omega_f}, \quad W_{ij}(x_s) = \langle\frac{\partial\Phi_i(x_s)}{\partial x_s}|\frac{\partial\Phi_j(x_s)}{\partial x_s}\rangle_{\Omega_f}.$$

Solutions of discrete spectrum were subjected to the following boundary conditions:

$$\lim_{x_s\to x^{\min}=0} g_{2s}(x_s)\frac{d\boldsymbol{\chi}^{(i)}(x_s)}{dx_s} = 0, \ \boldsymbol{\chi}^{(i)}(x_s^{\max}) = 0; \quad \text{and} \quad \boldsymbol{\chi}^{(i)}(x_s^{\min}) = 0, \quad \boldsymbol{\chi}^{(i)}(x_s^{\max}) = 0, \tag{9}$$

for $x_s = \rho$ or $x_s = r$, and $x_s = z$, respectively, and orthonormalization conditions

$$\int_{x_s^{\min}}^{x_s^{\max}} (\boldsymbol{\chi}^{(i)}(x_s))^T\boldsymbol{\chi}^{(j)}(x_s)g_{1s}(x_s)dx_s = \delta_{ij}. \tag{10}$$

For Model B we have potentials $\hat{V}_s(x_s) = 0$ and $\hat{V}_f(x_f) = 0$ and use: conditions (9) for $r_{\max} = r_0$ if (7), (10) for the case of QD, and the last conditions (9) for $s \leftrightarrow f$, $z_{\min} = -z_0/2$, $z_{\max} = z_0/2$ if (6) for the case of QW. Presented below results of solving the BVPs (4)–(10) in the SC and CC are found by means of complex of programs ODPEVP, POTHMF and KANTBP [15, 16, 17].

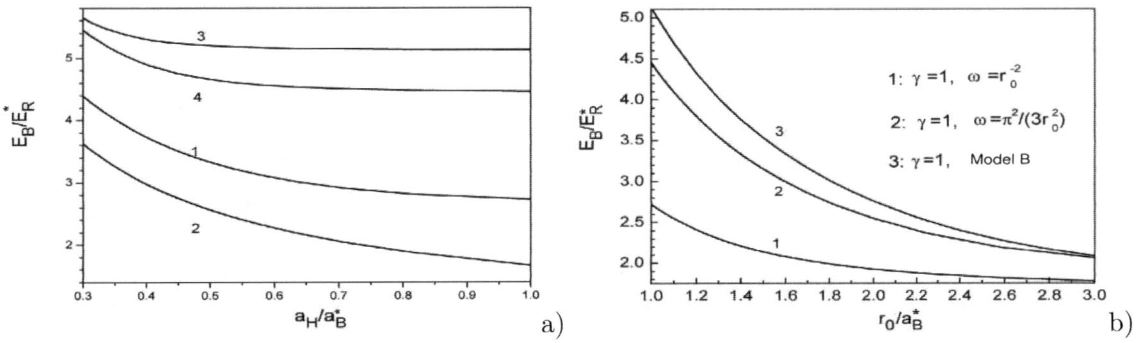

Figure 1. a) Binding energy of ground state electron E_B/E_R^* vs magnetic length $a_H^*/a_B^* = 1/\sqrt{\gamma}$ ($1 \le \gamma \le 11$) for $q = 1$, $m = 0$ and fixed radius of QD $r_0 = a_B^*$: 1 and 4 impurity electron (Model A) with confined potential frequencies $\omega = 1/r_0^2$ and $\omega = \pi^2/(3r_0^2)$; 3 is impurity electron in QD (Model B); 2 is impurity electron (Model A) without confined potential ($\omega = 0$). b)Binding energy of ground state electron E_B/E_R^* vs radius of QD at fixed magnetic field $\gamma = 1$.

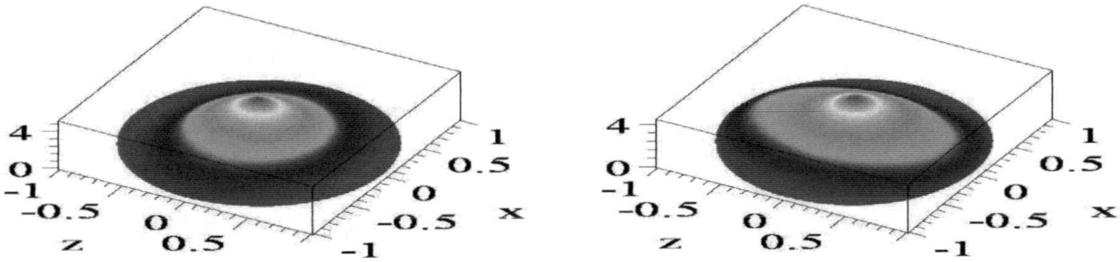

Figure 2. The profile of wave functions of the ground impurity states of QD with radius $r_0 = 1$ (Model B) in plane zx at magnetic field: $\gamma = 1$(left panel) and $\gamma = 10$(right panel).

3. Results and discussions

Determining binding energy of impurity as difference between electron without impurity center in QD E_0 ($q = 0$) and with one $E_1(q = 1)$, we can write $E_B = E_0 - E_1$. On Figs. 1a and 1b we show dependencies of the binding energies $E_B\,(a_H)$ at fixed radius of QD $r_0 = a_B^*$ and $E_B\,(r_0)$ at fixed magnetic field $H = H_0^* = 6$ T. As one can see from Fig. 1a with increasing H (decreasing magnetic length $a_H = a_B^*/\sqrt{\gamma}$) binding energy is increased, because the problem transforms to one-dimensional one. From Fig. 1b one can see, that with decreasing r_0 binding energy $E_B(r_0)$ is increased, because of reducing domain of localization of the electron probability density around the impurity center. From comparison of curves 1, 4 3 on Fig. 1a, and curves 1, 2 and 3 on Fig. 1b follows, that choosing value of parameter $\gamma_{r_0} = \pi^2/3$ provides quality agreement of models A and B. This fact connected with transforming of the discussed problem for QD as well as QWr to the problem of one dimensional hydrogen system. It is illustrated by the distributions of the wave functions of the ground state impurity electron in QD and QWr on Figs. 2 and 3, respectively.

On Fig. 4a we show the dependencies of binding energy of of the ground state impurity electron $E_B/(2E_R^*)$ in QWr without confinement potential ($\omega = 0$), versus magnetic field $\gamma = H/H_0^*, H_0^* = 6$T, that were calculated in different approximations and parametrizations of the problem (1) in the SC and CC. Curve 1 show values of E_B, i.e., $\mathcal{E}_1 = \gamma - 2E_1$, that where calculated by solving the BVP for (7)–(10) in the SC with $j_{max} = 10$ at a given accuracy (GA), which here was chosen as $\sim 10^{-8}$. Note, in interval $1 \le \gamma \le 11$ curve 1 corresponds to curve 2 on Fig. 1. Curves 4 and 5 show upper and lower estimations of binding energy, $E_{AD;c}^C$

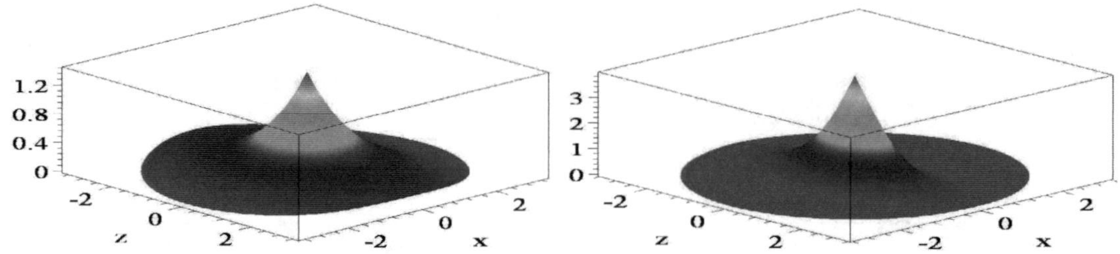

Figure 3. The profile of wave functions of ground impurity state in QWr without confinement potential($\omega = 0$) (Model A)in plane zx at magnetic field: $\gamma = 1$(left panel) and $\gamma = 10$(right panel).

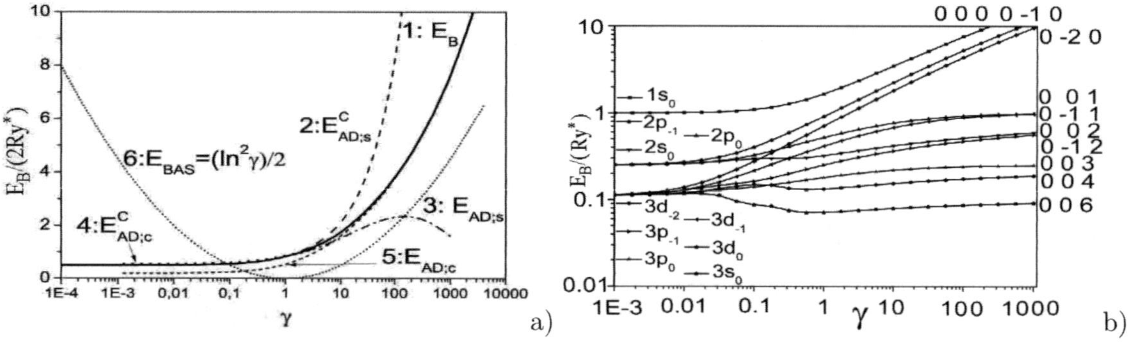

Figure 4. (a) The ground binding energy $E_B/(2E_R^*)$ of QWr (Model A) vs magnetic field $\gamma = H/H_0^*, H_0^* = 6$T in different approximation. (b) The first ten values of binding energy (in Ry*) levels $E_B/E_R^* = \mathcal{E}_i = \gamma - 2E_i$ classified by adiabatic sets of quantum numbers: spherical (N, l, m) for small γ and cylindrical (N_ρ, m, N_z) for large γ ones at $m \leq 0$, vs magnetic filed γ.

and $E_{AD;c}$, that where calculated by solving the BVP for (7) in the CC at $j_{max} = 1$ in a crude adiabatic approximation (CAD) at $W_{11} = 0$ and conventional adiabatic approximation (AD)at $W_{11} \neq 0$. One can see that these estimations go to curve 1 of the GA starting with $\gamma \geq 50$. For the interval of values $\gamma < 1$ the convergence rate of expansion (5) in the CC is slow, because conditions of the Kato type in a vicinity of point $z = 0$ does not fulfill [11].

Correspondingly, curves 2 and 3 show the upper and lower estimations of binding energy, $E^C_{AD;s}$ and $E_{AD;s}$, that were calculated by solving BVP for (7) in the SC at $j_{max} = 1$ in the CAD and the AD. One can see that these estimations approach to the values calculated with the GA (curve 1) starting from $\gamma < 1$ and remove from the GA values starting from $\gamma > 1$. However, for expansion (5) in the SC conditions of the Kato type near a vicinity of point $r = 0$ are satisfied, that leads to the high rate of convergence of such expansion and allows us to limit the same $j_{max} = 10$ for $1 < \gamma \leq 1000$ too [17]. For example, on Fig. 4b we show a behavior of the first ten values of binding energy (in Ry*) levels $\mathcal{E}_i = \gamma - 2E_i$ calculated with the GA in the SC and classified by adiabatic sets of quantum numbers: spherical (N, l, m) for small γ and cylindrical (N_ρ, m, N_z) for large γ at $m \leq 0$, vs magnetic field γ.

Note, if one describes the impurity states in QD, QWr and QW in strong magnetic field, on the first glance, it is possible to consider Coulomb interaction in problem (1) as small perturbation in the CC, and to use a well known estimation with logarithmic accuracy $E_{BAS} \simeq (\ln^2 \gamma)/2$ like in the case of ground state of a hydrogen atom in a strong magnetic field [18]. However from Fig. 4a one can see that there is three a times difference of the logarithmic estimation (curve 6) with respect to curve 1 of the GA. This question is discussed in details in [19].

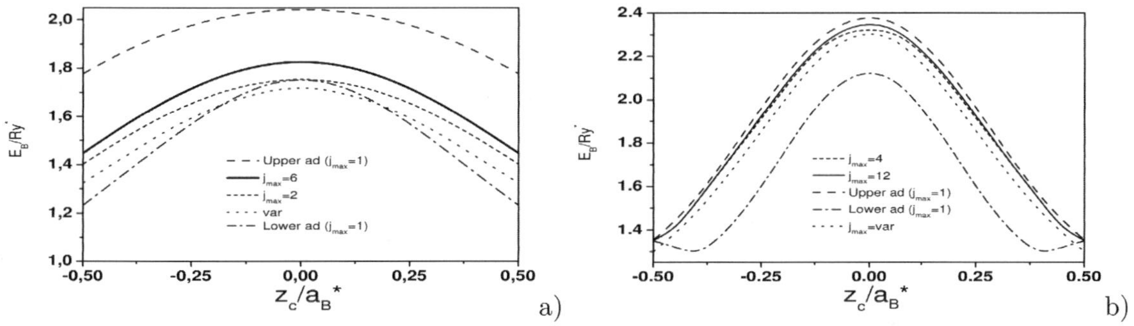

Figure 5. Binding energy $E_B/Ry^* = E_1^{th} - 2E_1(z_c)$ of QW vs position of Coulomb impurity z_c and number of basis functions $j_{\max}$, and crude adiabatic approximation (upper ad), adiabatic approximation (lower ad), and variational calculation (var): (a) Model A at $q = 1$, $m = 0$, $\omega = 3$, $E_1^{th} = \omega$ in the SC, and (b) Model B $q = 1$, $m = 0$, $z_0 = 1$, $E_1^{th} = \pi^2/z_0^2$ in the CC.

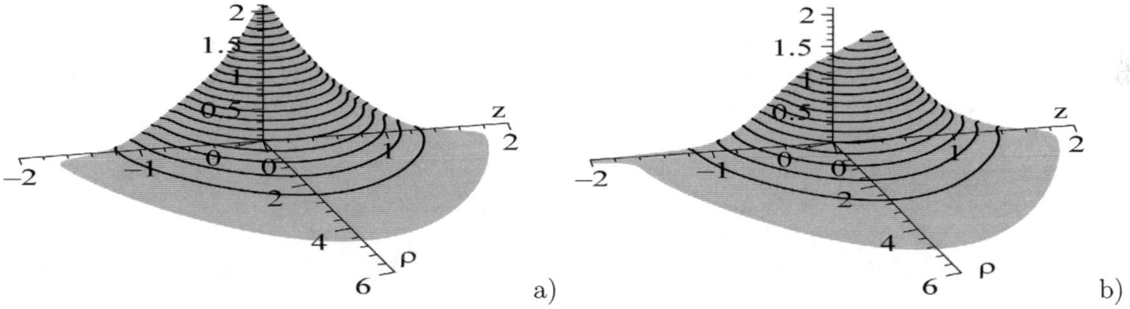

Figure 6. The profile of wave function of ground impurity state in QW (Model A) in plane zx at $q = 1$, $m = 0$, $\omega = 3$ and shift Coulomb center: (a) $z_c = 0$ and (b) $z_c = 0.4$.

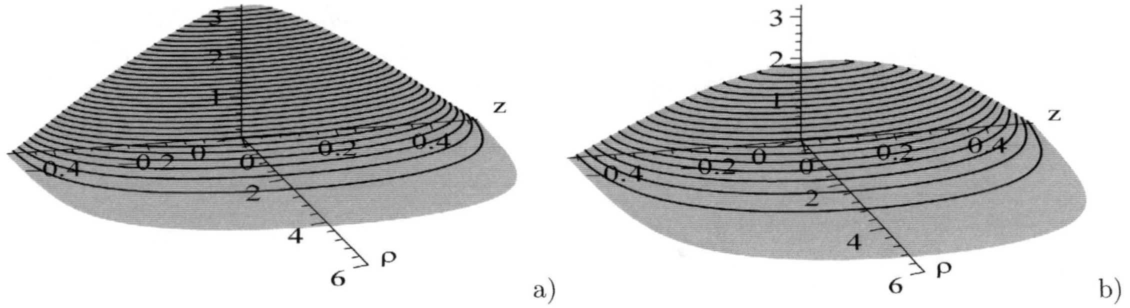

Figure 7. The profile of wave function of ground impurity state in QW (Model B) in plane zx at $q = 1$, $m = 0$, $z_0 = 1$ and shift Coulomb center: (a) $z_c = 0$ and (b) $z_c = 0.4$.

On Fig. 5 the upper and lower estimations of binding energy E_B of the ground state impurity electron in QW without magnetic field, correspondingly, for model A in the SC and model B in the CC, connected with impurity position z_c, as well as convergence by number j_{max} of basis functions in expansion (5) and their comparison with variational estimation are demonstrated. As it follows from Fig. 5, expansion (5) in the CC has a lower convergence rate with respect to one in the SC by the same reason [10] as for the QWr mentioned above. Fig. 6 and Fig. 7 show the wave functions of ground state of impurity electron in QW, for model A and B, respectively.

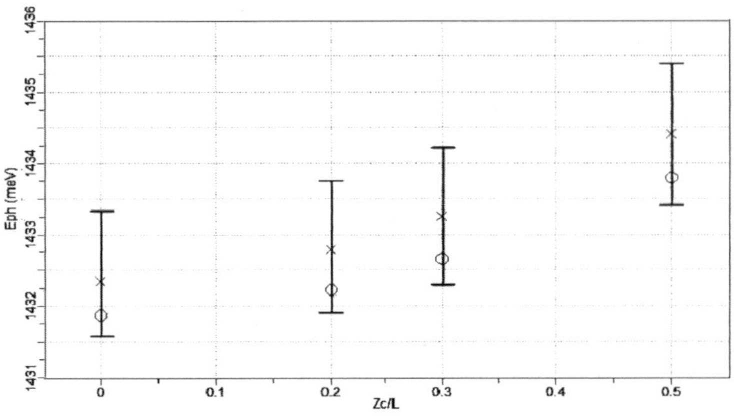

Figure 8. The interval of the resonance photon energy (in meV) $E_{ph} = \hbar\omega_{ph} = E_f - E_i$ determines the peak values of the absorption coefficient $K(\omega_{ph})$ vs the position of the impurity center z_c/L, $L = 102$Å: the vertical intervals are upper and lower adiabatic estimates, the crosses are variational estimates, the circles are estimates with given accuracy.

For the case of QW with parabolic confinement the coordinate of the maximum value of wave function coincide with the coordinate of impurity center. On the other hand, for Model B, due to Dirichlet boundary conditions in points $z_{\min} = -1/2$, $z_{\max} = 1/2$, the maximum of wave function is in a vicinity of the central point. Differences in the behavior of these wave functions allows us to verify models A and B.

Obtained estimations allow us to reveal the possible boundary of intervals of the photon resonance energies $E_{ph} = \hbar\omega_{ph}$ presented. For example, on Fig. 8, it is illustrated the maximal value of absorption coefficient $K(\omega_{ph})$ dependent upon the position of impurity in QW [9]. Based on the results for wave functions and eigenvalues it is possible to calculate impurity and inter impurity absorption coefficient for QD, QWr and QW. The impurity absorption coefficient $K(\omega)$ in parabolic GaAs QW is calculated with the help of variational functions [9]:

$$K(\omega_{ph}) = \frac{4\pi^2 e^2}{\kappa m_0^2 c\omega_{ph} V} \sum\nolimits_{i,f} \left|\vec{e}\vec{M}_{if}\right|^2 \delta\left(E_f - E_i - \hbar\omega_{ph}\right), \tag{11}$$

where V is the volume of a sample, $\vec{e}$ is the light polarization vector, $\vec{M}_{if}$ is the matrix element of the transition from the initial state i to the final one f, E_i and E_f are the energies of the initial and final states, delta-function provides the conservation of energy in the course of the transitions. Fig. 9a shows the absorption curves for transitions related to acceptors and donors. We can see, the strong difference exists between the shape of absorption curves, corresponding to donor and acceptor levels. During transitions between the acceptor level and conductive band the denominator of Eq. (11) at small values of wave vectors is close to unity, as far as for GaAs the ratio $m_c/m_v \approx 0.1$, therefore the absorption coefficient depending on the energy of incident photon has step-like feature, thus repeating the behavior of density of states. The absorption scenario related to donors is different. While increasing the denominator in Eq. (11) at a quite small interval from the absorption threshold the absorption rapidly decreases ($m_c/m_v \approx 9.25$).

Figure 9b shows the dependence of the blue shift on different values of two-dimensional concentrations of dominant impurity (acceptor). The difference between acceptor and donor energy levels can be presented as $E_{DAP} = E_{gap} + E_{bD} + E_{bA} + e^2/R$, where E_{bD} and E_{bA} are donor and acceptor binding energies, respectively; the fourth item in E_{DAP}, i. e. e^2/R, is the Coulomb term. In the frameworks of our calculations the blue shift $\Delta E_{blueshift}$ is proportional to the Coulomb term, i. e. inversely proportional to the distance between donor and acceptor $\Delta E_{blueshift} = e^2/R$. When the acceptor concentration is increasing (e. g. the concentration of Si atoms [20]), donors (e. g. residual C atoms [20]) and acceptors become spatially closer, the blue shift in the acceptor–donor transition peak takes place as a result of the increase of the Coulomb term [22, 20]. So, the growth of doping level is the reason for the increase of the blue shift.

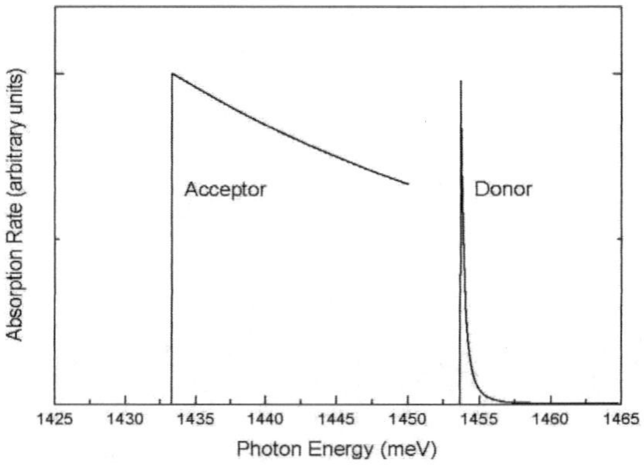

(a)

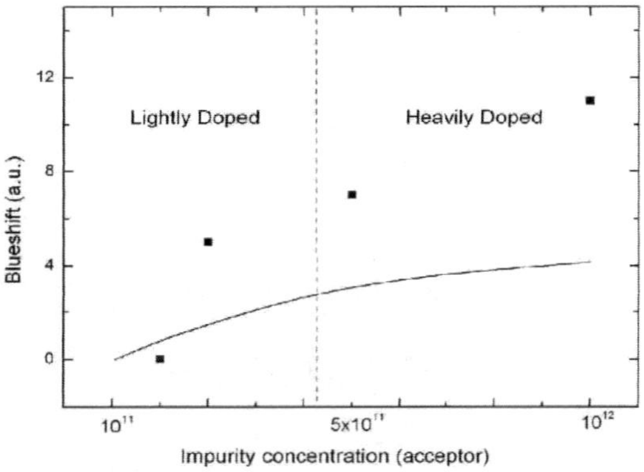

(b)

Figure 9. a) The difference of absorbtion for transitions "conduction band → acceptor band" (Acceptor) and "valence band → donor band" (Donor) at L=100Å, $m_v/m_c = 9.\ 2$ (in GaAs QW). b) The dependence of the blue shift (in a.u.) for transitions donor-acceptor vs concentration of dominant (acceptor) impurity (after the averaging between D–A distance): squares are experimental results [20], solid line is result of calculations [21], the left panel shows the domain of low concentration, the right one shows the domain of high concentration.

4. Conclusion

In this paper we demonstrate effective methods of calculation of impurity states in the present magnetic field for cases of QW, QWr and QD. As result of our investigation we can say that adiabatic approach is good mechanism for describing the Coulomb systems in quantum nanostructures. There are good agreement between adiabatic and variational procedures of calculation of energy spectra and wave functions of impurities in QW, QWr, QD.

Authors thank Profs. V. L. Derbov, V. P. Gerdt, A. A. Kostanyan and V. V. Serov for collaboration. The work was supported partially by RFBR (grants 10-01-00200 and 08-01-00604) and by the grant No. MK-2344.2010.2 of the President of Russian Federation.

[1] Miura N 2008 *Physics of semiconductors in high magnetic fields* (New York, Oxford Univ. Press Inc)
[2] Maksym P A and Chakraborty T 1990 *Phys. Rev. Lett.* **65** 108.
[3] J -L Zhu 1989 *Phys. Rev. B* **39** 8780
[4] J -L Zhu, Y Cheng and J -J Xiong 1990 *Phys. Rev. B* **41** 10792.
[5] Zhu J -L, Li Z -Q, Yu J -Z, Ohno K and Kawazoe Y 1997 *Phys. Rev. B* **55** 15819
[6] Elliott R and Loudon R 1959 *J. Phys. Chem. Sol.* **8** 382
[7] Avetisyan A A et al 1999 *Phys. Stat. Sol. B* **214** 91
[8] Kazaryan E M, Petrosyan L S and Sarkisyan H A 2001 *Int. J. Mod. Phys. B* **15** 4103
[9] Kazaryan E M et al 2005 *Physica E* **28** 423
[10] Gusev A A et al 2010 *Phys. At. Nucl.* **73** 331
[11] Chuluunbaatar O et al 2009 *Phys. At. Nucl.* **72** 811
[12] Gusev A A et al 2010 *Math. Comput. in Simulation* (accepted); arXiv:1005.2089v1 [cond-mat.mes-hall]
[13] Gusev A A et al 2010 *Lect. Notes in Comput. Sci.* **6244** 106; arXiv:1004.4202v1 [cond-mat.mes-hall].
[14] Vinitsky S I et al 2009 *Lect. Notes in Comput. Sci.* **5743** 334
[15] Chuluunbaatar O et al 2009 *Comput. Phys. Commun.* **180** 1358
[16] Chuluunbaatar O et al 2008 *Comput. Phys. Commun.* **178** 301
[17] Chuluunbaatar O et al 2008 *Comput. Phys. Commun.* **179** 685
[18] Landau L D and Lifshitz E M 1977 *Quantum Mechanics, 3rd ed* (Oxford, Pergamon)
[19] Karnakov B M and Popov V S 2003 *J. Exp. Theor. Phys.* **97** 890
[20] Guzman A et al 1999 *IEE Proceedings Optoelectronics* **146** 89
[21] Kazaryan E M, Kostanyan A A and Sarkisyan H A 2007 *J. Phys. Cond. Mat.* **19** 046212
[22] Ding Y J, Korotkov R, Khurgin J B, Rabinovich W S and Katzer D S 1998 *Appl. Phys. Lett.* **72** 534

International Conference on Theoretical Physics Dubna-Nano 2010 IOP Publishing
Journal of Physics: Conference Series **248** (2010) 012048 doi:10.1088/1742-6596/248/1/012048

TWO ELECTRONIC STATES IN A QUANTUM RING: MATHIEU EQUATION APPROACH

N.G. Aghekyan[1], **E.M. Kazaryan**[1], **L.S. Petrosyan**[1], **H.A. Sarkisyan**[1,2]
1 Russian-Armenian (Slavonic) University, 123 H. Emin, Yerevan, Armenia
2 Yerevan State University, 0025 Al. Manukyan, Yerevan, Armenia
E-mail: shayk@ysu.am

Abstract. The problem of two-electronic states is considered for the case of thin GaAs quantum ring in the framework of adiabatic approximation. It is assumed that electrons are localized on a circle, and the Coulomb interaction between the electrons depends only on angular coordinates. The real Coulomb interaction is replaced by function which makes it possible the separation of relative motion and the motion of the center of mass. The Mathieu equation is obtained for the relative motion of electrons. The harmonic, anharmonic and the libration regimes of oscillations are discussed.

1. Introduction

Quantum dots (QD) are very perspective objects in respect of the application in innovative semiconducting devises as an active media [1]. There is an opportunity of fabricating structures with desirable properties due to the wide range of methods of flexible manipulation of energetic spectrum of QD electron [1, 2]. The properties of the energetic spectrum of charge carriers in QD are defined by the range of characteristics of a sample such as the geometry and the component structure of the materials of QD. The close relationship between the above mentioned factors with the properties of QD have been investigated theoretically and experimentally in many works which are devoted to spherical, cylindrical, ellipsoidal, pyramidal and other QD-s (see for example [3]-[10]). Recently ring-shaped nanostructures have been fabricated [11], where the radial motion of carriers is confined both with the inner and outer borders. The importance of detailed theoretical study of physical properties of such systems is in the generalizing character of obtained results as one can realize quantum wells and wires as well as QDs by the limiting processes. The construction of the realistic Hamiltonian of a system, as far as it is possible, is necessary for the description of processes in qantum rings (QR), which allows us to calculate certain characteristics of QR. One and two electronic states in QR have been first investigated theoretically in pioneering work of Chakraboty and Pietilainen [12], where authors have modelled the confining potential of QR by the off-center two-dimensional oscillator. In the framework of this model, Halonen and others [13] have discussed the optical absorption in QR in a magnetic field by the same off-center oscillator approximation, in the presence of repulsive scattering centers. In [14] authors have studied QR as electron spin beam splitters. They have showed that quantum interference and spin-orbit interaction, in a one-dimensional mesoscopic semiconductor ring with one input and two output leads, can act as a spin beam splitter. The effects of the Coulomb interaction on the energy spectrum and the magnetization of two electrons in a strained

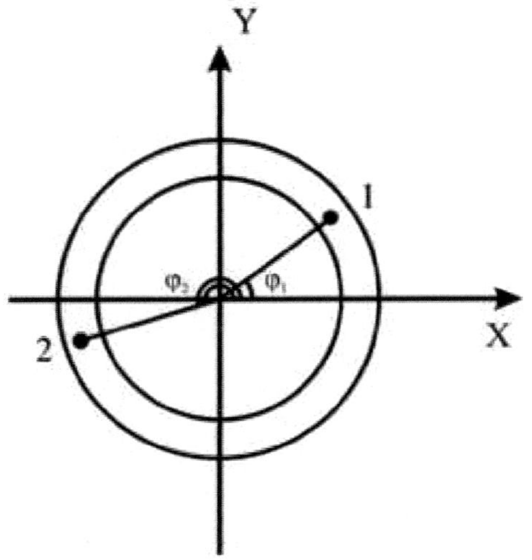

Figure 1. Quantum ring with two electrons.

$In_xGa_{1-x}As/GaAs$ ring-like nanostructures are analyzed with realistic parameters inferred from the cross-sectional scanning-tunneling microscopy data in [15]. Thereby, the theoretical study of two-electron states in QR has both academic and implicational value. It stimulates the further observations of this concept. In this work we present the theoretical model of two-electron states in a thin GaAs QR with the rectangular infinitely high confining potential. The condition of the small thickness of QR allows us to use the adiabatic approximation as a result of which the interaction of electrons becomes a function on angular coordinates only.

2. Theory

Let us consider a GaAs QR with a small width $R_1 - R_2 \ll \{R_1, R_2\}$, which contains two electrons. The influence of QR borders on electron states in radial direction is so strong that we can describe the radial states of each electron by one particle approximation. This assumption allows us to consider the electrons located on the same distance from the QR center and the energy of Coulomb interaction between the electrons as a function of polar angles φ_1 and φ_2 (Fig. 1). In this case the two-particle Hamiltonian is the following:

$$\hat{\mathcal{H}}(1,2) = \sum_{i=1}^{2} \left\{ -\frac{\hbar^2}{2\mu} \left(\frac{1}{\rho_i} \frac{\partial}{\partial \rho_i} \left(\rho_i \frac{\partial}{\partial \rho_i} \right) + \frac{1}{\rho_i^2} \frac{\partial^2}{\partial \varphi_i^2} \right) + V_{\mathrm{conf}}(i) \right\} + V_{\mathrm{Coul}}(\varphi_1, \varphi_2), \tag{1}$$

where μ is the effective mass of electron (for GaAs $\mu = 0.067 m_e$), V_{conf} is the confining potential in radial direction and can be expressed as follows:

$$V_{\mathrm{conf}} = \begin{cases} 0, \ R_1 \le \rho \le R_2, \\ \infty, \ [\rho < R_1, \ \rho > R_2]. \end{cases}$$

The expression of the Coulomb interaction will be given later. Since the radial wave functions of electrons are expressed by the Bessel function of the first and the second kinds, then in the adiabatic approximation the total wave function can be presented as the following product

$$\Psi(r_1, r_2, \varphi_1, \varphi_2) = f_0(r_1) f_0(r_2) \Phi(\varphi_1, \varphi_2), \tag{2}$$

where

$$f_0(r_i) = C_1 I_0(\kappa_0 r_i) + C_2 Y_0(\kappa_0 r_i), \tag{3}$$

$I_0(x), Y_0(x)$ are the Bessel functions of the first and the second orders respectively [16], $\kappa_0 = \sqrt{2\mu E_0/\hbar^2}$, E_0 is the energy of one particle of the ground state, which is determined from the boundary conditions

$$\begin{vmatrix} I_0(\kappa_0 R_1) & Y_0(\kappa_0 R_1) \\ I_0(\kappa_0 R_2) & Y_0(\kappa_0 R_2) \end{vmatrix} = 0. \tag{4}$$

Note, that function $f_0(r)$ satisfies the radial equation

$$-\frac{\hbar^2}{2\mu}\left(\frac{1}{\rho}\frac{d}{d\rho}\left(\rho\frac{d}{d\rho}\right)\right) f_0(\rho) + V_{\text{conf}}(\rho) = E_0 f_0(\rho). \tag{5}$$

Substituting (2) into (1) and using (5), for the angular wave function we obtain the following equation:

$$\begin{gathered} -\frac{\hbar^2}{2\mu R_{\text{eff}}^2}\left(\frac{\partial^2}{\partial\varphi_1^2} + \frac{\partial^2}{\partial\varphi_2^2}\right)\Phi(\varphi_1,\varphi_2) + V_{\text{Coul}}(\varphi_1,\varphi_2)\Phi(\varphi_1,\varphi_2) = \\ = (E - 2E_0)\Phi(\varphi_1,\varphi_2) = \varepsilon\Phi(\varphi_1,\varphi_2), \end{gathered} \tag{6}$$

where it is assumed, that the distances of the particles from the QR center are fixed:

$$R_{\text{eff}} = \frac{R_1 + R_2}{2}. \tag{7}$$

and the following notation is made: $\varepsilon = E - 2E_0$.

Let us turn to the definition of the expression of the potential $V_{\text{Coul}}(\varphi_1,\varphi_2)$ of electrons interaction. We deal with two particles on a circle interacting with the repulsive potential. On the other hand it is clear that this potential should depend only on the mutual location of the particles, i.e. on $\varphi_1 - \varphi_2$, and it should be an even function, as the replacement of particles should not affect on the Hamiltonian. Meanwhile if $\varphi_1 - \varphi_2 = \pi$ then $V_{\text{Coul}}(\varphi_1,\varphi_2)|_\pi = e^2/(2\varepsilon_d R_{\text{eff}})$, and if $\varphi_1 - \varphi_2 = \pi/2$ then $V_{\text{Coul}}(\varphi_1,\varphi_2)|_{\pi/2} = e^2/(\sqrt{2}\varepsilon_d R_{\text{eff}})$ (ε_d is dielectric constant and for GaAs $\varepsilon_d \approx 13$).

With the increase of $\varphi_1 - \varphi_2$ from $\pi/2$ to π the function $V_{\text{Coul}}(\varphi_1,\varphi_2)$ should decrease monotonically.

In simple approximation this function can be presented in the form:

$$\mathrm{V}_{\text{Coul}}^{\text{appr.}}(\varphi_1,\varphi_2) = \frac{\mathrm{e}^2}{\mathrm{R}_{\text{eff}}\varepsilon_d}\left\{\frac{1}{\sqrt{2}} + \frac{\sqrt{2}-1}{2}\cos(\varphi_1 - \varphi_2)\right\}. \tag{8}$$

Note, that the exact Coulomb function is as follows:

$$\mathrm{V}_{\text{Coul}}^{\text{ex.}}(\varphi_1,\varphi_2) = \frac{\mathrm{e}^2}{\sqrt{2}\varepsilon_d \mathrm{R}_{\text{eff}}\sqrt{1 - \cos(\varphi_1 - \varphi_2)}}.$$

The graphical representations of $\mathrm{V}_{\text{Coul}}^{\text{appr.}}(\varphi_1,\varphi_2)$ and $\mathrm{V}_{\text{Coul}}^{\text{ex.}}(\varphi_1,\varphi_2)$ functions in the region $(\pi/2,\pi)$ are presented in (Fig. 2). As follows from the figure the qualitative features of both functions are the same, and the maximal shift of the values of the two potentials is 0,027 (Fig. 3). Thereby, for the radial wave function we obtain the following equation:

$$\begin{gathered} -\frac{\hbar^2}{2\mu R_{\text{eff}}^2}\left(\frac{\partial^2}{\partial\varphi_1^2} + \frac{\partial^2}{\partial\varphi_2^2}\right)\Phi(\varphi_1,\varphi_2) + \\ + \frac{e^2}{\varepsilon_d R_{\text{eff}}}\left\{\frac{1}{\sqrt{2}} + \frac{\sqrt{2}-1}{2}\cos(\varphi_1 - \varphi_2)\right\}\Phi(\varphi_1,\varphi_2) = \varepsilon\Phi(\varphi_1,\varphi_2). \end{gathered} \tag{9}$$

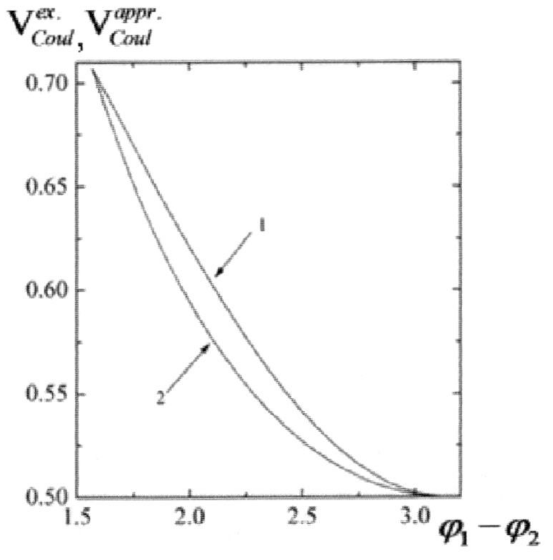

Figure 2. Approximated (1) and exact (2) functions of Coulomb interaction $\left(\text{in units of } \frac{\mathrm{e}^2}{\varepsilon_\mathrm{d} \mathrm{R}_\mathrm{eff}}\right)$.

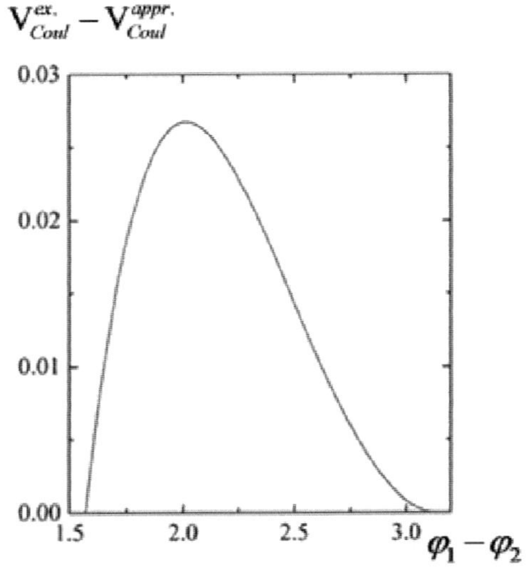

Figure 3. Difference between approximated and exact functions of Coulomb interaction $\left(\text{in units of } \frac{\mathrm{e}^2}{\varepsilon_\mathrm{d} \mathrm{R}_\mathrm{eff}}\right)$.

Let us insert new variables α and β, related to φ_1 and φ_2 by the expressions:

$$\alpha = \varphi_1 - \varphi_2, \; \beta = (\varphi_1 + \varphi_2)/2. \tag{10}$$

In this case the variables in (9) separate and the wave function, can be expressed in the form of the following product:

$$\Phi\,(\alpha,\,\beta) = u\,(\alpha)\,v\,(\beta)\,. \tag{11}$$

Applying the separation procedure one can define the functions $u\,(\alpha)$ and $v\,(\beta)$ from the equations

$$-\frac{\hbar^2}{\mu R_{\text{eff}}^2}\frac{d^2u\,(\alpha)}{d\alpha^2} + \frac{e^2}{\varepsilon_\text{d} R_\text{eff}}\left\{\frac{\sqrt{2}-1}{2}\cos\alpha\right\}u\,(\alpha) \;=\; E_\alpha u\,(\alpha)\,, \tag{12}$$

$$-\frac{\hbar^2}{4\mu R_{\text{eff}}^2}\frac{d^2v\,(\beta)}{d\beta^2} \;=\; E_\beta v\,(\beta)\,. \tag{13}$$

Here the following relation is taken into account:

$$(E_\alpha + E_\beta) = \varepsilon - \frac{e^2}{\sqrt{2}R_\text{eff}}. \tag{14}$$

Inserting in the equation (13) the electron total moment of inertia $J_\text{eff} = 2\mu R_\text{eff}^2$ for the eigenfunctions and the eigenvalues of equation (13) we can write:

$$v\,(\beta) = \frac{\exp\{im\beta\}}{\sqrt{2\pi}}, \quad E_\beta = \frac{\hbar^2 m^2}{2J_\text{eff}} \quad (m = 0, \pm 1, \pm 2, \ldots)\,. \tag{15}$$

Let us now consider equation (12). Making the following notations:

$$\alpha = 2\phi,\ \lambda = \frac{4\mu R_{\text{eff}}^2}{\hbar^2} E_\alpha,\ 2q = \left(\frac{\sqrt{2}-1}{2}\right) \frac{4\mu e^2 R_{\text{eff}}}{\varepsilon_{\text{d}} \hbar^2},$$

we can transform 12 to the Mathieu equation [17]:

$$\frac{d^2 u(\phi)}{d\phi^2} + (\lambda - 2q\cos 2\phi)\, u(\phi) = 0. \tag{16}$$

As the one dimensional equation (16) contains symmetrical potential energy with respect to the point $\phi = 0$, their solutions have to be either even or odd periodic functions. One can write these solutions in the form of the Fourier expansion

$$u_+(\phi) = \sum_{j=0}^{\infty} A_{2j} \cos(2j\phi), \quad u_-(\phi) = \sum_{j=1}^{\infty} B_{2j} \sin(2j\phi), \tag{17}$$

where A_{2j}, B_{2j} are the expansion coefficients. It is known from the Mathieu equation theory [17] that even and odd solutions and hence the energy levels alternate. If $\{a_{2j}\}_{j=0}^{\infty}$ is the ensemble of the even, and $\{b_{2j}\}_{j=1}^{\infty}$ is the ensemble of the odd solutions with respect to , hence the following inequalities take place:

$$a_0 < b_2 < a_2 < b_4 < a_4 < \ldots. \tag{18}$$

Finally for the even and odd energy levels we obtain the expressions:

$$E_{2j}^+ = \frac{\hbar^2 a_{2j}}{4\mu R_{\text{eff}}^2}, \quad E_{2j}^- = \frac{\hbar^2 b_{2j}}{4\mu R_{\text{eff}}^2}. \tag{19}$$

Taking into account electron spin, we obtain the singlet states corresponding to the even levels, as the spin part of the wave function is odd. In $z-$presentation by the Pauli matrix

$$\hat{\sigma}_z = \begin{pmatrix} 1 & 0 \\ 0 & -1 \end{pmatrix},$$

the asymmetric spin wave function can be written as

$$\chi_{\text{A}} = \frac{1}{\sqrt{2}} \{\xi(1)\varsigma(2) - \xi(2)\varsigma(1)\}, \tag{20}$$

where

$$\xi = \begin{pmatrix} 1 \\ 0 \end{pmatrix} \text{ and } \varsigma = \begin{pmatrix} 0 \\ 1 \end{pmatrix}.$$

For the odd levels spin part of the wave function is symmetric and hence, the total wave function is triplet with the spin wave functions

$$\chi_{\text{S}} = \left\{ \xi(1)\xi(2),\ \varsigma(1)\varsigma(2),\ \frac{1}{\sqrt{2}} \{\xi(1)\varsigma(2) + \xi(2)\varsigma(1)\} \right\}. \tag{21}$$

Thereby in the considered system the singlet and triplet levels alternate and we finally obtain for the energy spectrum the following expressions:

$$E_{2j,m}^+ = \left(2E_0 + \frac{\hbar^2 a_{2j}}{4\mu R_{\text{eff}}^2} + \frac{\hbar^2 m^2}{2J_{\text{eff}}} + \frac{e^2}{\sqrt{2}\varepsilon_{\text{d}} R_{\text{eff}}}\right). \tag{22}$$

$$E_{2j,m}^- = \left(2E_0 + \frac{\hbar^2 b_{2j}}{4\mu R_{\text{eff}}^2} + \frac{\hbar^2 m^2}{2J_{\text{eff}}} + \frac{e^2}{\sqrt{2}\varepsilon_{\text{d}} R_{\text{eff}}}\right). \tag{23}$$

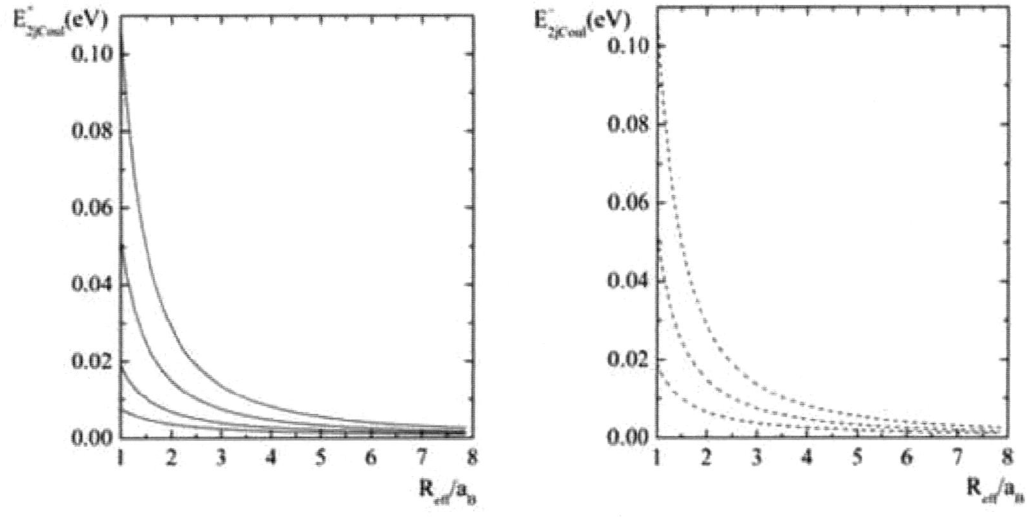

Figure 4. Dependence of energy from the effective radius of QR.

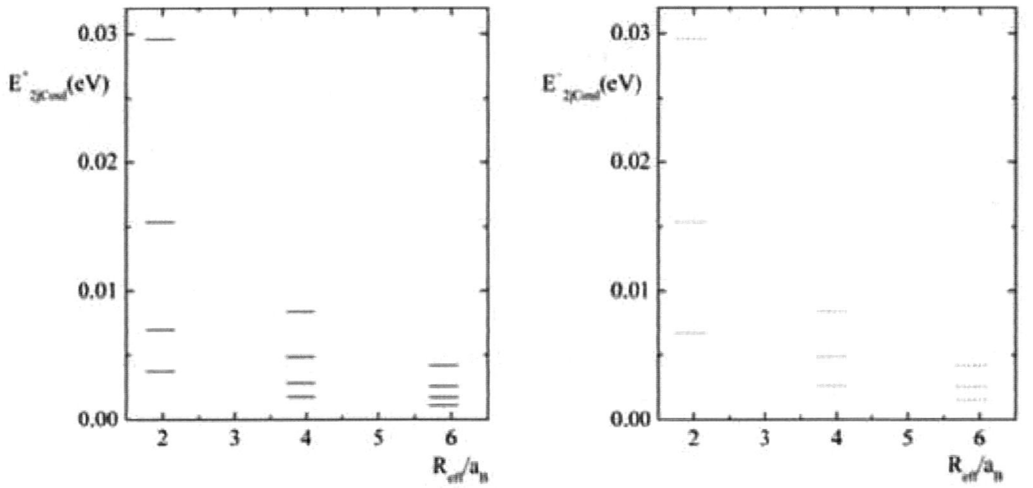

Figure 5. Energy diagrams of Coulomb levels.

3. Discussions of results

Now let us consider the behaviour of the system depending on the effective radius of QR. According to (16) the effective "depth" of the one dimensional well increases with the increase of $R_{\rm eff}$ as $q \sim R_{\rm eff}$. It can be shown from the Mathieu equation theory, that for $\alpha = \pi$ Mathieu equation reduces to the harmonic oscillator equation, with the eigenvalue:

$$E_n = \frac{\hbar^2 \sqrt{q}}{\mu R_{\rm eff}^2}\left(n + \frac{1}{2}\right) - \frac{q\hbar^2}{2\mu R_{\rm eff}^2}. \tag{24}$$

It is due to the small oscillations of diametrical electrons, as a result of the decrease of repulsive force, which is in its turn, the result of the increase of QR effective radius. That is why

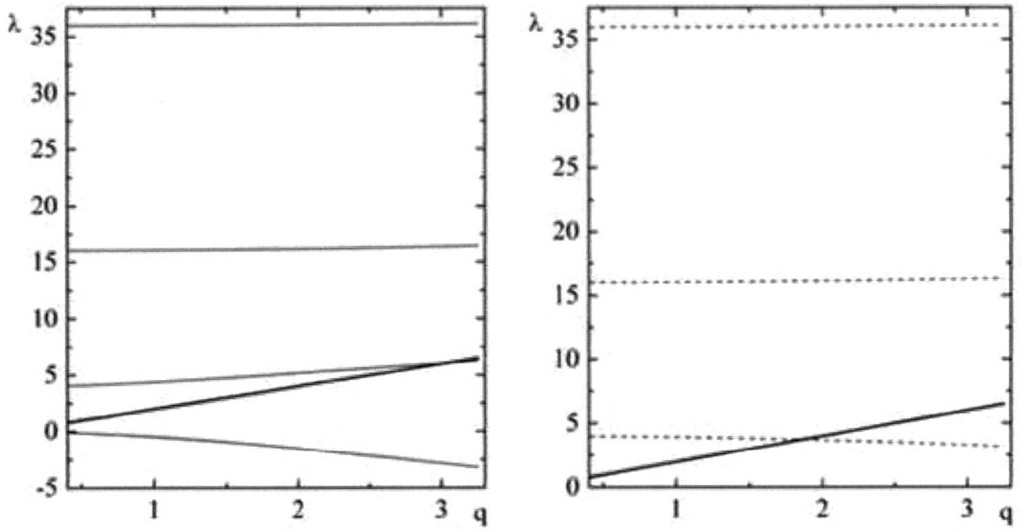

Figure 6. Dependence of λ parameter on q.

these oscillations will be harmonic. As follows from (24) the effective frequency of oscillations decreases with the increase of R_{eff} due to the weakening of the oscillation force. Only the Coulomb part of the spectra of E_{2jCoul} as a function of effective radius R_{eff} (the second and the forth terms in (22) and (23), are shown on (Fig. 4). Here the solid lines correspond to the even states, when the dashed lines correspond to the odd states. Note that the curves corresponding to the different parities do not cross and they decrease monotonically with the increase of R_{eff}, because the distance between electrons becomes larger, leading to the weaker Coulomb interaction. Meanwhile it follows from the behaviour of the solutions of the Mathieu equation, the levels E^+_{2Coul} and E^-_{2Coul} are very close to each other. The diagram of the discrete levels for $R_{\text{eff}} = 200\text{Å}, 400\text{Å}, 600\text{Å}$ is shown on (Fig. 5). The dependence of the parameter λ on q for the R_{eff} changing in the range from 100Å to 800Å is shown on (Fig. 6). The levels which have the energy with value smaller than $2q$ correspond to the anharmonic oscillations that transform to harmonic ones with the increase of R_{eff}. Concerning to the levels that lie above the value $2q$, we can insist that libration oscillations correspond to them. In this case electrons move all over the QR with small oscillations and feel each other weakly. For the levels with high energy the energy of Coulomb interaction is negligible and we deal with almost free electrons.

4. Conclusion

A model of two electrons in GaAs thin QR is presented, which allows us to apply the adiabatic approximation and reduce the initial Schrödinger equation to the angular one. The Schrödinger equation with Coulomb potential is not analytically solvable and, therefore, a new approximated potential is introduced, which has the qualitative behavior similar to the exact function in the region between 90 and 180 degrees. This approximation enables us to separate the variables in (9) and to reduce the the equation to the Mathieu equation. The singlet and triplet states of energy for two electronic system is discussed. Finally, three classes of electron oscillations are discussed: harmonic, anharmonic and libration.

References

[1] Z.I. Alferov, Semiconductors 32 (1998) 3.
[2] D. Bimberg, M. Grundman, N. Ledentsov, Quantum Dot Heterostructures, Wiley, New York, 1999.
[3] D. Chuu et al., Phys. Rev. B 46 (1992) 3898.
[4] Zhu, Jia-Lin; Chen, Xi, Physical Review B (Condensed Matter), Volume 50, Issue 7, August 15, 1994, pp.4497-4502.
[5] Tadić, M.; Peeters, F. M.; Janssens, K. L.; Korkusiński, M.; Hawrylak, P., Journal of Applied Physics, Volume 92, Issue 10, pp. 5819-5829 (2002).
[6] A.K. Atayan, E.M. Kazaryan, A. V. Meliksetyan, and H. A. Sarkisyan, Eur. Phys. J. B 63, 485 (2008).
[7] Dvoyan, K. G.; Kazaryan, E. M.; Petrosyan, L. S., Physica E: Low-dimensional Systems and Nanostructures, v. 28, iss. 4, p. 333-338 (2005).
[8] Barati, M.; Rezaei, G.; Vahdani, M. R. K., Physica status solidi (b), vol. 244, issue 7, pp. 2605-2610 (2007).
[9] Grundmann, M.; Stier, O.; Bimberg, D., Physical Review B (Condensed Matter), Volume 52, Issue 16, October 15, 1995, pp.11969-11981.
[10] Chang, K., Solid State Communications, vol. 104, issue 6, pp. 351-354 (1997).
[11] A. Lorke, R.J. Luyken, A.O. Govorov, J. Kottaus, Phys. Rev. Lett. 84 (2000) 2223.
[12] T. Chakraborty, P. Pietilainen, Phys. Rev. B 50 (1994) 8460.
[13] V. Halonen, P. Pietilainen, T. Chakraborty, Europhys. Lett. 33 (1996) 337.
[14] Foldi Pe ter, Kalman Orsolya, G. Benedict Miha ly, F.M. Peeters, Phys. Rev. B 73 (2006) 155325.
[15] Fomin et al., Physical Review B, vol. 77, Issue 20, id. 205326 (2008).
[16] Koshlyakov N.S., Gliner E.B., Smirnov M.M., Equations in partial derivatives of mathematical physics, Vysshaya Shkola, Moscow, 1970.
[17] M. Abramovitz, I.A. Stegun, Handbook of mathematical functions, Dover, New York, 1972.

Journal of Physics: Conference Series **248** (2010) 012049 doi:10.1088/1742-6596/248/1/012049

New stochastic equation for a harmonic oscillator: Brownian motion with adhesion

M Gitterman
Department of Physics, Bar-Ilan University, Ramat-Gan 52900 Israel

September 20, 2010

Abstract

In addition to the usually considered stochastic harmonic oscillator with an external random force (Brownian motion) or with random frequency and random damping, we consider an oscillator with a random mass for which the particles of the surrounding medium adhere to the oscillator for some (random) time after the collision, thereby changing the oscillator mass. We have calculated the first two moments and the Lyapunov exponent, which describes the stability of the fixed point. This model can be useful for the analysis of chemical and biological solutions as well as for nano-technological devices.

A simple, general and widely used model is that of a harmonic oscillator of unit mass [1], which is described by the following equation

$$\frac{d^2x}{dt^2} + 2\gamma\frac{dx}{dt} + \omega^2 x = 0 \tag{1}$$

Thousands of articles have been devoted to applications of this model to different phenomena in physics, chemistry, biology, economics, etc. However, all phenomena in Nature are subject to random forces, and for the adequate description of these phenomena, one has to supplement the dynamic equation (1) wih a random force [2]. This can be done in different ways.

1. Brownian motion

$$\frac{d^2x}{dt^2} + 2\gamma\frac{dx}{dt} + \omega^2 x = \xi(t) \tag{2}$$

This is universally known equation (with $\omega^2 = 0$) [3] describes the motion of nano-particle subject to a systematic viscous force, $2\gamma dx/dt$, and a random force $\xi(t)$, having its origin from non-equal (random) numbers of molecular collisions from opposite sides and is responsible for the zigzag motion of a Brownian particle. The $\omega^2 x$ term in Eq. (2) describes a Brownian motion in an external quadratic potential.

2. Random frequency.

$$\frac{d^2x}{dt^2} + 2\gamma\frac{dx}{dt} + \omega^2\left[1 + \xi\left(t\right)\right]x = 0 \tag{3a}$$

The many applications of this model include different fields in physics, such as wave propagation in a random medium [4], spin precession in a random external field [5], turbulent flow on the ocean surface [6], as well as biology (population dynamics [7]), economics (stock market prices [8]) and so on.

3. Random damping

$$\frac{d^2x}{dt^2} + 2\frac{\gamma}{m}\left[1 + \xi\left(t\right)\right]\frac{dx}{dt} + \omega^2 x = 0 \tag{4}$$

The many applications of this model include water waves influenced by a turbulent wind field [9], the Ginzburg-Landau equation with a convective term [10], mean flow passing through a region under study, such as the phase transitions under shear [11], open flows of liquids [12], dendritic growth [13], chemical waves [14], motion of vortices [15], etc.

In response to an external periodic field, the second moment of a harmonic oscillator with random frequency or random damping shows non-monotonic dependence on the noise strength and correletion length (stochastic resonance in linear systems [16]).

4. Here we study still another possibility for introducing randomness in the oscillator equation (1) by introducing a random mass, which is described by the following equation,

$$\left[1 + \xi\left(t\right)\right]\frac{d^2x}{dt^2} + 2\gamma\frac{dx}{dt} + \omega^2 x = 0 \tag{5}$$

or, for a concurrent action with the "Brownian" random force,

$$\left[1 + \xi\left(t\right)\right]\frac{d^2x}{dt^2} + 2\gamma\frac{dx}{dt} + \omega^2 x = \eta\left(t\right) \tag{6}$$

with

$$\left\langle \eta\left(t_1\right)\eta\left(t_2\right)\right\rangle = D_1\delta\left(t_2 - t_1\right) \tag{7}$$

The latter equation describes Brownian motion with adhesion, which means that the molecules surrounding a Brownian particle not only collide with it, but also adhere to the Brownian particle for some random time interval. As a result of this adhesion, the mass of a Brownian particle becomes a random quantity. There are many biological or chemical solutions of particles of different sizes which perform random collisions and adhesion. A comprehensive list of references can be found in [17]. Recent applications of this model include a nano-mechanical resonator which randomly absorbs and desorbs molecules [18].

We restrict our consideration to colored noise $\xi\left(t\right)$ with an exponential correlator,

$$\left\langle \xi\left(t_1\right)\xi\left(t_2\right)\right\rangle = \sigma^2\exp\left[-\lambda\left|t_2 - t_1\right|\right] \equiv D\lambda\exp\left[-\lambda\left|t_2 - t_1\right|\right] \tag{8}$$

where the random variable $\xi(t)$ may take the values $\pm\sigma$, so that $\langle \xi^2(t) \rangle = \sigma^2$, with mean waiting time $(\lambda/2)^{-1}$ in each of these two states. This noise is called dichotomous noise (random telegraph signal). The transition to white noise occurs when $\lambda \to \infty$.

Since an arbitrary function of dichotomous noise is linear, one can write

$$1 + \xi(t) = \frac{1-\sigma^2}{1-\xi(t)}, \tag{9}$$

which transforms Eq. (6) into the following form

$$\left(1-\sigma^2\right)\frac{d^2x}{dt^2} + (1-\xi(t))\left(2\gamma\frac{d}{dt} + \omega^2\right)x = 0 \tag{10}$$

Contrary to all other types of random forces in Eq. (1), the strength of the mass fluctuations is restricted. Due to the positivity of the oscillator mass, the condition $\xi^2 < 1$, and hence $\sigma^2 < 1$, is satisfied in Eq. (10). For the asymptotic case of small oscillations of the mass, $\sigma^2 << 1$, this equation can be rewritten as follows

$$\frac{d^2x}{dt^2} + 2\gamma\frac{d}{dt} + \omega^2 x = \xi(t)\left(2\gamma\frac{d}{dt} + \omega^2\right)x \tag{11}$$

Therefore, small dichotomous fluctuations of mass are equivalent to simultaneous fluctuations of frequency and damping.

In order to find the first moment $\langle x \rangle$, described by the second-order differential equation (6), it is convenient to rewrite this equation in the form of two first-order differential equations,

$$\frac{dx}{dt} = y; \qquad \frac{dy}{dt} = -\xi\frac{dy}{dt} - 2\gamma y - \omega^2 x \tag{12}$$

Multiplying the last equation in (12) by ξ, averaging the original and obtained equations and using the Shapiro-Loginov splitting procedure [19], one obtains [20] a fourth - order differential equation for $\langle x \rangle$.The use of dichotomous noise with $\left\langle \xi(t)^2 \right\rangle = \sigma^2$ allows us to split the hierarchy of higher - order correlation functions and get the expression of $\langle x \rangle$. In the case of white noise, one can compare the results for different types of noise. For additive noise, there is no change to the deterministic case. For the case of random frequency, one obtains a slight change (replacing ω by a renormalized frequency $\sqrt{\omega^2 - \gamma^2}$). For the case of random damping, the damping coefficient γ is replaced by $\gamma(1-2\gamma D)$, and, finally, for the case of random mass, ω and γ are replaced by $\omega\sqrt{1-2\gamma D}$ and $\gamma(1-2\gamma D)$, respectively..

In a similar way, one can find [20] the second moments $\langle x^2 \rangle$ associated with Eq. (6), which is transformed to six equations in six variables, $\langle x^2 \rangle$, $\langle y^2 \rangle$, $\langle xy \rangle$, $\langle \xi x^2 \rangle$, $\langle \xi y^2 \rangle$ and $\langle \xi xy \rangle$. We bring here only the results for $\langle x^2 \rangle$, obtained for some limit cases of equation (6). 1. Stationary solutions ($d/dt ...= 0$): $\langle x^2 \rangle = \frac{D_1}{2\eta\omega^2}$ as for pure Brownian motion. 2. White noise $\xi(t)$: One obtains two equations for $\langle x^2 \rangle$ and $\langle y^2 \rangle$. 3. In the absence of multiplicative noise $(\xi = 0)$, one obtains the same results as those of Brownian motion in a harmonic potential.

The problem of stability of the solutions obtained for the stochastic equation (5) is much more complicated than the calculation of the first moments [21]. Indeed, for a deterministic equation, the stability is defined by the sign of α, found from the solution of the form $\exp(\alpha t)$ of a linearized equation. The situation is quiet different for a stochastic equation. The first moment $\langle x(t) \rangle$ and higher moments become unstable for some values of the parameters. However, the usual linear stability analysis, which leads to instability thresholds, turns out to be different for different moments, making them unsuitable for the stability analysis. The rigorous mathematical analysis of random dynamic systems shows [22] that, similar to the order–deterministic chaos transition in nonlinear deterministic equations, the stability of a stochastic differential equation is defined by the sign of Lyapunov exponents. This means that for stability analysis, one has to go from the Langevin - type equations (2)-(6) to the associated Fokker-Planck equation, which describe properties of statistical ensembles. Here, we restrict our analysis to the case of the small fluctuations of the mass. Introducing the dimensionless time $\tau = \omega t$, Eq. (11) becomes

$$\frac{d^2x}{d\tau^2} + \frac{2\gamma}{\omega}\frac{d}{d\tau} + x = \xi(t)\left(\frac{2\gamma}{\omega}\frac{d}{d\tau} + 1\right)x \tag{13}$$

For stability analysis, it is convenient to introduce into Eq. (13) the variable $z = (dx/dt)/x$, yielding

$$\frac{dz}{d\tau} = \frac{d^2x/d\tau^2}{x} - \frac{(dx/d\tau)^2}{x^2} \equiv \frac{d^2x/d\tau^2}{x} - z^2 \tag{14}$$

In Eq. (13), replacing the variable x by the variable z leads to

$$\frac{dz}{d\tau} + z^2 + \frac{2\gamma}{\omega}z + 1 = \xi\left(\frac{2\gamma}{\omega}z + 1\right) \tag{15}$$

The general identity which connects the Lyapunov exponent $\lambda = \lim_{t\to\infty}[(1/t)\langle \log E \rangle]$ with the stationary average of the solution of Eq. (15), is [23]

$$\lambda = \int_{-\infty}^{\infty} zP_{st}(z)dz \tag{16}$$

where $P_{st}(z)$ is the stationary value of the distribution function $P(z,t)$, and the integral is to be understood to mean the principal value. We continue with the Fokker-Planck equation associated with Eq. (15). The Langevin equation

$$\frac{dz}{d\tau} = A(z) + B(z)\xi \tag{17}$$

is associated with the Fokker-Planck equation of the following form (Stratonovich interpretation) [24],

$$\frac{\partial P(z,\tau)}{\partial \tau} = -\frac{\partial}{\partial z}[A(z)P] + \frac{\Delta}{2}\frac{\partial}{\partial z}B(z)\left[\frac{\partial}{\partial z}B(z)P\right] \tag{18}$$

where J is the constant probability current,

$$B(z) = 1 + \frac{2\gamma}{\omega}; \qquad A(z) = -z^2 - B(z) \tag{19}$$

and

$$\langle \xi(\tau_1) \xi(\tau_2) \rangle = 2\Delta\delta(\tau_1 - \tau_2) \tag{20}$$

The stationary solution of Eq. (18) has the following form [21]

$$\begin{aligned} P_{st} &= P_{sr}(w) = \frac{J\omega}{N\Delta\gamma} (w)^{-c-1} \exp\left[-g\left(w - \frac{1}{w}\right)\right] \\ &* \int_{-\infty}^{w} dx\ x^{c-1} \exp\left[g\left(x - \frac{1}{x}\right)\right] \end{aligned} \tag{21}$$

where

$$c = \frac{\omega}{\Delta\gamma}\left(1 - \frac{\omega^2}{2\gamma^2}\right); \qquad g = \frac{\omega^3}{4\Delta\gamma^3} \tag{22}$$

There is no need to perform an analysis of Eq. (21) since the analogous calculation has been performed for the case of random damping (Eq. (4) yielding [25]

$$\begin{aligned} P_{st}(z) &= \frac{2J\omega_1}{ND} (w)^{-a-1} \exp\left[-b\left(w - \frac{1}{w}\right)\right] \\ &* \int_{-\infty}^{w} dx\ x^{a-1} \exp\left[b\left(x - \frac{1}{x}\right)\right] \end{aligned} \tag{23}$$

with

$$a = 4\gamma D; \qquad b = \frac{\omega}{D} \tag{24}$$

and

$$\lambda = \frac{8\int_0^\infty du\ K_1(8b \sinh u) \sinh[(1-a)u]}{\pi^2 \left[J^2_{2\gamma/D}\left(\frac{4}{D}\right) + Y^2_{2\gamma/D}\left(\frac{4}{D}\right)\right]} \tag{25}$$

where K_1 is a modified Bessel function of the second kind, and J and Y are Bessel functions of the first and second kind, respectively. The Bessel functions are always positive, and the sign of the Lyapunov exponent λ is the same as the sign of the hyperbolic sinh, i.e., the sign of $1 - a$.

Returning to Eq. (23), we conclude that an oscillator with fluctuating mass becomes unstable when $\mid c \mid = (\omega/\Delta\gamma) \mid 1 - \frac{\omega^2}{2\gamma^2} \mid < 1$, i.e., the instability of the fixed point $z = 0$ occurs for the noise strength $\Delta > \omega/\gamma \mid 1 - \frac{\omega^2}{2\gamma^2} \mid$, which reduces to $\Delta > \frac{\omega}{\gamma}$ for $\omega << \sqrt{2}\gamma$ and $\Delta > \omega^3/2\gamma^3$ for $\omega >> \sqrt{2}\gamma$.

We expect that the model of an oscillator with a random mass will find many applications in modern science.

References

[1] Moshinski M and Smirnov Yu F 1996 *The Harmonic Oscillator in Modern Physics* (Harwood, The Netherlands)

[2] Gitterman M 2005 *The Noisy Oscillator: the First Hundred Years, from Einstein until now* (World Scientific)

[3] Mazo R M 2008 *Brownian Motion : Fluctuations, Dynamics and Applicastions* (Oxford Science Publication)

[4] Ishimaru A 1999 *Wave Propagation and Scattering in Random Media* (IEEE Press)

[5] Kubo R 1969 in *Stochastic Processes in Chemical Physics,* ed. Shuler K E (New York: Wiley)

[6] Phillips J M 1977 *The Dynamics of the Upper Ocean* (Cambridge: Cambridge University Press)

[7] Turelli M 1997 *Theoretical Population Biology*(New York: Academic)

[8] Takayasu H, Sato A-H and Takayasu M 1997 *Phys. Rev. Lett.* **79** 966

[9] West B and Seshadri V 1981 *J. Geophys. Res.* **86** 4293

[10] Gitterman M 2003 *Phys. Rev. E* **70** 036116

[11] Onuki A 1997 *J. Phys.: Condens. Matter* **9** 6119

[12] Chomaz J M and Couairon A 1999 *Phys. Fluids* **11** 2977

[13] Hestol F and Libchaber A 1985 *Phys. Scr.* **T9** 126

[14] Saul A and Showalter K 1985 in *Oscillations and Traveling Waves in Chemical Systems* eds Field R J and Burger M (New York: Wiley)

[15] Gitterman M, Shapiro B Ya and Shapiro I 2002 *Phys. Rev. B* **65** 174510

[16] Gitterman M 2005 *Physica A* **352** 309

[17] Lambiotte R and Ausloos M 2005 *Phys. Rev. B* **73** 011105

[18] Portman J, Khasin M, Shaw S W and Dykman M I 2010 *Bulletin of the APS, March Meeting*

[19] Shapiro V E and Loginov V M 1978 *Physica A* **91** 563

[20] Gitterman M and Klyatskin V I 2010 *Phys. Rev. E* **81** 051139

[21] Gitterman M *Physica A* submitted.

[22] Arnold L 1998 *Random Dynamic Systems* (Berlin: Springer)

[23] Mallick K and Peyneau P-E 2006 *Physica D* **221** 72

[24] van Kampen N G 1992 *Stochastic Processes in Physics and Chemistry* (North-Holland)

[25] Leprovost N, Aumaitre S and Mallick K 2006 *Eur. Phys. J. B* **49** 453

Journal of Physics: Conference Series **248** (2010) 012050 doi:10.1088/1742-6596/248/1/012050

Cavity soliton mobility in semiconductor microresonators above laser threshold

Reza Kheradmand, Mansour Eslami

Photonics Group, Research Institute for Applied Physics and Astronomy, Tabriz University, Tabriz, Iran

E-mail: r_kheradmand@tabrizu.ac.ir

Abstract. In this paper we investigated cavity soliton switching in a VCSEL type semiconductor laser which has been driven by homogeneous holding beam. Using quadratic fitting of the gain curve we studied the incoherent switching technique for the controlling of cavity solitons in a VCSEL-type laser and we show the cavity soliton spontaneous mobility in this type of injection.

1. Introduction

Cavity Solitons (CS), which are classified as localized structures, are isolated intensity peaks formed on a homogeneous dark background of radiation [1]. In optical systems that can exhibit bistability in some parameter range, one can find a region in which the homogeneous solution is present simultaneously with pattern solution. In this case the CS can be created by applying a transient Gaussian beam [2], or similarly by local carrier injection [3]. These bright spots are then self-sustained, in that they are stable once created by either method. Their ability to remain stationary in time and space, and the possibility of switching them on and off, make them a suitable candidate for applications in information technology. When making use of them in optical storage, they appear as a bit of information, and when carrying information processing tasks they appear as logic gates [4-6]. In account of being able to control the position and mobility, performance of these structures can be parallel. For instance, in the case of logic gates, it is possible to locate any number of CS arrays in which logical operations can be performed independently.

Theoretical prediction of CSs in optical resonators is mainly based on the interplay among dispersive/absorptive nonlinearities, paraxial diffraction, and dissipation/feedback [7]. In most cases, and ours as well, the energy of the cavity is supplied by an external plane-wave field. In addition, we make use of an electric current as the pump source to provide laser threshold, beyond which the dynamics of the system is explored. Experimentally, CSs have been realized both above and below laser threshold in VCSEL (Vertical Cavity Surface Emitting Laser) [8, 9].

In this paper, by inserting a parameter which represents the quadratic fitting of the gain curve, we proposed a more realistic description for these kinds of systems, besides; we were able to write CSs in this system by two methods, i.e. both coherent and incoherent. However, we observed a spontaneous motion of the CSs which has been analyzed due to its novelty.

In section 2, full description of the system and the governing dynamical equations are presented. Section 3.1, is devoted to the solutions of the system and instability analysis of them. In section 3.2, the results obtained from simulations will be considered, including the switching process and description of movement of the CS.

2. Model

Cavity Solitons are usually produced by means of optical resonators containing nonlinear medium. But as to miniaturization purposes and efficient use in circuits due to fast response, it is desired to use semiconductor materials as the nonlinear medium. The VCSEL type resonator which is considered in this paper contains GaAs/AlAs multiple quantum well nanostructure as the active medium which is shown schematically in figure 1. Bragg reflectors are used at both ends of the cavity and the energy is provided to the system by a broad-area and stationary holding beam (HB) and also by an electric current.

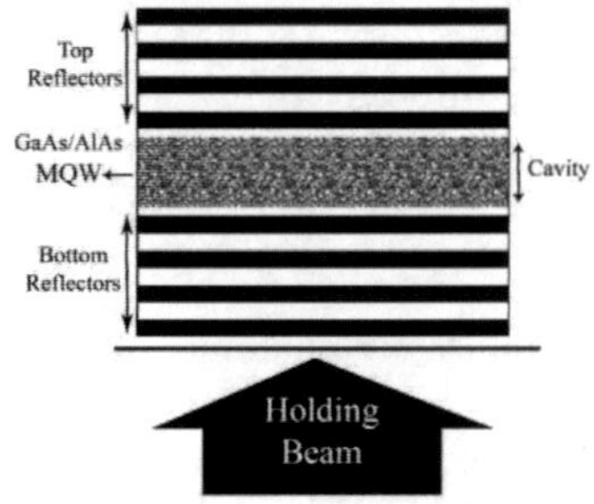

Figure 1. Schematic illustration of the microresonator with multiple quantum-well nanostructure active material filling the cavity.

We used full set of Maxwell-Bloch equations to describe the dynamics of the system, however, what makes this system distinguished from previous models is the introduction of a parameter to take into account the quadratic fitting of the gain curve, β, in order to make the model more realistic. So, the full set of effective Maxwell-Bloch equations describing the system will be:

$$\begin{cases} \partial E/\partial t = \sigma[P + E_I - (1 + i\theta)E + i\nabla^2 E] \\ \partial D/\partial t = -b[1/2\,(E^*P + P^*E) + D - J - d\nabla^2 D] \\ \partial P/\partial t = \Gamma(1 + i\Delta)[(1 - i\alpha)D(1 - \beta D)E - P] \end{cases} \tag{1}$$

where E, P are the slowly varying envelopes of electric field and of the effective macroscopic polarization, and D is the population variable proportional to the excess of carries with respect to transparency. It has to be mentioned that time is scaled to the dephasing rate τ_d of microscopic dipoles. The decay rates are defined as $\sigma = \tau_d/\tau_p$, $b = \tau_d/\tau_c$ where τ_p, τ_c are the photon life time and carrier recombination time, respectively. The difference between the cavity longitudinal mode frequency and of the injected field multiplied by τ_p, detuning, is represented by θ, which can be one of the important parameters. Amplitude of the injected field, holding beam, is denoted by E_I, and J is the pump current which is normalized in such a way that the threshold for the system can be obtained from $J_{th} = \frac{1-\sqrt{1-4\beta}}{2\beta}$, so for our model it will be $J_{th} = 1.171$ with $\beta = 0.125$. Finally d is the diffusion coefficient for carriers.

The two real parameters, Γ and Δ, determines the shape of the effective susceptibility and assumed to depend on the population variable, that can be derived phenomenologically. Taking into consideration the quadratic fitting of the gain curve, we set $\Gamma(D) = \frac{2.308D+1.206}{\sqrt{1+\alpha^2}}$, $\Delta = -\alpha + \frac{2\delta}{\Gamma}$. $\Gamma(D)$ is related to the gain line width, $\delta(D)$ is the detuning between the reference frequency and the frequency where the gain is maximum. The values used for physical parameters throughout this paper are summarized in table.1.

Table 1. The values for physical parameters which are kept fixed in calculations and simulations.

d	α	b	σ	τ_d	τ_p
0.052	4	10^{-4}	4×10^{-2}	$100\,fs$	$2.5\,ps$

Thus, the free parameters of the system will be the intensity of the HB, the pump parameter J and the detuning θ.

3. Results and discussion

3.1. *Homogeneous solution and linear stability analysis*

Homogeneous stationary solutions of the equations are obtainable by setting $\partial/\partial t = 0$ and $\nabla^2 = 0$, we arrive at:

$$|Ei|^2 = |Es|^2\{(1 - Ds + \beta Ds^2)^2 + (\theta + \alpha Ds - \alpha\beta Ds^2)^2\}, \quad (2)$$

$$Ds = \frac{(1 + |Es|^2) - \sqrt{(1 + |Es|^2)^2 - 4|Es|^2\beta J}}{2|Es|^2\beta}. \quad (3)$$

As can be seen in the figure 2, the homogeneous stationary solution exhibits bistability in some parameter region, and this allows the system to perform switching operations.

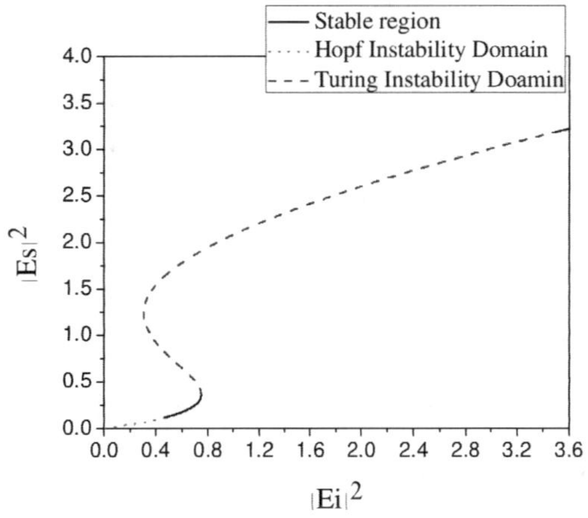

Figure 2. Stationary intensity $|E_s|^2$ versus the injected intensity $|E_I|^2$, which exhibits bistability for some specific range of parameters. $J = 1.288, \theta = -2, \alpha = 4$.

The linear stability of the homogeneous stationary solution is analyzed by studying the response of the system to small fluctuations around the steady state. We carry linear stability analysis by adding an ansatz of the form below and determine the instability domains for Turing and Hopf instabilities.

$$\begin{cases} E = Es + \delta Es \; e^{\lambda t - i(k_x x + k_y y)} \\ P = Ps + \delta Ps \; e^{\lambda t - i(k_x x + k_y y)} \\ D = Ds + \delta Ds \; e^{\lambda t - i(k_x x + k_y y)} \end{cases} \quad (4)$$

The set of equations which are obtained from applying such a fluctuation gets a nontrivial solution only if the eigenvalue λ satisfies equation below:

$$\lambda^5 + a_4\lambda^4 + a_3\lambda^3 + a_2\lambda^2 + a_1\lambda + a_0 = 0$$

λ in general is a complex number, however, we put Real part of λ zero and separate Real and Imaginary parts of the resultant equation. Thus we obtain the Turing and Hopf instability conditions. The instability curves for pump current of about 10% above threshold are shown in figure 3, 4.

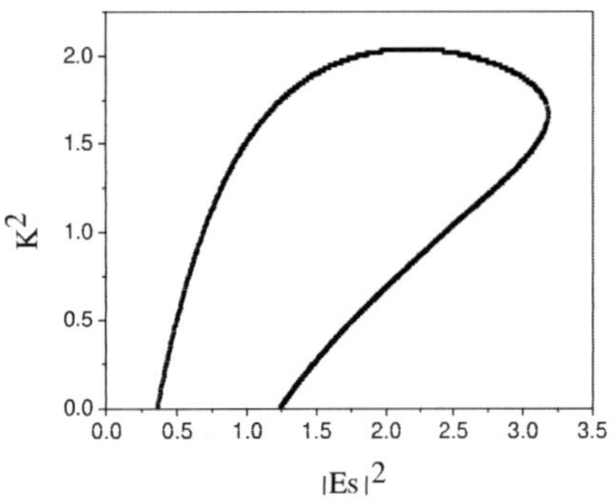

Figure 3. Instability domain for the Turing instability, *K* is the modulus of the transverse wave vector against which the stability of homogeneous solution is analyzed. $J = 1.288, \theta = -2,\ \alpha = 4.$

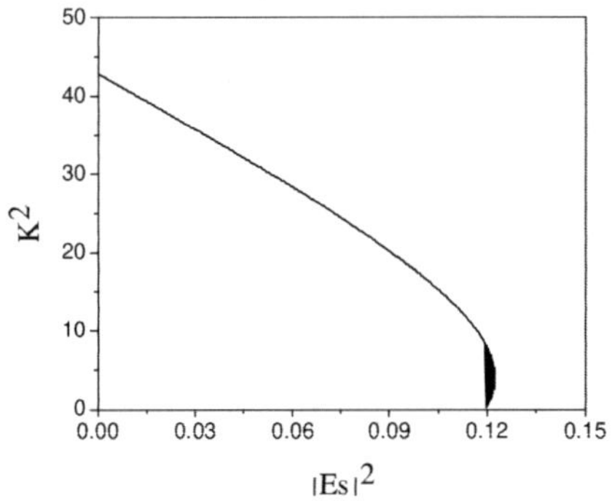

Figure 4. Instability domain for the the Hopf instability, *K* is the modulus of the transverse wave vector against which the stability of homogeneous solution is analyzed. $J = 1.288, \theta = -2,\ \alpha = 4.$

3.2. *Simulations*

As it is apparent from the set of equations used for describing the system, they involve both time derivative and Laplacian terms. Thus it is needed to use a specific method of integration, named split-step method. We used split-step method with periodic boundary conditions which consists in separating the algebraic and the Laplacian terms on the right-hand part of the equations; the algebraic term is integrated using a Runge-Kutta algorithm, while for the Laplacian operator a fast Fourier transform is adopted. This implies that the number of points for each side of the grid must be a power of 2 and we mostly assumed a 64 × 64 grid.

The next step is to switch on a CS and then analysing its behaviour in transverse plane. To do this we have several alternatives including Coherent and Incoherent switching methods. In Coherent method one uses a Gaussian beam as the writing beam in phase with holding beam. In contrast, the Incoherent method uses local injection of carriers to enhance the light emission rate. In this regard we preferred incoherent switching technique to coherent one because of several reasons. First of all, this technique requires shorter time interval compared with coherent switching. Second, and more important one, is the advantage of possibility of switching on a CS with almost any energy able to generate carriers in that specific region in transverse plane.

Figure 4 shows a typical switching on of a CS with injection field intensity of about $|E_I|^2$=0.6. The process of injection started at time 4 tu and continued until 4.92 tu. This is done through using a Gaussian function of the form $E_{inj} = -e_{inj} \exp -(\frac{Re(i-i_{loc})^2+Re(j-j_{loc})^2}{2Re(w)^2})$, where e_{inj} is the amplitude, i_{loc}, j_{loc} is the coordinate of injection point and w is the width of injection. As can be seen in Figure 5, simulations show that the cavity soliton is created and is stable even in absence of injection process.

Figure 6 shows the bistability curve together with solitonic branch which is indicated by filled triangles.

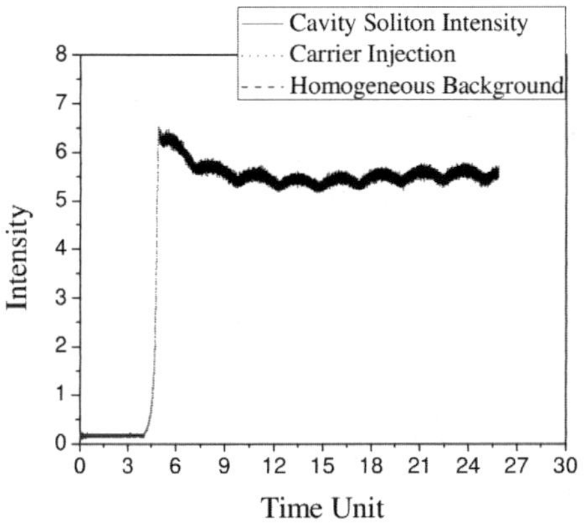

Figure 5. Switching on a cavity soliton in presence of holding beam intensity of about $|E_I|^2$=0.6.

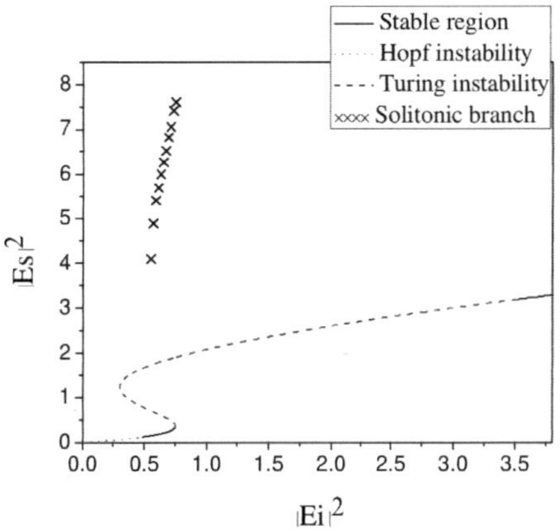

Figure 6. Bistability curve and solitonic branch for the same parameters as the previous graphs.

Of all the phenomena which are associated with CSs, mobility of them in transverse direction is of high interest. Cavity Solitons can have some instabilities, e.g. start moving, breathing, and oscillating. Moving of the CSs in some specific systems has been appointed to instabilities such as Hopf which, in the case of a system with thermal effects, arises as the result of presence of thermal effects [10]. For this system simulations show a spontaneous motion of CS along vertical direction with a velocity of about $\mathbf{0.442}$, in terms of time unit. Results implies a roughly periodic motion involving an interval of time in which the CS is stable in its position and then it undergoes a back and forth motion between its current position and the lower position which is due to the noise, and eventually it rests back on the lower position. This fact is shown in figure 7.

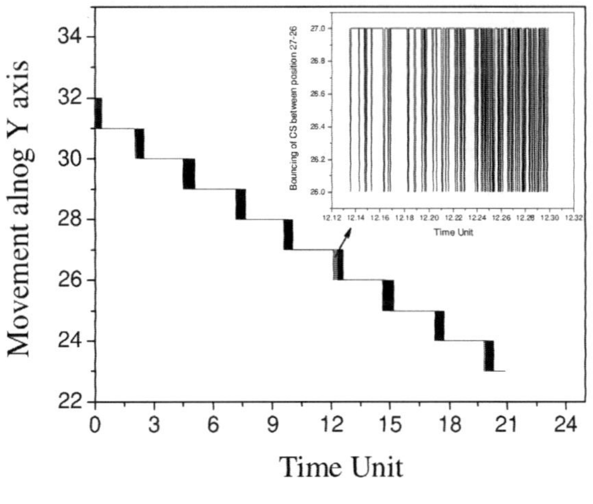

Figure 7. Movement of CS along vertical direction in presence of noise. The inset shows the effect of noise which results in bouncing of CS between two places before coming a step down.

Journal of Physics: Conference Series **248** (2010) 012050 doi:10.1088/1742-6596/248/1/012050

4. Conclusion

In this paper, first the electric field, carrier density and polarization equations have been investigated by quadratic fitting of the gain curve in semiconductor microresonator above laser threshold. The existence of cavity solitons in a driven vertical cavity semiconductor lasers has been demonstrated theoretically. We have studied the current injection technique for the control of a CS, which is called incoherent injection. We have analyzed the switching dynamics and its behaviour which have been observed. Furthermore, a spontaneous mobility in arbitrary direction has been found.

References

[1] S. Barland, J. R. Tredicce , M. Brambilla, L. A. Lugiato, S. Balle, M. Giudici, T. Maggipinto, L. Spinelli, G. Tissoni, T. Knodl, M. Miller, And R. Jager, "Cavity solitons as pixels in semiconductor microcavities," *Nature*, vol. 419, pp. 699–702 , London, 2002.

[2] X. Hachair, L. Furfaro, J. Javaloyes, M. Giudici, S. Balle, and J. Tredicce, "Cavity-solitons switching in semiconductor microcavities," *Phys. Rev. A* 72,013815, 2005.

[3] S. Barbay, Y. Ménesguen, X. Hachair, L. Leroy, I. Sagnes, and R. Kuszelewicz, 'Incoherent and coherent writing and erasure of cavity solitons in an optically pumped semiconductor amplifier,' OPTICS LETTERS, Vol.31, No.10, 2006.

[4] D. Gomila, M. A. Matias, and P. Colet *Phys. Rev. Lett.* 94, p. 063905, 2005.

[5] D. Gomila, A. Jacobo, M. A. Matias, and P. Colet *Phys. Rev. E* 75, p.026217, 2007.

[6] A. Jacobo, D. Gomila, M. A. Matias, and P. Colet, *Phys. Rev. A* 78, 053821, 2008.

[7] X. Hachair, F. Pedaci, E. Caboche, S. Barland, M. Giudici, R. Tredicce, F. Prati, G. Tissoni, R. Kheradmand, L. A. Lugiato , I. Protsenko and M. Brambilla, *IEEE* J, vol 12, No 3, 2006.

[8] Barland, S., et al., 'Cavity solitons as pixels in semiconductor microcavities," *Nature*, Vol.419, 699-702, 2002.

[9] Hachair, X., et al., 'Cavity solitons in a driven VCSEL above threshold," *Journal of Selected Topics in Quantum Electronics*, Vol.12, 339-351, 2006.

[10] L. Spinelli, G. Tissoni, L. A. Lugiato and M. Brambilla, "Thermal effects and transverse structures in semiconductor microcavities with population inversion," Phys. Rev. A 66, 023817, 2002.

[11] S. Barland, M. Brambilla, L. Columbo, L. Furfaro, M. Giudici, X. Hachair, R. Kheradmand, L. A. Lugiato, T. Maggipinto, G. Tissoni, and J. Tredicce, "Cavity solitons in a VCSEL: Reconfigurable micro pixel arrays," *Euro-phys. News*, vol. 34, pp. 136–139, 2003.

[12] Giovanna Tissoni, Lorenzo Spirielli, Luigi A. Lugiato, Massimo Brambilla, 'Thermal and electronic nonlinearities in semiconductor cavities,' *Proceedings of SPIE*, Vol. 4283, 2001.

[13] S. Barbay, Y. Ménesguen, X. Hachair, L. Leroy, I. Sagnes, and R. Kuszelewicz, 'Incoherent and coherent writing and erasure of cavity solitons in an optically pumped semiconductor amplifier *OPTICSLETTERS*, Vol.31, No.10, May15, 2006.

[14] S. Barbay and R. Kuszelewicz, 'Physical model for the incoherent writing/erasure of cavity solitons in semiconductor optical amplifiers' *OPTICS EXPRESS,* Vol. 15, No. 19, 2007.

Journal of Physics: Conference Series **248** (2010) 012051 doi:10.1088/1742-6596/248/1/012051

Charge transport in the α-helix proteins

D. Čevizović†, S. Galović†, S. Zeković†, Z. Ivić†

† Institute of Nuclear Sciences "'Vinča"', University of Belgrade, Serbia

E-mail: cevizd@vinca.rs

Abstract. In this paper we investigate charge transport in the α- helix proteins. It is considered that charge carriers are partially dressed small polarons, that are formed by electron self-trapping in the weak coupling limit. Polaron properties are considered by applying mean-field theory based upon the modified Lang-Firsov transformation and the Bogolyubov theorem. The transport properties have been calculated on the basis of the Holstein MC model. The obtained results are compared with the results of previous models.

1. Introduction

In the late fifties of the 20th century, Albert Szent-Gyorgy referred to demand of employment of ideas and methods of quantum physics in the problem of biochemistry of cancer. He believed that the cancer was related to the changes of electron transport properties at the molecular scale. The first serious examination of charge and energy transport in protein macromolecules at the quantum level was proposed in mid seventies by Davydov and co–workers [1]. The essence of their theory is the assumption that an extra electron or energy quanta released in the hydrolysis of adenosine triphosphate can be captured by the protein molecules and then transported along the polypeptide chain in a soliton form. However, due to the lack of direct experimental evidence of soliton existence in these substances, Davydov's idea has been for a long time considered just an interesting theoretical concept.

The situation changed at the beginning of the eighties, when G. Careri and A. Scott suggested [2] that appearance of the so–called unconventional amide–I band, observed in a previous year experiment by Careri in crystalline acetanilide (ACN) [3], might be explained in terms of Davydov's soliton theory. However, according to the general theory of self–trapping (ST) phenomenon, the original Davydov's concept should be substantially modified, and as shown by Alexander and Krumhansl [4], the whole concept must be reinterpreted in terms of the small–polaron (SP) model. Because of the ACN hydrogen–bonded structure that is similar to the structure of the α–helix proteins, it is believed that it provides a good model for studing ST phenomena in protein macromolecules. It means that transport processes in hydrogen–bonded biological macromolecules could be investigated within the standard SP models based on the Lang–Firsov (LF) transformation. As a consequence, these concepts rely on the assumption of fully dressed excitations. However, the values of basic energy parameters in hydrogen–bonded macromolecular chains fall into nonadiabatic weak coupling regime [5]. It means that the true states of system should be partially dressed SP states rather than the fully dressed ones. This fact refers to the necessity of using some kind of partial dressing strategy, or variational approach [5, 6, 7]. Another indication that it is necessary to approve the classic SP model in order to apply it to hydrogen–bonded macromolecules is the fact that the position of the anomal amide–I peak in IR spectra of ACN is temperature dependent. This fact cannot be explained within the scope of standard SP theories and the LF approach [7].

In this paper we offer a charge transport model in hydrogen–bonded biological macromolecules. It is based on the famous Holstein molecular crystal model (MCM) [8], and improved treatment of the ST, based on the modified Lang–Firsov (MLF) transformation and the Bogolyubov theorem [6, 9, 10, 11]. The results that our model predicts are compared with the results predicted by transport model based on the standard LF transformation.

2. Theoretical model

We base our further analysis on the results of the paper [7, 11] where, employing the nonadiabatic weak coupling approximation, we obtained the following expression for the optimal form of variational parameter:

$$f_q = \frac{F_q^*}{\hbar\omega_q + 2J\mathrm{e}^{-S(T)}(2\bar{n}_q + 1)(1 - \cos qR_0)}. \tag{1}$$

Its magnitude defines the degree of dressing and the character of ST states. In the above equation, $S(T) = \frac{1}{N}\sum_q |f_q|^2 (2\bar{n}_q + 1)(1 - \cos qR_0)$ is the electron band narrowing factor, J is the intersite transfer integral, ω_q is the phonon frequency, R_0 is the lattice constant, $F_q = F_{-q}^*$ is the electron–phonon coupling parameter, and $\bar{n}_q = 1/(\mathrm{e}^{\hbar\omega_q/k_BT} - 1)$ is the phonon average number. Incoherent neutron scattering studies of ACN [12] indicate that acoustic modes are not involved in vibron ST, and that the source of vibron ST in ACN is its coupling with optical phonons only. Additionally, recent pump–probe experiments seem to show that a unique phonon of frequency approximately equal to $\omega_0 \approx 50$ cm^{-1} is responsible for the self–trapping phenomenon [13]. In that case, the vibron (electron)–phonon coupling parameter is q–independent, and has the value $F \approx 25$ cm^{-1} [5, 14, 15]. Numerical values of the intersite transfer integral in ACN for amide–I mode is about $J \approx 5$ cm^{-1} [4, 5, 14, 15]. In the case of electron ST in proteins, this parameter ranges from $\sim$ 1cm^{-1} to $\sim$ 10cm^{-1} [16], depending on protein types, as well as on model based experimental estimations. In this paper we adopt numerical value that is characteristic of the vibron ST in ACN.

The character of the ST states is determined by relationships among the values of the basic energy parameters of the system: characteristic phonon energy $\hbar\omega_0$, quasiparticle energy bandwidth $2J$, and SP binding energy $E_b = \frac{1}{N}\sum_q \frac{|F_q|^2}{\hbar\omega_q}$. In particular, just two parameters characterize the ST in these media: adiabatic parameter $B = \frac{2J}{\hbar\omega_0}$ and the coupling constant $S = \frac{E_b}{\hbar\omega_0}$. In nonadiabatic limes (when $B << 1$) where quasiparticle and lattice deformation form new entity, we have "'dressed"' quasiparticle with new effective mass, and with reduced energy band. The above mentioned system parameter values that characterize ACN suggest that the vibron (electron) ST would result in the creation of the weakly dressed (S=0,25) nonadiabatic (B=0,16) small polaron.

In the next step we shall explore how the above result affects charge conductivity calculated within the conventional theories of SP transport [8, 10, 17, 18]. Conventional theories of SP transport predict substantially different character of SP motion in the low and high temperature regimes [8, 10, 18]. The first one is the coherent (band–like) SP motion appearing at low temperatures. The increase in temperature gradually destroys the polaron band and SP motion attains the character of random jumps between the neighbouring sites (incoherent, hopping motion).

Improving the Holstein results [8, 19, 20] using (Eq.1), in nondispersive phonon case we have the following relations for the hopping and band conductivities:

$$\frac{\sigma_{\text{hopping}}}{\sigma_0} = \frac{B^2}{4\tau}\sqrt{\frac{\pi\cosh\left(\frac{1}{2\tau}\right)}{S(\tau)}}\,\mathrm{e}^{-2S(\tau)\tanh\left(\frac{1}{4\tau}\right)\tanh\left(\frac{1}{2\tau}\right)} \tag{2}$$

$$\frac{\sigma_{\text{band}}}{\sigma_0} = \frac{1}{\tau}\sqrt{\frac{S(\tau)}{\pi\cosh\left(\frac{1}{2\tau}\right)}}\,\mathrm{e}^{-2\frac{S(\tau)}{\cosh\left(\frac{1}{2\tau}\right)}} \tag{3}$$

where $\sigma_0 = \frac{n_e e^2 R_0^2}{\hbar}$, n_e is the carrier concentration, $\tau = \frac{k_B T}{\hbar\omega_0}$ is normalized temperature, and the band narrowing factor has the form $S(\tau) = S\mathrm{cotanh}\left(\frac{1}{2\tau}\right)\left\{1 + 2B\mathrm{e}^{-S(\tau)}\mathrm{cotanh}\left(\frac{1}{2\tau}\right)\right\}^{-3/2}$

3. The obtained results

Figs. 1a and 1b graphically present the temperature dependence of the polaron band narrowing factor, for a few values of S and B.

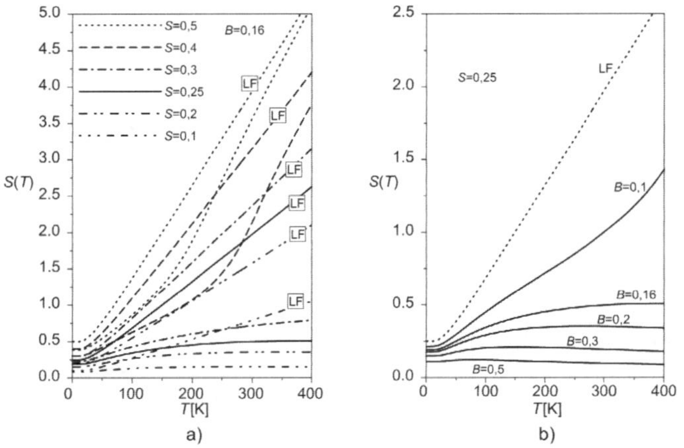

Figure 1. Temperature dependence of the SP band narrowing factor versus temperature for a few values of S and B = const. (Fig. 1a), and a few values of B and S = const. (Fig. 1b).

As it can be noticed, a variational approach based on MLF transformation predicts that the temperature dependence of $S(\tau)$ is quite different from $S(\tau)$ predicted by models based on the LF approach. The obtained values are significantly lower than the values predicted by standard SP theories, especially for the lower values of the coupling constant S (weak coupling limit) (Fig.1a), higher values of adiabatic constant B (Fig.1b) and at higher system temperature. For lower values of S there is a different temperature behaviour of $S(\tau)$, where, with increase of temperature after abrupt increasing of the band narrowing factor, a slow increase occurs. The obtained results suggest that the SP dressing in nonadiabatic weak coupling limit is lower than those predicted by the LF approach, and consequently SP must have lower values of effective mass (which is proportional to $\mathrm{e}^{S(\tau)}$), and higher mobility and electric conductivity. This prediction is confirmed by the results showed in Figs. 2a and 2b.

System parameter values that characterize ACN and hydrogen–bonded macromolecular chains imply that variational approach based on the MLF transformation can provide better description of SP in these systems [7, 11]. As a consequence, SP that may be formed in these structures are partially dressed, and have lower values of effective mass compared with the values that predict traditional SP theories. This fact implies that the SP mobility and electric

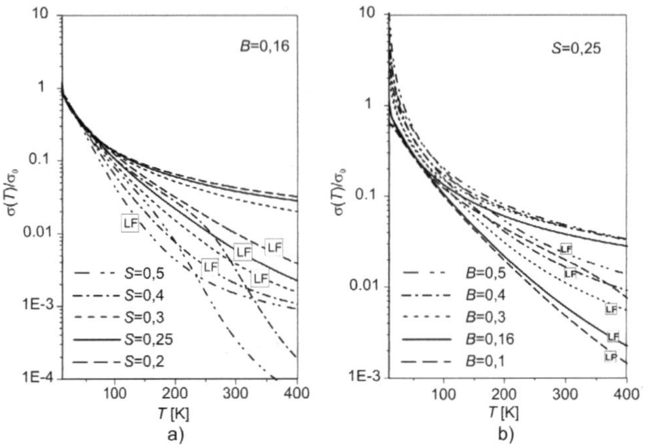

Figure 2. Temperature dependence of electric conductivity for a fixed value of B (Fig. 2a), and a fixed value of S (Fig. 2b).

conductance are greater than the values predicted by standard SP theories based on the LF approach, esspecially in the high temperature regime, and points to a possible significant role of self–trapping mechanism in charge and energy transport in biological macromolecules.

References

[1] Davydov A. S., *J. Theor. Biol.*, **38.**, pp.559, (1973); Davydov A. S., *Phys. Scr.*, **20.**, pp.387, (1979); Davydov A. S., *Phys. D*, **3.**, pp.1, (1981);
[2] Careri G., Buonotempo U., Carta F., Gratton E., Scott A. C., *Phys. Rev. Lett.*, **51.**, pp.304, (1983); Careri G., Buonotempo U., Galluzzi F., Scott A. C., Gratton E., Shyamsunder E., *Phys. Rev. B*, **30.**, pp.4689, (1984);
[3] Careri G. in: *Haken H., Wagner M. (Eds.), Cooperative Phenomena*, Springer-Verlag, Berlin, pp.391, (1973);
[4] Alexander D. M. and Krumhansl J. A., *Phys. Rev. B*, **33.**, pp.7172, (1986);
[5] Brown D. W., Ivić Z., *Phys. Rev. B*, **40.**, pp.9876, (1989);
[6] Emin D., *Adv. Phys.*, **22.**, pp.57, (1973;)
[7] Željko Pržulj, Dalibor Čevizović, Slobodan Zeković, Zoran Ivić, *Chem. Phys. Lett.*, **462**, pp.213, (2008);
[8] T. Holstein, *Ann. Phys.* (N.Y.), **8**, pp.343, (1959);
[9] Davydov A. S., *Teoriya tverdogo tela* Nauka, Moskva, 1976;
[10] Lang I. G., Firsov Yu. A., *Zh. Eksp. Teor. Fiz.*, **43.**, pp.1843, (1962), (Eng. Transl.: *Sov. Phys. JETP*, **16**, pp.1301, (1963));
[11] D. Čevizović, S. Zeković, Z. Ivić, On the possible role of small polarons in the charge and energy transport in the α–helix proteins, *International Bogolyubov Conference*, (2009); D. Čevizović, S. Zeković, Z. Ivić, *Physics of Elementary Particles and Atomic Nuclei, to be published*;
[12] M. Barthes, *J. Mol. Liquid*, **41**, pp.143-163, (1989);
[13] J. Edler, P. Hamm, A. C. Scott, *Phys. Rev. Lett.*, **88**, pp.067403, (2002);
[14] A.C.Scott, *Phys. Rep.*, **217**, pp.1, (1992);
[15] Peter Hamm, Julian Edler, *Phys. Rev. B*, **73**, pp.094302, (2006);
[16] Ji–Hoon Park, Han–Yong Choi, and Esther M. Conwell, *J. Phys. Chem. B*, **108**, pp.19483, (2004); E. M. Conwell, P. M. McLaughlin, and S. M. Bloch, *J. Phys. Chem. B*, **112**, pp.2268, (2008); Victor M. Kucherov, Colin D. Kinz-Thompson, and Esther M. Conwell, *J. Phys. Chem. C*, **114**, pp.1663, (2010);
[17] L. Friedman, *Phys. Rev. A*, **135**, pp.A233, (1964);
[18] I. J. Lang, Yu. A. Firsov, *Fiz. tverdogo tela*, **5**, pp.2799, (1963), (Eng. Transl.: *Sov. Phys. Solid State*, **5**, pp.2049, (1964));
[19] Dalibor Čevizović, Slobodanka Galović, Slobodan Zeković, Zoran Ivić, *Phys. B*, **404**, pp.270, (2009);
[20] Zoran Ivić, Slobodan Zeković, Dalibor Čevizović, Dragan Kostić, Gradimir Vujičić, *Phys. B*, **362**, pp.187, (2005);

International Conference on Theoretical Physics Dubna-Nano 2010 IOP Publishing
Journal of Physics: Conference Series **248** (2010) 012052 doi:10.1088/1742-6596/248/1/012052

Mobility in epitaxial GaN: Limitation of electron transport due to dislocation walls

S. Krasavin

Joint Institute for Nuclear Research, Bogoliubov Laboratory of Theoretical Physics, 141980 Dubna, Moscow Region, Russia

E-mail: krasavin@theor.jinr.ru

Abstract. A theoretical model to explain the sharp transition to low-mobility regime in n-type GaN layers is proposed. We show that this peculiarity of the mobility versus free carrier concentration can be interpreted with the well-known Read model when the scattering due to charged dislocation walls is included in consideration. An agreement with experiments is found above the critical value of carrier density.

1. Introduction

It is well known that the extended defects of different types (mainly dislocations and dislocation substructures) frequently occur in thin semiconducting films and layers. In this context, the problem of their effect on carriers transport is of considerable present interest, both for theoretical and practical reasons [1].

Recently, these investigations have been stimulated additionally by the reason of new promising applications of GaN epitaxial films in optoelectronics [2, 3] . However, the performance of the devices based on GaN is limited by threading dislocations (exclusive of recently synthesized materials [4, 5]) with large densities $10^8 - 10^{11}$ cm^{-2}, which are result from the large lattice mismatch between epilayer and substrate [6, 7, 8]. To improve the device characteristics, the effect of dislocations on the mobility should be studied.

In this paper we show that some peculiarities in electron mobility at room temperature in application to GaN epitaxial films can be explained in the framework of the well-known Read approach [9, 10] . Namely, it is explained the sharp mobility drop at some critical carrier density which, in turn, depends on the dislocation substructure [11, 12]. The total transverse mobility μ^* versus carrier density n is calculated when additional scattering due to electrostatic potential of row of dislocations or dislocation wall exists along with other mechanisms. The reason to consider dislocation walls is the fact found in most experiments that threading dislocations concentrate at subgrain boundaries lining up in row [11, 13]. In this picture, at some conditions (See discussion below) the mobility can be thermally activated because of the charged dislocation barriers present. The equation for the potential barriers is obtained here based on the model proposed in [14].

2. Electrostatic potential of dislocation wall

Let the wall of negatively charged dislocations be oriented along the y axis with the distance between them equal to D, and the lines of each dislocation are directed along z axis. The negative charge in the dislocation line is compensated by a cylindrical positive space-charge region, with radius R around any dislocation. Thus, we have the periodically ordered cylinders with radius R. The Poisson's equation in this case in the Schottky approximation can be written as

$$\frac{\partial^2\phi(x,y)}{\partial x^2}+\frac{\partial^2\phi(x,y)}{\partial y^2}=-\frac{e}{\epsilon\epsilon_0}\Big[(N_d^+-N_a^-)\theta(L(y)-|x|)-\frac{f\delta(x)}{c}\sum_{n=-\infty}^{\infty}\delta(y-nD)\Big], \quad (1)$$

$$\theta(\xi)=\begin{cases}1, & \xi>0\\ 0, & \xi\leq 0\end{cases}$$

where N_d^+, N_a^- are the volume doping and unintentional acceptor densities, respectively. $D=b/2\sin\frac{\alpha}{2}$, b is the absolute value of the Burgers vector, f is the fraction of sites of dislocation that are occupied, ϵ is the dielectric constant, $L(y)=\sqrt{R^2-y^2}$ is a periodic function along the y axis which is the screening length where potential $\phi(x,y)$ is not equal to zero (by definition, $\phi(x,y)=0$ when $|x|\geq L(y)$). Notice that $\phi(x,y)$ is even function on y with the period D, and can be represented as a Fourier series

$$\phi(x,y)=\phi_0(x)+\sum_{n=1}^{\infty}\phi_n(x)\cos(\frac{2\pi ny}{D}), \quad (2)$$

where the expansion coefficients by Eq.(1) satisfy the equations

$$\frac{d^2\phi_0(x)}{d^2x}=-\Big(\frac{e(N_d^+-N_a^-)}{\epsilon\epsilon_0}a_0(x)-\frac{2ef\delta(x)}{cD\epsilon\epsilon_0}\Big), \quad (3)$$

$$\frac{d^2\phi_n(x)}{d^2x}-A_n^2\phi_n(x)=-\Big(\frac{e(N_d^+-N_a^-)}{\epsilon\epsilon_0}a_n(x)-\frac{2ef\delta(x)}{cD\epsilon\epsilon_0}\Big), \quad (4)$$

$$A_n=\Big(\frac{2\pi n}{D}\Big)\Big(\frac{2R}{D}+\frac{\sin(\frac{4\pi nR}{D})}{2\pi n}\Big)^{1/2},$$

$$a_n(x)=\frac{2}{D}\int_{-R}^{R}d\xi\theta(L(\xi)-|x|)\cos(\frac{2\pi n\xi}{D}).$$

The solutions for $\phi_0(x)$ and $\phi_n(x)$ can be found using Green functions for Eqs.(3),(4) when $R=\sqrt{f/\pi c(N_d^+-N_a^-)}$ (this equation can be obtained from $\phi(x,y)=0$ when $|x|\geq L(y)$).

$$\phi_0(x)=-\frac{e(N_d^+-N_a^-)}{D\epsilon_0\epsilon}\Big(2Rx^2+\frac{4R^3}{3}\Big)+\frac{ef}{cD\epsilon_0\epsilon}|x|, \quad (5)$$

$$\phi_n(x)=\frac{2e(N_d^+-N_a^-)}{A_n^2\epsilon_0\epsilon\pi n}\sin\Big(\frac{2\pi nR}{D}\Big)-\frac{e(N_d^+-N_a^-)\alpha_n}{A_n\epsilon_0\epsilon D}\cosh(A_nx)-\frac{ef}{cDA_n\epsilon_0\epsilon}\exp(-A_n|x|), \quad (6)$$

$$\alpha_n=\int_{-R}^{R}d\xi\cos\Big(\frac{2\pi n\xi}{D}\Big)\exp(-A_nL(\xi)).$$

3. Mobility

The barrier-limited mobility, generally, takes the form [15]

$$\mu_{GB} = \Big(\frac{eL}{\sqrt{8kT\pi m^*}}\Big)\exp(-V(0, y = R)/kT), \tag{7}$$

where e is absolute value of the electron charge, m^* its effective mass, k is the Boltzmann constant, T is the temperature, L is the average grain size, $V(0, y = R) = -e\phi(0, y = R)$ is the value of the two-dimensional potential barrier caused by charged dislocations at $x = 0$. The explicit form of this potential can be derived from Eqs.(2),(5),(6). Notice that the potential $V(0, y = R)$ has quite simple form when $2R \approx D$ [14]

$$V(0, y = D/2) = -e\phi(0, y = D/2) = \frac{4e^2(N_d^+ - N_a^-)R^3}{3\epsilon_0\epsilon D} - \frac{e^2 f}{2c\pi\epsilon_0\epsilon}\ln 2. \tag{8}$$

Thus, the total effective mobility takes the form [16, 15]

$$\frac{1}{\mu^*} = \frac{1}{\mu_1} + \frac{1}{\mu_{GB}}, \tag{9}$$

where μ_1 is the mobility caused by different mechanisms of scattering in GaN films except charged dislocation walls. We have used the explicit form for μ_1 and the main parameters for GaN from paper [17]. The total mobility μ^* vs free carrier density n is defined by the Eqs.(7)-(9) with $V(0, y = R)$ from Eqs.(5),(6). The charge balance equation should be included in calculations $n = N_d^+ - N_a^- - n_t$, where n_t is the density of electrons trapped at the dislocation line [12]

4. Results and discussion

Using the proposed here model, we are able quantitatively to reproduce the drop of the mobility μ^* at some critical meanings of carrier density as one can see from Figure 1. The sharp transition to the low mobility regime caused by domination of μ_{GB} as n falls.

The nature of this domination is the rise of the potential barrier as the dopant density decreases, since $V(0, y = R)$ is a function of the ratio $2R/D$, while $R \sim 1/\sqrt{N_d^+ - N_a^-}$. As the result, μ_{GB} decreases as $n = N_d^+ - N_a^- - n_t$ decreases ($\mu_1 \gg \mu_{GB}$), and from Eq.(9) we have $\mu^* \simeq \mu_{GB}$. Since the increase of the doping level, the bulk single-crystal mobility starts to dominate. Also, at fixed $2R$, $V(0, y = R)$ will be greater for smaller D (increase of the local dislocation density in the dislocation wall). This explains the correlation between the position of the minimum of n and the dislocation density. However, the model does not allow to reproduce the rise of the mobility below the minimum ($2R/D > 1$). Most likely, this increase can be explained in the framework of the percolation theory [15, 18].

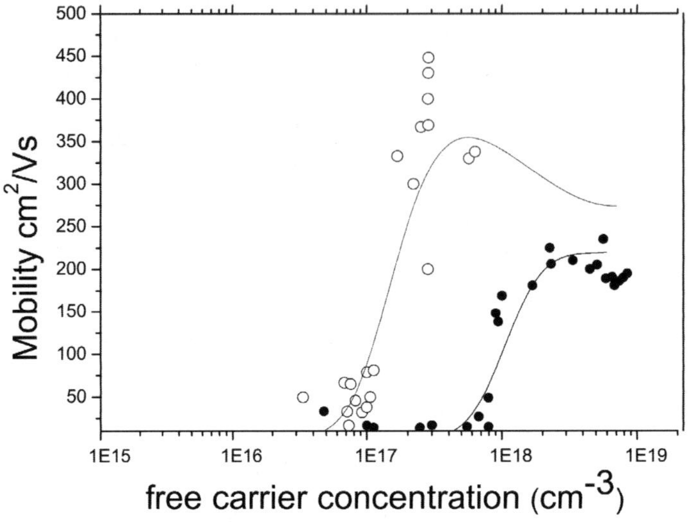

Figure 1. Room temperature mobility vs free carrier concentration for two samples with $N_{dis} = 5\times10^{9}\text{cm}^{-2}$ (open circles), $N_{dis} = 2\times10^{10}\text{cm}^{-2}$ (closed circles) from Ref.[12]. Solids lines are theoretical curves. A set of the model parameters used in calculations: (open circles) $f = 0.05$, $D = 1000\mathring{A}$, compensation ratio equal to 0.3, $L = 30000\mathring{A}$; (closed circles) $f = 0.05$, $D = 900\mathring{A}$, compensation ration equal to 0.4, $L = 30000\mathring{A}$ Other parameters are taken from Ref. [17]

References

[1] Matare H F 1971 *Defect Electronics in Semiconductors* (Wiley-Interscience, New York)
[2] Pearton S J, Zolper J C, Shul R J, Ren F 1999 *J.Appl.Phys.* **86** 1
[3] Edgar J, Strite S, Akasaki L, Amano H, Wetzel C 1999 *Properties, Processing and Applications of Gallium Nitride and Related Semiconductors* London: INSPEC, IEE 1999
[4] Huang D, Yun F, Reshchikov M A, Wang D, Morkoc H, Rode D L, Farina L A, Kurdak C, Tsen K T Park S S, Lee K Y 2001 *Sol State Electr.* **45** 711
[5] Look D C and Sizerlove J R 2001 *Appl.Phys.Lett* **105** 126104
[6] Kapolnec D, Wu X H, Heying B, Keller S, Keller B P, Mishra U K, DenBaars S P and Speck J S 1995 *Appl.Phys.Lett.* **67** 1541
[7] Lester S D, Ponce F A, Craford M G, and Steigerwald D A 1995 *Appl.Phys.Lett.* **66** 1249
[8] Heying B, Wu X H, Keller S, Li Y, Kapolnek D, Keller B P, DenBaars S P, and Speck J S 1996 *Appl.Phys.Lett.* **68** 643
[9] Read W T 1954 *Phil.Mag.* **45** 775
[10] Read W T 1955 *Phil.Mag.* **46** 111
[11] Fehrer M, Einfeldt S, Birkle U, Gollnik T, Hommel D 1998 *J.Cryst.Growth* **189/190** 763
[12] Farvacque J-L, Bougrioua Z, and Moerman I 2001 *Phys.Rev.* B **63** 115202
[13] Potin V, Ruterana P, and Nouet G, Pond R C, Morkoc H 2000 *Phys.Rev.* B **61** 5587
[14] Doschanov K M, Shamirzaev S X 1978 *Semicond.* **12** 2328
[15] Orton J W and Powell M J 1980 *Rep.Prog.Phys.* **43** 81
[16] Kazmerski L L, Berry W B, Allen C W 1972 *J.Appl.Phys.* **43** 3515
[17] Shur M, Gelmont B, and Asif Khan M 1996 *J.Elec.Mat.* **25** 777
[18] Kirkpatrick S 1973 *Rev.Mod.Phys.* **45** 574

Spin and orbital magnetism in 3d transition metal nanowires (Mn, Fe, and Co): a density functional theory study

Mahdi Sargolzaei[1,2], **Samaneh Ataee** [1]

[1] Department of Physics, Iran University of Science and Technology, Narmak 16345, Tehran, Iran.

[2] Computational Physical Science Laboratory, Department of Nano-Science, Institute for Research in Fundamental Sciences (IPM), Tehran, Iran.

E-mail: sargolzaei@iust.ac.ir

Abstract. In this work we study the electronic structure and magnetic properties of 3d transition metal nanowires (Mn, Fe, and Co). We have performed density functional calculations using the relativistic version of the full potential local orbital (FPLO) basis band structure method and optimized the equilibrium lattice parameters using LSDA total energy calculations. Relativistic LSDA produces orbital moments induced from spin polarization via spin-orbit coupling. It is found that the orbital moments of individual 3d atoms in low dimensional nanowires are very large compare to their bulk phases.

1. Introduction

In recent years a lot of experimental and theoretical studies [1, 2, 3, 4] have been devoted to the nanoscale materials, due to their interesting physical properties and many useful potential applications, such as electronic and optoelectronic devices, magnetic storage and spintronics. There are some experimental methods to provide nanowires using a scanning tunneling microscopy and a mechanically controllable break junction [5]. Beside these experimental methods many theoretical investigations like density functional calculations have been done on both finite and infinite chains such as linear or zigzag chains of transition metals [6, 7, 8]. The purpose of the present work is to study the electronic structure and magnetic properties (especially orbital magnetism) of 3d transition metal linear chains (Mn, Fe, and Co). Magnetic properties of the considered nanowires are mostly related to the partly filled d orbitals of transition metals.

2. Methodology

To study the electronic structure and magnetic properties of 3d transition metal (Mn,Fe, and Co) nanowires, we use the standard supercell approach to model an isolated linear atomic chain. We have carried out density functional calculations [9] using the relativistic version of the full potential local orbital method (FPLO) [10, 11, 12]. The site centered potentials and densities were expanded in spherical harmonic contributions up to l_{max}=12. The convergence of the total energies (10^{-6} Hartree) with respect to the k-space integrations was checked separately

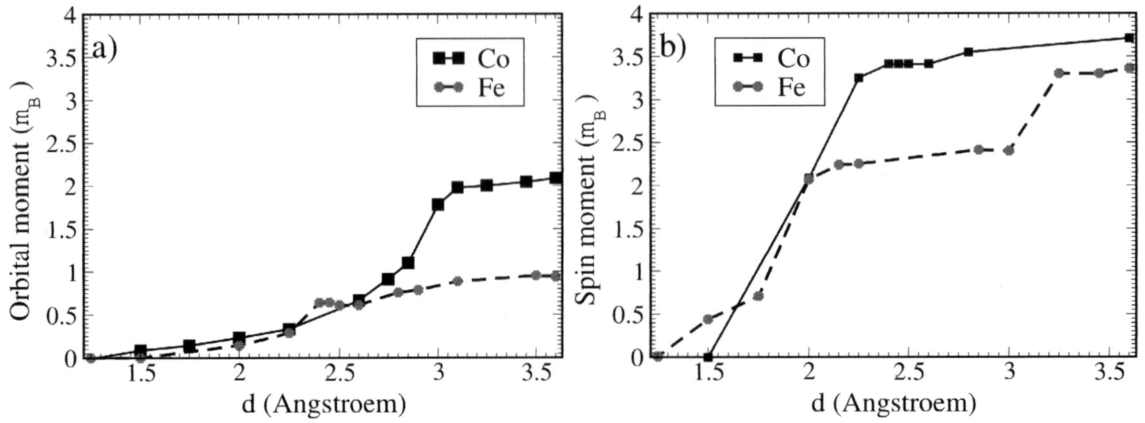

Figure 1. Calculated orbital (a) and spin (b) magnetic moments($\mu_B/atom$) of the ferromagnetic linear chains of Fe and Co as a function of interatomic distance (d) in LSDA approach.

for each of the considered nanowires. We performed numerical tests for up to 8×8×16 k-points in the Brilloin zone (BZ) to ensure that our results converged well with respect to BZ sampling. For linear chains the optimized equilibrium bond length of the linear atomic chain in the ferromagnetic or antiferromagnetic state is determined by locating the minimum in the calculated total energy as a function of the interatomic distance. We should use a wide enough separation between the nearest neighboring chains. To reach this aim we used a relatively large cubic supercell to avoid the wire-wire interactions (a=b=10 Å, c= distance of two pairs of atoms in the chain). The Perdew-Wang parameterization [13] of the exchange-correlation (XC) potential in the framework of LSDA was used.

3. Results and discussion

The d electrons of 3d transition metals usually determine the magnetic behavior of 3d transition metal nanowires. In an atomic picture, the Mn, Fe, and Co atoms are in configuration $3d^5$, $3d^6$ and $3d^7$, respectively. The calculated equilibrium bond lengths (d), total energy in ferromagnetic (FM) and antiferromagnetic (AFM) orderings (in units of Hartree), spin and orbital moments of Mn, Fe and Co linear chains in LSDA are tabulated in Table.I. The total energies of ferromagnetic and antiferromagnetic states in the linear chains of Mn, Fe, and Co were calculated to determine the magnetic ground state. It can be seen from Table.I that the magnetic ground state of Mn, Fe and Co linear chains are AFM, FM and FM, respectively . The ground state bond lengths for Mn, Fe and Co linear chains are 2.15Å, 2.18Å and 2.13Å, respectively. These bond lengths are generally small compare to their bulk structures. For example the calculated bond length for FM bcc Fe is 2.75Å. To study the orbital magnetism the relativistic spin-orbit coupling (SOC) plays an important role. Therefor we include the SOC in our self-consistent calculations. The spin magnetic moments for the atoms of the considered linear atomic chains in a full relativistic treatment are gained ± 3.61 μ_B/Mn, 3.25 μ_B/Fe and 2.24 μ_B/Co, respectively. However using the SOC we obtain a large significant orbital magnetic moment compare to fully quenched orbital moments in scalar relativistic approach by its nature [14]. As we see from Table.I for the magnetization along the chain axis, the calculated orbital magnetic moments of Mn, Fe and Co atoms in linear chains in their ground states are ± 0.01 μ_B, 0.25 μ_B and 0.28 μ_B. The orbital moments in the Fe and Co atomic chains are therefore considerably enhanced when compare to the bulk structure [12, 15]. The orbital magnetic moment of Mn is vanished because of Hund's second rule.

Table 1. Calculated spin and orbital magnetic moments in Bohr magnetons ($\mu_B/atom$) and total energy in ferromagnetic (FM) and antiferromagnetic (AFM) orderings (in unit of Hartree) using fully relativistic LSDA for 3d transition metal nanowires (Mn, Fe, and Co) in the equilibrium bond length (d).

	Mn (d=2.15 Å)	Fe (d=2.18 Å)	Co (d=2.13 Å)
m_s	±3.61	3.25	2.24
m_l	±0.01	0.25	0.28
E^{FM}(Hartree)	-2312.776	-2540.874	-2782.023
E^{AFM}(Hartree)	-2312.799	-2540.848	-2781.994
E^{FM}-E^{AFM} (meV)	626	-707	-789

To see how the magnetic properties of the atomic chains evaluate with the interatomic distances, the orbital and spin moments for the Fe and Co linear chains in the FM state are plotted in Fig.1 as a function of the bond length. It can be seen from Fig.1 as the bond length increases the atomic free state is more desirable. Therefor, the spin and orbital magnetic moments of the considered nanowires show values very closed to the atomic picture.

4. Conclusion

Density functional theory calculations were performed for the 3d transition metal ferromagnetic (Fe and Co) and antiferromagnetic (Mn) infinite linear chains. We have calculated the spin and orbital moments of each atom in the linear atomic chains by considering the spin-orbit interaction in full relativistic regime. Our calculations show that there is a smooth transition for the magnetism of the considered 3d atomic linear chains. Therefore one can conclude that the axial strain does not change drastically the magnetic behavior of the considered systems.

We are grateful to Manuel Richter, Ingo Opahle, Klaus Kopernik, and Helmut Eschrig for fruitful discussions. It made use of computing facilities of Computational Physical Science Laboratory, Department of Nano-Science, Institute for Research in Fundamental Sciences (IPM), Iran. This work has been financially supported by Research Devision at Iran University of Sceince and Technology.

References

[1] Heinze S, Bode M, Kubetzka A, Pietzsch O, Nie X, Blügel S, and Wiesendanger A, 2000 Science **288** 1805.
[2] Pietzsch O, Kubetzka A, Bode M, and Wiesendanger A, 2000 Phys. Rev. Lett. **84** 5212.
[3] Hernando A, Crespo P, Garcia M A, Fernandez Pinel A, Venta J, Fernndez A, and Penads S, 2006 Phys. Rev. B **74** 052403.
[4] Hernando A, Crespo P, and Garcia M A, 2006 Phys. Rev. Lett. **96** 057206.
[5] Ono T, Yamasaki H, Egami Y and Hirose K, 2003 Nanotechnology **14** 299-303.
[6] Ederer C, Komelj M and Fahnle M, 2003 Phys. Rev. B **68** 052402.
[7] Hong J and Wu R Q, 2003 Phys. Rev. B **67** 020406(R).
[8] Komelj M, Ederer C, Davenport J W and Fhnle M, 2002 Phys. Rev. B **66** 140407(R).
[9] Hohenberg P, and Kohn W, 1964 Phys. Rev. **136**, B864
[10] FPLO-9; improved version of the original FPLO code by K. Koepernik and H. Eschrig, Phys. Rev. B **59**, 1743 (1999); http://www.FPLO.de.
[11] Opahle I, Ph.D thesis, Technische Universität Dresden, 2001. H. Eschrig, M. Richter, and I. Opahle, in Relativistic Electronic Structure Theory-Part II: Applications, edited by P. Schwerdtfeger (Elsevier, Amsterdam, 2004), pp. 723776.
[12] Sargolzaei M, "Orbital Polarization in Relativistic Density Functional Theory", Lambert Academic Publishing, (2010).
[13] Perdew J P and Wang Y, 1992 Phys. Rev. B **45** 13244.

[14] Tung J C, and Guo G Y, 2007 Phys. Rev. B **76** 094413.
[15] Guo G Y, 1997 Phys. Rev. B **55** 11619.

Journal of Physics: Conference Series **248** (2010) 012054 doi:10.1088/1742-6596/248/1/012054

Study of all optical switching behaviour in semiconductor microresonator with nano-active layer

R. Kheradmand, H. Aryan

Photonics Group, Research Institute for Applied Physics and Astronomy, Tabriz University, Tabriz, Iran

E-mail: r_kheradmand@tabrizu.ac.ir , aryan86@ms.tabrizu.ac.ir

Abstract. In this paper the behaviour of carriers in spontaneous patterns formation and patterns switching has been studied. Results demonstrate that with increasing length of cavity the range of required input field amplitude for patterns formation increased slightly and also the minimum perturbation coefficient for switching decreased greatly. Increasing nonradiative recombination rate of carriers about ten percent appeared that required input field amplitude for patterns formation raised more than before, albeit the minimum perturbation coefficient for switching and switching and switching time dose not vary considerably.

1. Introduction

Recently, a switching technique that is based on sensitivity of transverse optical patterns has been developed in both atomic and semiconductor systems [1- 6]. A transverse optical pattern is the spatial structure of the electromagnetic field in the plane perpendicular to the propagation direction [3].In semiconductor micro resonator devices optical patterns emerge from the coupling of the nonlinear medium response, diffraction, and the feedback action of mirrors. In atomic system, Gauthier and his group have developed a switching technique experimentally and theoretically that uses a very weak beam of light to control a much stronger beam [3].

In this paper, we investigate a non-feedback technique [11-13], which allows selecting the orientation of spatial patterns in a semiconductor microresonator system. Our method is based on injecting a weak perturbation into the system to control patterns with a beam whose intensity is much smaller than the intensity of the pattern itself.

In reference [7-12], an all optical switch based on sensitivity of transverse optical patterns to tiny perturbation in a semiconductor microresonator in passive configuration had been discussed. In recent article for pattern generation an semiconductor optical cavity have been used that a holding beam has been imposed, they control patterns with a beam whose intensity is smaller than the intensity of the pattern itself and as fast as nanosecond scale for different values of perturbation coefficients. In semiconductor devices behavior of carriers determine the behavior of devices, also a powerful principle that could be explored to implement all-optical transistors, switches, logical gates, and memory is the concept of optical bistability. In systems that display optical bistability, the outgoing intensity is a strong nonlinear function of the input intensity, and might even display a hysteresis loop. Thus equations in homogeneous stationary state for field and carriers have been considered and with numerical calculations and by choosing proper physical parameters bistable behavior have been studied.

In each system some fundamental parameters have substantial rule in behavior of system, thus in this paper we perused the effects of some cavity parameters on the response manner of all optical switches. For this reason we studied the variation of cavity length on switching response time and required amplitude of perturbation for switching.

Journal of Physics: Conference Series **248** (2010) 012054 doi:10.1088/1742-6596/248/1/012054

2. Model

Similar with our previous model [4,5], we consider a broad area semiconductor device but in active configuration (i.e., with population inversion). The semiconductor microresonator is of the Fabry-Pe´rot type, with a MQW structure perpendicular to the direction z of propagation of the radiation inside the cavity. The dynamical equations for the slowly varying coherent field and carrier density, in the paraxial and mean field limit approximation are as in [4, 5].

2.1. Homogeneous Solution and Carriers Bistable Behavior

The homogeneous solution (E_s, N_s) of dynamical equations are obtained by setting equal to zero the time derivatives and neglecting the Laplacian. For appropriate choices of the parameters the curves of $|E_S|$ as a function of $|E_I|$ and also the curves of $|N_S|$ as a function of $|E_I|$ had been studied. As it has shown in Figure (1), the behavior of $|E_S|$ and $|N_S|$ as a function of $|E_I|$ is bistable. Also from their comparison at this Figure it is perspicuous that as well as mentioned with increasing injected field, output field increases with an S shape and carriers density decreases with an inverse S shape and the behavior of electric filed intensity and carrier density are on the contrary with each other.

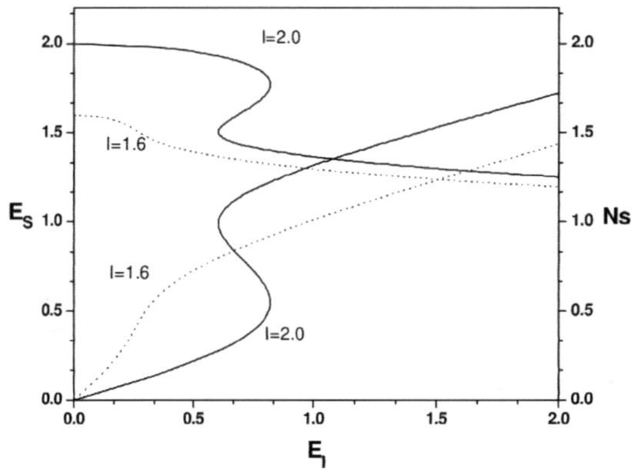

Figure 1. $(E_S - E_I)$ and $(N_S - E_I)$ curves for two different currents below threshold with these parameters " $\beta=0, \eta=0, \alpha=5, C=0.45, \theta=-2, d=0.052$", Ascendant curves depend on $(E_S - E_I)$ curves and falling curves depend on $(N_S - E_I)$ curves.

3. Simulations and numerical analysis

The results we will present here were obtained by numerically integration dynamical equations with the spatially perturbed injected field defined before. The numerical integration of dynamical equations was performed by using a split-step method with periodic boundary conditions. This method implies the separation of the algebraic and the Laplacian terms in the right-hand side of dynamical equations. The first part is integrated via a Runge-Kutta algorithm, while the linear operator (Laplacian) is integrated via a two-dimensional 2D-FFT. For the aim of achieving the initial condition of switching, what we typically do is to follow the pattern branch by changing the input intensity, starting from a case where the pattern was spontaneously formed.

3.1. The response time and sensitivity dependence on VCSEl parameters

Now in this section we investigate the effects of variation of physical parameters of microresonator on spontaneous patterns formation in system and also on switching performance. For the aim of switching performance and making comparison between different cases we just consider roll patterns switching in all cases. As mentioned before, physical characteristics that we study their variation in our system is: cavity length (L)

We study the effect of the cavity length by varying this parameter from $\pm 20\, to \pm 30$ percent. Some important parameters which on the cavity length: diffraction coefficient $a = v/2\kappa k_z$ which $v = c/n$, $k = vT/2L$ is the cavity half-width, and $k_z = \omega_0 n/c$; diffusion coefficient $d = l_D^2/a$ which $l_D = \sqrt{\tilde{D}\tau_r}$ is diffusion length; adimensional decay rate $\gamma = (\kappa\tau_r)^{-1}$ which τ_r is the nonradiative recombination rate of carriers; and the cavity detuning $\theta = \omega_c - \omega_0/\kappa$ which ω_0 being the frequency of the holding field and ω_c the longitudinal cavity frequency closest to ω_0. Thus with variation of the cavity length all of these parameters will change. Depends on different cavity length, bistable curve of $|E_S|$ as a function of $|E_I|$ and their Turing unstable region will be different which has shown in Figure (2). Thereupon region of injected field for patterns formation and also output patterns intensity will be different.

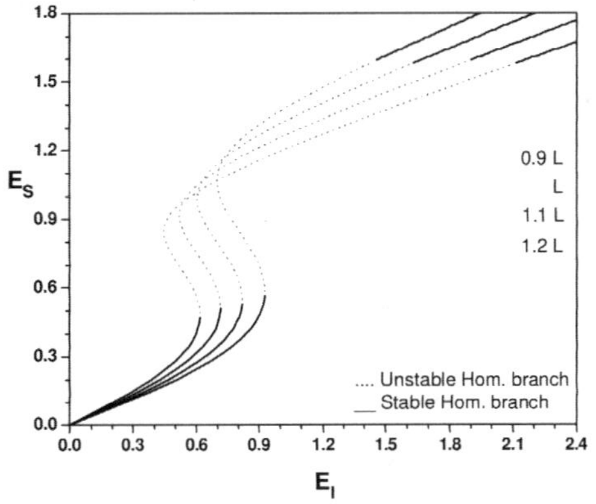

Figure 2. Steady-state curves of $|E_S|$ as a function of $|E_I|$ for different cavity length, dashed line refers to unstable Turing instability region.

Simulations show no regular patterns in proper times in system with decreasing cavity length. Thus we start to increase this parameter by 10, 20, and 30 percent and obtain patterns for each of these states. Then for each cavity length we switch one spontaneous roll pattern. In the case of 10 percent increment successful switching performed with $\alpha = 0.1$, $\alpha = 0.07$, $\alpha = 0.03$ and also $\alpha = 0.02$. As could be seen in this case required perturbation coefficient is very smaller than its value in the case of initial cavity length. Switching with $\alpha = 0.01$ was not successful thus $\alpha = 0.02$ is the minimum perturbation coefficient in this case. Then we increased cavity length up to 20 percent. Results of switching in this case demonstrate that minimum perturbation coefficient is $(\alpha = 0.002)$ too. This perturbation coefficient is 250 times smaller than minimum perturbation coefficient for switching in the case of initial cavity length. Also we switched roll pattern with perturbation that is 650 times weaker than injected field amplitude of roll pattern in this case $(E_I = 1.3)$. This consequence leads us to produce all optical switches that operate in ultra low light levels. It should be pointed out; however, increment in switching time is the cost of decreasing perturbation amplitude (Fig 3).

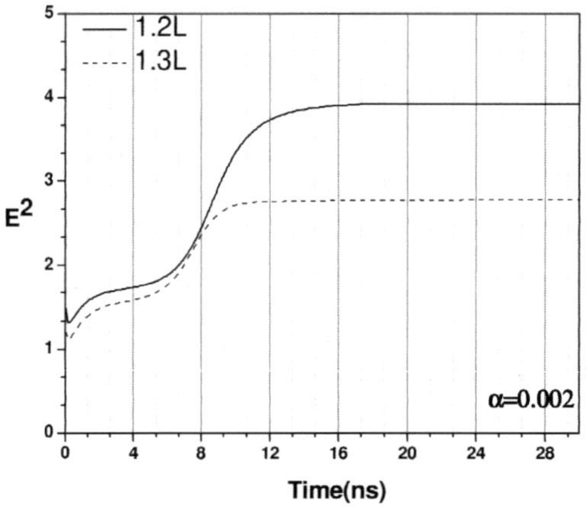

Figure 3. Evaluation of electric field intensity for 20 and 30 percent increment, for both cases injected field amplitude is $(E_I = 1.3)$ and perturbation coefficient is $(\alpha = 0.002)$.

Results of switching for 30 percent increment case show that minimum perturbation coefficient in this case is almost similar to 20 percent increment case. We could deduce from this result that increasing microcavity length for this purpose to some extent could be beneficial. Albeit in this case switching time is a little faster than previous case but their switching times are in a same range (Fig 3). According to Figure (3), here switching times are longer than initial case, while the used amplitude of perturbation here is 250 times smaller than its initial value, and switch could be faster with stronger perturbation coefficient.

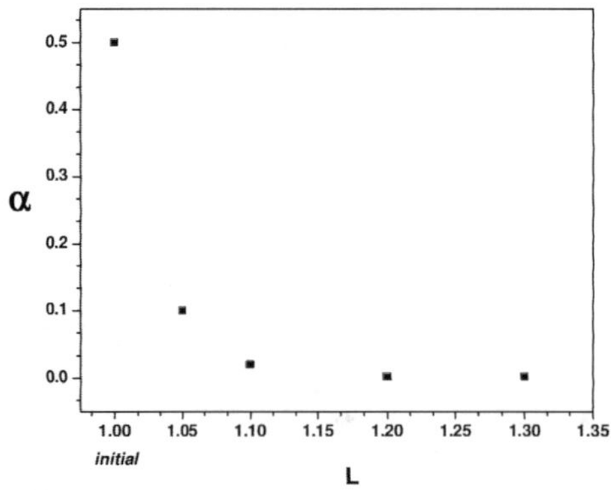

Figure 4. Comparing minimum perturbation coefficient for each cavity length.

In Figure (4) we plot the perturbation variation as a function of microcavity length. We defined the perturbation coefficient as the minimum coefficient needed for the pattern to rotate completely. It can be seen that, pattern can be controlled with very small coefficients of perturbations with enough larger length compared with initial length of the microresonator.

4. Conclusions

A spatial perturbation method has been used to select and control spontaneous optical pattern orientation in an active semiconductor microresonator. Our successful results have been demonstrated with a perturbation made by a tilted wave, corresponding in the Fourier space to only one of the spots of a roll pattern. The switching beam controls the orientation of output patterns with much stronger intensities: for example, we can rotate patterns with intensity 350 times stronger than the switching beam intensity. Also switching time is sensitive to the perturbation strength α and decreases by increasing α. Also for founding more regular spontaneous pattern after shorter time in system we could increase cavity detuning to some extent.

References

[1] Ramaswami, R. Sivarajan, K. N. "A Practical Perspective" *in Optical Networks* (Morgan Kaufmann, San Francisco, CA, ed. 2, 2002), ch. 12.
[2] H. M. Gibbs, Optical Bistability: Controlling Light with Light (Academic Press, Orlando, FL, 1985).
[3] A.M.C.Dawes,L.Illing,S.M.Clark,D.J.Gauthier,SCIENCE, 308, 672(2007).
[4] R. Kheradmand et al., *Eur. Phys. J. D*, Vol. 47, 107112 (2008).
[5] R. Kheradmand et al., *ETRI Journal*, Vol. 31, Number 5, October (2009).
[6] L. Spinelli, L. Tissoni, G. Brambilla,M. Prati, F. and Lugiato, L.A. " Spatial solitons in semiconductor microcavities", Phy. Rev. A VOL 58, N3 (1998).
[7] B. Sfez, J. L. Oudar, J. C. Michel, R. Kuszelewicz, and R. Azoulay, Appl. Phys. Lett. **57**, 324 (1990).
[8] B. Sfez, J. L. Oudar, J. C. Michel, R. Kuszelewicz, and R. Azoulay, Appl. Phys. Lett. **57**, 1849 (1990).
[9] B. Sfez, R. Kuszelewicz, and J. L. Oudar, Opt. Lett. **16**, 855 (1990).
[10] J. L. Jewell, J. P. Harbison, A. Scherer, Y. H. Lee, and L. T. Florez, IEEE J. Quantum Electron. **27**, 1332 (1991).
[11] Pellegrini, V. Fuso, F. Arimondo,E. Castelli,F. Lugiato, L.A. Bava, G.P. and Debernardi,P.L. Phys. Rev. A 50, 5219 (1994).
[12] C. H. Henry, IEEE J. Quantum Electron. **18**, 259 (1982).
[13] Dawes,A.M.C. Illing, L. Greenberg, J.A. and Gauthier, D.J . " All-Optical Switching with Transverse Optical Patterns"

Journal of Physics: Conference Series **248** (2010) 012055 doi:10.1088/1742-6596/248/1/012055

The effects of magnetic and nonmagnetic impurities on electronic properties of semiconductor multilayers

L Gharaee[1] and A Saffarzadeh[1,2]
[1] Department of Physics, Payame Noor University, Nejatollahi Street, 159995-7613 Tehran, Iran
[2] Computational Physical Sciences Laboratory, Department of Nano-Science, Institute for Research in Fundamental Sciences (IPM), P.O. Box 19395-5531, Tehran, Iran

E-mail: a-saffar@tpnu.ac.ir

Abstract. We present a theoretical study of electronic states for a semiconductor multilayer consisting of a binary alloy, by using the single-site coherent potential approximation (CPA) and the tight-binding model. In this study the self-energy depends on the layer number. This dependence was neglected in previous studies. Our results for both nonmagnetic semiconductor (NMS) and diluted magnetic semiconductor (DMS) multilayers are presented. We found that, with increasing the nonmagnetic (chemical) impurity concentration, some of the van-Hove singularities in local density of states (LDOS) disappeared. In addition, the applicability of CPA in appearance of a gap in LDOS has been investigated. In the presence of magnetic impurity, the bands are broadened due to the impurity spin fluctuation, and in the high impurity concentration, the van-Hove singularities in the LDOS may completely disappear and the curves become smooth.

1. Introduction

Many physical quantities of nanostructures such as multilayers behave thoroughly different from the same quantities in macro-sized structures. The main reason behind this is the change in the electron density of states (DOS). In this paper, the effect of substitutional disorder on the layer dependent DOS is investigated by using the CPA method and the tight-binding approximation. Both nonmagnetic semiconductor (NMS) and diluted magnetic semiconductor (DMS) multilayers are discussed here.

NMSs are composed of two semiconductors A and B with different band gap energies which are mixed together by concentration x as $A_{1-x}B_x$. On the other hand, DMSs are semiconductors of the general type $A_{1-x}M_xB$, where AB is either II-VI or III-V semiconductor and M is a magnetic element, most commonly Mn. [1,2]

2. Model and Formalism

In this study, under the assumption that the density of carriers is very low, we ignore the electron-electron interaction and write the Hamiltonian as follows:

$$H = \sum_{r,n} u_{r,n}(M,A) + \sum_{r,n,\sigma r',n',\sigma} \sum t_{rn,r'n'} \,|\, \mathrm{r}n,\sigma\rangle\langle \mathrm{r}n,\sigma \,| \tag{1}$$

where $\mathbf{r}$ denotes the positional vector in the x-y plane, $t_{\mathbf{r}n,\mathbf{r}'n'}$ is the hopping energy between nearest neighbours and the on-site potential $u_{\mathbf{r},n}$ depends on whether the site is nonmagnetic(A) or magnetic(M)

[1, 2]. For the A and M sites in turn we have:

$$u^{A}_{\mathbf{r},n} = \varepsilon_A \sum_{\sigma} | \mathbf{r}n,\sigma\rangle\langle \mathbf{r}n,\sigma | \tag{2}$$

$$u^{M}_{\mathbf{r},n} = \sum_{\sigma,\sigma'} |\mathbf{r},n,\sigma\rangle[\varepsilon_M \delta_{\sigma,\sigma'} - IS_{\mathbf{r}n}.\tau_{\sigma,\sigma'}]\langle \mathbf{r},n,\sigma'| \tag{3}$$

where, $|\mathbf{r},n,\sigma\rangle$ is an atomic orbital with spin $\sigma(=\uparrow \text{ or } \downarrow)$ and ε is the random site energy [1]. Furthermore, we have used the Bloch-Wannier wave functions to describe the system. According to the CPA, the multilayer Hamiltonian is replaced by the following effective Hamiltonian:

$$H_{eff} = \sum_{\mathbf{r},n,\sigma} \sum_{\mathbf{r}',n';\sigma'} [\Sigma_n(\omega)\delta_{n,n'}\delta_{\mathbf{r},\mathbf{r}'} - t_{\mathbf{r}n,\mathbf{r}'n'}] |\mathbf{r},n,\sigma\rangle\langle \mathbf{r}',n',\sigma'| \tag{4}$$

The physical properties of the real system can be obtained from the configurationally averaged Green's function $\langle G\rangle_{av} = \langle(\omega - H)^{-1}\rangle_{av}$ which is replaced in the CPA with an effective medium Green's function $\overline{G} = (\omega - H_{eff})^{-1}$. The matrix elements of this effective Green's function are calculated by the following Dyson equation [4]:

$$\overline{G}_{n,n'}(r,r';\omega) = G^0_{n,n'}(r,r';\omega) + \sum_{n''=1}^{N_z}\sum_{r''} G^0_{n,n''}(r,r'';\omega)\Sigma_{n''}(\omega)\overline{G}(r'',r';\omega) \tag{5}$$

where $\Sigma_{n''}(\omega)$ is the site-dependent self-energy and G^0 is the clean system Green's function and its matrix element is given by [3]:

$$G^0_{n,n'}(r,r';\omega) = \frac{1}{N_{\parallel}} \sum_{l=1}^{N_z}\sum_{\mathbf{k}_{\parallel}} G^0_{n,n'}(l,\mathbf{k}_{\parallel};\omega) e^{i\mathbf{k}_{\parallel}.(r-r')} \tag{6}$$

Here, $G^0_{n,n'}(l,\mathbf{k}_{\parallel};\omega)$ is the mixed Bloch-Wannier representation of G^0 and can be expressed as

$$G^0_{n,n'}(l,\mathbf{k}_{\parallel};\omega) = \frac{\frac{2}{N_z+1}\sin(\frac{l\pi}{N_z+1}n)\sin(\frac{l\pi}{N_z+1}n')}{\omega + i\eta - \varepsilon_l(\mathbf{k}_{\parallel})} \tag{7}$$

where $\varepsilon_l(\mathbf{k}_{\parallel})$ is the clean system band structure, $\mathbf{k}_{\parallel}$ is a wave vector parallel to the layer, l is the mode of the subband and $N_{\parallel}$ is the number of lattice sites in each layer [3]. The perturbation term is given as

$$V = H - H_{eff} = \sum_{r,n} v_{r,n} \tag{8}$$

where $v_{\mathbf{r},n} = v^{A}_{\mathbf{r},n}$ for the A site and $v_{\mathbf{r},n} = v^{M}_{\mathbf{r},n}$ for the M site are given by [1]

$$v^{A}_{\mathbf{r},n} = \sum_{\sigma} |\mathbf{r},n,\sigma\rangle[\varepsilon_A - \Sigma_{r,n}]\langle \mathbf{r},n,\sigma| \tag{9}$$

$$v^{M}_{r,n} = \sum_{\sigma,\sigma'} |r,n,\sigma\rangle[(\varepsilon_M - \Sigma_{r,n})\delta_{\sigma,\sigma'} - IS_{r,n}.\tau_{\sigma,\sigma'}]\langle r,n,\sigma'| \tag{10}$$

by writing the Green's function of the real system in terms of the effective Green's function and the total scattering matrix T as $G = \overline{G} + \overline{G}T\overline{G}$ and using the multiple scattering theory [4] and finally applying the CPA condition ; $\langle T\rangle_{av} = 0$, we obtain $\langle G\rangle_{av} = \overline{G}$ [1, 2, 4].

Consequently, the LDOS in each layer can be obtained from the following equation in which $F_n(\omega)$ is the diagonal element of the effective Green's function matrix [1,3]:

$$g_n(\omega) = -\frac{1}{\pi}\operatorname{Im} F_n(\omega) \tag{11}$$

4. Results and Conclusion

In this section we report the results of LDOS for two types of nonmagnetic and magnetic impurities for a multilayer with N_z =5. In case of nonmagnetic impurity, we characterize the scattering-strength parameter

Journal of Physics: Conference Series **248** (2010) 012055 doi:10.1088/1742-6596/248/1/012055

by $\Delta = (\varepsilon^A - \varepsilon^B)/W$, where $W \cong 6t$ is the half-bandwidth of either constituent. Also we measure the energies in units of t and we set $\varepsilon_A = 0, \varepsilon_B = -3, -8$ and $\varepsilon_M = -0.75$. Figures 1-4 show that the LDOS clearly depends on the layer number and sharp peaks and step-like futures (Van Hove singularities) vary in different layers. Moreover, because of symmetry for a five-layered system, diagrams of the layer 1 and 5 as well as those of 2 and 4 are similar. To demonstrate the impurity effects in NMS multilayer, the LDOS as a function of energy and impurity concentration is plotted in figures 1 and 2.

At zero impurity concentration, the band is the same as the clean system but as the impurity increases, for $\varepsilon_B < 0$ (donor impurity), the curves are being relocated to the lower energy levels, sharpness of peaks decreases and the curves become smooth until at x=0.5, the LDOS becomes completely smooth and symmetric. With further increasing x, the spectrum tends to create the sharp peaks again, and in the case of x=1, the spectrum of B atom is created similar to the A atom with only a shift in energy equal to ε_B.

By drawing and comparing the figures 1 and 2 one would realize that in the case of nonmagnetic impurity no band broadening happens. In addition, enhancement of Δ for a given concentration creates an energy gap in the LDOS diagrams. Therefore, the efficiency of CPA is seen.

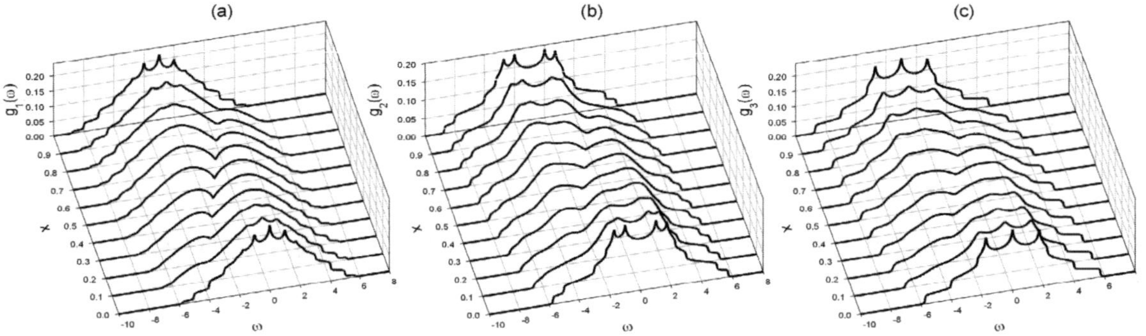

Figure 1. The LDOS for a NMS multilayer with $\varepsilon_B = -3$ as functions of energy and concentration for the layers (a) n=1,5, (b) n=2,4, and (c) n=3.

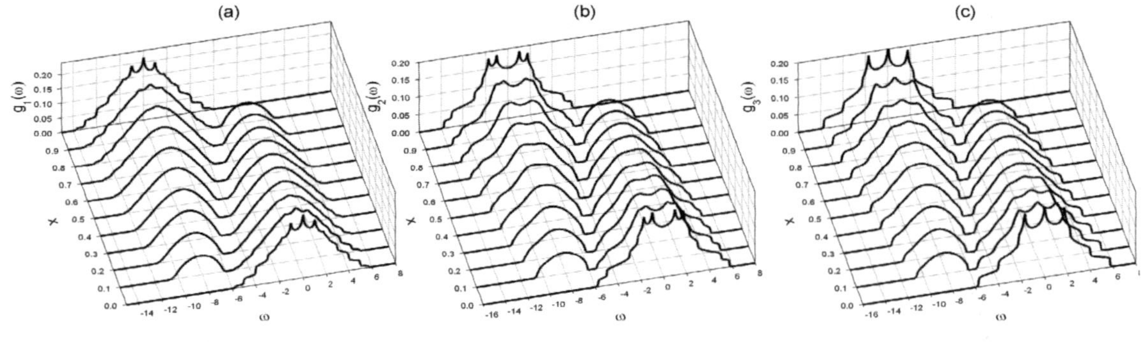

Figure 2. The LDOS for a NMS multilayer with $\varepsilon_B = -8$ as functions of energy and concentration for the layers (a) n=1, 5, (b) n=2, 4, and (c) n=3.

When the magnetic impurity is considered the plots become smoother with increasing the impurity concentration. In the case of x=1 we have the most smooth and symmetric graphs (Figures 3 and 4). Also, enhancement of magnetic impurity concentration broadens the band due to the impurity spin fluctuation. We have exhibited the results for both weak (IS=−0.4), and strong (IS= −1.2) exchange interaction. Results clearly prove that when the exchange interaction is stronger, the band broadening is more obvious. For this type of impurity, with increasing x the Van-Hove singularities decrease continuously, and at the concentration x=1.0, they are entirely disappeared and the plot is completely being smooth and symmetric around the center of the band. Another interesting future is that, the curves of LDOS are the same for different signs of IS, because upon our assumption, the spin of atoms is treated classically. By comparing the figures 1 and 2 with 3 and 4, we can realize that the LDOS curves for a magnetic impurity are wider than the chemical one. Moreover, with further increasing the IS value, the spectra will be wider. For large values of IS, the plots clearly tend to show a gap.

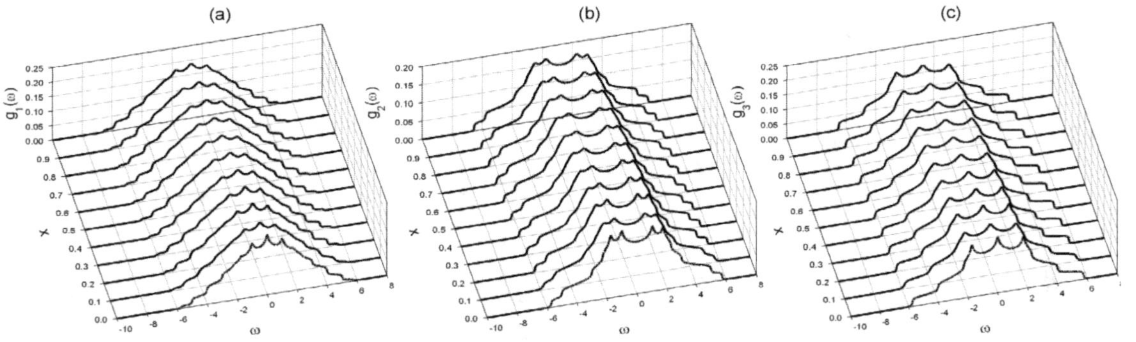

Figure 3. The LDOS of a DMS multilayer with $\varepsilon_M = -0.75$ and IS=−0.4 as functions of energy and concentration, for the layers (a) n=1, 5, (b) n=2, 4, and (c) n=3, respectively.

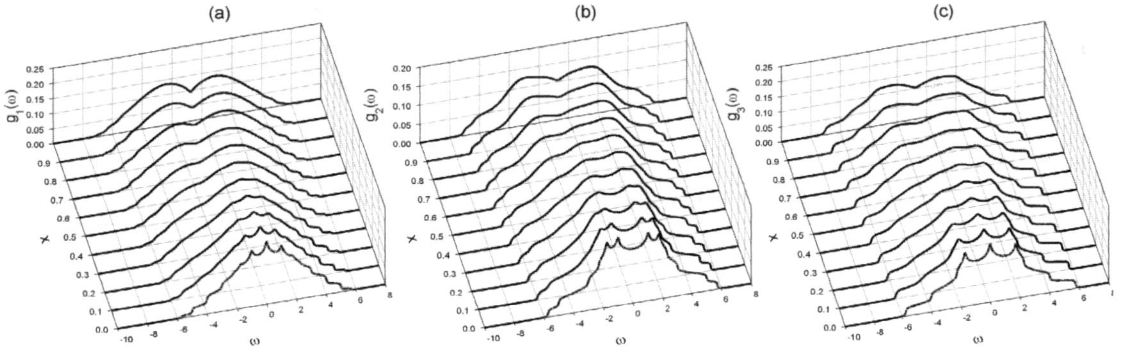

Figure 4. The LDOS of a DMS multilayer with $\varepsilon_M = -0.75$ and IS=−1.2 as functions of energy and concentration, for the layers (a) n=1, 5, (b) n=2, 4, and (c) n=3, respectively.

References

[1] Saffarzadeh A 2007 Chemical and magnetic impurity effects on electronic properties of semiconductor quantum wires *Phys. Rev.* B **76**, 214201
[2] Takahashi M 1999 Coherent-potential approach to magnetic and chemical disorder in diluted magnetic semiconductors *Phys. Rev.* B **60**, 15858
[3] Saffarzadeh A and Gharaee L 2010 Optical absorption spectrum in disordered semiconductor multilayers *J. Appl. Phys.* **107**
[4] Gonis A 1992 Green Functions for Ordered and Disordered Systems: Studies in Mathematical Physics (North-Holland, Amsterdam,), Vol. 4.

Journal of Physics: Conference Series **248** (2010) 012056 doi:10.1088/1742-6596/248/1/012056

Plasma Oscillations and Electron – Elecron Interactions in 2D Electron Systems

A B Dubois

Ryazan Institute of Open Education, 2 Bronnaya str., Ryazan, Russia, 390010

Ryazan State Radio Technical University, 59/2 Gagarin str., Ryazan, Russia, 39005

E-mail: abd@rioo.ru, abd-69@mail.ru

Abstract. Electron-electron interactions in a single highly doped heterojunction are considered taking into account both transitions. Expressions are derived for the time of electron-electron interaction, matrix elements of the full screening potential and dynamic dielectric function in 2D electron system with fine structure of energy spectrum, and for the electron density spatial distribution. The calculated dependences provide a good description of the experimental times of Landau levels collisional broadening.

1. Introduction

Starting from the pioneering works [1] and up to present time [2,3] the electron – electron (e-e) interactions are the subject of ever growing interest because of their fundamental role in kinetic phenomena. Among others, one should note the hot electron effects, quantum corrections to the conductivity, and damping (destruction) of Landau quantization in bulk and two-dimensional semiconductors with degenerate electrons. Also known are the anomalies in the low-temperature magnetotransport arising when 2D electrons fill several size-quantized subbands. In the present paper we report the results of the study of the e-e relaxation processes in a system of highly degenerate 2D electrons with the fine structure of energy spectrum and the electron density spatial distribution. Expressions are derived for the times of electron-electron intra- (τ_{ee}^{intra}) and intersubband (τ_{ee}^{inter}) interaction, matrix elements of the full screening potential and dynamic dielectric function in a 2D electron system.

2. Channels of electron interactions

Our analysis of electron – electron is based on the calculation of the conduction band energy structure $E_c(z)$ of strongly doped (charge density $n_S \simeq 10^{12}\,\text{cm}^{-2}$) heterojunction $Al_xGa_{1-x}/GaAs$ plotted in Figure 1. The system of equations which describe the model is the following: Schrödinger equation for the states of the conduction electrons, a constitutive equation giving the charge density in terms of the occupied states and the uniform-background charge density, and Poisson equation giving the potential in terms of the charge density. The potential well of this heterojunction was approximated by the triangular profile with the sharp bends $E_m = E_c(d_m)$ and $E_p = E_c(d_p)$ [4].

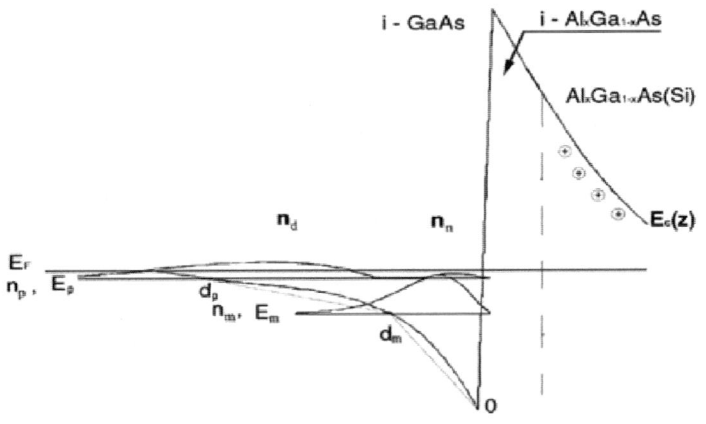

Figure. 1. Energy diagram of the conduction band $E_c(z)$ for a heterojunction with two filled quantized subbands.

Schematics of the e-e interactions in the 2D system is presented in Figure 2 which also shows the typical channels through which the perturbation affects the electron system. Here n_m is the 2D electron density in the ground (main) size-quantized subband while n_n and n_d are the satellites of the perturbed (exited) subband component $n_p = n_d + n_n$. Of all the channels of Landau quantization damping which we consider one should emphasize those three that reveal the major features. Since $n_m >> n_d, n_n$ and centres of gravity of $|\psi|^2$ at the levels E_m and E_p are spatially resolved, the perturbation is sensed (absorbed) during the time $\tau^{(1)}$ by the set of 2D electrons. Then through the intrasubband (τ_{mm}^{intra}) and intersubband (τ_{mn}^{inter}, τ_{md}^{inter}) e-e interactions mediated by the τ_{nn}^{intra} and τ_{dd}^{intra} intrasubband interactions the perturbation is extended to the entire 2D system. The path of this mechanism is shown in Figure 2 by the solid line. The major factor here is the interaction of n_m and n_d electrons, while n_n is a passive element.

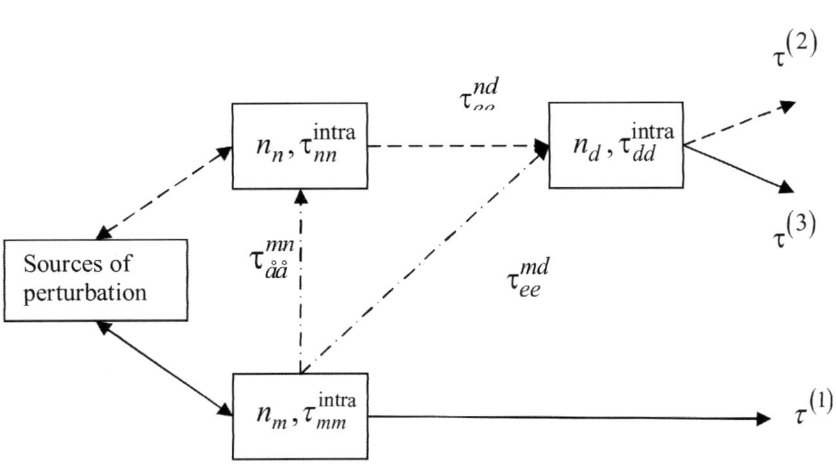

Figure 2. Schematic model illustrating the channels of perturbation

The second scenario corresponds to the situation where perturbation is first sensed by n-satellite. In that case the time needed to destroy the quantization is formed in the chain shown in Figure 2 by the dashed line. Here the system behaviour is governed by the intersubband τ_{mn}^{inter} and τ_{md}^{inter} and intrasubband τ_{nd}^{intra} interactions. The τ_{nd}^{intra} interaction is major one in this channel and n_m plays the role of a passive element. The third version is formed in the chain where the τ_{mn}^{inter} interaction dominates while n_d is a passive element. The corresponding path is shown by the dot-and-dash line. With respect to the nature of induced transitions the e-e interactions can be classified into three types: 1) interactions within a single subband limited to the transitions within the same band; 2) intrasubband interaction exiting the intersubband transitions, and 3) intersubband interactions resulting also in the intrasubband transitions.

3. Times of e-e interactions

One of the important points in the derivation of expression for τ_{ee}^{intra} and τ_{ee}^{inter} is the calculation of the full screening potential matrix elements which, within the perturbation theory approach, implies the transformation of the potential $V(\mathbf{r},t)$ into $V_{tot}(\mathbf{q},\omega)$. To within the second order in external potential $V_{tot}(\mathbf{q},\omega)$ in the perturbation theory expansion, the time required for the e-e interaction to change the state $\langle \mathbf{k}|\mathbf{p}\rangle$ into $\langle \mathbf{k}+\mathbf{q}|\mathbf{p}-\mathbf{q}\rangle$ is given by a expression [5]:

$$\frac{1}{\tau_{ee}} = \int_{-\infty}^{\infty} \frac{d\omega}{\pi^2 \operatorname{ch}^2(\hbar\omega/2k_B T)} \sum_{i,j,k,m} \sum_{\mathbf{q}} \left|V_{tot}^{ijkl}(\mathbf{q},\omega)\right| \chi_{ik}(\mathbf{q},\omega)\, \chi_{ik}^{*}(\mathbf{q},\omega),$$

with matrix elements $V_{tot}^{ijkl}(\mathbf{q},\omega)$ and dielectric function $\chi_{ik}(\mathbf{q},\omega)$ with $\gamma = 2m^{*}S/\pi\hbar$ (see below). S is 2D electron gas area. In the form convenient for calculations the relaxation times for the first and second type transitions are written as

$$\frac{1}{\tau^{\text{intra}}} = \frac{E_m^2 m^{*2} S}{16\pi^5 d_m^2 \hbar^4 n_s^2} P_{-n}(T)\varsigma(T,n_s), \tag{1}$$

while for the third type

$$\frac{1}{\tau^{\text{inter}}} = \frac{m^{*2} S}{16\pi^6 (d_n - d_p)^2 \hbar^4} \left\{ \frac{(E_n - E_m)^2}{8\pi} Q_{-n}(T)\varsigma(T,n_s) - \frac{(E_n - E_m)^{3/2}}{\sqrt{8\pi}} W_{-n}(T)\varsigma(T,n_s) \right\} \tag{2}$$

The polynomials are

$$P_{-n}(T) = \frac{B_1}{T} + \frac{B_2}{T^2} + \frac{B_3}{T^3} + \ldots \qquad Q_{-n}(T) = \frac{B_1}{T} + \frac{B_2}{T^2} + \frac{B_3}{T^3} + \ldots \qquad W_{-n}(T) = \frac{B_2}{T^2} + \frac{B_4}{T^4} + \ldots$$

with the coefficients B_n defined by the Riemann zeta-function $\varsigma(T,n_m,n_d,d_m,d_p)$. The calculations of the time e-e interactions were performed within the outlined schematic model of the Landau quantization destruction taking into account the paths corresponding to the channels (1-3) including both intra- and intersubband transitions according to (1) and (2) employing the Matthiessen rule $\tau^{-1} = \sum_i \tau_i^{-1}$ where the summation is performed over all intra- and intersubband components of the schematic model presented in figure 2.

4. Spectrum of collective exitation

Figures 3a and 3b show the experimental and calculated curves for the time of destruction of Landau quantization for two heterostructures with electron density sufficient for filling of two size-quantized subbands.

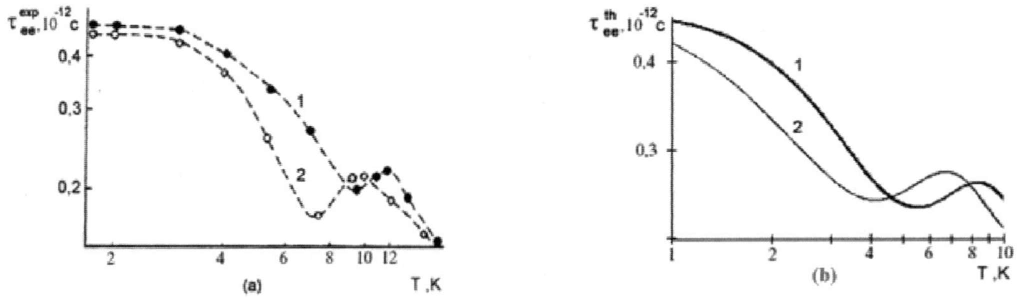

Figure 3. Comparison of experimental $\tau_{ee}^{\exp}(T)$ (a) and calculated $\tau_{ee}^{th}(T)$ (b) for different values of the m-subband electron density: 1 - $9{,}1\cdot10^{11}\,\mathrm{cm}^{-2}$, 2 - $10\cdot10^{11}\,\mathrm{cm}^{-2}$.

Bearing in mind the oscillations in $\tau_{ee}^{\exp}(T)$ and $\tau_{ee}^{th}(T)$, it is natural to expect the resonant response of the components of a complex 2D electron system to the external perturbation $V_{tot}^{ijkl}(\mathbf{q},\omega)$ at the plasma oscillations frequency. The 2D system responds to the spectrum of $V_{tot}^{ijkl}(\mathbf{q},\omega)$ by one of its n_m, n_d components, and in time $\tau_{ee}^{th}(T)$ the perturbation extends to the entire system system resulting in the destruction of quantum states. The latter is formally equivalent to the rise of temperature at which the measurements are taken. Therefore the resonant response featuring the Landau quantization destruction corresponds to a minimum in the curves $\tau_{ee}^{th}(T)$ and $\tau_{ee}^{\exp}(T)$. We have performed a spectral analysis of the dispersion equations for dielectric function

$$\chi_{ik}(\mathbf{q},\omega)=\frac{\gamma}{q}\left\{\frac{q}{\sqrt{n_i}}-\left[\sqrt{2\pi}-\frac{1}{n_i}\left(\frac{q}{2}-\frac{m\omega}{\hbar q}\right)^2\right]^{\frac{1}{2}}-\left[\sqrt{2\pi}-\frac{1}{n_i}\left(\frac{q}{2}+\frac{m\omega}{\hbar q}\right)^2\right]^{\frac{1}{2}}\right\}+c.c.$$

for various channels of e-e interactions according to the scheme shown in Fig. 2 and various relative values of the densities in the situation where two size-quantized subbands are occupied. The plasma oscillation frequencies are found from the dispersion equation $\chi_{ik}(\mathbf{q},\omega)=0$, the minima in $\tau_{ee}^{th}(T)$ corresponding to the minima in $\operatorname{Re}\chi_{ik}(\mathbf{q},\omega)$ and $\operatorname{Im}\chi_{ik}(\mathbf{q},\omega)$ while the maxima in $\tau_{ee}^{th}(T)$ correspond to $\operatorname{Re}\chi_{ik}(\mathbf{q},\omega)=0$ and maxima in $\operatorname{Im}\chi_{ik}(\mathbf{q},\omega)$ [4,5].

[1] Landau L D and Kompaneets A S 1935 *ZhETF.* **5**, 276
[2] Gantmakher V F and Levinson I B 1984 Scattering of charge carriers in metals and semiconductors *Moscow, "Nauka"*
[3] Slutzky M, Entin-Wohlman O, Berk Y, and Palevsky A 1996 *Phys. Rev.* B **53** 4065.
[4] Kadushkin V I, Dubois A B 2004 *Phys. Low-Dim. Struct.* **5/6** 107.
[5] Kadushkin V I, Dubois A B 2002 *Phys. Low-Dim. Struct.* **11/12** 27.

Journal of Physics: Conference Series **248** (2010) 012057 doi:10.1088/1742-6596/248/1/012057

Lattice Heat Capacity of Mesoscopic Nanostructures

B Gharekhanlou, S Khorasani, A Vafai

School of Electrical Engineering, Sharif University of Technology, PO Box 11365-9363, Tehran, Iran.

E-mail: khorasani@sharif.edu, vafai@sharif.edu

Abstract. We present a rigorous full quantum mechanical model for the lattice heat capacity of mesoscopic nanostructures in various dimensions. Model can be applied to arbitrary nanostructures with known vibrational spectrum in zero, one, two, or three dimensions. The limiting case of infinitely sized multi-dimensional materials are also found, which are in agreement with well-known results. As examples, we obtain the heat capacity of fullerenes.

1. Introduction

This paper presents a rigorous and unified quantization of bosonic fields, applicable to phononic, stress, strain, and electromagnetic fields. We then proceed to develop an exact relation for the lattice heat capacity of mesoscopic nanostructures, having finite physical dimensions. Extension of the model to the analysis of bulk media is shown to be agreement with known results. Numerical examples of this model for analysis of fullerenes and are shown and discussed.

2. Field and Energy

Suppose that a vector bosonic field state is $\langle \mathbf{r} | \mathbf{F} \rangle = \mathbf{F}(\mathbf{r}, t)$, connected to a scalar energy functional as

$$\Pi = \tfrac{1}{2}\mathbf{F} : \ddot{\mathbf{K}} : \mathbf{F} = \tfrac{1}{2}\sum_{i,j=1}^{3} F_i K_{ij} F_j \tag{1}$$

where $\ddot{\mathbf{K}} = \left[K_{ij}\right]$ is a second-rank tensor, describing the inherent properties of the medium. For a lossless medium $\ddot{\mathbf{K}}$ has to Hermitian $\ddot{\mathbf{K}} = \ddot{\mathbf{K}}^{\dagger} = \ddot{\mathbf{K}}^{t}$ and real-valued. Hence, isotropic media require $\ddot{\mathbf{K}} = K\ddot{I}$, which results in $\Pi = \frac{1}{2}K\mathbf{F}\cdot\mathbf{F} = \frac{1}{2}K|\mathbf{F}|^2$. Now, using Fourier transformation pair we get

$$\mathbf{F}(\mathbf{r},t) = \frac{1}{\sqrt{2\pi}}\int_{-\infty}^{+\infty}\mathbf{B}(\mathbf{r},\omega)e^{-j\omega t}d\omega, \quad \mathbf{B}(\mathbf{r},\omega) = \frac{1}{\sqrt{2\pi}}\int_{-\infty}^{+\infty}\mathbf{F}(\mathbf{r},t)e^{-j\omega t}dt \tag{2}$$

As $\mathbf{F}(\mathbf{r},t)$ is a real field we get $\mathbf{B}(\mathbf{r},-\omega) = \mathbf{B}^*(\mathbf{r},+\omega)$, and hence the energy density expands as $\Pi(\mathbf{r},t) = \Pi_1(\mathbf{r},t) + \Pi_2(\mathbf{r},t) + \Pi_3(\mathbf{r},t) + \Pi_4(\mathbf{r},t) =$

$$\frac{1}{4\pi}\left\{\int_0^{+\infty}\int_0^{+\infty}\mathbf{B}(\mathbf{r},\omega):\ddot{\mathbf{K}}:\mathbf{B}(\mathbf{r},\varpi)e^{-j(\omega+\varpi)t}d\omega d\varpi + \int_0^{+\infty}\int_0^{+\infty}\mathbf{B}(\mathbf{r},\omega):\ddot{\mathbf{K}}:\mathbf{B}^*(\mathbf{r},\varpi)e^{-j(\omega-\varpi)t}d\omega d\varpi \right.$$
$$\left. + \int_0^{+\infty}\int_0^{+\infty}\mathbf{B}^*(\mathbf{r},\omega):\ddot{\mathbf{K}}:\mathbf{B}(\mathbf{r},\varpi)e^{+j(\omega-\varpi)t}d\omega d\varpi + \int_0^{+\infty}\int_0^{+\infty}\mathbf{B}^*(\mathbf{r},\omega):\ddot{\mathbf{K}}:\mathbf{B}^*(\mathbf{r},\varpi)e^{+j(\omega+\varpi)t}d\omega d\varpi\right\} \tag{3}$$

Now if the medium allows orthogonal eigenstates $\mathbf{M}_{(n)}(\mathbf{r},t) = e^{-j\omega_{(n)}t}\mathbf{M}_{(n)}(\mathbf{r}) = e^{-j\omega_{(n)}t}\langle \mathbf{r}|(n)\rangle$ with $\langle (n)|(m)\rangle = e^{-j(\omega_{(m)}-\omega_{(n)})t}\delta_{(n)(m)} = \delta_{(n)(m)}$, and if the eigenkets $|(n)\rangle$ are complete, we get $|\mathbf{F}\rangle = \sum_{(n)} f_{(n)}|(n)\rangle$ for every bosonic field $|\mathbf{F}\rangle$ with $f_{(n)} = \langle (n)|\mathbf{F}\rangle$. Then

$$\mathbf{B}(\mathbf{r},\omega) = \frac{1}{\sqrt{2\pi}}\int_{-\infty}^{+\infty}\sum_{(n)} f_{(n)}e^{-j\omega_{(n)}t}\mathbf{M}_{(n)}(\mathbf{r})e^{+j\omega t}dt = \sqrt{2\pi}\sum_{(n)} f_{(n)}\mathbf{M}_{(n)}(\mathbf{r})\delta\left(\omega-\omega_{(n)}\right) \tag{4}$$

Hence, using (4) in (3)

$$\Pi_1(\mathbf{r},t) = \tfrac{1}{2}\sum_{(n)} f_{(n)}{}^2\mathbf{M}_{(n)}(\mathbf{r}) : \vec{\mathbf{K}} : \mathbf{M}_{(n)}(\mathbf{r})\exp\left(-j2\omega_{(n)}t\right) = \Pi_4^*(\mathbf{r},t) \tag{5.1}$$

$$\Pi_2(\mathbf{r},t) = \tfrac{1}{2}\sum_{(n)} \left|f_{(n)}\right|^2\mathbf{M}_{(n)}(\mathbf{r}) : \vec{\mathbf{K}} : \mathbf{M}_{(n)}^*(\mathbf{r}) = \Pi_2(\mathbf{r}) = \Pi_3^*(\mathbf{r}) \tag{5.2}$$

Integrating from (5), the total energy is obtained and given by

$$E(t) = \iiint \Pi(\mathbf{r},t)d^3r = E_1(t) + E_2(t) + E_3(t) + E_4(t) \tag{6.1}$$

$$E_2(t) = E_3(t) = \tfrac{1}{2}\sum_{(n)}\left|f_{(n)}\right|^2, \quad E_1(t) = E_4^*(t) = \tfrac{1}{2}\sum_{(n)} f_{(n)}{}^2 e^{-j2\omega_{(n)}t}\iiint \mathbf{M}_{(n)}(\mathbf{r}) : \vec{\mathbf{K}} : \mathbf{M}_{(n)}(\mathbf{r})d^3r \tag{6.2}$$

Here, the second and the third terms are time independent while the first and the latter ones vibrate with $\pm 2\omega_{(n)}$ frequency, and hence have zero contribution to the slowly time-varying components. Now, using (6) it is clear that the total time-average energy of the system is simply given by $E = \sum_{(n)}\left|f_{(n)}\right|^2 = \sum_{(n)}\tfrac{1}{2}\left(f_{(n)}^* f_{(n)} + f_{(n)}f_{(n)}^*\right)$, which shows that the total time-average energy of the system is equal to the sum of squared amplitudes of its eigen-modes.

2.1. Classical Hamiltonian

As bosons can occupy same level of energy with population more than one, this can produce a macroscopic quantum field $\mathbf{F}(\mathbf{r},t)$. We can define the functions $a_{(n)}(t) = \left(\hbar\omega_{(n)}\right)^{-\frac{1}{2}} f_{(n)}e^{-j\omega_{(n)}t}$ which satisfy $\frac{d}{dt}a_{(n)}(t) = -j\omega_{(n)}a_{(n)}(t)$. This allows us to rewrite the total energy in the different form of

$$E = \sum_{(n)} E_{(n)} = \sum_{(n)}\tfrac{1}{2}\hbar\omega_{(n)}\left[a_{(n)}^*(t)a_{(n)}(t) + a_{(n)}(t)a_{(n)}^*(t)\right] \tag{7}$$

For a stress field due to phonons, a tensor representation is needed instead of vector field $\mathbf{F}(\mathbf{r},t)$, such as $\boldsymbol{\sigma}(\mathbf{r},t) = \left[\sigma_{ij}(\mathbf{r},t)\right]$ [1]. The energy density will be given by $\Pi = \frac{1}{2}\boldsymbol{\sigma} : \vec{\mathbf{c}} : \boldsymbol{\sigma} = \frac{1}{2}\sum_{i,j,k,l}\sigma_{ij}c_{ijkl}\sigma_{kl}$, as well. Here, instead of $\vec{\mathbf{K}} = \left[K_{ij}\right]$ which is a second-rank tensor, we have to use the fourth-rank elasticity tensor $\vec{\mathbf{c}} = \left[c_{ijkl}\right]$, to get scalar energy density. The rest of the analysis will be exactly the same.

3. Transition to Quantum Mechanics

Similarity of (7) to that of a quantized harmonic oscillator suggests using a sum on the energies of oscillators with frequencies $\omega_{(n)}$, to construct the quantum mechanical Hamiltonian. So we define the ladder operators in a standard way as $\hat{a}_{(n)}(t) = \hat{a}_{(n)}(0)e^{-j\omega_{(n)}t}$ and $\hat{a}_{(n)}^\dagger(t) = \hat{a}_{(n)}^\dagger(0)e^{+j\omega_{(n)}t}$, with the algebra $\left[\hat{a}_n, \hat{a}_m^\dagger\right] = \delta_{nm}$ and $\left[\hat{a}_n, \hat{a}_m\right] = 0$. This would guarantee the independence of bosonic modes. Therefore, the quantum mechanical Hamiltonian will be

$$\mathbb{H} = \sum_{(n)}\mathbb{H}_{(n)} = \sum_{(n)}\frac{1}{2}\hbar\omega_{(n)}\left[\hat{a}_{(n)}^\dagger(t)\hat{a}_{(n)}(t) + \hat{a}_{(n)}(t)\hat{a}_{(n)}^\dagger(t)\right] \tag{8}$$

Furthermore, the ladder operators satisfy $\frac{d}{dt}\hat{a}_{(n)} = +\frac{j}{\hbar}\left[\mathbb{H}, \hat{a}_{(n)}\right]$ and likewise $\frac{d}{dt}\hat{a}_{(n)}^\dagger = +\frac{j}{\hbar}\left[\mathbb{H}, \hat{a}_{(n)}^\dagger\right]$.

4. Heat Capacity

Heat capacity is defined as

$$C = \partial E/\partial T \tag{9}$$

To obtain C, it is enough to know the number distribution of bosons at the given temperature T under thermodynamic equilibrium. Now, a phononic field under thermal equilibrium is given by [2]

$$|\psi\rangle = \sum_{\{m\}} \psi_{\{m\}} |\{m\}\rangle \tag{10}$$

Here $\psi_{\{m\}}$ are coefficients of the expansion and $|\{m\}\rangle = |m_{(0)}\rangle |m_{(1)}\rangle |m_{(2)}\rangle \cdots |m_{(n)}\rangle \cdots$ are eigenkets, meaning that the mode (n) can have $m_{(n)}$ bosons each having an energy of $\hbar\omega_{(n)}$. Now, application of Bose-Einstein statistics, requires that the probability of occupation of state $\hbar\omega_{(n)}$ with $m_{(n)}$ bosons is $P(m_{(n)}) \propto e^{-m_{(n)}\beta_{(n)}}$ with $\beta_{(n)} = \hbar\omega_{(n)}/kT$, in which k is the Boltzmann's constant. So, we get

$$|\psi\rangle = \sum_{\{m\}} \sqrt{P_{\{m\}}} |\{m\}\rangle = \sqrt{N} \sum_{\{m\}} \exp\left(-\tfrac{1}{2} \sum_{n} m_{(n)}\beta_{(n)}\right) |\{m\}\rangle \tag{11.1}$$

$$N = \prod_{(n)} N_{(n)} = \prod_{(n)} [1 - \exp(-\beta_{(n)})] \tag{11.2}$$

where N is the normalization constant to support $\langle\psi|\psi\rangle = 1$. A simple check shows that the Bose-Einstein distribution for bosons is readily obtained through this method as $f(E) = \left[\exp\left(\frac{E}{kT}\right) - 1\right]^{-1}$. Now, the energy of the system under the thermodynamic equilibrium at temperature T is

$$E = \langle\psi|\mathbb{H}|\psi\rangle = N \sum_{\{m\}} \exp\left(-\sum_{n} m_{(n)}\beta_{(n)}\right) \langle\{m\}|\mathbb{H}|\{m\}\rangle \tag{12}$$

which can be simplified through lengthy but straightforward calculations [3] as

$$E = \sum_{(n)} \hbar\omega_{(n)} \left[\exp(\beta_{(n)}) - 1\right]^{-1} + E_0 \tag{13}$$

where $E_0 = \frac{1}{2} \sum_{(n)} \hbar\omega_{(n)}$ is zero point energy. Using (9) we get the exact relation for heat capacity [3]

$$C = \frac{1}{kT^2} \sum_{(n)} \left\{\hbar\omega_{(n)} \left[\exp\left(\hbar\omega_{(n)}/kT\right) - 1\right]\right\}^2 \tag{14}$$

Small size of a nanostructure makes the energy spectrum both discrete and finite, so that (14) has a finite number of terms. This makes it possible to calculate the accurate value of (14) through numerical methods. At high temperatures, to a good approximation we have $\exp\left(\hbar\omega_{(n)}/kT\right) - 1 \approx \hbar\omega_{(n)}/kT \ll 1, \forall (n) \in \mathcal{R}^3$, so that $E \approx \sum_{(n)} kT + E_0 = LkT + E_0$, where L is number of modes of the system. Thus $C \approx Lk$, will be independent of T. On the other hand, at very low temperatures we have $\exp\left(-\hbar\omega_{(n)}/kT\right) \approx 0, \forall (n) \in \mathcal{R}^3$, so that we get $C \approx (kT)^{-2} \sum_{(n)} [\hbar\omega_{(n)}]^2 \exp\left(-2\hbar\omega_{(n)}/kT\right)$. Therefore at the limit of zero temperature we always have $C = 0$. Furthermore, when the size of structure approaches infinity, the relation for heat capacity reduces to

$$C = \left(\frac{\hbar}{kT}\right)^2 \int_0^\infty \left[\omega \exp\left(\hbar\omega/kT\right) - 1\right]^{-2} D(\omega)\, d\omega \tag{15}$$

with $D(\omega) = \sum_{(n)} \delta\left[\omega - \omega_{(n)}\right]$ being the density of states. But for an N-dimensional bulk medium, we have $D(\omega) = \alpha\omega^{N-1}$, and the corresponding lattice heat capacity will be given by direct integration as $C \propto T^N$, being in agreement with known results.

5. Examples

5.1. Fullerenes: C60, C70, C80

Fullerenes are zero-dimensional (0D) nanostructures of Carbon (figures 1,2), and their calculated heat capacities are calculated and plotted in figure 3. It is interesting to note that all heat capacities detach from the origin at $C(T=0)=0$, and furthermore they behave as $C \propto T^0 = \text{const}$ for temperatures roughly below $30\,\text{K}$.

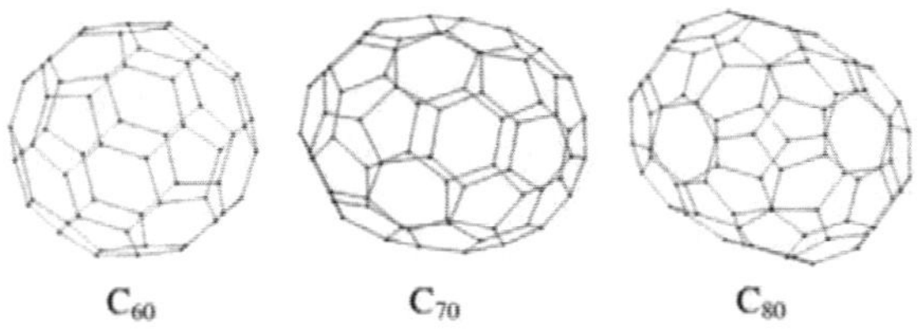

Figure 1. Fullerenes [4]

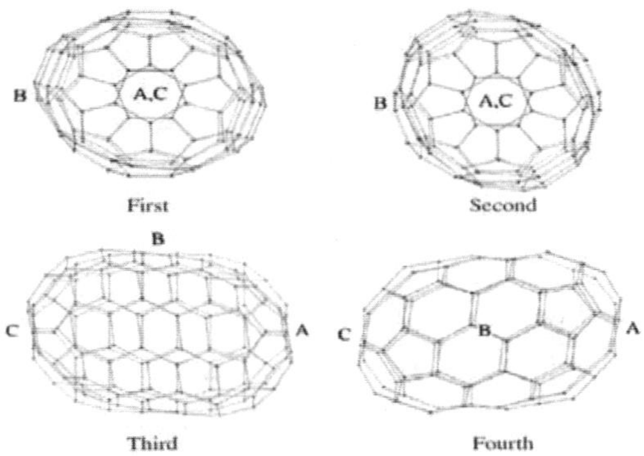

Figure 2. Vibrational modes of C80 [4]

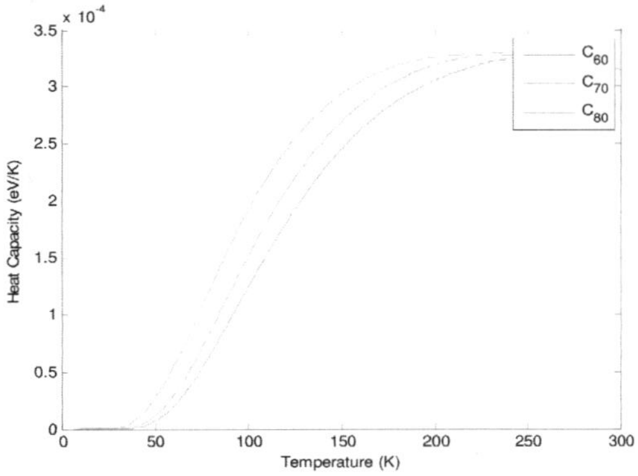

Figure 3. Calculated heat capacities for fullerenes.

6. Conclusions

We presented a rigorous and general method for quantization of bosonic fields, which was applicable to phononic fields as well. This enabled us to find the exact expression for heat capacity. We studied its behavior at high and low temperatures and also at the limit of infinite multi-dimensional structures. Agreement to well-known expressions was noticed. As application examples, we have calculated and plotted the heat capacities for C60, C70, and C80 fullerenes using their recently published spectrum.

References

[1] Stroscio M A and Dutta M 2001 *Phonons in Nanostructures* (Cambridge University Press)
[2] Schleich W 2001 *Quantum Optics in Phase Space* (Berlin: Wiley-VCH)
[3] Khorasani S 2009 *Applied Quantum Mechanics* (in Persian) (Sharif University of Technology)
[4] Sakhaee-Pour A and Vafai A 2010 *Appl. Phys. Lett.* **96** 021903

Journal of Physics: Conference Series **248** (2010) 012058 doi:10.1088/1742-6596/248/1/012058

"Spontaneous Movable Semifluxons "- New Phenomenon arising in Nano-Electronic Superconducting System

Hanaa Farhan-Hassan[1,2] and F V Kusmartsev[1]

[1] Department of Physics, Loughborough University, Leicestershire LE11 3TU, UK
[2] Department of Physics, King Abdulaziz University, Jeddah, KSA.

Email: hfarhan@kau.edu.sa, F.Kusmartsev@lbro.ac.uk

Abstract. Semifluxon is a very rare topological excitation. It is a spontaneous Josephson vortex carrying a half of the magnetic flux quantum ϕ_0 created on the boundary between 0 and π junctions. Semifluxons play an important role as qubits for quantum computing. Typically they are static and arise in a variety of Josephson 0- π nano-junctions, in particular. The properties of semifluxons are very different from the properties of fluxon. Typically, semifluxons are pinned at the phase discontinuity point in 0- π long Josephson junctions (LJJ) and may have two polarities carrying the flux $\pm\ \phi_0/2$, that was attracting to use them as qubits. In this paper, we report for a first time about a novel type of semifluxons arising in superconductors, which are movable, i.e. not pinned. We show that a spontaneous generation of movable semifluxons is created in conventional LJJ due to flux cloning in extended T-Josephson nano-junctions. We also show that there is a strong interaction between semifluxon and anti-semifluxon once they have been generated and that leads to their very short life time, i.e. to their fast annihilation and next recreation.

1. Introduction

Josephson nano-junctions are one of the most important elements in superconducting electronic circuits. Josephson vortices are stable topological excitations arising in long Josephson junctions (LJJs) subjected to magnetic field and to strong electric current pulses [1],[2]. They are also known as Sine-Gordon (SG) solitons. Magnetic flux trapped by these vortices or SG solitons is quantized with the single flux quantum ϕ_0. The study of fluxons in Josephson nano-junctions has attracted a lot of attention because of their potential applications in superconducting electronics [3]-[6], digital logic circuitry [7]-[11] and, in general, modern quantum circuits [2]. The very long Josephson junction may be used as Josephson transmission line (JTL) through which a SG soliton can freely move along.

The standard description of fluxon dynamics in JTL is based on the (2+1)-dimensional unperturbed sine-Gordon equation. With its boundary conditions [12], this description is extremely accurate:

$$\phi_{xx} + \phi_{yy} - \phi_{tt} - \sin(\phi) = 0 \tag{1}$$

where $\phi(x, y, t)$ is the superconducting phase difference, and the subscripts denote partial derivatives of x, y, which are normalized to the Josephson penetration depth λ_J, and time t, which is normalized to the inverse of the Josephson angular plasma frequency ω_J. The exact SG soliton solutions of the SG equation have been given as

$$\phi(x,y,t) = 4\,\tan^{-1}\left(\exp\left(\frac{x-x_0-ut}{\sqrt{1-u^2}}\right)\right) \tag{2}$$

With the SG soliton energy E as equal to

$$E = \int_{-\infty}^{\infty} dx \int_0^y dy \left[\frac{\phi_t^2}{2} + \frac{\phi_x^2}{2} + \frac{\phi_y^2}{2} + 1 - \cos(\phi)\right] = \frac{8W}{\sqrt{1-u^2}} \tag{3}$$

Where u is a velocity of soliton propagating in a straight two-dimensional strip of the width W. For our next consideration we assume that its initial position is located at $x = x_0$ for $t = 0$. The velocity u (normalized to the Swihart velocity $\bar{c}$, the maximum electromagnetic propagation velocity in the junction) may take values within the range $0 < u < 1$ [13]-[20].

During last years it was shown both theoretically [21]-[22] and experimentally [23]-[24] that there was also significant interest in vortex excitations carrying an arbitrary fraction of standard flux quantum as $\phi = \phi_0\ \kappa/2$, where κ is a the value of the discontinuity [25]. For π discontinuity, Josephson vortices carry only half of the magnetic flux quantum. This half-vortex has been studied theoretically by several authors [1],[2], [21],[25], [27], [28]-[34]. Because of containing one-half of the flux quantum, they are called ''semifluxon''. In other words, a 2π vortex is a fluxon carrying ϕ_0, while a π vortex is a semifluxon carrying $\phi_0/2$, see figure 1 [25]. A single semifluxon can have positive or negative polarity carrying the flux $+\phi_0/2$ (semifluxon) or $-\phi_0/2$ (anti-semifluxon), respectively [29],[27]. In addition, it has been conjectured that semifluxons can be utilized in superconducting memory and logic devices and accurate measurements of the Josephson penetration length of a superconductor (for details, see Refs [22], [30]).

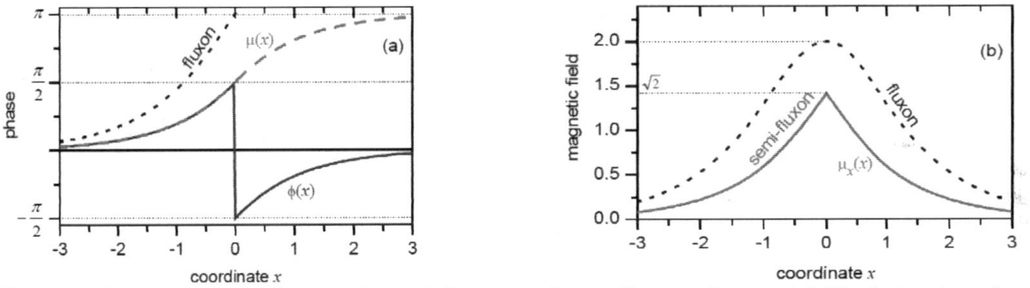

Figure 1 (color online). Comparison of fluxon and semifluxon shapes. (a) The behavior of total phase $\phi(x)$ and of magnetic component $\mu(x)$ only. (b) Magnetic-field profile $\mu_x(x)$, see references [25] and [34], for details.

Semifluxons exist in specially designed long Josephson junctions (LJJ), which are so-called "0-π" LJJ, see Fig. 2. This 0-π junction consists of two joint parts. The 0-junction which is a conventional Josephson junction (one may literally say, "with positive critical current") and the other second part is a π-junction, which is a standard junction "with a negative critical current", see for example, the Ref. [25]. Semifluxons can be created at the boundaries between the 0 and π regions provided the Josephson junction is long enough [31]. The 0-π boundaries are sometimes called *discontinuity points* [32].

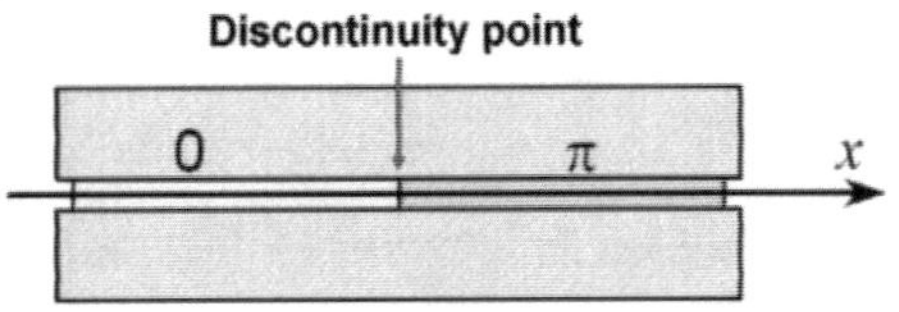

Figure 2. 0-π-tunnel long Josephson junction geometry.

Semifluxons are very interesting objects which are not studied in detail, especially experimentally, because of difficulties with a fabrication of 0-π junctions [29]. Although the classical properties of semifluxons are understood, their quantum behavior and their possible applications in the quantum domain still have to be studied [31]-[34]. The purpose of the present work is to demonstrate the existence of a new novel phenomenon for semifluxons carrying half-integer magnetic flux quantum in conventional extended Josephson junction. Whereas semifluxons and fluxons can be described by sine-Gordon-equation, their properties are very different; fluxons are topologically stable solitons with very long life time while the semifluxons created dynamically have a very short life time . We present the first evidence and observation of moving semifluxon which may arise in a manner similar to normal fluxon. In addition, we show here that semifluxon can be created and moved in extended Josephson junctions by means of flux cloning phenomenon arising in the T- shaped Josephson junctions.

2. Solitons in T- Josephson junctions

In Josephson electronic circuits, two Josephson transmission lines (JTLs) can be connected together to form a T-shaped junction. In general, T-junctions, have two different widths; a main JTL (MJTL) of width W_0 along the x −axis, and an additional JTL (AJTL) of width W along the y −axis (see figure 4). They can be connected to form two kinds of T junctions of the different widths (see figure 3 (a) and (b)) [36]-[38]. Flux cloning is a newly discovered phenomenon that occurs as a result of vortex passing through T-shaped junctions [36],[37]. In connection area, the dynamic behavior of flux is shown that there are two different types of behaviors - either reflection from the T junction (figure 4 (a)) or flux cloning (figure 4 (b)). When the mother vortex in the MJTL approaches T-junction and its speed satisfies certain conditions, it splits to generate two vortices [36]-[38]. These behaviors are dependent on the value of velocity of fluxon, less or greater than critical velocity, which has been theoretically calculated for T-junction in figure 3 (a) [36],[37].

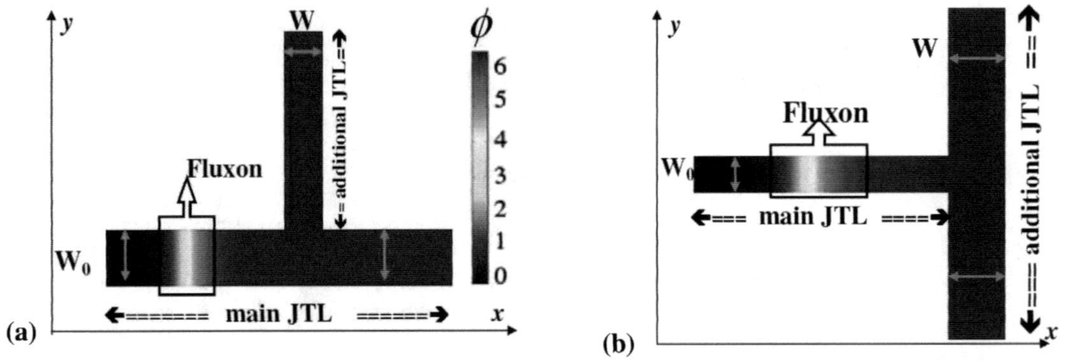

Figure 3 (color online). (a) the general shape of the Josephson T-junctions studied in the present paper, (b) The T-junctions rotated by - 90^0. The T-junction consists of two attached Josephson transmission lines. The colour scale represents the superconducting phase difference ϕ. The soliton-fluxon is represented by a bright colour strip.

In T-junction, the condition of critical velocity of soliton has to satisfy the flux cloning condition:

$$u_c = \frac{\sqrt{W(W+2W_0)}}{W+W_0} \quad (4)$$

Otherwise, the soliton will be reflected if the velocity (u) is less than u_c [36]-[37]. After cloning, a new baby vortex generate in the AJTL whereas the mother vortex will still propagate continuously in MJTL. In the case $W \geq W_0$, the behavior of soliton changes because the critical velocity becomes very high close to Swihart velocity. In addition, there is a discrepancy between analytical estimation based on the soliton energy arguments[36-37] and numerical evaluation of critical velocities when $W > W_0$, see figure 5. The theoretical prediction of critical velocity can be calculated by equation (4), while the numerical value has been evaluated when the soliton changes behavior from reflection to splitting [36], see figure 4. In this paper, we show that movable semifluxons can be excited in extended T-Josephson junctions when fluxon will be crossing T-junctions. In general, by extended T-Josephson junctions we mean that the widths of MJTL (W_0) and AJTL (W) are larger than the Josephson length (λ_J). Here, extended T-Josephson junctions are denoted to both $W_0, W, > \lambda_J$ and $W > W_0$.

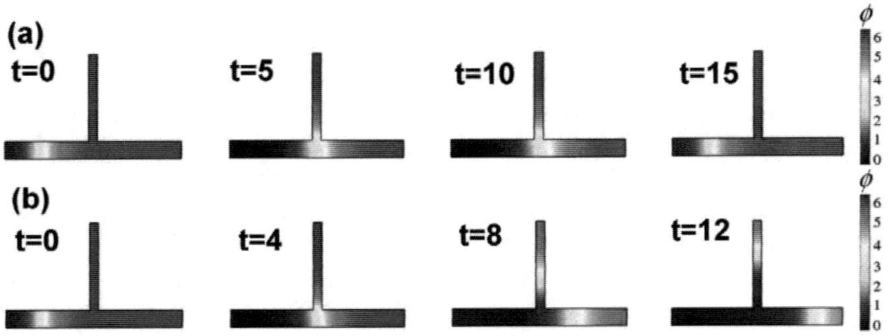

Figure 4 (color online). Numerical simulations of the time evolution of superconducting phase difference; the yellow-green stripe corresponds to the soliton-fluxon (a) The time snapshots of the reflection of an incident fluxon-soliton propagating without cloning and (b) Cloning of the fluxon-soliton propagating through the T -junction. After the time t=4 due to the cloning there are created two fluxon-solitons. The color scale on a right side represents the superconducting phase difference ϕ [36].

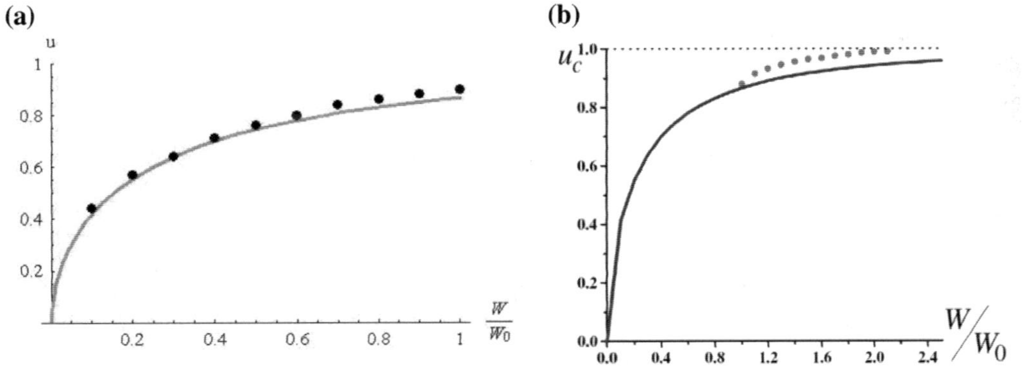

Figure 5 (color online). Schematic representation of the dependence of the critical velocity on the width of the additional JTL (a) The case when $W \leq W_0$ [36]-[37]. (b) The case when $W \geq W_0$. Dots represent the data obtained in numerical calculation made within 2D geometry, solid lines are related to qualitative analytic estimations obtained from eq. (4).

In special case, when the energy of incident fluxon is very large, the T-shaped Josephson junction allows vortex excitations carrying a half of the standard flux quantum. When the width W is too wide, mother vortex shows strange behavior in both MJTL and AJTL. In contrast to original prediction (u_c) indicating that fluxon moving with velocity high than critical will be cloned[36-37], we show here that in some cases the mother vortex is reflected from the T-junction while at the same time at the moment of the reflection some new excitations will be created and continuously propagate in MJTL. In connection area of T-junction, soliton starts to divide and to clone but it has no enough energy, ie it cannot generate a complete quantized fluxons (baby and mother vortex). Then, the fluxon-soliton will be trapped in the T-junction and semifluxons will be created. Then it becomes possible for them to propagate in AJTL and MJTL. In MJTL, the semifluxon after creation moves and its stability looks similar of the original fluxon-soliton. In AJTL, however, a massive shape wave (tsunamis) [39],[40] will be appeared together with semifluxon which becomes unstable and fast decays into plasma waves.. Because of the broad wideness of AJTL, W, the propagation of unstable semifluxon over a large distance requires a high energy and high velocity to move along the AJTL, which can be estimated from the equation (4). Therefore, unstable semifluxon will struggle to continue its motion due to the condition that its velocity is not large enough. Then, the unstable semifluxon, which moves along the AJTL, will be reflected and moved back to the MJTL in x and -x directions. In connection area of the T-junction, the trapped fluxon-soliton will be reflected as well and then starts propagating in - x directions. On the other hand, in x direction, unstable semifluxon starts moving and passing through the MJTL at the same time transforming as unstable anti-semifluxon. Typically unstable anti-semifluxon moves faster than stable semifluxon. Consequently, the semifluxon and anti-semifluxon are strongly attracting each other like quarkes in a high energy physics. This results typically in the fast enough annihilation between them. Hereafter, a generation of periodic stable of semifluxon and anti-semifluxon chains may be spontaneously appeared and propagate in MJTL. As a result, another and another annihilation and after consequent re-creation of such pairs will take place indicating on a complex dynamics existing between these new solitons, semifluxons and anti-semifluxons.

3. Numerical Studies of Semifluxon in T- Josephson junctions

Our numerical simulations of the flux cloning dynamics have been performed with the use of the finite element numerical analysis, in particular the technique package FEMLAB has been utilized. We have studied the time-dependent sine-Gordon-equation with boundary conditions depending on external magnetic field or a current. The dynamical behavior of Josephson vortices can be indicated by time dependent change of the value of the superconducting phase ϕ. The solution can be presented on the figures 7 – 9. It is convenient for each value of the phase difference to put in a correspondence a color. The dark blue color represents the minimum value of the phase ϕ and red color represents the maximum value of the phase ϕ. In general, semifluxons are presented as the phase difference change, $\Delta\phi = \pi$, while fluxon-solitons are presented as the phase difference change, $\Delta\phi = 2\pi$ (the intermediate color between these values represents the core of the vortex, see the color scale in figure 7). In the present paper we use the simplest boundary conditions $\phi_x(x = 0, L, y, t)= 0$ and $\phi_y(x, y = 0, W, t)=0$ (see reference [12] for details).

In all our numerical simulations, we have used initial conditions with the following form: for the phase $\phi(x,y,t)|_{t=0} = \phi_{soliton}(x)$ and $\partial\phi(x,y,t)/\partial t|_{t=0} = -u\ \partial\phi_{soliton}(x)/\partial t$ for its time derivative, where $\phi_{soliton}(x) = \phi(x,y,0)$. In addition, we limit our analyses to the case when the Josephson junction widths is W_0 =1 and consider many cases when W >1. First, we model 0-π–long Josephson junction discussed in references [21-34 using the finite element methods. The results, calculated here, are similar to previous results[30], see, for example, figures 1 and 6.

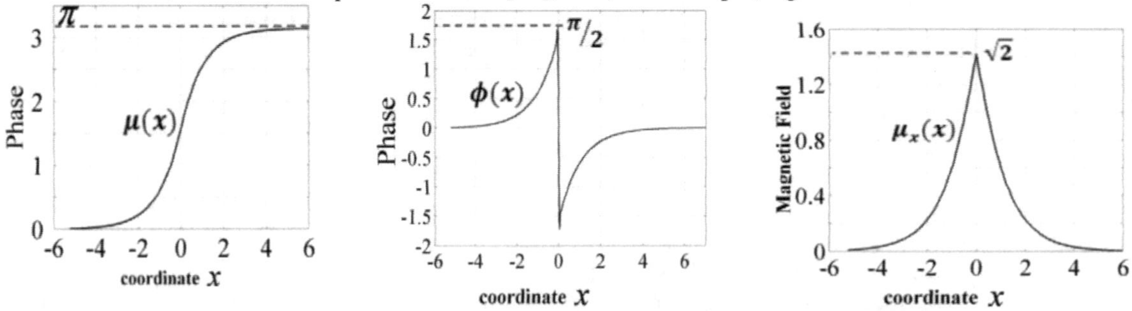

Figure 6 (color online). (a) The behavior of magnetic component $\mu(x)$ for a semi-fluxon pinned by 0-pi junction. It is stationary and does not change at any time. (b) The same, ie the behavior of semifluxon for total phase $\phi(x)$. (c) The stationary magnetic-field profile for the pinned semifluxon, $\mu_x(x)$.

Furthermore, we consider the case when the fluxon cannot be cloned, for example, when W=2.2. The critical velocity (u_c) is equal to 0.94. Note, although the fluxon-soliton is moving with the speed $u \cong 0.99$, the flux cloning does not occur. The numerical simulations show that in this case the soliton is reflected. In the figure 7, the mother vortex starts moving with high velocity (u= 0.985) and then it is reflected from the T-junction. Figure 7 shows that semifluxons can be created when soliton crosses T-shaped junction, ie at the time t=5. After the creation, this stable semifluxon is moved in MJTL, while at the same time another semifluxon created together with a massive shape wave, see, for example, the Refs [39],[40] is moved in AJTL (see the snapshot of the semifluxon dynamics presented in the figure 7 at the time t= 6). In addition, an unstable semifluxon in AJTL is reflected and moved in MJTL as unstable anti-semifluxon (presented on the same figure 7 at the time t = 8). Finally at the time t =12, semifluxon and anti-semifluxon are meet each other and annihilated. Then, spontaneous serious of movable pairs consisting of the semifluxon and anti-semifluxon are continuously generated and annihilated in MJTL. To simplify our result, we plot the phase difference ϕ along $x-$ axis; see figure 8 (a) and (b). The figure 8 (b) shows that at the time t =15, the series of pairs consisting of the semifluxon and anti-semifluxon appear and in the next moment their annihilation event takes place.

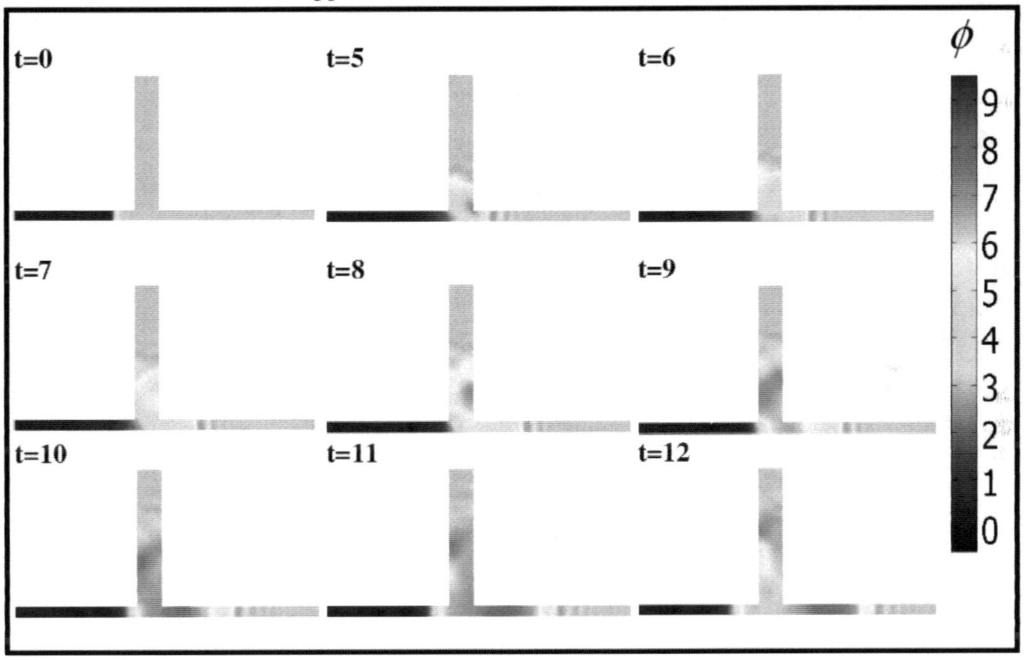

Figure 7 (color online). Numerical simulations of the creation of movable semifluxons. In this case no flux cloning occurs. An incident fluxon-soliton (created at t=0) is propagating with the velocity u= 0.985 when W= 2.2. For a comparison in this case the estimated critical velocity is u_c= 0.94[36]. A movable semifluxon is created in MJTL after fluxon-soliton crosses the T-junction (at t=-5). In a short time after this event, the anti-semifluxon begin to propagate in MJTL. The semifluxon and anti-semifluxon strongly interact. The color scale represents the superconducting phase difference ϕ. A fluxon-soliton is represented by 0 to 2π phase differences whereas (anti-)semifluxons are represented by phase differences as $\Delta\phi = \pi$.

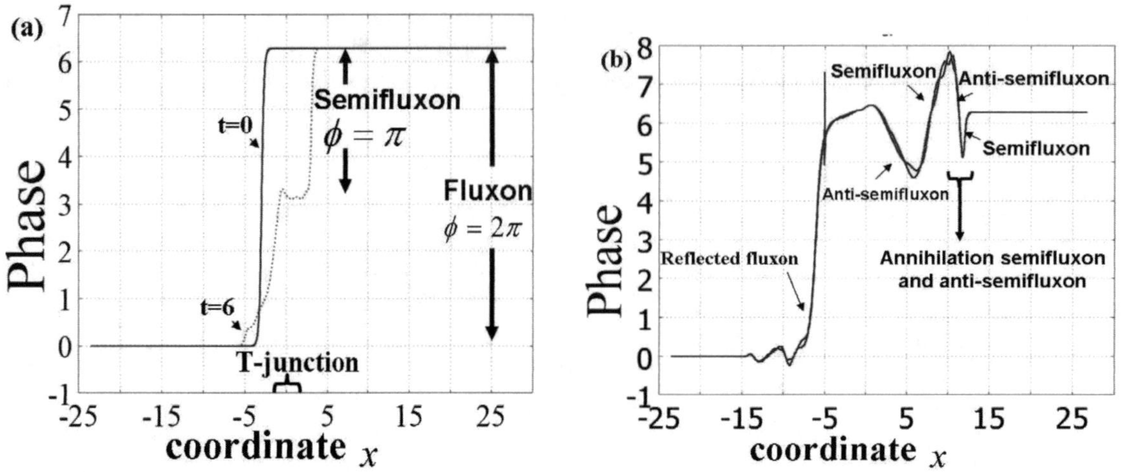

Figure 8 (color online). The change of the phase ϕ along $x-$ axis. An incident fluxon is propagating with a velocity higher than the critical one [36] (a) At the time t=0 (solid line), the change of the phase difference of incident fluxon-soliton as usual, is equal to 2π. At t=6 (dotted line), after fluxon has just crossed the T-junction, the change of the phase difference is equal to π as it is needed for a semifluxon, which is located at approximately $x = 4$. It is movable semifluxon and at t=15, there the series of the semifluxon and anti-semifluxon pairs appears. Each semifluxon has the change of the phase difference equal to π. The semifluxon and anti-semifluxon are colliding at approximately x = 12 and then again are annihilated by each other.

Eventually, we consider the case when the fluxon-soliton can be cloned. In the case W=1.7, the critical u_c= 0.93[36] while the numerical evaluation gives u_c= 0.979. A movable semifluxon can be temporarily formed inside the T-junction only if its velocity belongs to a specific critical range. The values of such velocities should be slightly lower than the critical velocity ($u \cong u_c$) obtained in our numerical experiments. In the figure 9, an incident fluxon-soliton propagates with the velocity u= 0.97. The numerical simulations show that movable semifluxons in MJTL and AJTL are created at the time t=5, see figure 9 (a) and (b). An annihilation of the semifluxon and a reflection of the incident fluxon are shown in figure 9 (a) at the time t =13. In summary, we show that there are three types of the dynamical behavior for the fluxon-solitons in extended T-Josephson junctions: 1) reflection (see figure 4 (a)), 2) flux cloning (see figure 4 (b)) and 3) a spontaneous generation of movable semifluxon and anti-semi-fluxons that arises during a reflection from the T-junction of the high energy fluxon-solitons (see figure 9 (a)).

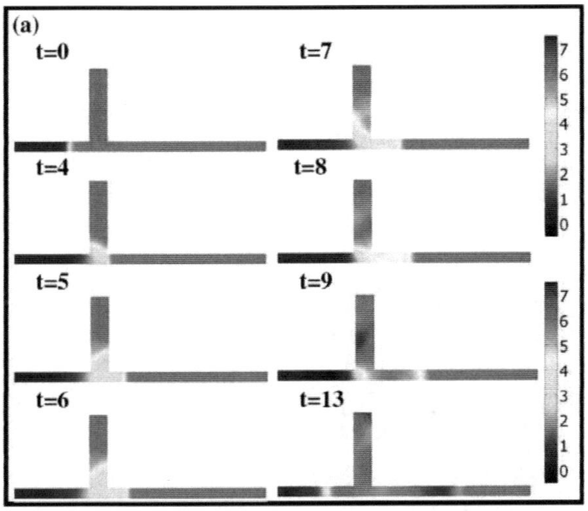

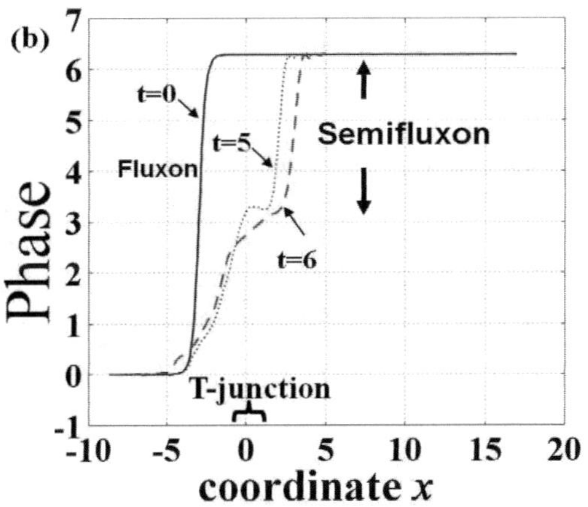

Figure 9 (color online). Numerical simulations of spontaneous movable semifluxon generation arising during a reflection of an incident high energy fluxon-soliton propagating with the velocity u= 0.97, W= 1.7. The critical velocity is here equal to u_c= 0.979. A movable semifluxon is here temporarily formed after the initial fluxon-soliton has crossed the T-junction. Then it is annihilated after a very short time. The color scale represents the superconducting phase difference ϕ. A soliton is represented by 0 to 2π phase differences whereas (anti-)semifluxons are represented by phase differences as $\Delta\phi = \pi$. (b) The phase difference ϕ along $x -$ axis at t=0 (solid line), t=5 (dotted line), and t=6 (dashed line). A movable semifluxon has been temporarily formed after the fluxon has crossed the T-junction. For a movable semifluxon the change of the phase difference is equal to π .

4. Conclusions

We demonstrate here numerically that in conventional T-Josephson junctions it is possible the existence of movable semifluxons. In all previous studies only localised, ie unmoveable semifluxons pinned by topological defects have been discussed[29-35]. In the case of the very large extended T-shaped Josephson junctions, when $W > W_0$, our numerical calculations show clear evidence of three different interesting and important features for the semifluxon generation: (a) (anti-)semifluxons can be spontaneously appeared at the T-shaped junctions when a high energy fluxon-soliton is crossing this junction, (b) (anti-)semifluxon is not localized but it may be freely moved along the long branches of the Josephson junction, (c) there a spontaneous generation of periodic movable (anti-)semifluxon chains may occur. Also there a spontaneous series of annihilation and creation between anti- and semi-fluxons takes place. The finding of these semifluxon and their dynamics is very useful to understand the behavior and decay of the high elementary particles. If the fluxon is equivalent to such particle, then the semifluxon will be equivalent to a quark. Thus, the T-shaped Josephson junction may be considered as a very economic laboratory to study such a quark creation and other high energy phenomena.

Acknowledgments
The work was supported by the ESF network-program AQDJJ, and by EPSRC grant, EP/F005482/1. We are very grateful to Dima Gulevich and Reinhold Kleiner for very useful discussions.

References

[1] Dewes A, *et al.* 2008, *Phys. Rev. Lett.,* **101**, 247001
[2] Gaber T *et al.* 2007, *Appl. Phys.* A **89** 587–592
[3] Tahara S *et al.* 2001, *IEEE Trans.on Appl. Super.* **11** 463
[4] Raissi F 2005, *Appl. Phys. Lett.* **86**, 263503
[5] Raissi F, and Nordman J 1994, *Appl. Phys. Lett.* **65** 1838
[6] Olsen O, Ustinov A and Pedersen N 1993, *Phys. Rev.* B **48**, 13133
[7] Koshelets V 1991, *Supercand. Sci. Technol.* **4** 55S560
[8] Shaju P, Kuriakose V 1999, *Physica* C **322** 163
[9] Guéret P 1975, *IEEE Trans. Magn.* MAG-**11** 751
[10] Nakajima K, Onodera Y and Ogawa Y 1976, *J. of Appl. Phys.* **47** 1620
[11] Nakajima K and Onodera Y 1987, *J. of Appl. Phys.* **49**, 2958
[12] Barone A. and Paterno G 1982, *Physics and Applications of the Josephson Effect* (John Wiley: New York)
[13] Schmidt V V 1997, *The Physics of Superconductors: Introduction to Fundamentals and Applications*, ed P. Müller , A. V. Ustinov (Springer:New york)
[14] Doderer T *et al. 1993*, *Phys. Scr.* **T49**A 172-175
[15] Fulton T and Dynes R 1973, *Solid State Commun.* **12**, 57
[16] Duzer T and Turner C 1990, *Principles of Super Divices & circuts* (Prentice-Hall PTR: New Jersy)
[17] Pedersen N and Ustinov A 1995, Supercond. Sci. Technol. **8** 389
[18] Li H, Shen S, Liang J and Liu W 2005, *Phys. Rev.* B **72** 014546
[19] Ustinov A 1998, *Physica* D **123** 315
[20] Lachenmann S *et al.* 1993, *Phys. Rev.* B **48** 22
[21] Xu J, Miller J, Jr and Ting C 1995, *Phys. Rev.* B **51** 11958 – 11961
[22] Kuklov A, Boyko V and Malinsky J 1995, *Phys. Rev.* B **51** 11965
[23] Kirtley J, Tsuei C and Moler K 1999, *Science* **285** 1373 -1375
[24] Sugimoto A, Yamaguchi T and Iguchi I 2002, *Physica* C **367** 28
[25] Kusmartsev F V, 1992 Phys. Rev. Lett. **69** 2268
[26] Goldobin E, Koelle D and Kleiner R 2004, *Phys. Rev.* B **70** 174519
[27] Susanto H *et al.* 2003, *Phys. Rev.* B **68** 104501
[28] Walser R *et al.* 2008, *New J. of Phys.s* **10** 045020
[29] Goldobin E, Stefanakis N, Koelle D and Kleiner R 2004, *Phys. Rev.* B **70** 094520
[30] Susanto H and Gils S 2004, *Physica* C **408** 579-580
[31] Goldobin E *et al.* 2010, *Phys. Rev.* B **81** 054514
[32] Goldobin E *et al.* 2005, *Phys. Rev.* B **71** 104518
[33] Goldobin E, Sterck A, Gaber T, Koelle D and Kleiner R 2004, *Phys. Rev. Lett.* **92** 057005
[34] Goldobin E, Koelle D and Kleiner R 2003, *Phys. Rev.* B **67** 224515
[35] Goldobin E, Koelle D and Kleiner R 2002, *Phys. Rev.* B **66** 100508
[36] Gulevich D R and Kusmartsev F V 2006, *Phys Rev. Lett.* **97** 017004
[37] Gulevich D R and Kusmartsev F V 2007, *Supercond. Sci. Technol.* **20** S60-S67
[38] Farhan-Hassan H and Kusmartsev F V 2009, *Int. J. of Modern Phys.* B **23** 4345-4351
[39] Gulevich D *et al.* 2008, *Phys. Rev. Lett.* **101** 127002
[40] Gulevich D *et al.* 2008, *J of Appl. Phys.* **10** 064507

Physical limits for scaling of integrated circuits

Waldemar Nawrocki

Poznan University of Technology, Faculty of Electronics and Telecommunications, ul. Piotrowo 3A, PL 60965 Poznan, Poland

E-mail: nawrocki@et.put.poznan.pl

Abstract. In this paper we discuss some physical limits for scaling of devices and conducting paths inside of semiconductor integrated circuits (ICs). Since 40 years only a semiconductor technology, mostly the CMOS and the TTL technologies, are used for fabrication of integrated circuits in the industrial scale. Miniaturization of electronic devices in integrated circuits has technological limits and physical limits as well. In 2010 best parameters of commercial ICs shown the dual-core Intel Core i5-670 processor manufactured in the technology of 32 nm. Its clock frequency in turbo mode is 3.73 GHz. A forecast of the development of the semiconductor industry (ITRS 2009) predicts that sizes of electronic devices in ICs circuits will be smaller than 10 nm in the next 10 years. The physical gate length in a MOSFET will even amount 7 nm in the year 2024. At least 5 physical effects should be taken into account if we discuss limits of scaling of integrated circuits.

1. Introduction

Scaling of electronic devices in integrated circuits has technological limits and physical limits as well. Since 40 years only a semiconductor technology, mostly the CMOS and the TTL technologies, are used for fabrication of integrated circuits in the industrial scale. Probably the CMOS technology will be used at least in the next 10-15 years. In 2010 best parameters of commercial ICs shown the dual-core Intel Core i5-670 processor manufactured in the technology of 32 nm. Its clock frequency in turbo mode is 3.73 GHz. The other example is the quad-core Intel Core i7-975 manufactured in the technology of 45 nm. The last forecast of the development of the semi-conductor industry (ITRS 2009) predicts that sizes of electronic devices in ICs circuits will be smaller than 10 nm in the next 10 years [1]. The physical gate length in a MOSFET will even amount 7 nm (see Table 1) in the year 2024. One can notice that the previous forecast (ITRS 2007) gave different values of the predicted length of a physical gate in ICs: 9 nm in 2016 and 4 nm in 2022.

At least 5 physical effects should be taken into account if we discuss limits of miniaturization of integrated circuits:

- spread of doping atoms in a semiconductor material; each dopant would induce a relatively high potential bump;
- quantization of both electrical and thermal conductance in narrow and thin transistors`channels and in conducting paths;
- propagation time of electromagnetic wave along and across a chip (IC);
- electrostatics; a loss of electrostatic control of the drain current vs the gate voltage;

- electron tunneling between a source and a drain inside a MOSFET through a insulation (oxide).

Table 1. Data of integrated circuits (IC) according to the Report of The International Technology Roadmap for Semiconductors (Edition 2009) [1]

Year		2009	2012	2015	2018	2021	2024
Physical gate length in a FET transistor inside of ICs (microprocessor unit – MPU)	nm	29	22	17	12.8	9.7	7.4
Clock frequency (on chip MPU)	GHz	5.45	6.82	8.52	10.6	13.3	16.6
Functionality of IC (number of transistors)	mln	2212	4424	8848	17696	35391	70782
Supply voltage	V	1.0	0.9	0.81	0.73	0.66	0.60
Dissipated power (cooling on)	W	143	158	143	136	133	130

We discuss spread of doping atoms in a semiconductor material and quantization of both electrical and thermal conductance in nanostructures. We mention only the other physical effects important for scaling of integrated circuits.

2. Spread of doping atoms in a semiconductor material

Classical theories of electrical and thermal conductance assume a huge number of atoms and free electrons. Let`s assume a silicon cube with one side dimension of a and with common doping of 10^{16} cm^{-3}. In a n-doped silicon cube with the size of $(100\ nm)^3$ there are 5×10^7 atoms and 10 free electrons at 300 K, but in the Si cube with the size of $(10\ nm)^3$ there are 5×10^4 atoms and 1% chance only to find **one** free electron. Free electrons are necessary for electrical conductance as charge carriers. It means not only that classical theories of conductance are not valid for nanostructures. It means that common doping in semiconductor material is not sufficient for electronic devices of nanometric size.

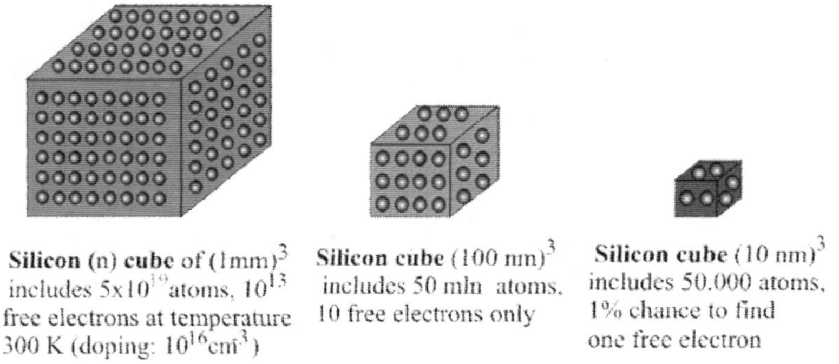

Figure 1. Free electrons in a silicon cube

In order to keep the conductive properties of the semiconductor material one should apply more intensive doping, eg. 10^{20} cm^{-3}. However such intensive doping decreases resistivity of the material dramatically from 2×10^{-3} Ωm to 10^{-5} Ωm, respectively (for n-type Si, at 300 K). Low number of free electrons should be scattered evenly in whole volume of a material.

3. Quantization of electrical conductance in nanostructures

Electric and thermal proprieties of electronic devices or paths with nanometer sizes are not more described by a classical theory of conductance but by quantum theories. The theoretical quantum unit of electrical conductance $G_0 = 2e^2/h$ was predicted by Landauer in his theory of electrical conductance [2]. Parameters characterizing the system are a Fermi wavelength λ_F ($\lambda_F = 2\pi/k_F$, where k_F is the Fermi wavevector). and a mean free path Λ. For metals like gold $\lambda_F \approx 0.5$ nm is shorter then free electron path Λ ($\Lambda_{Au} = 14$ nm). If a length of the system is shorter than the free electron path, the impurity scattering is negligible, so the electrons transport is ballistic. If a wire has outside diameter comparable with the Fermi wavelength λ_F, and its length L is less than Λ, the system can be regarded as one-dimensional (1-D), the electron – as a wave, and one can expect quantum effects – see Fig 2.

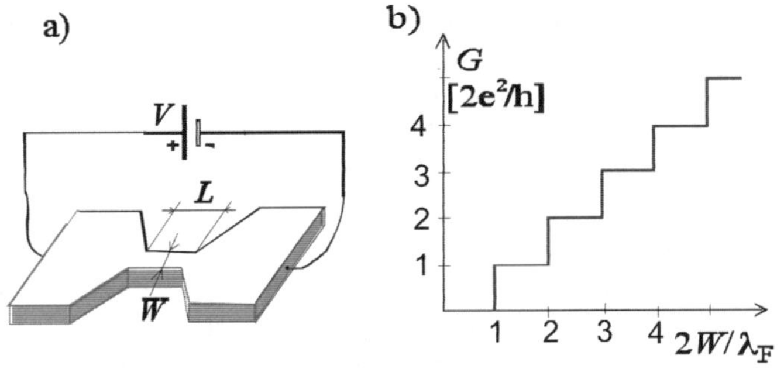

Figure 2. Conductance quantization in a nanowire (conductor with length $L< \Lambda$ and width W comparable with the length of Fermi wave λ_F): a) nanowire outline (the third dimension is not considered); b) conductance quantization G versus width W

The total electrical conductance of a nanowire is given by formula (1).

$$G = \frac{2e^2}{h} N \tag{1}$$

where N is the number of transmission channels. For 1-D system, with thickness $H \leq \lambda_F$, N depends on the width of the wire, $N = \text{int}\,(2W/\lambda_F)$. For 2-D system, with H, $W \geq \lambda_F$, $N = \text{int}\,(W \times H/\lambda_F^2)$, where int ($A$) means the integer of A. However, defects, impurities and irregularities of the shape of the conductor can induce scattering, then conductivity is given by the Landauer equation:

$$G = \frac{2e^2}{h} \sum_{i,j=1}^{N} t_{ij} \tag{2}$$

where t_{ij} denotes probability of the transition from j^{th} to i^{th} state. In the absence of scattering $t_{ij} = \delta_{ij}$ thus Eq. (2) is reduced to Eq. (1). Figure 1 presents a picture of a path (nanowire) – the constriction in an electrical conductor with dimension W (width), H (thickness) and L (length) W.

A set-up for measurements of electrical conductance in nanowires formed by mechanical contact between two microwires is shown in Figure 3. The experimental setup consisted of a pair of metallic wires (they formed a nanowire), a digital oscilloscope, a motion control system (doesn`t show on the picture) and a PC. Instruments are connected in one system using the IEEE-488 interface. There was the resistor $R_p = 1$ kΩ in series to the connected wires. The circuit was fed by the constant voltage V_s and measurements of current $I(t)$ have been performed.

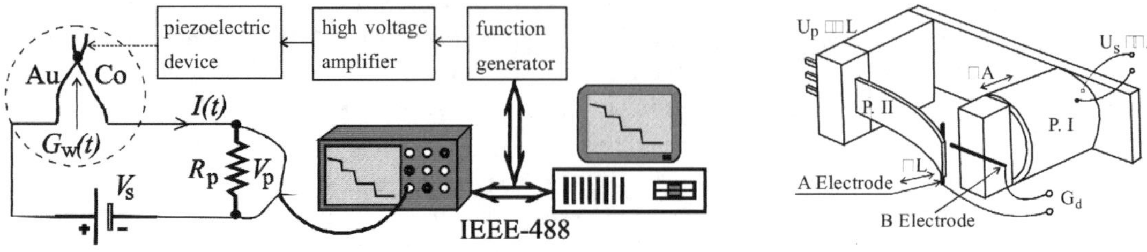

Figure 3. A system for measurements of conductance quantization: an electrical circuit (left) and a piezoelectric actuator for precise moving of electrodes (right)

Transient effects of making contact or breaking the contact give time dependent current. The voltage V_p on the resistor R_p was measured with computer controlled oscilloscope. The piezoelectric device is used to control the backward and forward movement of the macroscopic wires between which nanowires are formed. A high voltage amplifier controlled by a digital function generator supplies the piezoelectric device. Both electrodes (macroscopic wires) are made of wire 0.5 mm in diameter. The conductance was measured between two metallic electrodes, moved to contact by the piezoelectric tube actuator. The oscilloscope was triggered by a single pulse. All experiments were performed at room temperature and at ambient pressure. The quantization of electric conductance depends neither on the kind of element nor on temperature. For conductors and semiconductors the conductance quantization in units of $G_0 = 2e^2/h = (12.9\ \text{k}\Omega)^{-1}$ was measured in many experiments.

The quantization of electric conductance depends neither on the kind of metal nor on temperature. However, the purpose of studying the quantization for different metals was to observe how the metal properties affect the contacts between wires. For nonmagnetic metals, the conductance quantization in units of $G_0 = 2e^2/h = 7.75\times10^{-5}$ [A/V] $= (12.9\ \text{k}\Omega)^{-1}$ was previously observed for the following nanowires: Au-Au, Cu-Cu, Au-Cu, W-W, W-Au, W-Cu. The quantization of conductance in our experiment was evident. All characteristics showed the same steps equal to $2e^2/h$. We observed two phenomena: quantization occurred when breaking the contact between two wires, and quantization occurred when establishing the contact between the wires. The characteristics are only partially reproducible; they differ in number and height of steps, and in the time length. The steps can correspond to 1, 2, 3 or 4 quanta. It should be emphasised that quantum effects were observed only for some of the characteristics recorded. The conductance quantization has been so far more pronouncedly observable for gold contacts. Figure 4 and 5 show example plots of conductance vs. time during the process of drawing Au- and Cu-nanowire, respectively, for the bias voltage $V_s = 0.42$ V [3].

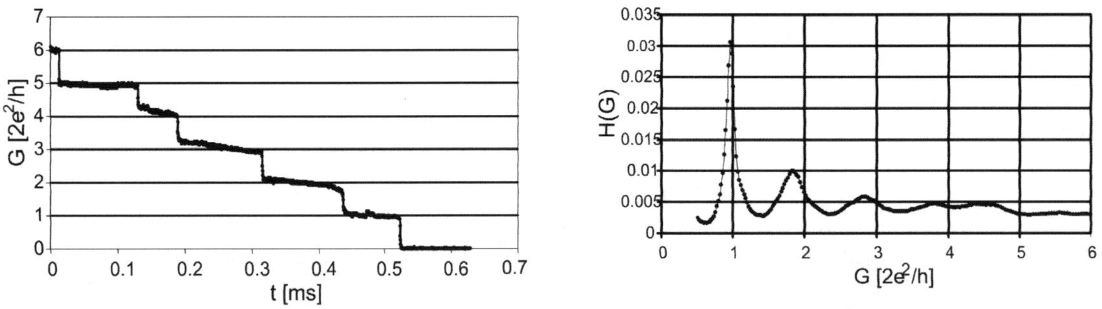

Figure 4. Conductance quantization in gold nanowires: a time plot (left) and a histogram from 6000 consecutive formations of a nanowire (measurements made by M. Wawrzyniak, PTU, Poland)

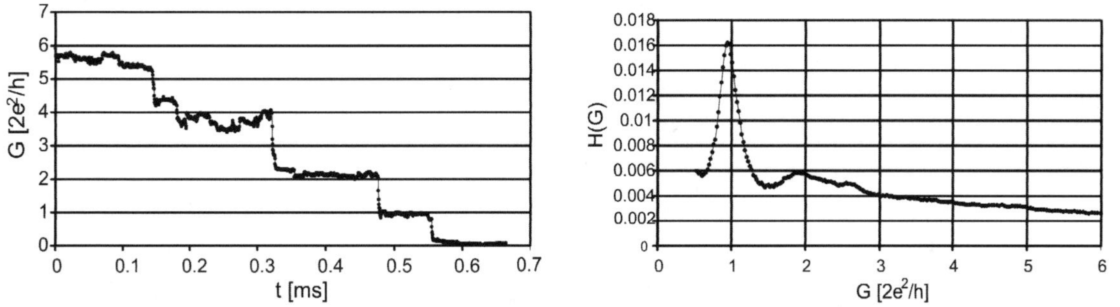

Figure 5. Conductance quantization in copper nanowires: a time plot (left) and a histogram from 6000 consecutive formations of a nanowire (measurements made by M. Wawrzyniak)

Let`s consider a silicon path with length L = 20 nm ($L < \Lambda_{\mathrm{Si}}$), width W = 4 nm, thickness H = 4 nm and λ_F = 1 nm. Such path is a nanowire with the conductance: $G_P = (2e^2/h)\times N$; $N = \mathrm{int}\,(W\times H/\lambda_F^2) = 16$; $G_P = 7.75 \times 10^{-5} \times 16 = 12.4 \times 10^{-4}$ [A/V]. A resistance of the path $R_P = 1/G_P = 800\ \Omega$, thus it is surprisingly high. An electrical capacity of a path is 2 pF/cm [4], so for L = 20 nm $C_P = 4\times10^{-18}$ F. The path inside an IC forms a low-pass RC filter with the upper frequency $f_u = 1/(2\pi R_P C_P) \approx 50\times10^{12}$ Hz. Thus the path-nanowire filters of signals.

4. Quantization of thermal conductance in nanostructures

It is generally known that limits for speed-up of digital circuits, especially microprocessors, are determined by thermal problems. Both electrical G_E and thermal G_T conductance of a nanostructure describe the same process: electron transport in nanostructures. Therefore there are several analogues between the two physical quantities. Electron transport in a nanowire does two effects: an electrical current $I = G_E \times \Delta V$ and a heat flux density $Q_D = G_T \times \Delta T$, where G_E – electrical conductance of a sample, ΔV – difference of electrical potentials, G_T – thermal conductance of a sample, ΔT – temperature difference.

$$G_E = \sigma \times A/l, \qquad G_T = \lambda \times A/l \tag{3}$$

where σ – electrical conductivity, λ – thermal conductivity, l – length of a sample (e.g. nanowire), A – area of a cross-section of a sample.

Quantized thermal conductance in one-dimensional systems (e.g. nanowires) was predicted theoretically by Rego [11] using the Landauer theory. The thermal conductance is considered in a similar way like the electrical conductance. In one-dimension systems are formed conductive channels. Each channel contributes to a total thermal conductance with the quantum of thermal conductance G_{T0}. Quntized thermal conductance and its quantum (unit) G_{T0} was confirmed experimentally by Schwab [7]. The quantum G_{T0} of thermal conductance

$$G_{T0}\ [\mathrm{W/K}] = (\pi^2 k_B^2/3h)T = 9.5\times10^{-13}T \tag{4}$$

depends on the temperature (10). At T = 290 K value of $G_{T0} = 2.8\times10^{-10}$ [W/K]. This value is determined for an ideal ballistic transport (without scattering) in a nanowire, with the transmission coefficient t_{ij} = 100%. It means that in all practical cases (for t_{ij} < 100%) the thermal conductance is below the limit given by formula (10). However an analyze of thermal conductance is more complex than electrical conductance because of contribution either phonons or electrons in heat exchange. Quantization of thermal conductance in one-dimensional systems was predicted theoretically by

Greiner [5] and by Rego [6] for ballistic transport of electrons and phonons. Quantized thermal conductance G_T and its quantum (unit) G_{T0} was confirmed experimentally by Schwab [7].

Table 2. Electrical resistance R and thermal conductance G_T of gold nanowire with diameter D (λ_F = 0.5 nm)

Diameter D	nm	0.5	1.0	1.5	2.0	2.5	3.0
Number of channels	–	1	4	9	16	25	36
$R = 1/G_E$	Ω	12 903	3226	1434	806	516	358
G_T (at 300 K)	10^{-9} [W/K]	0.285	1.14	2.56	4.56	7.12	10.26

The thermal conductance of a nanowire is very small. We can compare the thermal conductance in a nanowire with the thermal conductance of Au microwire which connects a silicon part with a metallic terminal (pin) in a transistor. For a microwire with a length of L = 1.5 mm and a diameter D = 25 μm is the thermal conductance is much larger: $G_T \approx 10^{-4}$ [W/K]

One can obtain a similar (but not the same) value of the quantum of thermal conductance G_{T0} using the Wiedemann-Franz law. The law describes the relation between the thermal conductivity λ, and electrical conductivity σ of a sample for macroscopic objects (5).

$$\frac{\lambda}{\sigma} = \frac{1}{3}\left(\frac{\pi k_B}{e}\right)^2 T = L \times T = 2.35 \times 10^{-8} T \tag{5}$$

where e – electron charge, k_B – Boltzmann constant, L – Lorenz number, T – temperature.

The Wiedemann-Franz law is valid for the relation between conductances G_T/G_E as well. The value of G_{T0} can be obtained directly from (6).

$$G_{T0} = L \times T \times G_{E0} = \frac{1}{3}\left(\frac{\pi k_B}{e}\right)^2 T \times \frac{2e^2}{h} = \frac{2}{3}\frac{(\pi k_B)^2}{h} \times T = 1.89 \times 10^{-12} T \tag{6}$$

The value of G_{T0} obtained from the Wiedemann-Franz law is twice larger than this from (4) and it is not correct.

A single nanowire should be consider together with its terminals (Figure 6). They are colled reservoirs of electrons. A single nanowire should be consider together with its terminals. Electron transport in the nanowire is ballistic itself, it means the transport without scattering of electrons and without energy dissipation. The energy dissipation occurs in terminals. Because of the energy dissipation the local temperature T_{term} in terminals is higher then the temperature T_{wire} of nanowires itself. A heat distribution in terminals of a nanostructe should be analyzed.

In small structures a dissipated energy is quite large. For the first step of conductance quantization, $G_E = G_{E0} = 7.75 \times 10^{-5}$ [A/V], and at the supply voltage V_s = 1,4 V the current in the circuit I = 100 μA (I = 190 μA for the second step of quantization). The power dissipation in terminals of nanowires is $P = I^2/G_{E0}$ = 130 μW for the first step and P = 230 μW for the second step. One ought to notice that the density of electric current in nanowires is extremely high. The diameter of the gold nanowire on the first step of quantization can be estimated to D = 0.4 nm, so for I = 100 μA the current density $J \approx 8 \times 10^{10}$ [A/cm^2].

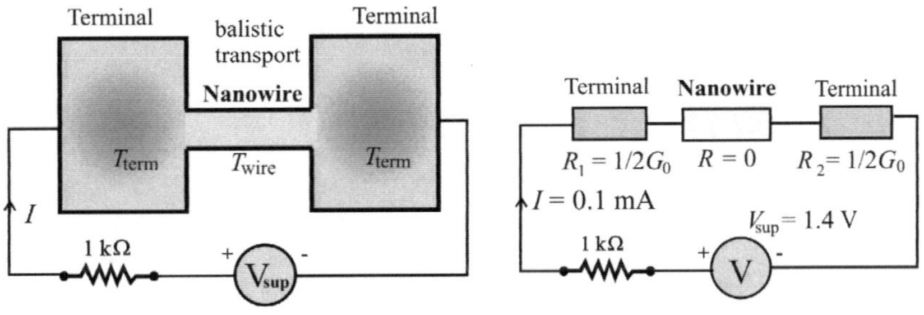

Figure 6. Conductance distribution in a nanowire with a ballistic transport

5. Other physical limits

The channel length L_E of Si MOSFET, limited by a degradation of electrostatic control in the transistor, was analyzed by Frank [4] and Likharev [8]. The minimal channel length L_E (see Fig. 5) depends on thickness of channel H_{ch}, thickness of insulation layer H_i, dielectric constants of a channel ε and insulation ε_i – formula (7) [8].

$$L_E = \left(\frac{\varepsilon H_{ch} H_i}{2\varepsilon_i} \right)^{1/2} \tag{7}$$

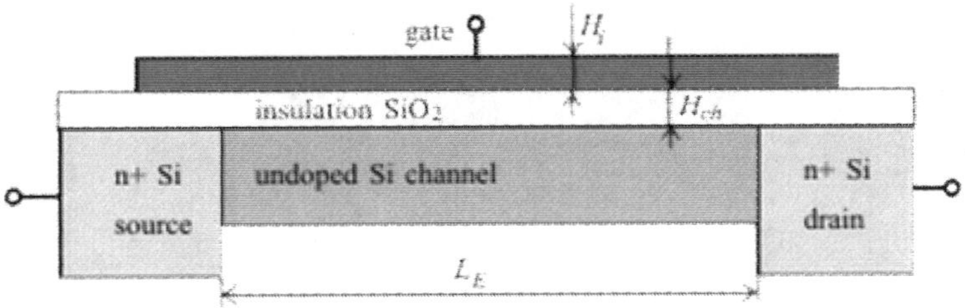

Figure 7. MOSFET transistor – the simple model

If we take the ratio $\varepsilon_i/\varepsilon \approx 0.3$ (for silicon oxide and silicon), thickness $H_{ch} = 2$ nm, $H_i = 1.5$ nm – the estimated minimal length of channel is $L_E \approx 3$ nm. The channel length L_E can be shorter if a better insulator than silicon oxide (SiO_2) would be applied. Following materials are tested as an insulator for MOSFETs: silicon nitride, hafnium oxide and zirconium silicate.

The next limit to miniaturization of electronic devices comes from source to drain tunneling through the potential barrier along the channel. The tunneling effect depends on the channel length L and the supply voltage. Because of tunneling the minimal channel length for silicon device is around 2 nm [8].

High-tech microprocessors have dimensions over 30 mm. A new Intel product the quad-core Intel Core i5-670 has a width $a = 37.5$ mm and a length $b = 37.5$ mm. The pins of the IC are distributed on the bottom plate, along its four sides too. The period of the highest clock frequency (3.73 GHz) is $T_{ck} =$ 270 ps. But the propagation time for an electromagnetic wave on the way of $(a + b)$ is: $T_p = (a + b)/v$ $= 75$ mm/2.5×10^8 m/s $= 300$ ps (v – speed of an electromagnetic wave in silicon). Thus, the propagation time of the clock signal is longer than the period of the clock signal.

Journal of Physics: Conference Series **248** (2010) 012059 doi:10.1088/1742-6596/248/1/012059

6. Conclusions

Conductance quantization has proved to be observable in a simple experimental setup, giving opportunity to investigate subtle quantum effects in electrical conductivity. The energy dissipation in nanowires takes part in their terminals. Because of the energy dissipation the local temperature in terminals is higher then the temperature of a nanowire itself. According to the state of the art the minimal length of gate in MOSFET in silicon integrated circuits is around 3 nm (thus – technology of 3 nm!). However technological limits allow to apply only the 10 nm-technology in the next 10 years.

References

[1] International Technology Roadmap for Semiconductors, 2010, Edition 2009, www.itrs.net/reports
[2] Landauer R 1989 *J. Phys.: Cond. Matter* vol. 1 p 8099
[3] Nawrocki W 2008 *Electron transposrt in nanosystems*, eds Bonca J, Kruchinin S (Springer: Dordrecht) p 17
[4] Frank D J, Dennard R H, Nowak E, Solomon P M, Taur Y and H.S.P. Wong H S P 2001 *Proc. IEEE* vol. 89 p 259
[5] Greiner A, Reggiani L, Kuhn T and Varani L 1997 *Phys. Rev. Lett.* vol. 78 p 1114
[6] Rego L G C and Kirczenow G 1998 *Phys. Rev. Lett.* vol. 81 p 232
[7] Schwab K, Henriksen E A, Worlock J M and Roukes M L 2000 *Nature* vol. 404 p 974
[8] Likharev K 2003 *Advanced Semiconductor and Organic Nano-techniques*, ed Markoç H. (Amsterdam: Elsevier) chapter 4

Journal of Physics: Conference Series **248** (2010) 012060 doi:10.1088/1742-6596/248/1/012060

The effect of surfactants on chemical development of ion track nanopores in polymer

Yu. Yamauchi[1], **P. Yu. Apel**

FLNR, JINR, Dubna, Moscow Region, 141980, Russia

E-mail: yukari_yamauchi@yahoo.co.jp

Abstract. The use of surfactants in etching of the track membranes (TMs) enables one to control the pore shape in TMs. This technique is useful to improve flow rate, and to optimize the retention and permeation properties of TMs. The addition of an anionic surfactant, sodium dodecyl diphenyloxide disulphonate (SDDD), to etching solutions leads to the highly tapered pore shape in TMs of the polyethylene terephthalate (PET). To understand the mechanism of the surfactant effect on track etching in the nanometer range, we study the diffusion and adsorption of SDDD on non-etched non-porous, etched non-porous and etched porous PET films. The comparison of non-etched non-porous and etched non-porous films shows the effect of negatively charged surface on the adsorption of surfactant. The comparison of etched non-porous and etched porous PET films shows the different adsorption on film surface and inner wall of nanopore, thus the influence of curvature of the surface on the adsorption of surfactant molecules.

1. Introduction

The nano and micro pores with controlled size and shape are important for variety of applications, such as the micro and ultra filtration membranes, molecular sensors, optical filters, template synthesis of nanotubes and nanowires, nano and microfluidics, etc. [1]. For usual production of TMs, a polymer film is irradiated with the heavy ion beams, and gets the linear traces of damage. Then chemical etching is performed with alkaline aqueous solution to get the pores along the traces. TMs of polymer film usually give cylindrical pore channels. However addition of an anionic surfactant, sodium dodecyl diphenyloxide disulphonate (SDDD) to the alkali solution leads the pore channels to the highly tapered pore profile [2].

Previous works suggest a model as follows [2]-[4]. The chemical etching performed with surfactant decreases the etching rate of the film surface because the adsorption layer of SDDD protects the film surface from hydroxide ions (alkali). However the etching rate of inner wall of pores does not decrease at the first stage because the molecule of SDDD is too large to enter inside of pore and to construct the adsorption layer on inner wall of pores. As the result, highly tapered pores ("cigar-like") are produced.

For better understanding of these phenomena, we carry out the present work. Goals of the present work are, to measure adsorption isotherms of SDDD on PET films, to study effect of electrolyte

[1]Corresponding author

Journal of Physics: Conference Series **248** (2010) 012060 doi:10.1088/1742-6596/248/1/012060

concentration on the adsorption isotherms, to compare adsorption of SDDD on flat surface and on the pore walls of different curvature.

2. Experimental

PET films (thickness 20μm) were used as the substrate. The three kinds of pre-treatment were performed on the PET films: non-etched non-pore film (NENP, pristine PET film), etched non-pore film (ENP, PET film etched with alkali), etched pore film (EP, PET film irradiated with heavy-ion and etched with alkali). In fig.1 we show the SEM image of an etched pore film which has the pore diameter of 260 nm and pore number of $4.8*10^8 cm^{-2}$. Sodium dodecyl diphenyloxide disulphonate (SDDD) was used for adsorption experiments. The chemical structure of SDDD is shown in Fig. 2. This is an anionic surfactant which has two negatively charged groups in hydrophilic head. The experimental procedure is also shown in Fig.2. Adsorption experiments were performed at room temperature. Small pieces of PET film were put into the SDDD aqueous solution of the concentration C_o. After 24 hours, the PET films were removed and the concentration of SDDD solution C_{eq} was measured by UV spectrophotometer. From the difference between C_{eq} and C_o, we found the adsorption of SDDD on the PET films. For NENP and ENP firm, the surface area means two sides of film surfaces (up and bottom sides). For EP film, the surface area is sum of two film surfaces and inner wall of pores.

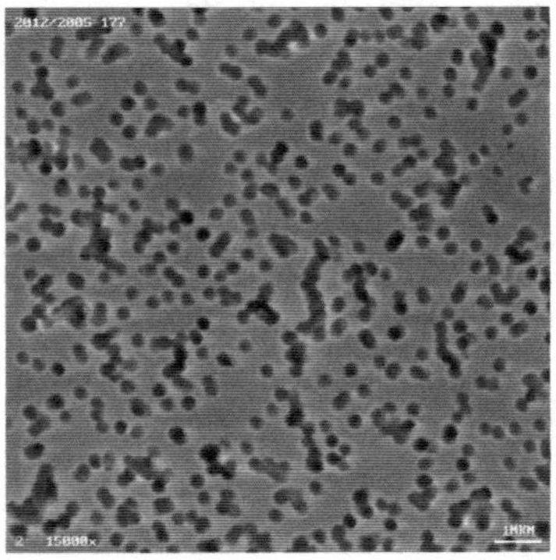

Fig.1 SEM image of the etched-pore film (EP-b) surface. The pore diameter is 260 nm.

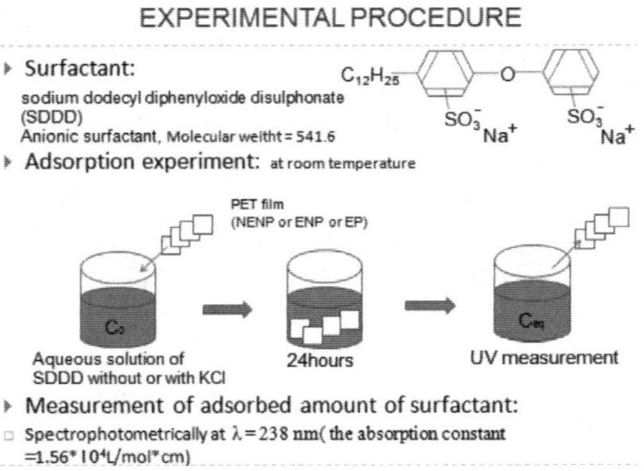

Fig.2 The chemical structure of SDDD and the experimental procedure.

3. Results

Fig. 3 left shows the influence of the electrolyte KCl concentration on the adsorption isotherm of SDDD on NENP films. Fig.3 right shows the average area occupied by SDDD molecule on NENP film surface. The lines in Figures show the results of fitting by Langmuir formula $V= K\,Ceq\,V_{max} / (1+K\,C_{eq})$. Here, K - Langmuir equilibrium constant, C_{eq} - the aqueous concentration, V - the amount adsorbed, and V_{max} - the maximum amount adsorbed as C increases. It shows the higher concentration of KCl leads to the higher adsorption of SDDD on NENP films.

Fig.4 left shows the influence of the electrolyte KCl concentration on the adsorption isotherm of SDDD on ENP film. Fig.4 right shows the average area occupied by SDDD molecule on ENP film. The same as the case of NENP film, the higher concentration of KCl leads to the higher adsorption of SDDD on ENP films.

Fig.5 left shows the influence of the electrolyte KCl concentration on the adsorption isotherm of SDDD on EP-b film. Fig.5 right shows the average area occupied by SDDD molecule on EP-b film. We observed that the adsorption of SDDD on EP-b film is significantly lower than the cases of NENP and ENP films at the same concentration of KCl. It suggests that the adsorption onto the inner walls of pores is lower than the adsorption onto the film surface. We also found that there is practically no adsorption of SDDD in the absence of electrolyte. And this result was same even for EP films which have pores as large as 1000 nm in diameter. (The result is not shown here.)

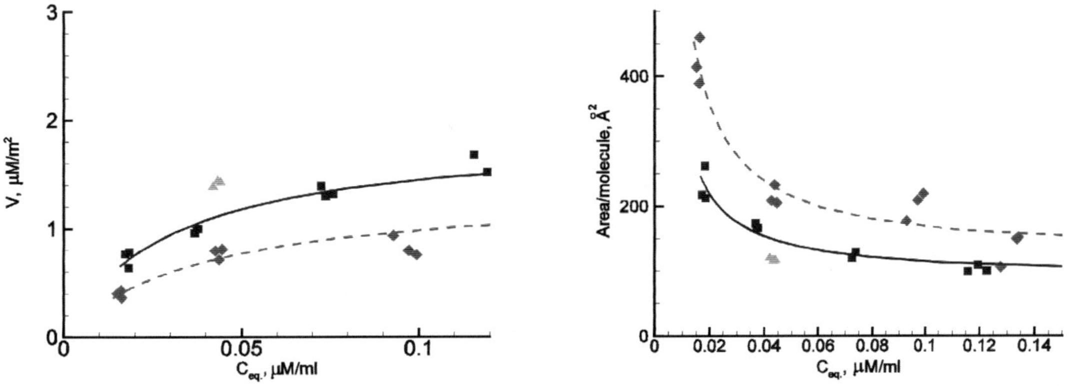

Fig.3 Left - The influence of the electrolyte KCl concentration on the adsorption isotherm of SDDD on NENP film. No KCl - diamonds, 0.1M KCl -squares, 1.0M KCl –deltas. Right - The average area occupied by SDDD molecule on NENP film surface. The lines in Figures show the results of fitting by Langmuir formula $V= K\ Ceq\ V_{max} / (1+K\ C_{eq})$. K - Langmuir equilibrium constant, C_{eq} - the aqueous concentration, V - the amount adsorbed, and V_{max} - the maximum amount adsorbed as C increases.

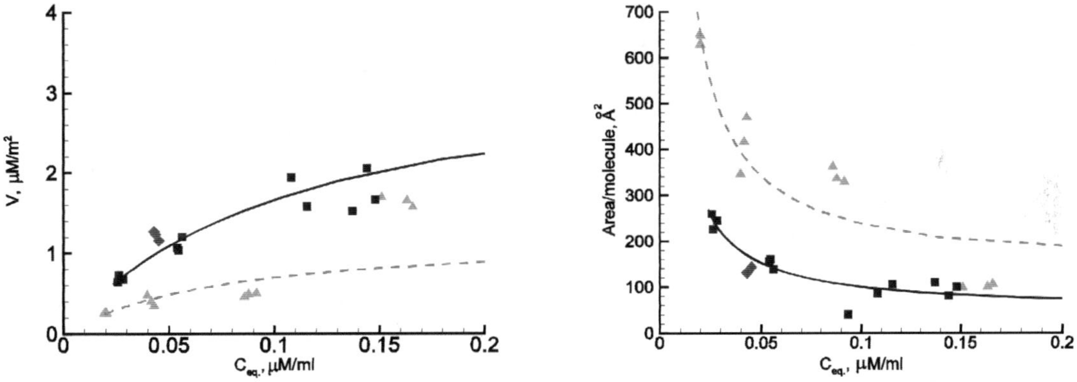

Fig.4 Left - The influence of the electrolyte KCl concentration on the adsorption isotherm of SDDD on ENP film. No KCl - deltas, 0.1M KCl -squares, 1.0M KCl - diamonds. Right - The average area occupied by SDDD molecule on ENP film.

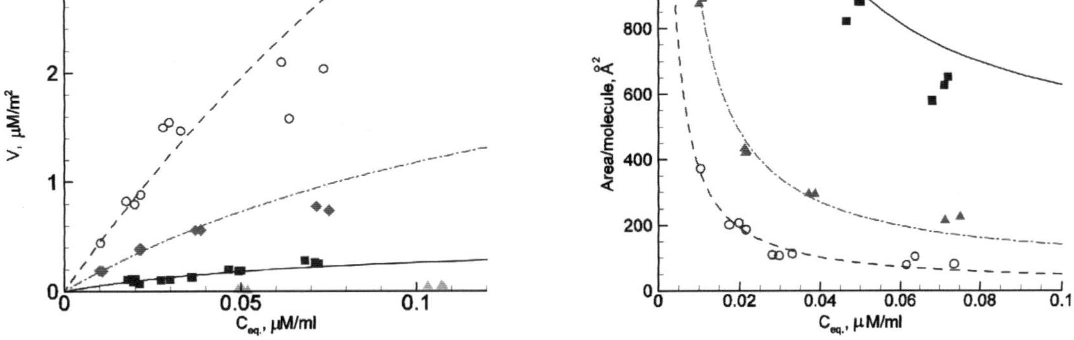

Fig.5 Left - The influence of the electrolyte KCl concentration on the adsorption isotherm of SDDD on EP-b film. No KCl - deltas, 0.1M KCl -squares, 1.0M KCl – diamonds, 3.0M KCl – circles. Right - The average area occupied by SDDD molecule on EP-b film.

Fig.6 shows the adsorption isotherms of SDDD on EP films with different sizes and pore shapes. The electrolyte KCl concentration of SDDD aqueous solution is 0.1M. EP-a has the highly tapered pores which are produced by the chemical etching with SDDD. The effective pore diameter found from gas flow is 280 nm, while the pore diameter at the film surface is about 100 nm. EP-b, EP-c and EP-d have almost cylindrical pores which are slightly tapered towards the depth. Their pore diameters are 260 nm, 480 nm and 1020 nm, respectively. Fig.6 shows that all porous substrates have almost same dependence of SDDD adsorption at 0.1M KCl.

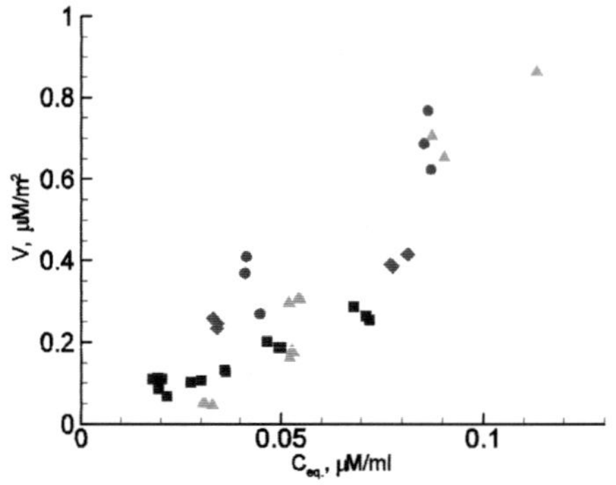

Fig.6 The adsorption isotherm of SDDD on EP films with different sizes and pore shapes. The electrolyte KCl concentration of SDDD aqueous solution is 0.1M. EP-a has the highly tapered pores which are produced by the chemical etching with SDDD. The effective pore diameter found from gas flow is 280 nm, while the pore diameter at the film surface is about 100 nm. EP-b, EP-c and EP-d have almost cylindrical pores which are slightly tapered towards the depth. Their pore diameters are 260 nm, 480 nm and 1020 nm, respectively. EP-a -diamonds, EP-b –squares, EP-c –deltas and EP-d –circles.

4. Conclusions

The experiments clearly show that the adsorption of anionic surfactant in 260 nm pores is significantly lower than that on flat surface of both etched and non-etched PET film. This phenomenon is quite surprising and needs further investigation. It was also shown that the higher concentration of KCl leads to the higher adsorption of anionic surfactant on all three kinds of PET films (non-etched, etched non-porous, etched porous). Repulsion between the negatively charged sulfonate groups of surfactant molecules seems to be neutralized by the addition of electrolyte KCl. As a result, the anionic surfactant forms denser adsorption layer.

The results obtained are important: (i) for efficient control over the structural characteristics of track membranes; (ii) for better understanding of the processes of adsorption and diffusion of ions and molecules in nanocapillaries with electrical charge on the walls.

References

[1] R.Spohr, Radiation Measurements 40, 191-202 (2005).

[2] P. Apel, I. V. Blonskaya, S.D. Dmitriev, T.I. Mamonova, O.L. Orelovitch, B. Sartowska, Yu. Yamauchi, Radiation Measurements, 43, S552-559 (2008).

[3] P. Apel, I. V. Blonskaya, A. Yu. Didyk, S.D. Dmitriev, O.L. Orelovitch, L.I. Samoilova, V.A. Vutsadakis, Nucl. Instrum. and Meth. in Phys. Res. B179, 55- 62 (2001).

[4] P. Apel, I. V. Blonskaya, S.D. Dmitriev, O.L. Orelovitch, B. Sartowska, Journal of Membrane Science, 282, 393-400 (2006).

Bipolar Transistor Based on Graphane

B Gharekhanlou, S B Tousaki, S Khorasani

School of Electrical Engineering, Sharif University of Technology,
PO Box 11365-9363, Tehran, Iran.

E-mail: khorasani@sharif.edu

Abstract. Graphane is a semiconductor with an energy gap, obtained from hydrogenation of the two-dimensional grapheme sheet. Together with the two-dimensional geometry, unique transport features of graphene, and possibility of doping graphane, p and n regions can be defined so that p-n junctions become feasible with small reverse currents. Our recent analysis has shown that an ideal I-V characteristic for this type of junctions may be expected. Here, we predict the behavior of bipolar juncrion transistors based on graphane. Profiles of carriers and intrinsic parameters of the graphane transistor are calculated and discussed.

1. Introduction

Graphane is an extended two-dimensional (2D) covalently bonded hydrocarbon. It is a fully saturated hydrocarbon derived from a single graphene sheet with formula C_6H_6. All of the carbon atoms are in sp^3 hybridization forming a hexagonal network and the hydrogen atoms are bonded to carbon on both sides of the plane in an alternating manner. It has a relatively large direct energy gap of about 3.6eV [1-6]. The numerical estimate for the intrinsic carrier density of graphane turns out to be too low for practical applications [7]. One of the simplest ways to reduce the energy gap is to bring up hydrogen deficiency in graphane structure [7]. So we use C_6H_3 Structure. In this paper, we analyze a 2D bipolar transistor based on doping of a modified graphane sheet. The graphane structure is supposed to encompass hydrogen deficiency, with a typical stoichiometric composition of C_6H_{6-x}. We have used the tight-binding method to show how hydrogen deficiency may be exploited to reduce the band gap of graphane from 3.6eV down to about 1.1eV [7]. This increases the intrinsic carrier density and thereby the conductivity of doped graphane. Analysis of the junctions with the aid of Shockley's law is done to obtain the profiles and behaviour of the corresponding bipolar transistor.

2. Analysis

Necessary parameters are calculated at 300K [7]. The density of states for conduction and valence bands are respectively N_c=2.1×10^{17}m^{-2} and N_v=2.6×10^{17}m^{-2}, with the intrinsic carrier density of n_i=4.5×10^{7}m^{-2}. Now, we proceed with a p^+np bipolar junction transistor, with the dopant-concentrations in the emitter, base and collector respectively given by N_E=2.0×10^{15}m^{-2}, N_B=10^{14}m^{-2}, and N_C=10^{13}m^{-2}. One can immediately use the relation $V_b=V_T\ln(N_A N_D/n_i^2)$ with $V_T=K_BT/q$, to readily get V_{jBE}=0.8340V and V_{jBE}=0.6968V for emitter-base and base-collector junctions, respectively.

As we deal with 2D junctions, the analysis the electric field is different to that of bulk step junctions [7]. For this purpose, we need to divide the charge sheets into infinitely many thin wire line charges,

Journal of Physics: Conference Series **248** (2010) 012061 doi:10.1088/1742-6596/248/1/012061

with constant charge densities λ_i [7]. The partial electric field due to each line charge is $E_i(r)=\lambda_i/2\pi\varepsilon r_i$, where r_i is the distance from the charge wire. This model is illustrated in figure 1. Integrating from the electric feild gives the potential distribution. Calculated total electric field over the surface of the transistor is demonstrated in figure 2. We also find the width of the deplation regions in the emitter, base and collector to be in the order of micrometer.

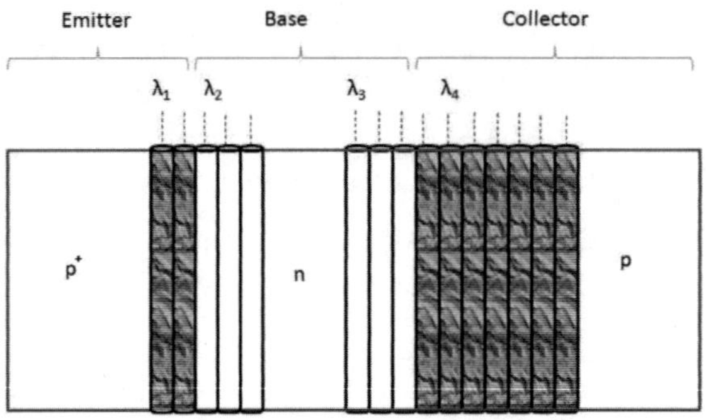

Figure 1. Schematic diagram of a pnp bipolar transistor.

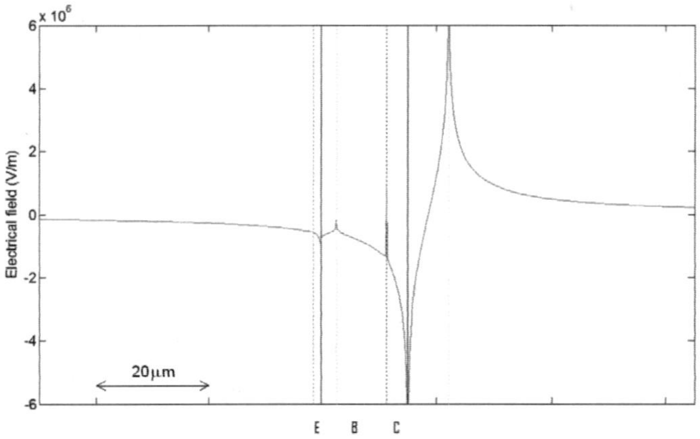

Figure 2. The electric field distribution in the graphane bipolar transistor.

3. Current-voltage characteristics

Figure 3 represents a schematic of a pnp transistor biased in the common-base configuration. It also indicates all the relevant current components of the transistor in the circuit.

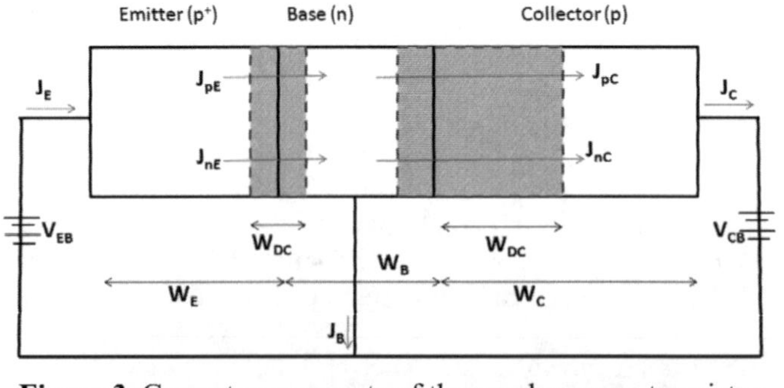

Figure 3. Current components of the graphane pnp transistor.

In the neutral base region, the injected minority-carriers distribution (holes) is governed by the continuity equation, with the boundary conditions at the two edges of neutral region

$$-\frac{p_n - p_{n0}}{\tau_p} - \mu_p E\frac{dp_n}{dx} + D_p\frac{d^2 p_n}{dx^2} = 0 \tag{1.1}$$

$$p_{nB}(x = x_{nE0}) = n_{nB0}\exp\left(\frac{qV_{EB}}{K_B T}\right) \tag{1.2}$$

$$p_{nB}(x = w - x_{nC0}) = n_{nB0}\exp\left(\frac{qV_{CB}}{K_B T}\right) \tag{1.3}$$

The general solution to the above equation is given by

$$\begin{aligned} p_n(x) = {} & p_{n0}\exp\left(\frac{x+x_{nC}}{L_{p1}}\right)\left[-\exp\left(\frac{w}{L_{p2}}\right)+\exp\left(\frac{qV_{EB}}{K_BT}+\frac{w}{L_{p2}}\right)+\exp\left(\frac{x_{nE}+x_{nC}}{L_{p2}}\right)-\exp\left(\frac{qV_{CB}}{K_BT}+\frac{x_{nC}+x_{nE}}{L_{p2}}\right)\right]\times \\ & \left[\exp\left(\frac{w}{L_{p2}}+\frac{x_{ne}+x_{nC}}{L_{p1}}\right)-\exp\left(\frac{w}{L_{p1}}+\frac{x_{ne}+x_{nC}}{L_{p2}}\right)\right]^{-1} \\ & + p_{n0}\exp\left(\frac{x}{L_{p2}}\right)\left\{-\exp\left(\frac{x_{ne}}{L_{p1}}\right)\left[1-\exp\left(\frac{qV_{CB}}{K_BT}\right)\right]+\exp\left(\frac{w-x_{nC}}{L_{p2}}\right)\left[1-\exp\left(\frac{qV_{EB}}{K_BT}\right)\right]\right\}\times \\ & \left[\exp\left(\frac{w-x_{nC}}{L_{p2}}+\frac{x_{nE}}{L_{p1}}\right)-\exp\left(\frac{w-x_{nC}}{L_{p1}}+\frac{x_{nE}}{L_{p2}}\right)\right]^{-1} + p_{n0} \end{aligned} \tag{2}$$

The hole density in the base region is used for calculation of hole current due to the collector and emitter depletion regions. The electron currents at the emitter edge J_{nE} and the collector edge J_{nC} are given by

$$\begin{aligned} J_{nE} = -qD_n\left.\frac{dn_{pe}}{dx}\right|_{x=-x_{pe0}} &= \frac{qD_n n_{pe0}}{L_{n2}(x)}\left[\exp\left(\frac{qV_{eb}}{K_BT}\right)-1\right]\exp\left[\frac{x+x_{pe0}}{L_{n2}}\right]\Bigg|_{x=-x_{pe0}} \\ &= \frac{qD_n n_{pe0}}{L_{n2}(x=-x_{pe0})}\left[\exp\left(\frac{qV_{eb}}{K_BT}\right)-1\right] \end{aligned} \tag{3.1}$$

$$L_{n2}(x) = \left\{\frac{\mu_n}{2D_n}E(x) + \sqrt{\left[\frac{\mu_n}{2D_n}E(x)\right]^2 + \frac{1}{\tau_n D_n}}\right\}^{-1} \tag{3.2}$$

$$\begin{aligned} J_{nC} = -qD_n\left.\frac{dn_{pC}}{dx}\right|_{x=-x_{pC0}} &= \frac{qD_n n_{pC0}}{L_{n2}(x)}\left[\exp\left(\frac{qV_{CB}}{K_BT}\right)-1\right]\exp\left[\frac{x-(w+x_{pc0})}{L_{n1}}\right]\Bigg|_{x=-x_{pC0}} \\ &= \frac{qD_n n_{pC0}}{L_{n1}(x=-x_{pC0})}\left[\exp\left(\frac{qV_{CB}}{K_BT}\right)-1\right] \end{aligned} \tag{3.3}$$

$$L_{n1}(x) = \left\{\frac{\mu_n}{2D_n}E(x) - \sqrt{\left[\frac{\mu_n}{2D_n}E(x)\right]^2 + \frac{1}{\tau_n D_n}}\right\}^{-1} \tag{3.4}$$

The emitter and collector currents can be then obtain from

$$J_E = J_{nE} + J_{pE} \tag{4.1}$$

$$J_C = J_{nC} + J_{pC} \tag{4.2}$$

Here J_{pE} and J_{pC} are minority-carriers current densities in the base region, flowing to emitter and collector regions, respectively. Figure 4 shows a typical collector current characteristics, which confirms the exponential current-voltage dependence of the graphane transistor. Here, the collector-base junction is biased at V_{CB} =−1V. Figure 5 shows a representative set of output characteristics for the common base configuration.

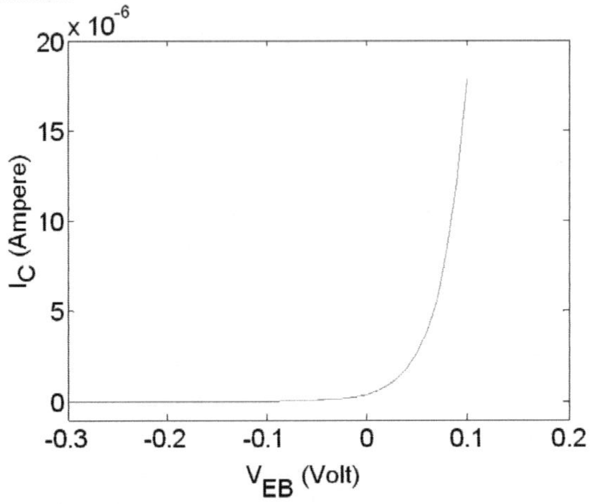

Figure 4. Collector current characteristics versus emitter-base voltage.

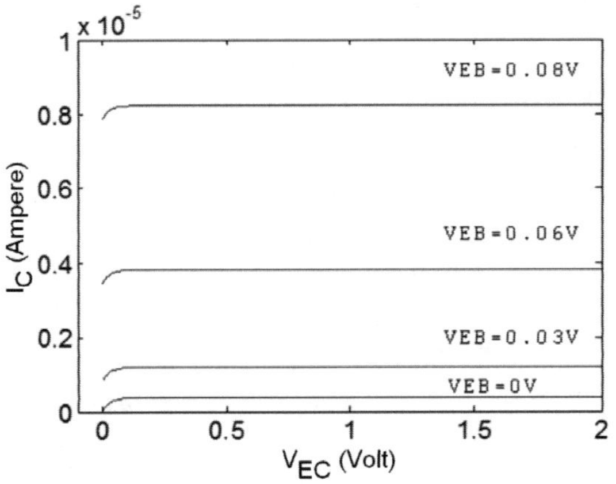

Figure 5. Output characteristics for common base configuration.

4. Conclusions

We have introduced a 2-D p^+np bipolar transistor based on graphane with hydrogen deficiency to reduce the band gap effectively. Using a basic analysis, we have shown that, within the approximation of the Shockley's law of junctions, an exponential ideal I–V characteristic is expectable. Furthermore, we have calculated the surface distribution of electric field in the vicinity of the junction. Curvature of collector current characteristics shows good agreement with an ideal bipolar transistor.

References

[1] Sofo J O, Chaudhari A S and Barber G D 2007 *Phys. Rev. B* **75** 153401
[2] Elias D C *et al.* 2009 *Science* **323** 610
[3] Cudazzo P, Attaccalite C, Tokatly I V, and Rubio A 2010 *Phys. Rev. Lett.* **104** 226804
[4] Şahin H, Ataca C, and Ciraci S 2010 *Phys. Rev. B* **81** 205417
[5] Singh A K, Penev E S, and Yakobson B I 2010 *ACS Nano* **4** 3510
[6] Tozzini V and Pellegrini V 2010 *Phys. Rev. B* **81** 113404
[7] Gharekhanlou B and Khorasani S 2010 *IEEE Trans. Electron Dev.* **57** 209

Nanoscale characterization of martensite structures in copper based shape memory alloys

O Adiguzel[1]
Firat University Department of Physics, 23169 Elazig, Turkey

E-mail: oadiguzel@firat.edu.tr

Abstract. Martensitic transformations are first order displacive transitions and occur in the materials on cooling from high temperature. Shape memory effect is an unusual property exhibited by certain alloy systems, and leads to martensitic transition. Copper-based alloys exhibit this property in beta phase field which possess simple bcc- structures, austenite structure at high-temperatures. As temperature is lowered the austenite undergoes martensitic transition following two ordering reactions, and structural changes in nanoscale govern this transition. Atomic movements are also confined to interatomic lengths in sub-μm or angstrom scale in martensitic transformation. The formation of the layered structures in copper based alloys consists of shears and shear mechanism. Martensitic transformations occur in a few steps with the cooperative movement of atoms less than interatomic distances by means of lattice invariant shears on a {110} - type plane of austenite matrix which is basal plane or stacking plane of martensite. The lattice invariant shears occurs, in two opposite directions, <110 > -type directions on the {110}-type plane. These shears gives rise to the formation of layered structure.

1. Introduction

Shape memory alloys take place in a class of functional materials by exhibiting a peculiar property called shape memory effect. This property is characterized by the recoverability of a certain shape of material at different conditions. Shape memory effect is associated with martensitic transformation which is a solid state phase transformation occurring with the cooperative movements of atoms in the alloy on cooling from high temperature austenite phase region. Shape memory effect refers to the shape recovery of materials resulting from martensite to austenite transformation when heated above reverse transformation temperature after deforming in the martensitic phase. These alloys also cycle between two certain shapes with changing temperature. Copper based shape memory alloys exhibit this property in metastable β-phase field. High temperature β-phase bcc-structures martensitically undergo the non-conventional structures following two ordered reactions on cooling, and structural changes in nanoscale level govern this transition cooling [1-3]. Atomic movements are also confined to interatomic lengths in sub-μm or angstrom scale.

In the shape memory alloys, the austenite lattice has a higher order of symmetry than that of martensite. More than one martensite variant can be induced from one austenite. Martensite variants have identical crystal lattice, but are oriented in different directions [2].

[1] To whom any correspondence should be addressed.

The relationship of microstructures in shape memory alloys is essential for a better understanding of the mechanism using the basic characteristic of shape memory alloys.

The basic characteristic of shape memory alloys is the occurrence of different variants in the low temperature martensitic phase. The martensitic transformation is a shear-dominant diffusionless solid-state phase transformation, and when a shape memory alloy undergoes a martensitic phase transformation, it transforms from the parent phase to one or more of the different variants of the martensitic phase [4, 5].

In the absence of applied stresses or mechanical loading, the variants of the martensitic phase usually arrange themselves in a self-accommodating manner through twinning, resulting in no observable macroscopic shape change [5]. By applying mechanical loading the martensitic variants are forced to reorient (detwin) into a single variant leading to large macroscopic inelastic strains. The multiple martensite variants begin to convert to single variant, the preferred variant determined by alignment of the habit planes with the axis of loading [4-8].

Martensitic transformations occur in a few steps with the cooperative movement of atoms less than interatomic distances by means of lattice invariant shears on a {110} - type plane of austenite matrix which is basal plane of martensite. These shears give rise the formation of unusual complex structures called long period layered structures such as 3R, 9R or 18R depending on the stacking sequences on the close-packed planes of the ordered lattice. The complicated long-period stacking ordered structures mentioned above can be described by different unit cells.

All of these martensite phases are long-period stacking ordered structures that is the underlying lattice is formed by stacks of close-packed planes. In case the parent phase has a B2-type superlattice, the stacking sequence is ABCBCACAB(9R) [2, 9]. The stacking of $(110)_\beta$ -planes in DO_3-type structure and formation of layered structures are shown in Figure1. Martensitic transformation is characterized by a change in the crystal structure of the material at the nano-level rather than micro-level, and the transformed region consists of parallel bands containing alternately two different variants.

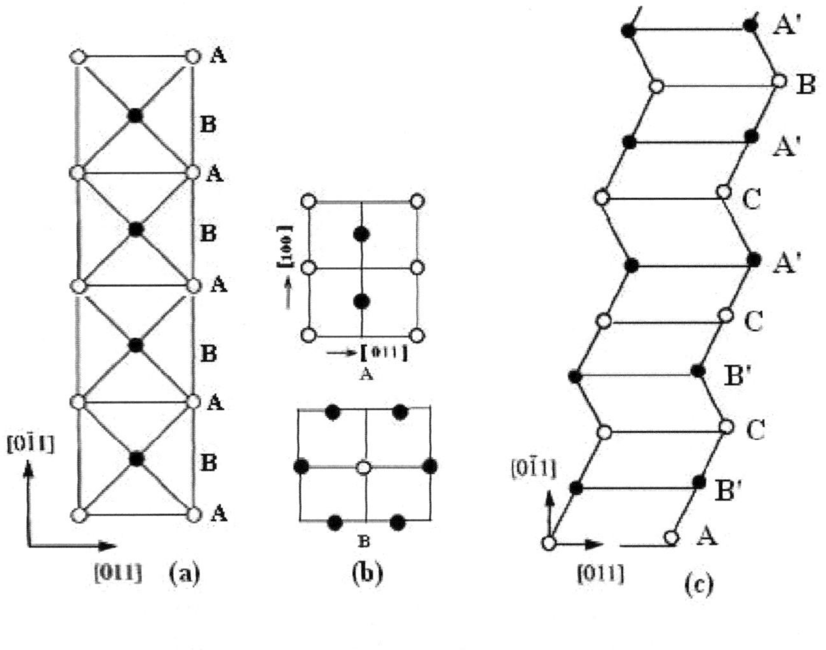

Figure 1. a) Stacking of $(110)_\beta$ planes viewed from $[001]_\beta$ direction in DO_3-type structures, b) atomic configuration on first and second layers of (110) β - plane in DO_3-type structures, c) inhomogeneous shear and formation of layered structures.

All of these martensite phases have the long-period stacking ordered structures, and microstructural evaluation provides a mechanism by which the transformation from the high temperature austenite phase to the low temperature martensite phase takes place.
The fundamental structures of the beta-type martensites are orthorhombic close-packed structures, and monoclinic distortion takes place in some cases by means of microstructural evaluation depending on the atomic distribution in nanoscale or angstrom level, and 18R structure is modified as M18R

2. Experimental

In the present contribution, two copper based ternary alloys were selected for investigation: a CuZnAl alloy with a nominal composition by weight of 26.1%zink, 4%aluminium, the balance copper, while the other was a CuAlMn alloy with a nominal composition by weight of 11% aluminium, 6% manganese and the balance copper. Powder specimens for X-ray examination were prepared by filling the alloys. Specimens for TEM examination were also prepared from 3mm diameter discs and thinned down mechanically to 0.3mm thickness. These specimens were heated in evacuated quartz tubes in the β-phase field (15 minutes at 830°C for CuZnAl and 20 minutes at 700°C for CuAlMn) for homogenization and quenched in iced-brine.
These specimens were also given different post-quench heat treatments and aged at room temperature. TEM and X-ray diffraction studies carried out on these specimens. TEM specimens were examined in a JEOL 200CX electron microscope, and X-ray diffraction profiles were taken from the quenched specimens using Cu-K_{α} radiation with wavelength 1.5418 Å.

3. Results and Discussion

An x-ray powder diffractogram taken from the quenched CuZnAl alloy samples is shown in Figure 2. An electron micrograph and two electron diffraction patterns taken from CuZnAl and CuAlMn alloy samples are also shown in Figure 3a, b and c, respectively. X-ray powder diffractograms and electron diffraction patterns reveal that this alloys exhibit superlattice reflections. As seen from the electron micrograph, the main characteristics martensite in copper-based β-phase alloys are the prevalence of groups of essentially parallel-sided plates [10]. X-ray powder diffractograms and electron diffraction patterns taken from the specimens in a large time interval were compared with each other.

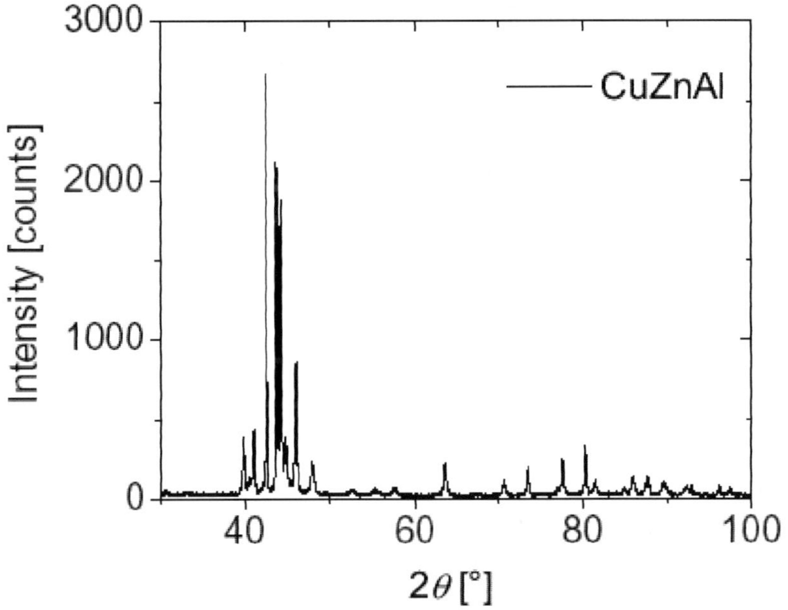

Figure 2. An x-ray powder diffractogram taken from CuZnAl alloy sample aged at room temperature for more than ten years.

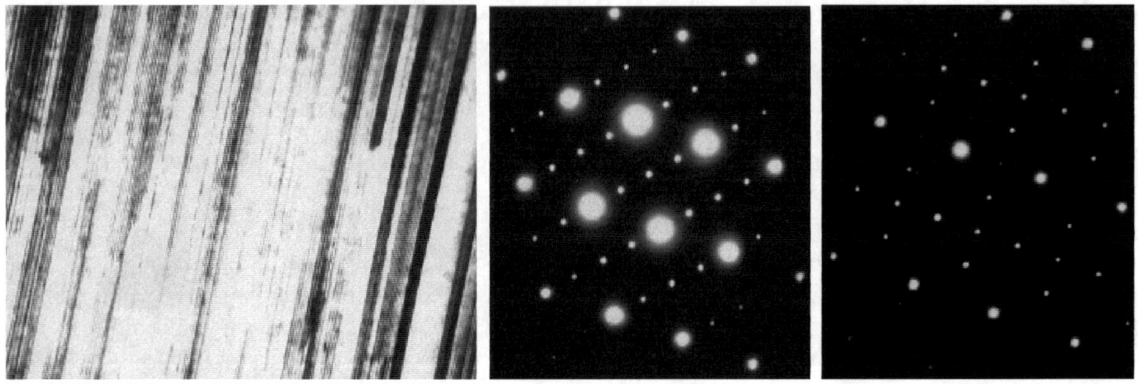

Figure 3. a) An electron micrograph showing the fine martensite structure in CuZnAl alloy1 (x 90k), (b) electron diffraction patterns taken from CuZnAl alloy, (c) electron diffraction patterns taken from CuAlMn alloy sample.

It has been observed that electron diffraction patterns exhibit similar characteristics, but some changes have been occurred in the locations and intensities of diffraction peaks on the x-ray diffractograms with aging duration. In particular, some peak pairs satisfying a specific relation between miller indices move and come close each other with ageing [10]. These changes imply new transitions which have diffusive character. It means that some neighbour atoms change their locations. The monoclinic distortion of 18R-type structure contributes to the martensite stabilization which proceeds by a diffusion-controlled process [11].

Metastable phases of copper-based shape memory alloys are very sensitive to the ageing effects, and heat treatments can change the relative stability and the configurational order of crystal planes. The parent phase has highly symmetric structure and the product phase has internally twinned and complex structures. Also, several types of microscopic deformation involving changes can occur in the stacking sequence of close-packed planes of material with martensite formation [2, 12]. This change gives rise the increase in the complexity of crystal structure. Atom locations in the lattice sites in the crystal unit cell are very important for the analysis and process of transformation.

It can be concluded from the above results that the copper-based shape memory alloys are very sensitive to the ageing treatments, and heat treatments can change the relative stability and the configurational order of atoms in the material. This result attributes to rearrangement of atoms in diffusive manner.

Key Words: Martensitic transition, shape memory effect, layered structures, stacking sequence.

References

[1] Adiguzel O 1995 *Materials Research Bulletin* **30** 755
[2] Zhu J J Liew K M 2003 *Acta Materialia* **51** 2443
[3] Sutou Y et al 2005 Acta *Materialia* **53** 4121
[4] Otsuka K and Kakeshita T 2002 *MRS Bulletin* **27**(2) 91
[5] Popov P and Lagoudas D C 2007 *International Journal of Plasticity* **23**(10-11) 1679
[6] Rahman M A 2008 *Recent Patents on Mechanical Eng.* **1** 65
[7] Otsuka K and Ren X 2005 *Prog. Mater. Sci.* **50**, 511
[8] Lee E Sand and Aln S 1998 *Acta materialia* **46** (12) 4357
[9] Pelegrina J L and Romero R 2000 *Materials Science and Engineering A* **282** 16
[10] Adiguzel O 2007 *Journal of Materials Processing Technology* **185**(1-3) 120
[11] Li Z Gong S and Wang M P 2008 *Journal of Alloys and Compounds* **452** 307
[12] Paidar V 2008 *Materials Science and Engineering A* **481–482** 243

International Conference on Theoretical Physics Dubna-Nano 2010 IOP Publishing
Journal of Physics: Conference Series **248** (2010) 012063 doi:10.1088/1742-6596/248/1/012063

The origin of multibranch IV-characteristics of shunted Josephson junction arrays

Alexander Grib[1,2] and Paul Seidel[1]

[1] Institut für Festkörperphysik, Friedrich-Schiller-Universität Jena, D-07743 Jena, Germany
[2] Physics Department, V. N. Karazin Kharkiv National University, 61077 Kharkiv, Ukraine

E-mail: "Paul Seidel" <Paul.Seidel@uni-jena.de>

Abstract. To model the IV-characteristics of shunted intrinsic Josephson junctions we calculated phase dynamics of a chain of two identical junctions covered by a resistive shunt and with a resistive crosspiece between the shunt and the junctions. The coherent solution of dynamic equations for this system is unstable at some bias currents above critical currents of junctions. We have shown that this instability leads to the multibranch structure of IV-curves in the hysteretic region. The multibranch structure appears due to the random switching of junctions to the superconducting state above their critical currents. We found that the resistance of the interface between a covering shunt and the junctions can regulate the multibranch behaviour.

1. Introduction

Recently, intensive investigations are made to study electrical properties of intrinsic Josephson junctions in high temperature superconductors (HTSC) [1, 2, 3]. It was found that the IV-characteristic of a chain of Josephson junctions loaded by a resistive shunt can show the multibranch behaviour in some regions of bias currents *above* the critical currents of junctions [4]. Such a behaviour of IV-characteristics show that there are instabilities at some bias currents above critical currents. This behaviour is not investigated until the present time. The origin of unstable dynamic states in the resistively shunted Josephson junctions is as follows. When the chain of junctions is loaded by a resistive shunt, the ac current $\tilde{i}$ produced by the voltages over junctions is flowing in the circuit consisted of all the chain and the load. To fulfill the current conservation conditions, this current must be subtracted from the bias current i [5], and the total current through the chain $i_{tot} = i - \tilde{i}$ decreases. In some regions of bias currents the values of i_{tot} become lower than the critical currents of junctions, so some of junctions can switch to the superconducting state. However, in the superconducting state these junctions do not produce their ac current, so the total current increases again. Junctions then switch back to the voltage state etc. This unstable dynamical state of the whole system is the subject of our present investigation. The analyses of stability by means of Floquet exponents [6] allows to reveal unstable regions of the IV-characteristics but it does not give the information about the shape of IV-characteristics. The aim of the present work is to investigate the shape of IV-characteristics within unstable regions. For this we applied the same approach as for the stability analyses, i.e. we added random perturbations to solutions of dynamic equations and then solved the dynamic equations without any additional random current terms and found IV-characteristics. Because the equations of phase dynamics are differential equations of the second order, we can introduce perturbations in two ways. One of them is the addition of random perturbations of phase differences over junctions and another is the addition of random voltage drops over junctions. We added random perturbations to voltages over junctions because these additions have the straight physical meaning of noise produced by random electromagnetic fields. To refine the effects connected with the instability, we considered only two shunted *identical* junctions and decreased voltage perturbations to the lowest magnitude of 10^{-15} in units of voltages normalized to the critical voltage of the junction.

2. The Model

The high-frequency scheme of the system is shown in Fig. 1 a. For the shunted chain of Josephson junctions investigated in the experiment [4], the resistance R_1 represents the gold shunting cover and R_2 is resistance of the interfaces between the shunt and the chain of junctions. We suppose all critical currents of junctions are equal to each other and all the resistances of junctions are the same. The dynamic equations within a capacitively extended model of a resistively shunted junction [5] read

$$\beta_C \ddot{\varphi}_k + \dot{\varphi}_k + \sin\varphi_k + \tilde{i}_k(\tau) = i,$$
$$\tilde{i}_k(\tau) = \frac{1}{r_1(1+2\gamma)}\left[\dot{\varphi}_k + \gamma\left(\dot{\varphi}_1 + \dot{\varphi}_2\right)\right], \qquad (1)$$

where $k = 1, 2$, φ_k, $\dot{\varphi}_k$ and $\ddot{\varphi}_k$ are the phase difference across the k-th junction and its first and second derivatives used with respect to dimensionless time $\tau = \frac{2\pi RI_c t}{\Phi_0}$ with Φ_0 is the quantum of magnetic flux, I_c is a critical current, R is the resistance of the junction, $\beta_C = \frac{2\pi I_c CR^2}{\Phi_0}$ is the McCumber parameter of the junction with C as the capacitance of the junction, i and $\tilde{i}_k$ are the bias current and the current flowing through the k-th junction normalized with respect to I_c, r_1 and r_2 are resistances of the load normalized with respect to R and $\gamma = \frac{r_2}{r_1}$. Voltages across the k-th junction measured in units of RI_c are $\nu_k = \dot{\varphi}_k$. Equations (1) were solved by the standard method of Runge-Kutta. Initial conditions for each bias current were values of variables obtained during the last time step at the previous bias current (conditions with the memory). For the investigation of IV-characteristics of junctions in the perturbed state, we perturbed *only initial conditions* of voltages, i.e. we added random values of the order of magnitude 10^{-15} to ac voltages $\dot{\varphi}_k$ of junctions. We applied also initial conditions without the memory (i. e. voltages were randomly perturbed around zero values at every new bias current).

3. Results and Discussion

Considering currents in the shunt for the system of equivalent junctions without voltage perturbations, we conclude that in the coherent state the IV-characteristics of junctions should be the same for all possible values of r_2 because the equal currents flow through this resistance in opposite directions. IV-characteristics of systems with arbitrary values of r_2 should coincide with each other and IV-characteristics of both junctions should be equal to each other.

However, calculations with randomly perturbed initial conditions for voltages show quite different results. To make the particularities of the curves clear we plot curves for the increase of the bias current only.

IV-characteristics of both junctions are the same for $r_2 = 0.1$ (Fig. 1 b) and coincide with those for the coherent solution (the value $\overline{\nu}$ in the plot is the averaged over time voltage across the junction). The increase of the resistance r_2 to 1 and further to 10 leads to the appearance of branches of the voltage state (the active state) and the zero voltage (passive) state (Figs. 1 c, d) above the critical currents of the junctions. If we apply the initial conditions without the memory, the switching of junctions between active and passive states becomes randomly (Fig. 1 e). The

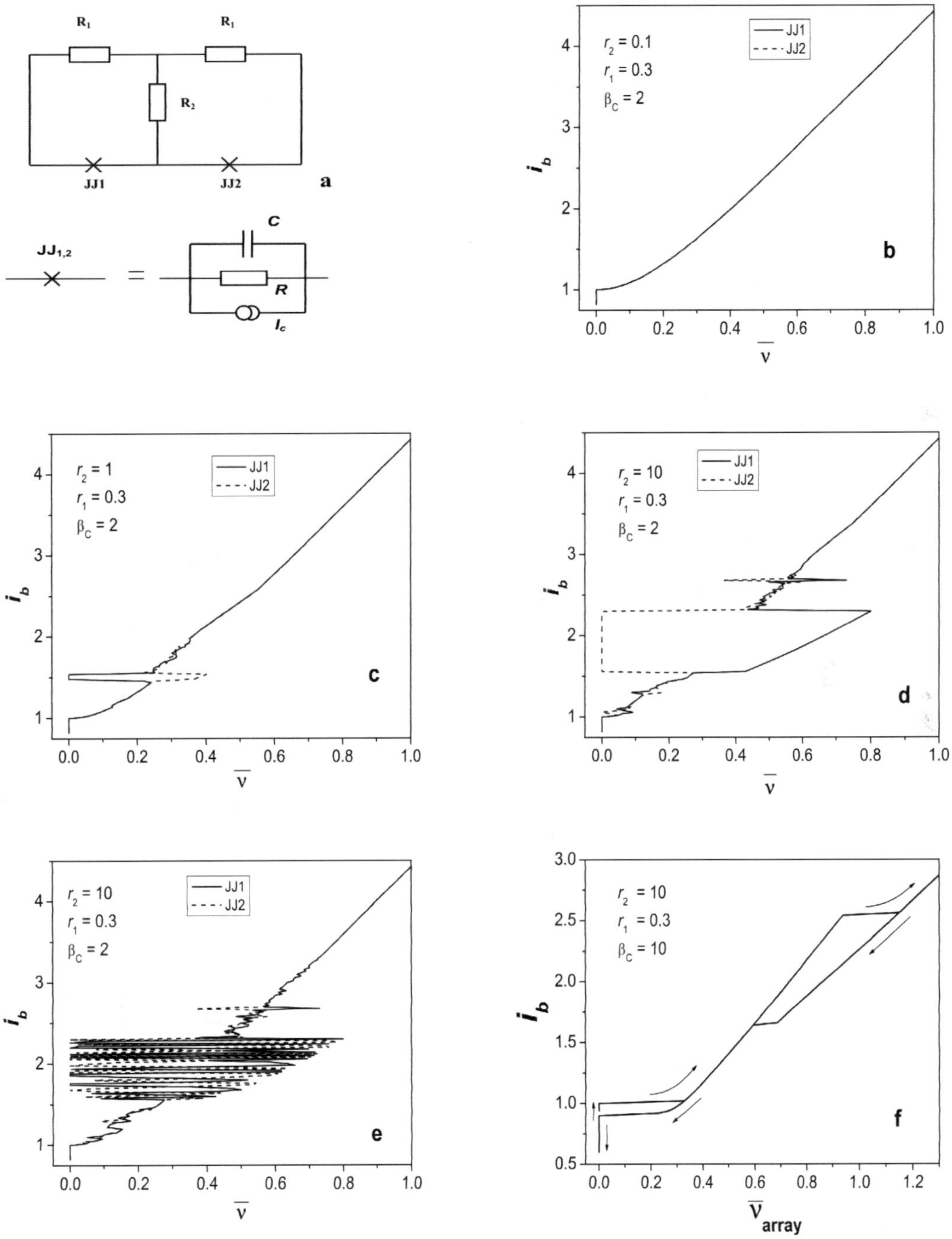

Figure 1. The high-frequency scheme of the system (a), individual IV-characteristics of two junctions with equal critical currents and $\beta_C = 2$ loaded by the shunt with $r_1 = 0.3$ and $r_2 = 0.1$ (b), $r_2 = 1$ (c), $r_2 = 10$ (d), $r_2 = 10$ and with initial conditions without memory (e); the IV-characteristic of the whole system for $\beta_C = 10$, $r_2 = 10$ and with initial conditions with memory (f). Arrows in (f) show the change of the movement of the bias current.

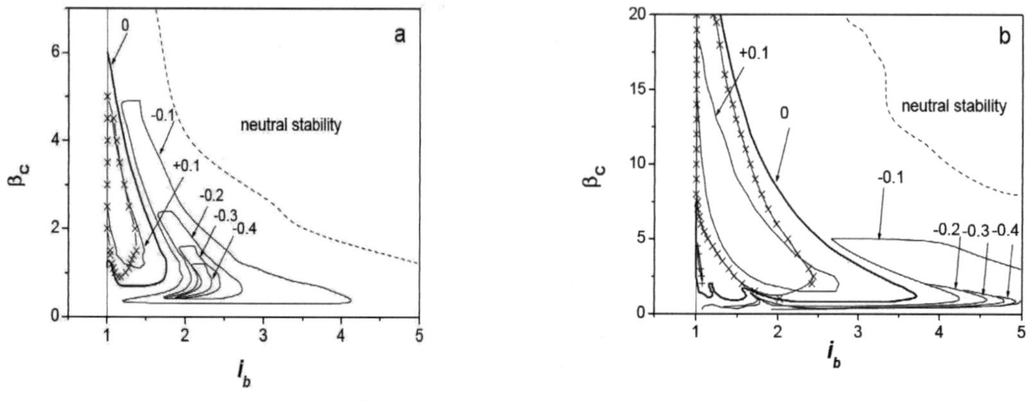

Figure 2. Contours of the largest real part of the Floquet exponents λ on planes $i_b - \beta_C$ for the array of two junctions loaded by a matched resistor (a) and the shunt shown in Fig. 1 a with $r_1 = 0.3$ (b). Contours of regions of multibranch IV-characteristics are marked by the line with crosses. Numbers on the plots are values of λ.

regions of the random switching are placed inside the regions of the unstable coherent solutions (see Fig. 2, where we marked the areas of random switching by crosses). This proves that the origin of the random switching is the instability of the coherent solutions. The ambiguity of switching leads to the appearance of the additional branch on the IV-characteristic of the whole array (Fig. 1 f, initial conditions with the memory).

The resistance r_2 plays the role of the interaction coefficient. Due to the resistance r_2 junctions exchange perturbations with each other. If $r_2 < r_1$, the perturbations are small. The strongest interaction between junctions provides the shunt with $r_2 >> r_1$.

4. Conclusions

We showed that in the resistively shunted array of junctions random perturbations of voltages across junctions lead to the switching of their IV-characteristics from the voltage state to the superconducting state above critical currents that can result in the multibranch structure of IV-characteristics of the whole system. Thus, the multibranch structure appears due to instability of the coherent solution of dynamic equations. The resistance r_2 between the shunt and junctions is the parameter which regulates the splitting of the IV-curves. If $r_2 >> r_1$ the splitting is largest and if $r_2 = 0$ there is no splitting. To avoid the multibranch and switching behaviour the resistance r_2 has to be as small as possible.

Acknowledgments

This research was supported by German DFG.

References

[1] Koyama T, Matsumoto H, Machida M and Kadowaki K 2009 *Phys. Rev.* B **79** 104522
[2] Kurter C, Ozyuzer L, Proslier T, Zasadzinski J F, Hinks D G, and Gray K E 2010 *Phys. Rev.* B **81**, 224518
[3] Mans M, Schneidewind H, Büenfeld M, Schmidl F and Seidel P 2006 *Phys. Rev.* B **74**, 214514
[4] Grib A, Mans M, Scherbel J, Büenfeld M, Schmidl F, Seidel P 2006 *Supercond. Sci. Technol.* **19** 200
[5] Likharev K K 1991 *Dynamics of Josephson Junctions and Circuits* (Philadelphia: Gordon and Breach).
[6] Hadley P, Beasley M R, Wiesenfeld K, 1988 *Phys. Rev.* B **38** 8712

AUTHOR INDEX

AUTHOR INDEX

Institute of Physics
Dirac House, Temple Back
Bristol BS1 6BE UK

CURRAN ASSOCIATES INC
proceedings.com